Compendium of Organic Synthetic Methods

Compendium of Organic Synthetic Methods

IAN T. HARRISON

and

SHUYEN HARRISON

SYNTEX RESEARCH
PALO ALTO, CALIFORNIA

QD
262
H32
v. 1

WILEY - INTERSCIENCE
A Division of John Wiley & Sons, Inc.
New York • London • Sydney • Toronto

Library of Congress Catalog Card Number: 71–162800

ISBN O–471–35550–X

Printed in the United States of America.

10 9 8 7 6 5 4

PREFACE

Compendium of Organic Synthetic Methods is a systematic listing of functional group transformations designed for use by bench chemists, persons planning syntheses, students attending courses on synthetic chemistry, and teachers of these courses. The idea for this compilation came from the observation that organic chemists spend a large proportion of their time searching a formidable original literature for these hard-to-find synthetic methods.

A key feature of this book is the classification of reactions on the basis of the functional group of the starting material and of the product, without reference to the reaction mechanism. We wished to produce as comprehensive a set of reactions as possible, covering all branches of organic chemistry. Reactions giving low yields or requiring exotic reagents are not omitted. Consequently reactions included cover the full range of methods from boiling in oil to treatment with fluorine or orange-peel enzymes.

The presentation of each synthetic method in the form of representative reactions without discussion follows the plan used successfully in *Steroid Reactions* (Djerassi, Holden-Day). The limitations of such compilations containing much information but few words are obvious; there is, however, a great need for a comprehensive one-volume listing of synthetic methods as an intermediary between the chemist and the literature. The reader interested in a detailed discussion of synthetic methods should consult *Reagents for Organic Synthesis* (Fieser and Fieser, Wiley) and *Survey of Organic Syntheses* (Buehler and Pearson, Wiley).

We apologize to authors for the space-saving anonymity of references and for referring in many instances not to papers by the originators of a reaction but rather to subsequent articles by other authors. We make no apology, however, for omitting unnecessary reference punctuation, and avoiding the use of *ibid.* and other sources of confusion.

Ian T. Harrison
Shuyen Harrison

Palo Alto, California
March 1971

CONTENTS

ABBREVIATIONS

Ac	acetyl
Bu	butyl
DCC	dicyclohexylcarbodiimide
DDQ	2,3-dichloro-5,6-dicyanobenzoquinone
DMA	dimethylacetamide
DMF	dimethylformamide
Et	ethyl
HMPA	hexamethylphosphoramide
Me	methyl
Ms	methanesulfonyl
NBA	*N*-bromoacetamide
NBS	*N*-bromosuccinimide
NCS	*N*-chlorosuccinimide
Ni	Raney nickel
Ph	phenyl
Pr	propyl
Pyr	pyridine
THF	tetrahydrofuran
THP	tetrahydropyranyl
Ts	*p*-toluenesulfonyl

INDEX

Sections—heavy type
Pages—light type

PREPARATION OF →

FROM ↓

Sections are shown in heavy type, page numbers in light type. Each cell lists section / page.

FROM ↓ \ PREPARATION OF →	Acetylenes	Carboxylic acids, acid halides, anhydrides	Alcohols, phenols	Aldehydes	Alkyls, methylenes, aryls	Amides	Amines	Esters	Ethers, epoxides	Halides, sulfonates, sulfates	Hydrides (RH)	Ketones	Nitriles	Olefins
Acetylenes	1 / 1	16 / 16	31 / 75	46 / 132	61 / 177	76 / 203	91 / 230	106 / 271		136 / 329		166 / 379	181 / 457	196 / 479
Carboxylic acids, acid halides, anhydrides	2 / 2	17 / 18	32 / 76	47 / 132	62 / 178	77 / 204	92 / 230	107 / 272	122 / 309	137 / 329	152 / 357	167 / 380	182 / 457	197 / 482
Alcohols, phenols	3 / 3	18 / 26	33 / 73	48 / 137	63 / 180	78 / 208	93 / 232	108 / 280	123 / 310	138 / 331	153 / 359	168 / 386	183 / 459	198 / 484
Aldehydes	4 / 3	19 / 31	34 / 80	49 / 144	64 / 181	79 / 209	94 / 233	109 / 287	124 / 316	139 / 338	154 / 363	169 / 396	184 / 460	199 / 489
Alkyls, methylenes, aryls		20 / 36		50 / 146	65 / 182	80 / 210				140 / 338	155 / 365	170 / 400	185 / 464	200 / 493
Amides		21 / 39	36 / 85	51 / 148	66 / 184	81 / 211	96 / 236	111 / 289		141 / 338	156 / 366	171 / 403	186 / 464	201 / 495
Amines	7 / 5	22 / 41	37 / 86	52 / 150	67 / 184	82 / 213	97 / 240	112 / 290	127 / 318	142 / 339	157 / 367	172 / 404	187 / 465	202 / 496
Esters	8 / 6	23 / 42	38 / 87	53 / 152	68 / 185	83 / 218	98 / 249	113 / 291	128 / 318	143 / 342	158 / 368	173 / 406	188 / 467	203 / 498
Ethers, epoxides		24 / 46	39 / 92	54 / 154	69 / 186	84 / 220	99 / 249	114 / 293	129 / 320	144 / 343	159 / 369	174 / 408	189 / 467	204 / 501
Halides, sulfonates, sulfates	10 / 6	25 / 47	40 / 102	55 / 156	70 / 186	85 / 220	100 / 250	115 / 295	130 / 320	145 / 345	160 / 370	175 / 411	190 / 468	205 / 504
Hydrides (RH)		26 / 53	41 / 107	56 / 162	71 / 191	86 / 222	101 / 255	116 / 299	131 / 322	146 / 349	161 / 375	176 / 417	191 / 471	206 / 512
Ketones	12 / 10	27 / 56	42 / 110	57 / 164	72 / 193	87 / 223	102 / 258	117 / 302	132 / 323	147 / 353	162 / 375	177 / 419	192 / 473	207 / 513
Nitriles		28 / 62		58 / 166	73 / 198	88 / 225	103 / 262	118 / 304			163 / 376	178 / 433	193 / 474	208 / 520
Olefins	14 / 13	29 / 64	44 / 119	59 / 168	74 / 198	89 / 227	104 / 264	119 / 305	134 / 325	149 / 354	164 / 377	179 / 435	194 / 475	209 / 520
Miscellaneous compounds	15 / 13	30 / 68	45 / 122	60 / 172	75 / 202	90 / 229	105 / 266	120 / 307	135 / 328	150 / 356	165 / 377	180 / 442	195 / 47f.	210 / 526

PROTECTION

	Sect.	Pg.
Carboxylic acids	30A	71
Alcohols, phenols	45A	124
Aldehydes	60A	174
Amines	105A	266
Ketones	180A	449

Blanks in the table correspond to sections for which no examples were found in the literature

INTRODUCTION

Classification and Organization. *Compendium of Organic Synthetic Methods* contains approximately 3000 examples of published chemical transformations classified according to the reacting functional group of the starting material and the functional group formed. Those reactions that give products with the same functional group form a chapter. The reactions in each chapter are further classified into sections on the basis of the functional group of the starting material. Within each section reactions are listed in a somewhat arbitrary order although an effort has been made to put chain lengthening processes before degradations.

The classification is unaffected by allylic, vinylic, or acetylenic unsaturation, which appears in both starting material and product, or by increases or decreases in the length of carbon chains. For example, the reactions t-BuOH → t-BuCOOH, PhCH$_2$OH → PhCOOH, and PhCH = CHCH$_2$OH → PhCH = CHCOOH are all found in Section 18 on carboxylic acids from alcohols.

The terms hydrides, alkyls, and aryls classify compounds containing reacting hydrogens, alkyl groups, and aryl groups, respectively; for example, RCH$_2$-H → RCH$_2$COOH (carboxylic acids from hydrides), RMe → RCOOH (carboxylic acids from alkyls), RPh → RCOOH (carboxylic acids from aryls). Note the distinction between R$_2$CO → R$_2$CH$_2$ (methylenes from ketones) and RCOR′ → RH (hydrides from ketones).

The following examples illustrate the application of the classification scheme to some potentially confusing cases:

RCH=CHCOOH → RCH=CH$_2$	(hydrides from carboxylic acids)
RCH=CH$_2$ → RCH=CHCOOH	(carboxylic acids from hydrides)
ArH → ArCOOH	(carboxylic acids from hydrides)
ArH → ArOAc	(esters from hydrides)
RCHO → RH	(hydrides from aldehydes)
RCH=CHCHO → RCH=CH$_2$	(hydrides from aldehydes)
RCHO → RCH$_3$	(alkyls from aldehydes)
R$_2$CH$_2$ → R$_2$CO	(ketones from methylenes)
RCH$_2$COR → R$_2$CHCOR	(ketones from ketones)
RCH=CH$_2$ → RCH$_2$CH$_3$	(alkyls from olefins)
RBr + RC≡CH → RC≡CR	(acetylenes from halides, also acetylenes from acetylenes)

ROH + RCOOH → RCOOR   (esters from alcohols, also esters from carboxylic acids)

Sulfonic esters are grouped with halides. Hydrazines are listed with amines and hydrazides with amides.

Yields quoted are overall, reduced to allow for incomplete conversion and impurities in the product.

Trivial reactions not described in the given references but required to complete a sequence are indicated by a dashed arrow.

How to Use the Book. Examples of the preparation of one functional group from another are located via the index on p. xi, which gives the corresponding section and page. Thus Section 1 contains examples of the preparation of acetylenes from other acetylenes; Section 2, acetylenes from carboxylic acids; Section 3, acetylenes from alcohols; etc.

Sections giving examples of the reactions of a functional group are found in horizontal rows of the index. Thus Section 1 gives examples of the reactions of acetylenes forming other acetylenes; Section 16, reactions of acetylenes forming carboxylic acids; Section 31, reactions of acetylenes forming alcohols; etc.

Examples of alkylation, dealkylation, homologation, isomerization, transposition, etc. are found in Sections 1, 17, 33, and so forth, which lie close to a diagonal of the index. These sections correspond to the preparation of acetylenes from acetylenes, carboxylic acids from carboxylic acids, alcohols and phenols from alcohols and phenols, etc.

Examples of the protection of carboxylic acids, alcohols, phenols, aldehydes, amines, and ketones are also indexed on page xi.

Examples of name reactions can be found by first considering the nature of the starting material and product. The Wittig reaction, for example, is to be found in Section 199 on olefins from aldehydes and Section 207 on olefins from ketones.

The pairs of functional groups, alcohol–ester, carboxylic acid–ester, amine–amide, carboxylic acid–amide, can be interconverted by quite trivial reactions. When a member of these groups is the desired product or starting material, the other member should of course also be consulted in the text.

The original literature must be used to determine the generality of reactions. A reaction given in this book for a primary aliphatic substrate may in fact also be applicable to tertiary or aromatic compounds.

The references given usually yield a further crop of references to previous work. Subsequent publications can be found through Science Citation Index.

Reactions Included in the Book. Interconversions of monofunctional compounds form the major part of this compilation. Reactions of bifunctional compounds in which the two functions are identical and which give monofunctional products are also included; for example, $R_2C(COOH)_2 \rightarrow R_2CO$ (ketones from carboxylic acids).

Examples of the removal of one functional group from bifunctional compounds and the preparation of functional groups from groups not listed on the index are included in the miscellaneous sections; for example, $RCH(Br)COR \rightarrow RCH_2COR$, $RCH(OH)COR \rightarrow RCH_2COR$ and $RCH=CHCOR \rightarrow RCH_2CH_2COR$ (ketones from miscellaneous compounds), $RCH_2NO_2 \rightarrow RCHO$ (aldehydes from miscellaneous compounds).

Reactions are included even when full experimental details are lacking from the given reference. In some cases the quoted reaction is a minor part of a paper or may have been investigated from a purely mechanistic aspect. When several references are given, the first refers to the reaction illustrated; others give further examples, related reactions, or reviews.

Reactions Not Included in the Book. Reactions forming bifunctional products are not included, for example, $RCH=CH_2 \rightarrow RCH(OH)CH_2OH$. Chain lengthening processes via unsaturated intermediates, for instance, $RCHO \rightarrow [RCH=CHCOOEt] \rightarrow RCH_2CH_2COOEt$, are only partially covered. Ring forming reactions and reactions that involve several functional centers (e.g., the Diels-Alder reaction) are represented by very few examples. These gaps will be filled by a second volume, presently under consideration, which will include reactions forming unsaturated and other bifunctional products.

Reactions published after early 1971 are not included.

Compendium of Organic Synthetic Methods

Chapter 1 PREPARATION OF ACETYLENES

Section 1 <u>Acetylenes from Acetylenes</u>

Review: The Synthesis of Acetylenes   Org React (1949) <u>5</u> 1

$C_5H_{11}C{\equiv}CH$ $\xrightarrow[\text{2 BuI HMPA}]{\text{1 i-PrMgCl Et}_2\text{O}}$ $C_5H_{11}C{\equiv}CBu$ 66%

Bull Soc Chim Fr (1964) 2000

$BuC{\equiv}CH$ $\xrightarrow[\text{2 Et}_2\text{SO}_4\ \text{Et}_2\text{O}]{\text{1 EtMgBr Et}_2\text{O}}$ $BuC{\equiv}CEt$ 70%

JACS (1936) <u>58</u> 796
JOC (1959) <u>24</u> 840

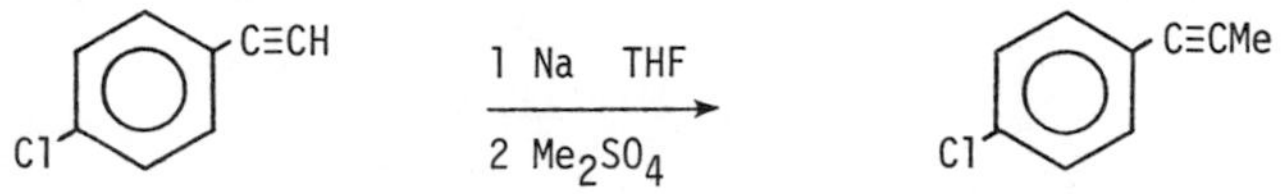

Cl—C$_6$H$_4$—C≡CH $\xrightarrow[\text{2 Me}_2\text{SO}_4]{\text{1 Na THF}}$ Cl—C$_6$H$_4$—C≡CMe   36%

Bull Soc Chim Fr (1965) 1525

Further examples of the reaction $RC\equiv CH + R'X \longrightarrow RC\equiv CR'$ are included in section 10 (Acetylenes from Halides, Sulfonates and Sulfates)

$PhC\equiv CH$ $\xrightarrow{\text{CH}_2\text{I}_2 \quad \text{Zn-Cu} \quad \text{Et}_2\text{O}}$ $PhC\equiv CMe$ 37%

Bull Soc Chim Fr (1965) 1525
Tetrahedron (1958) <u>3</u> 197

$EtCH_2C\equiv CH$ $\xrightarrow{\text{KOH} \quad \text{EtOH} \quad 170\text{-}180°}$ $EtC\equiv CMe$ ~70%

JACS (1951) <u>73</u> 1273
Quart Rev (1970) <u>24</u> 585

$C_5H_{11}C\equiv CCH_3$ $\xrightarrow[150°]{\text{NaNH}_2 \quad 1,2,4\text{-trimethylbenzene}}$ $C_5H_{11}CH_2C\equiv CH$ 80%

Org React (1949) <u>5</u> 1

$BuC\equiv C(CH_2)_3CH_3$ $\xrightarrow{\text{NaNH}_2 \quad \text{mineral oil}}$ $Bu(CH_2)_4C\equiv CH$

Org React (1949) <u>5</u> 1

Section 2   <u>Acetylenes from Carboxylic Acids and Acid Halides</u>
ooooooooooooooooooooooooooooooooooooooooooooooooooooooo

$BuC\equiv CCOCl$ $\xrightarrow[2 \ \sim280°]{1 \ \text{Ph}_3\text{P=CHPh} \quad \text{Et}_3\text{N} \quad \text{C}_6\text{H}_6}$ $BuC\equiv C\text{-}C\equiv CPh$

JCS (1964) 543

$$PhCH_2COOH \xrightarrow[\text{2 PhMgBr-CdCl}_2]{\text{1 SOCl}_2} PhCH_2COPh \xrightarrow[\text{2 NaNH}_2]{\text{1 PCl}_5} PhC \equiv CPh \qquad 34\%$$

Helv (1938) <u>21</u> 1356

Section 3   Acetylenes from Alcohols

$$BuOH \xrightarrow{\text{Na} \quad \text{TsCl}} BuOTs \xrightarrow{\text{NaC} \equiv \text{CH} \quad \text{NH}_3} BuC \equiv CH \qquad 37\text{-}47\%$$

JACS (1937) <u>59</u> 1490
Org React (1949) <u>5</u> 1

Section 4   Acetylenes from Aldehydes

$$PhCHO \xrightarrow[\text{Et}_2O]{\text{Ph}_3P=CHO-\bigcirc-Me} PhCH=CHO-\bigcirc_{Me} \xrightarrow{\text{PhLi}} PhC \equiv CH \qquad 29\%$$

Ber (1962) <u>95</u> 2514

$$AcOCH_2CH=\underset{Me}{C}(CH_2)_2CHO \xrightarrow[\substack{\text{lithium piperidide} \\ \text{2 MeONa} \quad \text{MeOH} \\ \text{3 BuLi} \quad \text{Et}_2O}]{\overset{+}{1} \; Ph_3PCH_2Cl \; \overset{-}{Cl} \quad Et_2O} HOCH_2CH=\underset{Me}{C}(CH_2)_2C \equiv CH \qquad 55\%$$

JACS (1969) <u>91</u> 4318

$C_6H_{13}CHO$ $\xrightarrow{\begin{array}{l}1\ Ph_3P{=}CHMe\quad THF\\ 2\ BuLi\quad THF\quad pentane\\ 3\ N\text{-}Chlorosuccinimide\end{array}}$ $C_6H_{13}CH{=}\underset{Cl}{C}Me$ $\xrightarrow{\begin{array}{c}NaNH_2\\ NH_3\end{array}}$ $C_6H_{13}C{\equiv}CMe$

Tetr Lett (1970) 447

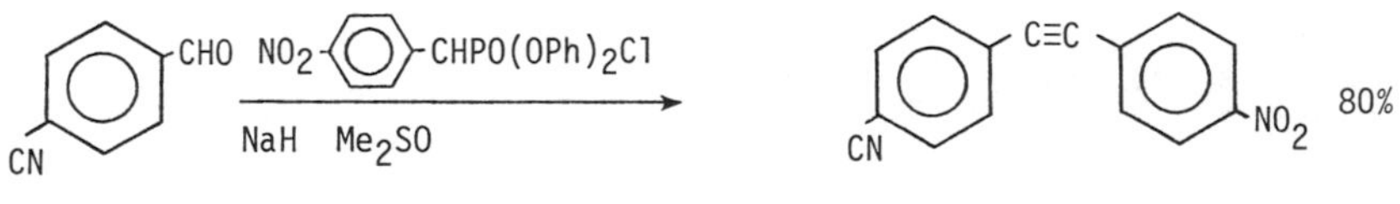

JACS (1965) <u>87</u> 2777

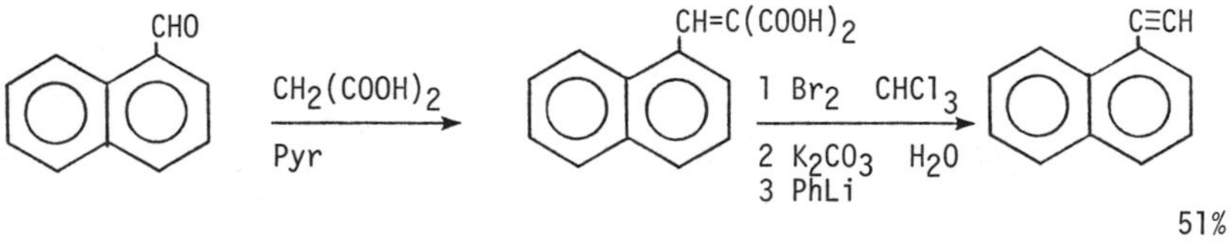

51%

Compt Rend (1949) <u>229</u> 660

PhCHO $\xrightarrow{\begin{array}{c}BrCH_2COOEt\\ \hline Zn\quad C_6H_6\end{array}}$ $Ph\underset{OH}{C}HCH_2COOEt$ $\xrightarrow{\begin{array}{l}1\ N_2H_4\\ 2\ NaNO_2\quad HCl\\ \quad H_2O\quad pet\ ether\\ 3\ NOCl\quad Pyr\\ 4\ KOH\quad H_2O\end{array}}$ $PhC{\equiv}CH$

JACS (1951) <u>73</u> 4199

Section 5 Acetylenes from Alkyls, Methylenes and Aryls

No examples

Section 6 Acetylenes from Amides

No examples

Section 7 Acetylenes from Amines

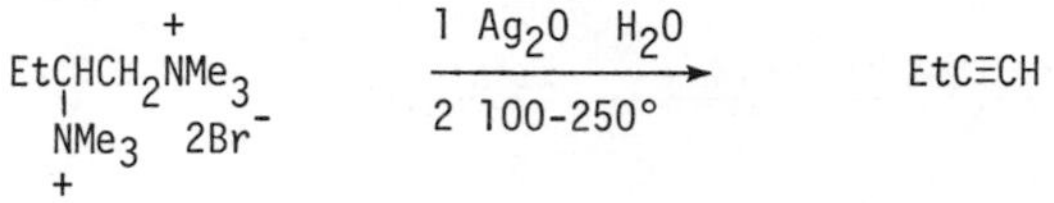
EtCHCH2NMe3 +
NMe3 + 2Br-
1 Ag2O H2O
2 100-250°
EtC≡CH

JACS (1939) 61 1943

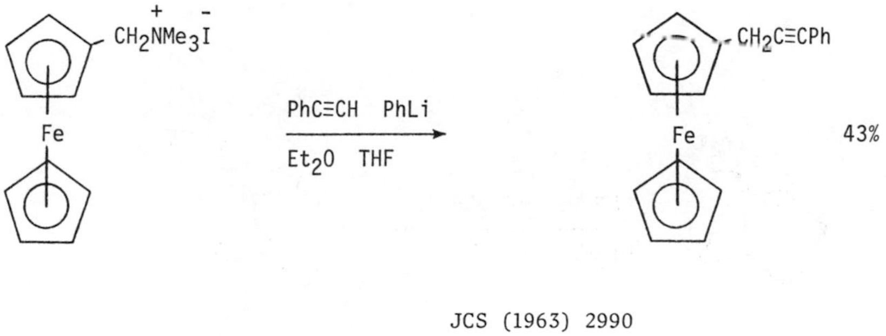
CH2NMe3I
+ -
Fe
PhC≡CH PhLi
Et2O THF
CH2C≡CPh
Fe
43%

JCS (1963) 2990

Section 8   Acetylenes from Esters
$\circ\circ\circ\circ\circ\circ\circ\circ\circ\circ\circ\circ\circ\circ\circ\circ\circ\circ\circ\circ\circ\circ\circ\circ$

$$\underset{\underset{OAc}{|}}{MeCHCH_2OAc} \xrightarrow{\ 450°\ } MeC\equiv CH$$

Izv (1959) 43
(Chem Abs $\underline{54}$ 7547)

Section 9   Acetylenes from Ethers
$\circ\circ\circ\circ\circ\circ\circ\circ\circ\circ\circ\circ\circ\circ\circ\circ\circ\circ\circ\circ\circ\circ\circ\circ$

No examples

Section 10   Acetylenes from Halides, Sulfonates and Sulfates
$\circ\circ\circ\circ\circ\circ\circ\circ\circ\circ\circ\circ\circ\circ\circ\circ\circ\circ\circ\circ\circ\circ\circ\circ\circ\circ\circ\circ\circ\circ\circ\circ\circ\circ\circ\circ\circ\circ\circ\circ\circ\circ\circ\circ\circ\circ\circ\circ$

Review: The Synthesis of Acetylenes   Org React (1949) $\underline{5}$ 1

$$i\text{-PrBr} \xrightarrow[150°]{PhC\equiv CLi\quad dioxane} i\text{-PrC}\equiv CPh \qquad 65\%$$

Annalen (1958) $\underline{614}$ 37

$$\xrightarrow[Me_2SO]{LiC\equiv CH\cdot H_2NCH_2CH_2NH_2}$$

53%

JACS (1969) $\underline{91}$ 4771
Angew (1959) $\overline{71}$ 245

$Me_2C=CH(CH_2)_2\overset{|}{\underset{Me}{C}}=CHCH_2Br$ →[1 $LiCH_2C\equiv CSiMe_3$][2 $AgNO_3$
3 $NaCN$] $Me_2C=CH(CH_2)_2\overset{|}{\underset{Me}{C}}=CHCH_2CH_2C\equiv CH$

Tetr Lett (1968) 5041
(1970) 2247

BuOTs →[$PhC\equiv CNa$][Toluene] $PhC\equiv CBu$ 65-70%

Org React (1949) <u>5</u> 1
JACS (1937) <u>59</u> 1490

Et_2SO_4 →[$NaC\equiv CH$ NH_3] $EtC\equiv CH$ 60%

JACS (1931) <u>53</u> 289

$i-Pr_2SO_4$ →[$NaC\equiv CH$ NH_3] $i-PrC\equiv CH$ 29-50%

JACS (1937) <u>59</u> 1490

EtBr →[1 Mg Et_2O][2 $BrCH_2C\equiv CCH_2Br$] $EtCH_2C\equiv CCH_2Et$

JCS (1946) 1009

JOC (1963) <u>28</u> 3313
JACS (1964) <u>86</u> 4358

98%

$$PhCH=CHBr \xrightarrow{\quad PhC\equiv CCu \quad DMF \quad} PhCH=CHC\equiv CPh \qquad 75\%$$

Chem Comm (1967) 1259

$$PhI \xrightarrow[\substack{2\ \text{Acid hydrolysis} \\ 3\ \text{NaOH}\ \text{MeOH}\ H_2O}]{1\ CuC\equiv CCH(OEt)_2 \quad pyr} PhC\equiv CH$$

JCS C̲ (1969) 2173

Further examples of the reaction RC≡CH + R'X → RC≡CR' are included in
section 1 (Acetylenes from Acetylenes)

$$BuBr \xrightarrow[\substack{2\ PhC\equiv CBr\ \ CoCl_2}]{1\ Mg\ \ Et_2O} BuC\equiv CPh \qquad 32\%$$

JCS (1954) 1704

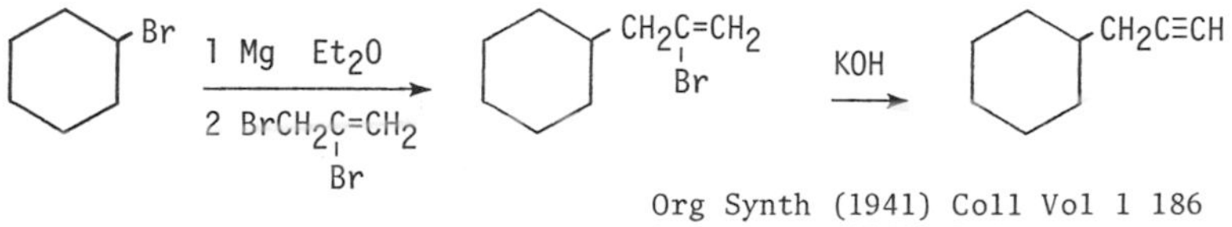

Org Synth (1941) Coll Vol 1 186
Org React (1949) 5̲ 1

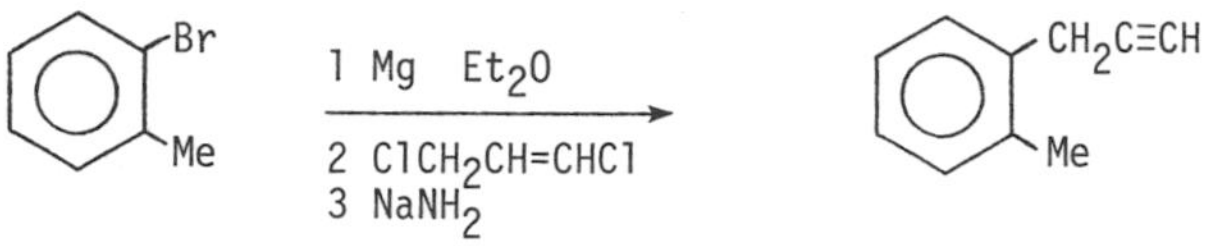

Compt Rend (1925) 181̲ 555

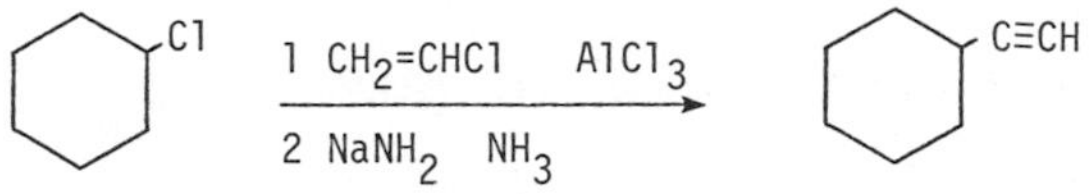

$$\text{1 } CH_2=CHCl \quad AlCl_3$$
$$\text{2 } NaNH_2 \quad NH_3$$

36%

Rec Trav Chim (1965) <u>84</u> 31

EtCHCH$_2$Br
 |
 Br

1 Me$_3$N
2 Ag$_2$O H$_2$O
3 100-250°

EtC≡CH   9%

JACS (1939) <u>61</u> 1943

C$_8$H$_{17}$CH-CH$_2$
          | |
          Br Br

NaNH$_2$ Me$_2$SO

C$_8$H$_{17}$C≡CH   54%

Tetrahedron (1970) <u>26</u> 2127
JACS (1934) <u>56</u> 2120

EtCH$_2$CH-CH$_2$
         | |
         Br Br

KOH   EtOH

EtCH$_2$C≡CH

Org React (1949) <u>5</u> 1

Further examples of the conversion of dibromides into acetylenes are
included in section 14 (Acetylenes from Olefins)

Section 11   Acetylenes from Hydrides (RH)
             °°°°°°°°°°°°°°°°°°°°°°°°°°°°°°°°°°°

No examples

Section 12   Acetylenes from Ketones
              ○○○○○○○○○○○○○○○○○○○○○○○○○

Review: The Synthesis of Acetylenes   Org React (1949) 5 1

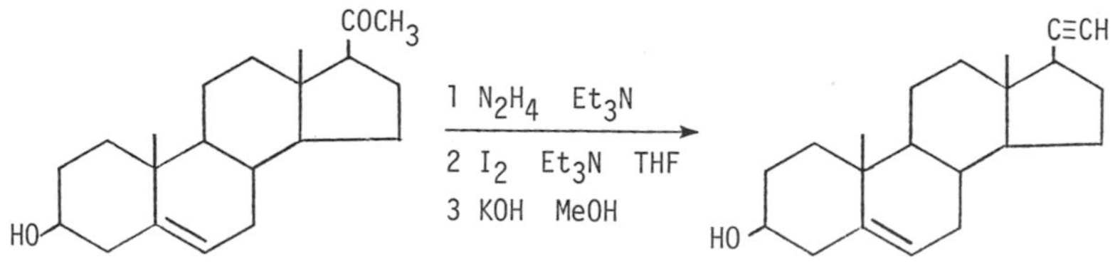

JOC (1969) 34 3502

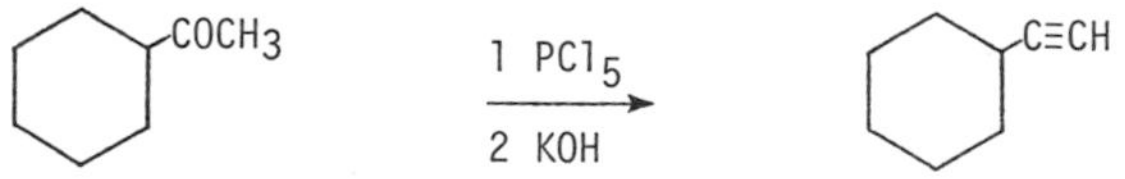

~35%

JACS (1932) 54 1184

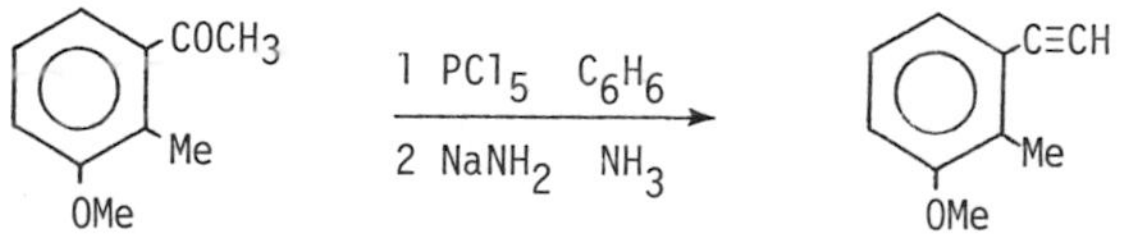

73%

Tetrahedron (1969) 25 4249

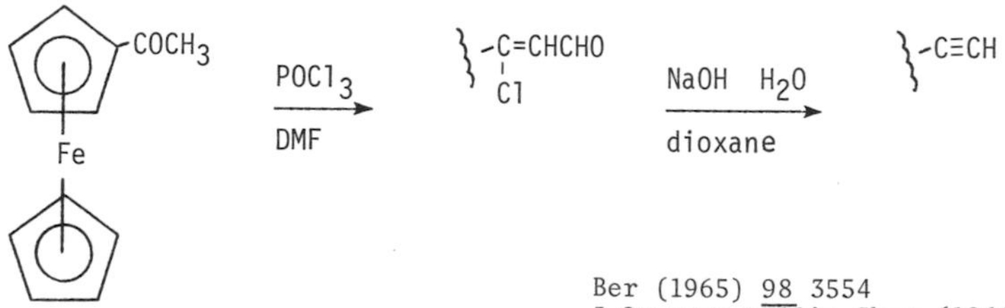

77%

Ber (1965) 98 3554
J Organometallic Chem (1966) 6 173

PhCH$_2$COMe $\xrightarrow[\text{2 (CF}_3\text{COO)}_2\text{Hg}_2 \quad \text{Et}_2\text{O}]{\text{1 N}_2\text{H}_4 \cdot \text{H}_2\text{O}}$ PhC≡CMe   48%

JOC (1966) <u>31</u> 624

PhCOCH$_3$ --→ PhCOCH$_2$Br $\xrightarrow{\text{(PhO)}_3\text{P}}$ PhC≡CH

Dokl (1965) <u>163</u> 656
(Chem Abs <u>63</u> 11338)

PhCOCH$_3$ --→ PhCOCH$_2$Cl $\xrightarrow[\text{2 NaNH}_2 \quad \text{Et}_2\text{O} \quad \text{NH}_3]{\text{1 (EtO)}_3\text{P}}$ PhC≡CH   ~75%

JCS (1963) 3712

PhCO
|
Et $\xrightarrow[\text{Zn} \quad \text{C}_6\text{H}_6]{\text{BrCH}_2\text{COOEt}}$ Et
                                                                    |
                                                                    PhCCH$_2$COOEt
                                                                    |
                                                                    OH $\xrightarrow[\substack{\text{2 NaNO}_2 \quad \text{HCl} \\ \text{H}_2\text{O} \quad \text{pet ether} \\ \text{3 KOH} \quad \text{EtOH} \quad \text{H}_2\text{O}}]{\text{1 N}_2\text{H}_4}$ PhC≡CEt   80%

JACS (1951) <u>73</u> 4199

$\xrightarrow{\substack{\text{1 N}_2\text{H}_4 \\ \text{2 MnO}_2 \\ \text{3 CHCl}_3 \quad \text{t-BuOK} \\ \text{4 MeLi}}}$

Chem Ind (1969) 1306

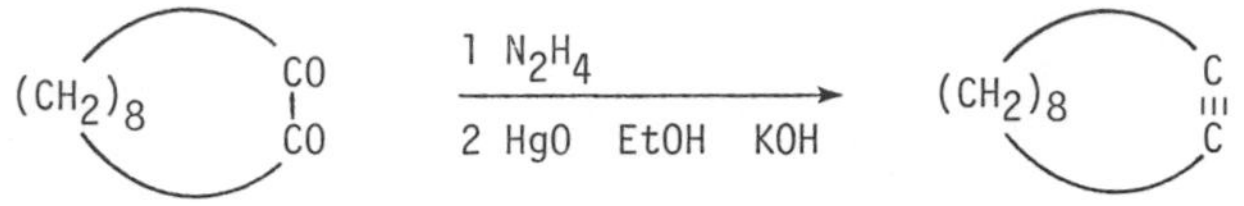

$(CH_2)_8$ ⟨CO / CO⟩ $\xrightarrow[\text{2 HgO EtOH KOH}]{\text{1 } N_2H_4}$ $(CH_2)_8$ ⟨C≡C⟩

JACS (1952) <u>74</u> 3636 3643

PhCOCOPh $\xrightarrow[\text{HgO } C_6H_6]{N_2H_4 \cdot H_2O \text{ PrOH}}$ PhC≡CPh   67-73%

Org Synth (1963) Coll Vol 4 377

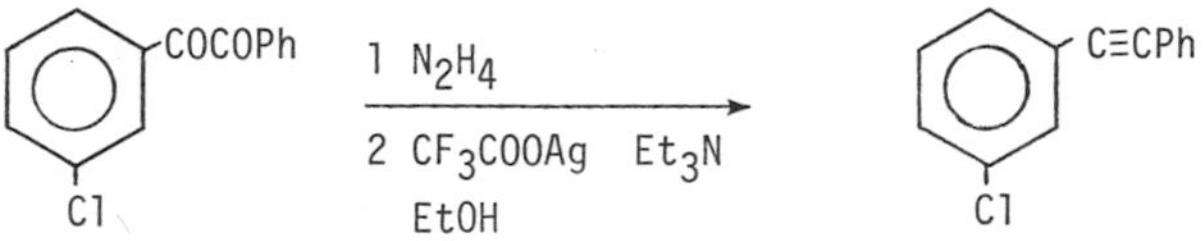

COCOPh (on Cl-substituted benzene) $\xrightarrow[\text{EtOH}]{\substack{\text{1 } N_2H_4 \\ \text{2 } CF_3COOAg \text{ Et}_3N}}$ C≡CPh (on Cl-substituted benzene)   80%

JOC (1958) <u>23</u> 665

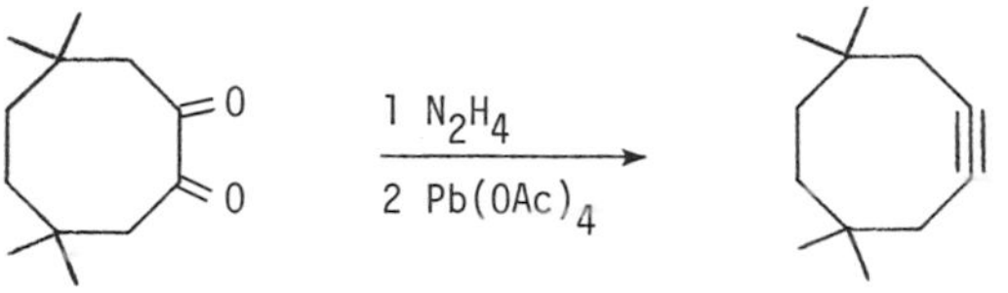

(dione ring) $\xrightarrow[\text{2 Pb(OAc)}_4]{\text{1 } N_2H_4}$ (alkyne ring)   26%

Tetr Lett (1968) 4511

PhCOCOPh $\xrightarrow{\text{(EtO)}_3P \text{ 215°}}$ PhC≡CPh   60%

JOC (1964) <u>29</u> 2243

Section 13   Acetylenes from Nitriles
              ⸰⸰⸰⸰⸰⸰⸰⸰⸰⸰⸰⸰⸰⸰⸰⸰⸰⸰⸰⸰⸰⸰⸰⸰

No examples

Section 14   Acetylenes from Olefins
              ⸰⸰⸰⸰⸰⸰⸰⸰⸰⸰⸰⸰⸰⸰⸰⸰⸰⸰⸰⸰⸰⸰⸰

$BuCH=CH_2$ $\xrightarrow{\text{Li \quad THF}}$ $BuC\equiv CLi$ $\xdashrightarrow{\text{H}_2\text{O}}$ $BuC\equiv CH$

                                    ~65%

                                             JOC (1967) 32 105

                      1 Br$_2$ CCl$_4$
    ───────────────────────►   48%
                      2 NaNH$_2$ HMPA

                                             Tetr Lett (1970) 41
                                             Tetrahedron (1970) 26 2127

                         1 Br$_2$
$C_8H_{17}CH=CH(CH_2)_7COOH$ ───────────────────────► $C_8H_{17}C\equiv C(CH_2)_7COOH$ 68%
                     2 NaNH$_2$ NH$_3$ Et$_2$O

                                             J Am Oil Chem Soc (1951) 28 27
                                             (Chem Abs 45 8449)
                                             Org Synth (1947) 27 76

Section 15   Acetylenes from Miscellaneous Compounds
              ⸰⸰

Review: The Synthesis of Acetylenes   Org React (1949) 5 1

PhCH=CHBr $\xrightarrow{\text{BuLi} \quad \text{Et}_2\text{O}}$ PhC≡CLi $\xdashrightarrow{\text{H}_2\text{O}}$ PhC≡CH

JACS (1940) <u>62</u> 2327

PhCH=CHCl $\xrightarrow{\text{PhLi} \quad \text{Et}_2\text{O}}$ PhC≡CH   100%

Ber (1941) <u>74</u>B 1474

PhCH=CHBr $\xrightarrow{\text{NaH} \quad \text{HMPA}}$ PhC≡CH   78%

Bull Soc Chim Fr (1966) 1293

$C_6H_{13}\underset{\text{Br}}{C}=CH_2$ $\xrightarrow{\text{NaH} \quad \text{HMPA}}$ $C_6H_{13}C\equiv CH$ 70%

Bull Soc Chim Fr (1966) 1293

$\underset{\text{Me}}{PhC}=CHBr$ $\xrightarrow{\text{NaNH}_2 \quad \text{HMPA} \quad C_6H_6}$ PhC≡CMe   40%

Bull Soc Chim Fr (1966) 1293

$\underset{\text{Cl} \quad \text{Cl}}{CF_3C=CCF_3}$ $\xrightarrow{\text{Zn} \quad \text{EtOH}}$ $CF_3C\equiv CCF_3$ 54%

JACS (1949) <u>71</u> 298

PhCH=CClF $\xrightarrow{\text{BuLi}}$ PhC≡CBu   ~20%

Angew (1963) <u>75</u> 638
(Internat Ed <u>2</u> 477)

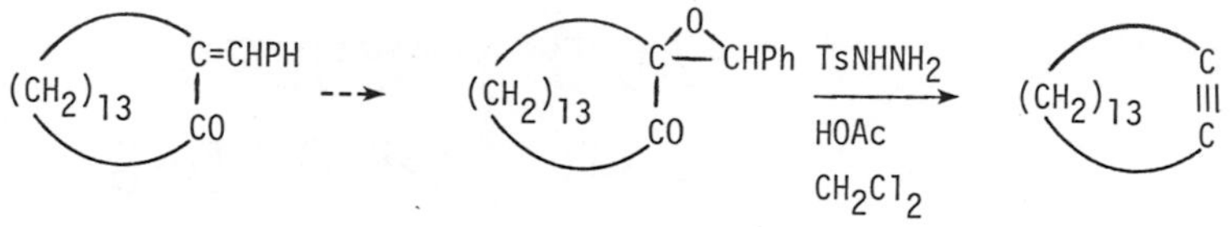

Helv (1967) <u>50</u> 2101

PhCOCHPh $\xrightarrow[\text{2 } C_6H_{13}COOH \quad 500°]{\text{1 } HC(OEt)_3 \quad HOAc}$ PhC≡CPh   40%
  |
 OH

Chem Comm (1970) 206

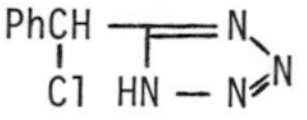

$\xrightarrow[\text{176°}]{\text{Mesitylene}}$ PhC≡CH   59%

Tetr Lett (1966) 1663

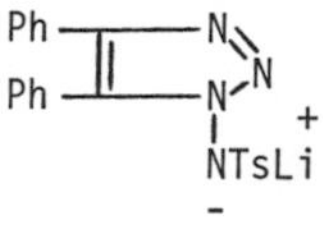

$\xrightarrow{\text{h}\nu \quad \text{dioxane}}$ PhC≡CPh   85%

Angew (1964) <u>76</u> 144
(Internat Ed <u>3</u> 138)

CH$_2$=CHCH$_2$NMe$_3$Br $\xrightarrow[\text{2 } 310\text{-}325°]{\text{1 } Ag_2O \quad H_2O}$ HC≡CMe   <34%

JOC (1949) <u>14</u> 1

Chapter 2   PREPARATION
OF
CARBOXYLIC ACIDS
ACID HALIDES
AND ANHYDRIDES

Section 16   <u>Carboxylic Acids from Acetylenes</u>

ooooooooooooooooooooooooooooooo

$BuC{\equiv}CH$ $\xrightarrow[\substack{\text{2 BuLi THF} \\ \text{3 } CO_2}]{\text{1 } (i\text{-}Bu)_2AlH}$ $BuCH_2CH(COOH)_2$ $\xrightarrow{\triangle}$ $BuCH_2CH_2COOH$

                                      62%

Tetr Lett (1966) 6021

$\begin{matrix} CH \\ {\ \ \ \ \ \ }\mathop{|||}\limits \\ C \\ | \\ (CH_2)_8 \\ | \\ COOH \end{matrix}$ $\xrightarrow[\text{2 } CO_2 \ \ Et_2O]{\text{1 } NaNH_2 \ \ NH_3}$ $\begin{matrix} CCOOH \\ \mathop{|||}\limits \\ C \\ | \\ (CH_2)_8 \\ | \\ COOH \end{matrix}$ $\xrightarrow{H_2 \ \ Ni \ \ H_2O}$ $\begin{matrix} CH_2COOH \\ | \\ CH_2 \\ | \\ (CH_2)_8 \\ | \\ COOH \end{matrix}$ 33%

JACS (1945) <u>67</u> 1171

$C_6H_{13}C{\equiv}CH$ $\xrightarrow[\text{2 TsCl}]{\text{1 Na}}$ $C_6H_{13}C{\equiv}CCl$ $\xrightarrow{\text{KOH} \ \ EtOH}$ $C_6H_{13}CH_2COOH$ 42%

Annales de Chimie (1931) <u>16</u> 309

16

BuC≡CH $\xrightarrow[\text{2 m-Chloroperbenzoic acid   THF}]{\text{1 Dicyclohexylborane   THF}}$ BuCH$_2$COOH

JACS (1967) <u>89</u> 291

PhC≡CH $\xrightarrow[\text{CH}_2\text{Cl}_2]{\text{CF}_3\text{COO}_2\text{H   Na}_2\text{HPO}_4}$ PhCOOH   +   PhCH$_2$COOH

25%   38%

JACS (1964) <u>86</u> 4866

C$_8$H$_{17}$C≡C(CH$_2$)$_7$COOH $\xrightarrow[\text{pH 12}]{\text{KMnO}_4\quad\text{H}_2\text{O}}$ HOOC(CH$_2$)$_7$COOH   80%

JOC (1952) <u>17</u> 1063

BuC≡CBu $\xrightarrow[\text{2 NaI}]{\text{1 O}_3\quad\text{CCl}_4\quad\text{HOAc}}$ BuCOOH   35%

Annalen (1953) <u>583</u> 29

$\xrightarrow[\text{2 H}_2\text{O}]{\text{1 O}_3\quad\text{CCl}_4}$ 41%

Carbohydrate Res (1966) <u>2</u> 315

Carboxylic acids may also be prepared by conversion of acetylenes into esters or amides, followed by hydrolysis. See section 106 (Esters from Acetylenes) and section 76 (Amides from Acetylenes)

Section 17   Carboxylic Acids, Acid Halides and Anhydrides
              from Carboxylic Acids and Acid Halides

$C_5H_{11}COCl$
  → 1 Cyclododecanone morpholine enamine Et_3N
     2 NaOH EtOH
     3 N_2H_4 KOH triethanolamine
$C_5H_{11}(CH_2)_{12}COOH$

~70%

Ber (1967) 100 4010

$$\begin{array}{c} COCl \\ | \\ (CH_2)_8 \\ | \\ CH \\ \| \\ CH_2 \end{array}$$
  → 1 MeCOCHCOOEt (Na) C_6H_6
     2 MeONa MeOH
     3 $I(CH_2)_{10}COOEt$ K_2CO_3
     4 KOH MeOH
$$\begin{array}{c} COOH \\ | \\ (CH_2)_{11} \\ | \\ CO \\ | \\ (CH_2)_8 \\ | \\ CH \\ \| \\ CH_2 \end{array}$$
  → N_2H_4
     KOH
     triethylene
     glycol
$$\begin{array}{c} COOH \\ | \\ (CH_2)_{11} \\ | \\ CH_2 \\ | \\ (CH_2)_8 \\ | \\ CH \\ \| \\ CH_2 \end{array}$$

Arkiv Kemi (1949) 1 99
(Chem Abs 43 7414)

$C_5H_{11}COCl$
  → 1 $NaC(COOCH_2Ph)_2$ [$(CH_2)_6COOCH_2Ph$]
     2 H_2 Pd-C
     3 78° (decarbox)
     4 H_2SO_4
$C_5H_{11}CO(CH_2)_7COOH$ → N_2H_4 / NaOH → $C_5H_{11}(CH_2)_8COOH$

56%

JCS (1950) 174

$$\begin{array}{c} COCl \\ | \\ (CH_2)_8 \\ | \\ CH \\ \| \\ CH_2 \end{array}$$
  → 1 $NaC(COO\text{-}\bigcirc\text{-}O)_2$ with $[(CH_2)_{10}COO\text{-}\bigcirc\text{-}O]$
     2 C_6H_6 reflux
$$\begin{array}{c} COOH \\ | \\ (CH_2)_{11} \\ | \\ CO \\ | \\ (CH_2)_8 \\ | \\ CH \\ \| \\ CH_2 \end{array}$$
  → N_2H_4 KOH
     diethylene
     glycol
$$\begin{array}{c} COOH \\ | \\ (CH_2)_{11} \\ | \\ CH_2 \\ | \\ (CH_2)_8 \\ | \\ CH \\ \| \\ CH_2 \end{array}$$

JCS (1952) 3945

$C_6H_{13}COCl$ $\xrightarrow[\text{2 } N_2H_4 \cdot H_2O \quad KOH]{\text{1 Thiophene} \quad SnCl_4}$ C_7H_{15}—⟨thiophene⟩—S $\xrightarrow[\substack{\text{anhydride} \\ AlCl_3 \\ \text{2 } N_2H_4 \quad KOH \\ \text{3 Ni} \quad Na_2CO_3}]{\text{1 Succinic}}$ $C_7H_{15}(CH_2)_7COOH$

JCS (1954) 4162
    (1962) 350

$EtCOOH$ $\xrightarrow[\substack{\text{2 } CH_2N_2 \quad Et_2O \\ \text{3 } CH_2=CO \quad toluene \\ \text{4 } N_2H_4 \quad NaOH \quad \text{diethylene glycol}}]{\text{1 } SOCl_2}$ $Et(CH_2)_3COOH$ 28%

Annalen (1964) <u>678</u> 113

MeO—⟨ring⟩—COCl $\xrightarrow[\substack{\text{2 Collidine} \quad PhCH_2OH \\ \text{3 KOH} \quad MeOH}]{\text{1 } EtCHN_2}$ MeO—⟨ring⟩—CHCOOH / Et   37%

JOC (1948) <u>13</u> 763

Cl—⟨ring⟩—COOH $\xrightarrow[\substack{\text{2 } MeCHN_2 \quad Et_2O \\ \text{3 } PhNEt_2 \quad PhCH_2OH \\ \text{4 Hydrolysis}}]{\text{1 DCC} \quad Et_2O}$ Cl—⟨ring⟩—CHCOOH / Me   17%

JCS <u>C</u> (1970) 971

PhCOCl
$\xrightarrow{\begin{array}{l}1 \ MeCHN_2 \quad Et_2O \\ \hline 2 \ Ag_2O \quad PhNH_2 \\ 3 \ Acid \ hydrolysis\end{array}}$
PhCHCOOH
|
Me

Chem Ind (1955) 1673

$C_{14}H_{29}CH_2COCl$ $\xrightarrow{\begin{array}{l}1 \ CH_2N_2 \quad Et_2O \\ \hline 2 \ HCl \quad Et_2O\end{array}}$ $C_{14}H_{29}CH_2COCH_2Cl$ $\xrightarrow{KOH}$ $C_{14}H_{29}CHCOOH$ 70%
|
Me

Chem Phys Lipids (1968) $\underline{2}$ 213

$\xrightarrow{\begin{array}{l}1 \ ClCOOEt \quad Et_3N \quad CH_2Cl_2 \\ \hline 2 \ CH_2N_2 \quad Et_2O \\ 3 \ h\nu \quad dioxane \quad H_2O\end{array}}$
CH$_2$COOH   ~34%

JCS $\underline{C}$ (1969) 1319
Tetr Lett (1969) 4517

$\xrightarrow{\begin{array}{l}1 \ CH_2N_2 \quad Et_2O \\ \hline 2 \ Ag_2O \quad Na_2CO_3 \quad Na_2S_2O_3 \\ dioxane \quad H_2O\end{array}}$
~74%

Org React (1942) $\underline{1}$ 38
JCS (1950) 926

$\xrightarrow{\begin{array}{l}1 \ Me_2\overset{+}{S}O\overset{-}{C}H_2 \quad THF \\ \hline 2 \ h\nu \quad H_2O\end{array}}$
~80%

JACS (1964) $\underline{86}$ 1640

$BuCH_2COOH$

1 NaH (i-Pr)$_2$NH THF
————————————————→
2 BuLi
3 C$_6$H$_{13}$Br

$BuCHCOOH$
$\overset{|}{C_6H_{13}}$

86%

JACS (1970) <u>92</u> 1397

$C_5H_{11}CH_2COOH$

1 BuLi (i-Pr)$_2$NH hexane THF
————————————————————→
2 HMPA
3 BuBr

$C_5H_{11}CHCOOH$
$\overset{}{Bu}$

93%

JOC (1970) <u>35</u> 262

$C_5H_{11}\overset{|}{\underset{Bu}{CHCOOH}}$

1 BuLi (i-Pr)$_2$NH hexane THF
————————————————————→
2 BuBr

$C_5H_{11}\overset{Bu}{\underset{Bu}{CCOOH}}$

50%

JOC (1970) <u>35</u> 262

$PhCH_2COOH$

1 Sodium-naphthalene THF
————————————————→
2 i-PrBr

$PhCHCOOH$
$\overset{}{i\text{-}Pr}$

88%

JOC (1967) <u>32</u> 2797

$PhCH_2COOH$

1 i-PrMgCl HMPA Et$_2$O
————————————————→
2 EtI

$PhCHCOOH$
$\overset{}{Et}$

Bull Soc Chim Fr (1964) 2000

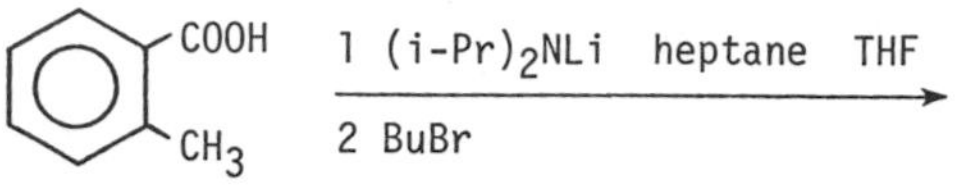

69-73%

JACS (1970) 92 1396

Alkylation of acids may also be accomplished via ester or amide intermediates. See section 113 (Esters from Esters) and section 81 (Amides from Amides)

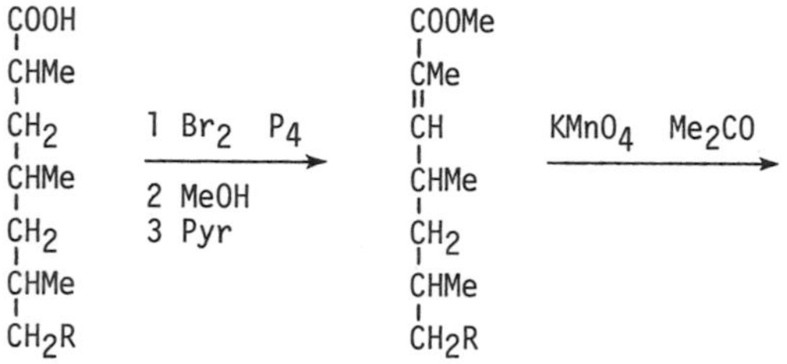

JCS (1963) 3081

$$PrCH_2CH_2COCl \xrightarrow[\text{2 i-PrCH}_2\text{OH} \quad \text{3 PhNEt}_2]{\text{1 Br}_2} PrCH{=}CHCOOCH_2Pr\text{-}i \xrightarrow[300\text{-}340^\circ]{KOH} PrCOOH$$

Biochem J (1951) 50 163

$$CCl_3COOH \xrightarrow[DMF]{SOCl_2} CCl_3COCl$$

89%

Helv (1959) 42 1653
Org Synth (1932) Coll Vol 1 12

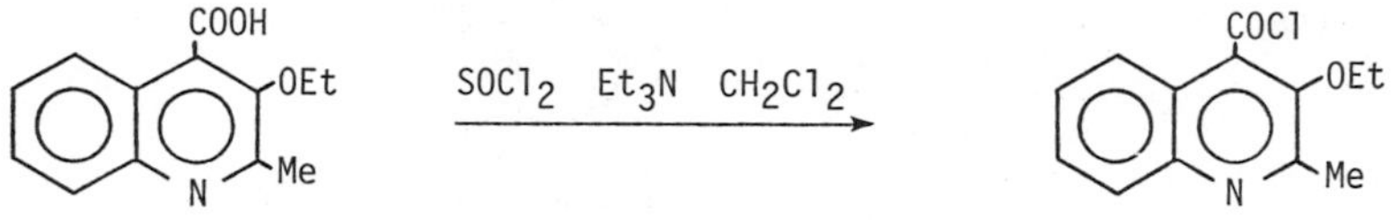

$$\xrightarrow{\text{SOCl}_2 \quad \text{Et}_3\text{N} \quad \text{CH}_2\text{Cl}_2}$$

JCS (1963) 491

$$\text{PhCH}_2\text{CH}_2\text{COOH} \xrightarrow{\text{(COCl)}_2 \quad \text{C}_6\text{H}_6} \text{PhCH}_2\text{CH}_2\text{COCl} \qquad 98\%$$

JACS (1920) 42 599
Can J Chem (1955) 33 1515

$$\text{t-BuCOOH} \xrightarrow{\text{PhCOCl}} \text{t-BuCOCl} \qquad 92\%$$

JACS (1938) 60 1325

$$\text{PhCH}_2\text{COOH} \xrightarrow{\text{PCl}_3} \text{PhCH}_2\text{COCl}$$

Org Synth (1943) Coll Vol 2 156

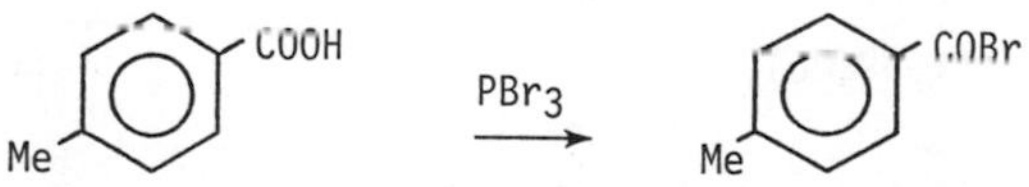

$$\xrightarrow{\text{PBr}_3}$$

65%

JCS (1934) 1406

$$\xrightarrow{\text{Ph}_3\text{P} \quad \text{CCl}_4}$$

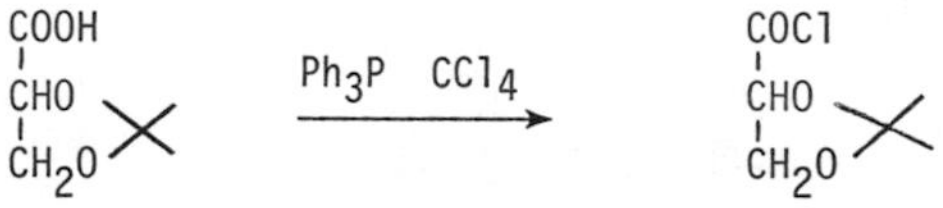

JACS (1966) 88 3440

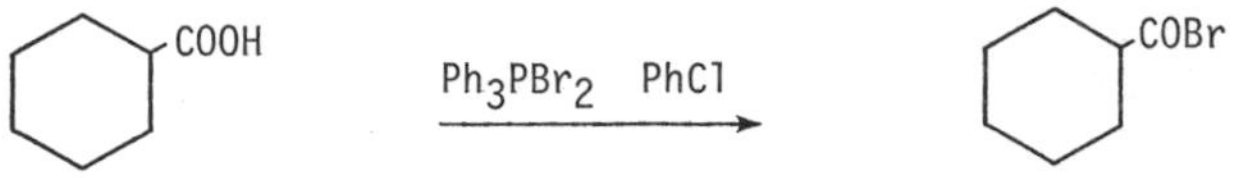

$$\text{(cyclohexane-COOH)} \xrightarrow[]{\text{Ph}_3\text{PBr}_2 \quad \text{PhCl}} \text{(cyclohexane-COBr)}$$

Annalen (1966) 693 132

$$C_{15}H_{31}COOH \xrightarrow[\text{HCl} \quad \text{CHCl}_3]{\text{NN'-Carbonyldiimidazole}} C_{15}H_{31}COCl \qquad 68\%$$

Annalen (1966) 694 78

$$PhCOOH \xrightarrow[]{\text{(benzene)} O,O-PCl_3} PhCOCl \qquad 77\%$$

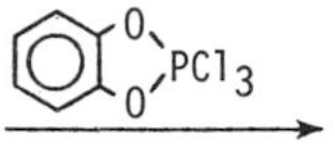

Ber (1963) 96 1387

$$MeCH=CHCOOH \xrightarrow[]{MeCCl_2OEt} MeCH=CHCOCl \qquad 79\%$$

Rec Trav Chim (1957) 76 969

$$C_{17}H_{35}COOH \xrightarrow[]{Ac_2O} (C_{17}H_{35}CO)_2O \qquad 50\text{-}80\%$$

JACS (1941) 63 699

$$PhCOOH \xrightarrow[]{TsCl \quad pyr} (PhCO)_2O \qquad 97\%$$

JACS (1955) 77 6214

$C_6H_{13}COONa$ $\xrightarrow{\quad C_6H_{13}COCl \quad H_2O \quad}$ $(C_6H_{13}CO)_2O$ 60%

JCS (1964) 755

$C_{15}H_{31}COOH$ $\xrightarrow{\quad DCC \quad CCl_4 \quad}$ $(C_{15}H_{31}CO)_2O$ 85%

J Lipid Res (1966) $\underline{7}$ 174

$Me[CH_2]_7\underset{OH}{CH}-\underset{OH}{CH}[CH_2]_7COOH$ $\xrightarrow[\text{Et}_3N \quad THF]{\quad ClCOOEt \quad}$ $(Me[CH_2]_7\underset{OH}{CH}-\underset{OH}{CH}[CH_2]_7CO)_2O$

JOC (1963) $\underline{28}$ 1905

$C_{15}H_{31}COOH$ $\xrightarrow{\quad \text{N-Trifluoroacetyl imidazole} \quad C_6H_6 \quad}$ $(C_{15}H_{31}CO)_2O$ 54%

Ber (1962) $\underline{95}$ 2073

$PhCOOAg$ $\xrightarrow{\quad CS_2 \quad}$ $(PhCO)_2O$ 98%

Proc Chem Soc (1957) 20

$C_6H_{13}COOH$ $\xrightarrow{\quad HC\equiv COMe \quad}$ $(C_6H_{13}CO)_2O$ 67%

JCS (1954) 1860

PhCOOH $\xrightarrow[\text{2 Distillation}]{\text{1 CH}_2\text{=CO}}$ $(PhCO)_2O$ 96%

JACS (1932) 54 3427

PrCOOH $\xrightarrow{\text{PhCOCH=CHCOPh Bu}_3\text{P C}_6\text{H}_6}$ $(PrCO)_2O$ 77%

JOC (1964) 29 1385

Section 18   Carboxylic Acids from Alcohols and Phenols
ooooooooooooooooooooooooooooooooooooooo

$\xrightarrow{\text{CH}_2\text{=CCl}_2 \text{ H}_2\text{SO}_4}$ 24%

Ber (1967) 100 978

PhCHOH $\xrightarrow{\text{CH}_2\text{=CCl}_2 \text{ H}_2\text{SO}_4 \text{ BF}_3}$ PhCHCH$_2$COOH   50%
| |
Me   Me

Angew (1965) 77 967
(Internat Ed 4 956)

t-BuOH $\xrightarrow{\text{HCOOH H}_2\text{SO}_4}$ t-BuCOOH   75%

Org Synth (1966) 46 72
JACS (1961) 83 3980

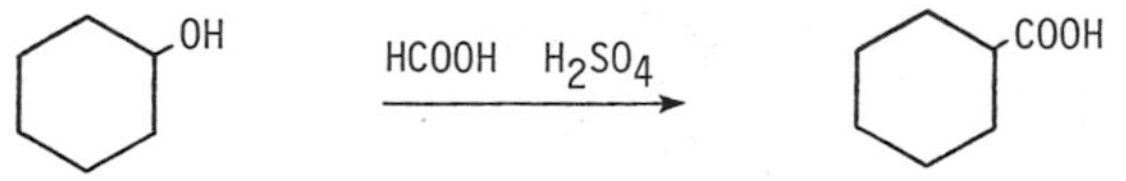

$$\text{cyclohexanol} \quad \xrightarrow{\text{HCOOH} \quad H_2SO_4} \quad \text{cyclohexane-COOH} \quad 75\%$$

Ber (1966) <u>99</u> 1149

$$\text{MeOH} \quad \xrightarrow[\text{HI} \quad H_2O \quad C_6H_6]{\text{CO (400 psi)} \quad RhCl_3 \cdot 3H_2O} \quad \text{MeCOOH} \quad 99\%$$

Chem Comm (1968) 1578

$$FCH_2(CH_2)_8CH_2OH \quad \xrightarrow{CrO_3 \quad HOAc \quad H_2O} \quad FCH_2(CH_2)_8COOH \quad 93\%$$

JACS (1956) <u>78</u> 2255
     (1960) <u>82</u> 6147

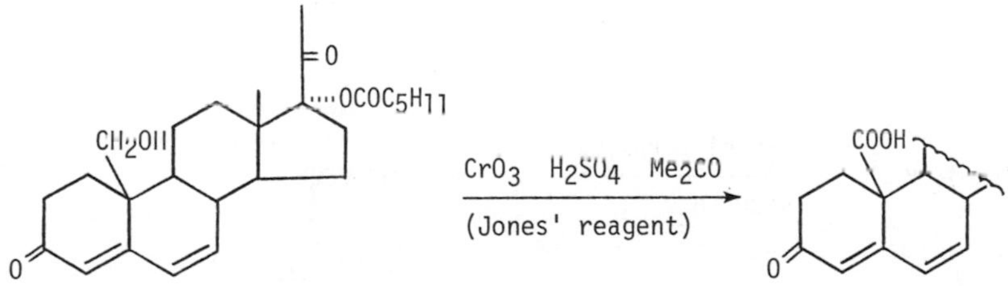

$$\xrightarrow[\text{(Jones' reagent)}]{CrO_3 \quad H_2SO_4 \quad Me_2CO}$$

77%

Helv (1967) <u>50</u> 269
J Med Chem (1970) <u>13</u> 926

$$Cl_3C(CH_2)_3CH_2OH \quad \xrightarrow{KMnO_4 \quad H_2O} \quad Cl_3C(CH_2)_3COOH \quad 92\%$$

Bull Chem Soc Jap (1963) <u>36</u> 1264

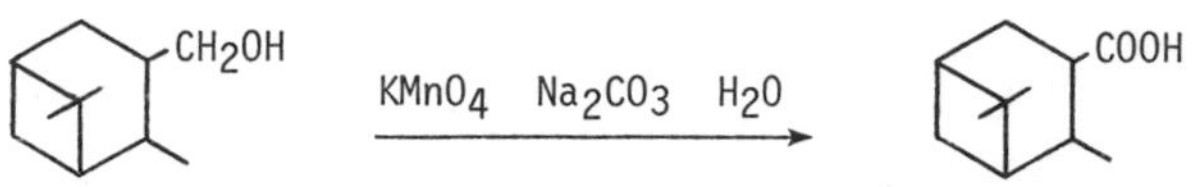
CH2OH
KMnO4 Na2CO3 H2O
COOH

66%

Bull Acad Polon (1964) 12 15
(Chem Abs 61 1895)
JACS (1950) 72 2953

$Br(CH_2)_5CH_2OH \xrightarrow{HNO_3} Br(CH_2)_5COOH$

JACS (1950) 72 5137

$C_9H_{19}CH_2OH \xrightarrow{N_2O_4} C_9H_{19}COOH$

Ber (1956) 89 202

$C_5H_{11}CH_2OH \xrightarrow{RuO_4 \quad CCl_4} C_5H_{11}COOH$ 10%

JACS (1958) 80 6682

$Ph(CH_2)_2CH_2OH \xrightarrow[NaOH \quad H_2O]{\text{Nickel peroxide}} Ph(CH_2)_2COOH$ 70%

JOC (1962) 27 1597

$PhCH{=}CHCH_2OH \xrightarrow[NaOH \quad H_2O]{\text{Nickel peroxide}} PhCH{=}CHCOOH$ 81%

JOC (1962) 27 1597

$$\underset{\underset{Me}{|}}{EtCHCH_2CH_2OH} \xrightarrow{\text{AgO}} \underset{\underset{Me}{|}}{EtCHCH_2COOH} \qquad 100\%$$

Tetr Lett (1968) 5685

$$(HOCH_2)_3CCH_2OH \xrightarrow{O_2 \quad Pt \quad NaHCO_3 \quad H_2O} (HOCH_2)_3CCOOH \qquad 50\%$$

Ber (1956) 89 1648
Angew (1957) 69 600

$$Me(CH_2)_{14}CH_2OH \xrightarrow{\text{t-Butyl chromate}} Me(CH_2)_{14}COOH \qquad 54\%$$

Bull Chem Soc Jap (1965) 38 893

$$PhCH_2OH \xrightarrow{CCl_4 \quad KOH \quad t\text{-}BuOH \quad H_2O} PhCOOH \qquad 75\%$$

JACS (1969) 91 7510

$$MeCH_2OH \xrightarrow{\text{Xenic acid} \quad H_2O} MeCOOH$$

JACS (1964) 86 2078

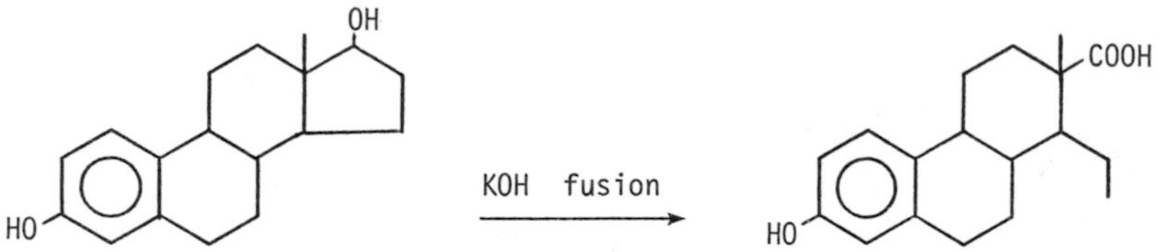

KOH   fusion

47%

Helv (1944) 27 1727
JACS (1948) 70 3485

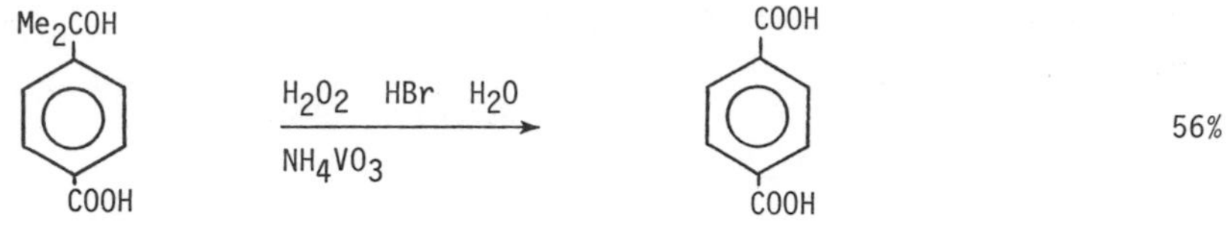
Me2COH
COOH
H2O2 HBr H2O
NH4VO3
COOH
COOH
56%

JCS (1961) 4082

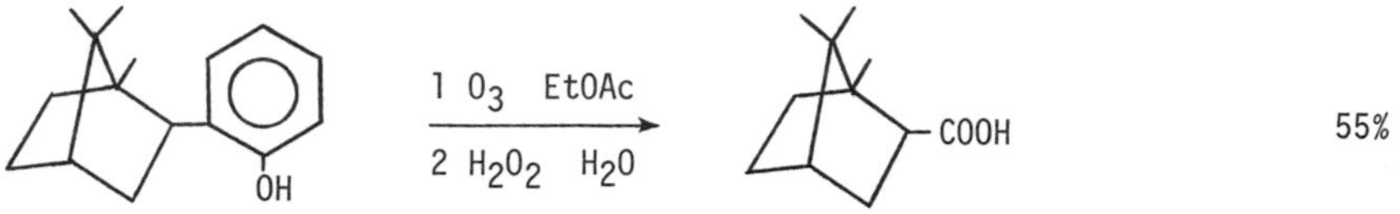
OH
1 O3 EtOAc
2 H2O2 H2O
COOH
55%

Zh Org Khim (1967) 3 1636
(Chem Abs 68 29874)
Tetr Lett (1967) 4729

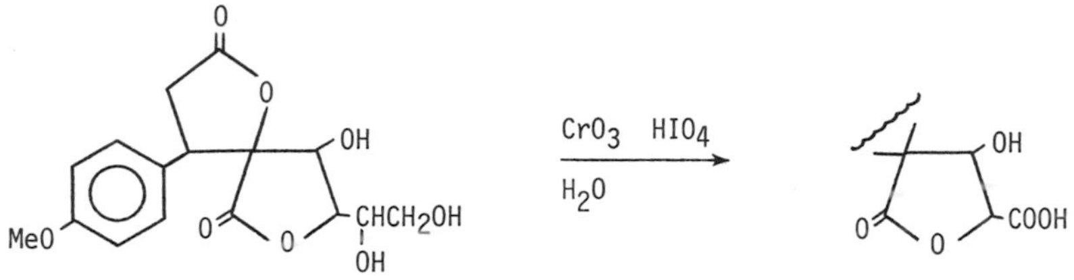
O
O
OH
MeO
O
O
CHCH2OH
OH
CrO3 HIO4
H2O
OH
O
O
COOH

JCS C (1966) 1918

$C_8H_{17}CH(OH)CH_2OH$ $\xrightarrow{O_2\ \text{cobalt laurate}\ \text{PhCN}}$ $C_8H_{17}COOH$ 70%

Tetr Lett (1968) 5689

Section 19   Carboxylic Acids and Acid Halides from Aldehydes

$$\xrightarrow{\begin{array}{c}1\ [(EtO)_2PO]_2CHNMe_2\quad NaH\quad dioxane\\ \hline 2\ HCl\quad H_2O\end{array}}$$

Angew (1968) 80 364
(Internat Ed 7 391)

i-PrCHO $\xrightarrow{\begin{array}{c}1\ (MeO)_3P=C\overset{S}{\underset{S}{\diagup}}\\ \hline 2\ Hydrolysis\end{array}}$ i-PrCH$_2$COOH

Tetr Lett (1967) 3201

$$\xrightarrow{\begin{array}{c}NaHSO_3\\ \hline NaCN\quad H_2O\end{array}} \qquad \xrightarrow{\begin{array}{c}HI\quad P_4\\ \hline H_2O\end{array}}$$

78%

JOC (1956) 21 1149

PhCHO $\xrightarrow{\begin{array}{c}1\ PhCONHCH_2COOH\\ \hline 2\ Base\ or\ acid\end{array}}$ PhCH$_2$COCOOH $\xrightarrow{H_2O_2}$ PhCH$_2$COOH

Org React (1942) 1 210

$$\xrightarrow{\begin{array}{c}1\ Rhodanine\quad NaOAc\quad HOAc\\ \hline 2\ NaOH\\ 3\ NH_2OH\\ 4\ Ac_2O\\ 5\ KOH\end{array}}$$

74%

JACS (1940) 62 1512
Org React (1942) 1 210

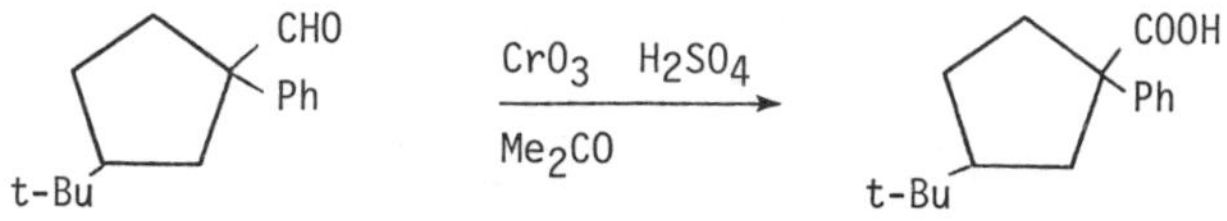

$$\xrightarrow[\text{Me}_2\text{CO}]{\text{CrO}_3 \quad \text{H}_2\text{SO}_4}$$

85%

JCS C (1970) 1168

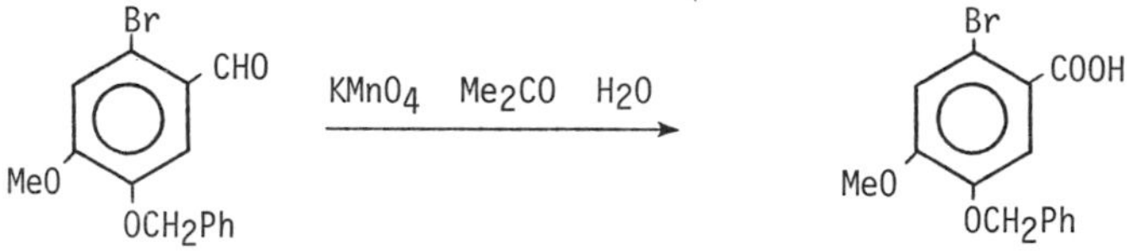

$$\xrightarrow{\text{KMnO}_4 \quad \text{Me}_2\text{CO} \quad \text{H}_2\text{O}}$$

JCS C (1970) 1208

$$\xrightarrow{\text{KMnO}_4 \quad \text{Pyr}}$$

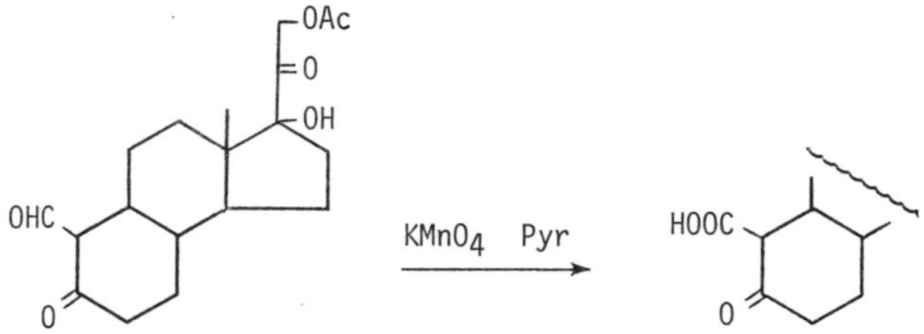

Steroids (1964) 3 639

$$C_6H_{13}CHO \xrightarrow{\text{KMnO}_4 \quad \text{H}_2\text{SO}_4 \quad \text{H}_2\text{O}} C_6H_{13}COOH \qquad 76\text{-}78\%$$

Org Synth (1943) Coll Vol 2 315

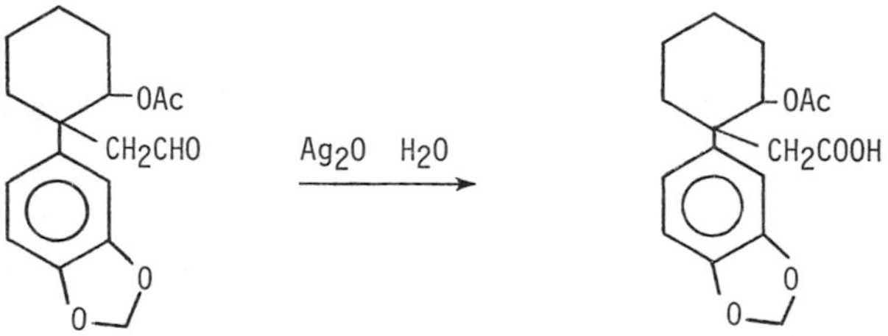

$$\xrightarrow{\text{Ag}_2\text{O} \quad \text{H}_2\text{O}}$$

93%

Tetrahedron (1968) 24 6583

$Me_2C=CH(CH_2)_2\underset{Me}{C}=CH(CH_2)_3CHO$ $\xrightarrow[\text{THF} \quad H_2O]{\text{AgO}}$ $Me_2C=CH(CH_2)_2\underset{Me}{C}=CH(CH_2)_3COOH$ 55%

Tetr Lett (1969) 1837
JACS (1968) 90 5617

PrCHO $\xrightarrow{\text{Argentic picolinate}}$ PrCOOH

Tetr Lett (1967) 415

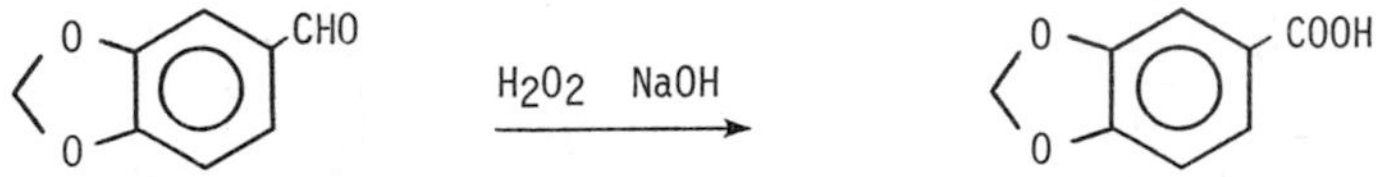

$\xrightarrow{H_2O_2 \quad NaOH}$ 90%

Monatsh (1955) 86 325

$C_6H_{13}CHO$ $\xrightarrow{MeCOO_2H}$ $C_6H_{13}COOH$ 88%

Org React (1957) 9 73

PrCHO $\xrightarrow[\text{PhCMe}_2 \quad OsO_4 \quad MeOH]{\overset{OOH}{|}}$ PrCOOH

JCS (1950) 2169

$\xrightarrow{SeO_2 \quad H_2O_2 \quad t\text{-BuOH}}$

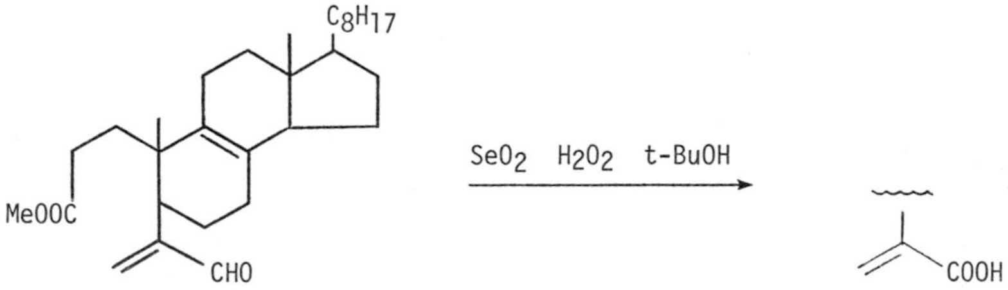

Chem Comm (1969) 945

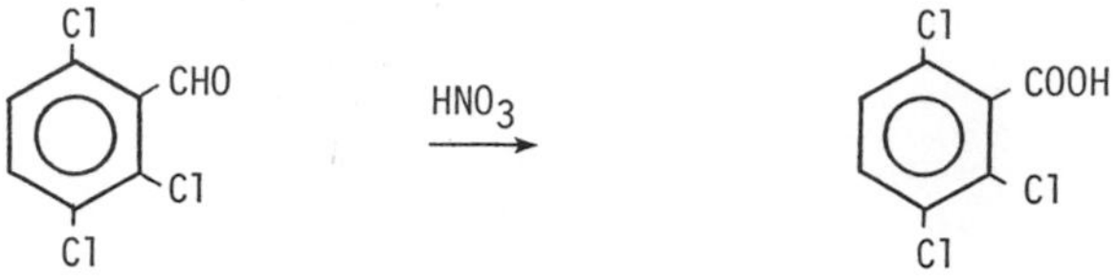

24%

JCS (1951) 1208

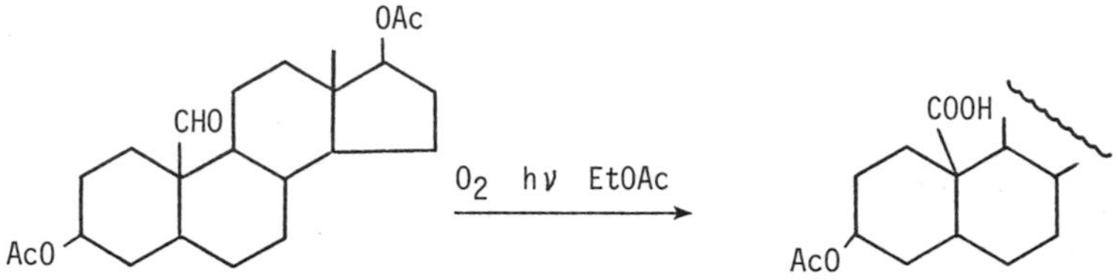

50%

Tetr Lett (1966) 2507

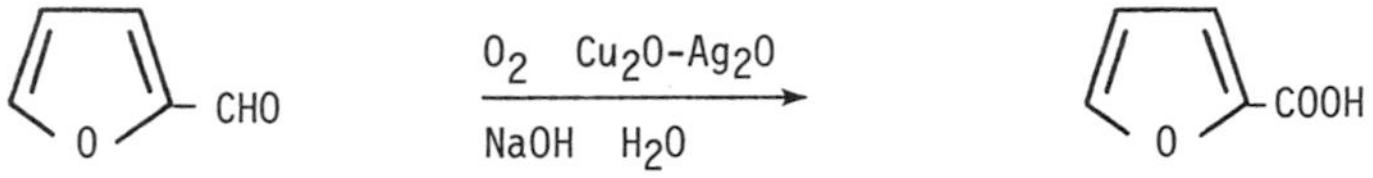

86-90%

Org Synth (1963) Coll Vol 4 493

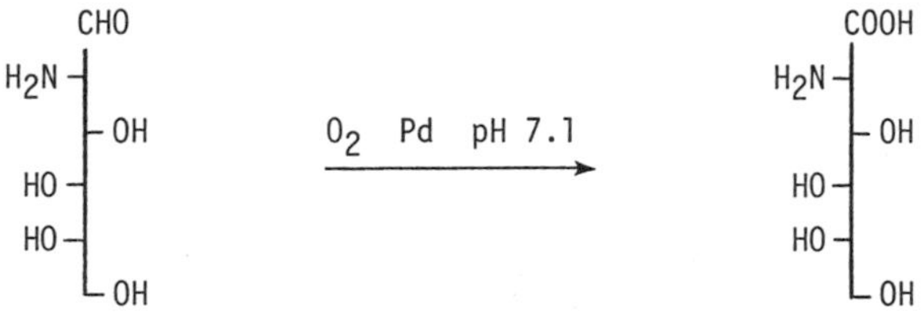

54-60%

Helv (1957) 40 2383

$\underset{\text{OMe}}{\overset{\text{CHO}}{\bigcirc}}$ $\xrightarrow{\text{KOH   H}_2\text{O}}$ $\underset{\text{OMe}}{\overset{\text{COOH}}{\bigcirc}}$ 34%

Org React (1944) <u>2</u> 94
Org Synth (1963) Coll Vol 4 974

BuCHCHO $\xrightarrow{\text{NaOH   H}_2\text{O}}$ BuCHCOOH   20%
 | |
 Et   Et

Helv (1951) <u>34</u> 1211

i-PrCHO $\xrightarrow[\text{2 KOH   diethylene glycol}]{\text{1 NH}_2\text{OH·HCl   NaOAc   EtOH   H}_2\text{O}}$ i-PrCOOH   88%

JOC (1962) <u>27</u> 629

MeOCHCHO   MeOCHCOOH
 | |
 O $\xrightarrow[\text{KHCO}_3\text{   H}_2\text{O}]{\text{I}_2\text{   KI   K}_2\text{CO}_3}$ O
 | |
MeOCHCHO   MeOCHCOOH

JACS (1954) <u>76</u> 3188

$\underset{\text{Cl}}{\overset{\text{CHO}}{\bigcirc}}$ $\xrightarrow{\text{Cl}_2}$ $\underset{\text{Cl}}{\overset{\text{COCl}}{\bigcirc}}$ 70%

Org Synth (1941) Coll Vol 1 155

i-PrCH$_2$CHO $\xrightarrow{\text{Benzoyl peroxide   CCl}_4}$ i-PrCH$_2$COCl   60%

JACS (1947) <u>69</u> 2916

Carboxylic acids may also be prepared by conversion of aldehydes into esters, followed by hydrolysis. See section 109 (Esters from Aldehydes). Some of the methods listed in section 27 (Carboxylic Acids from Ketones) may also be applied to the preparation of carboxylic acids from aldehydes

Section 20 Carboxylic Acids from Alkyls, Methylenes and Aryls

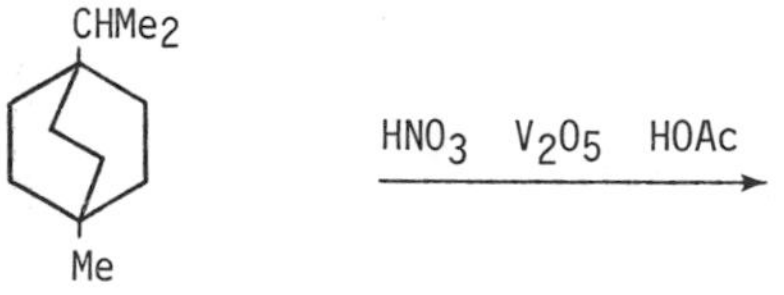
CHMe2
HNO3 V2O5 HOAc
COOH
Me
Me

25%

J Med Chem (1970) 13 254

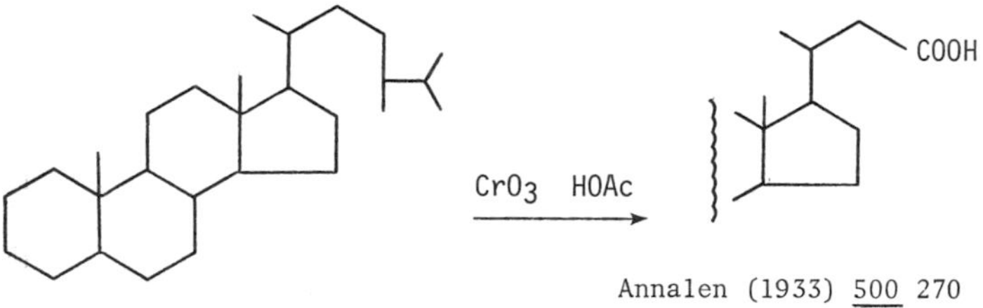
CrO3 HOAc
COOH

0.3%

Annalen (1933) 500 270

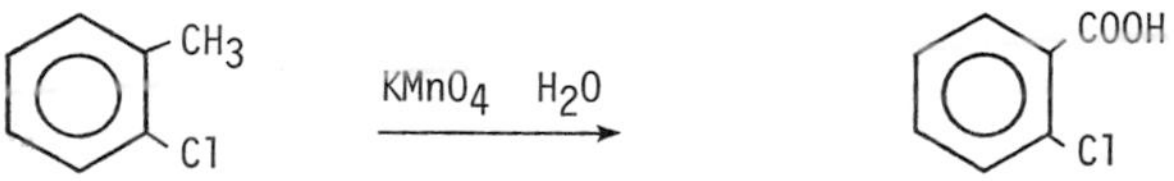
CH3
Cl
KMnO4 H2O
COOH
Cl

76-78%

Org Synth (1943) Coll Vol 2 135

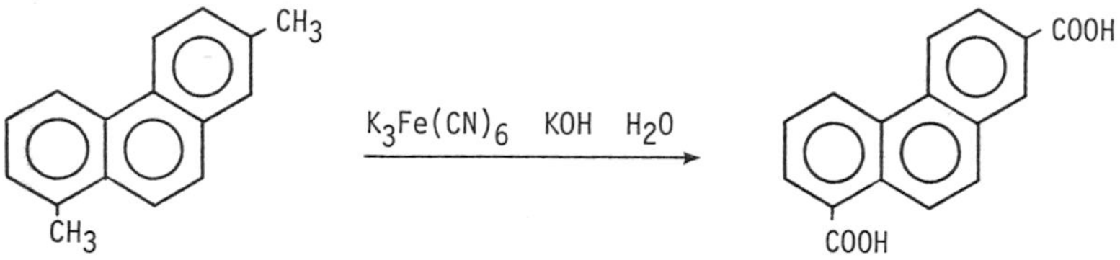
CH3
CH3
K3Fe(CN)6 KOH H2O
COOH
COOH

Helv (1931) 14 233

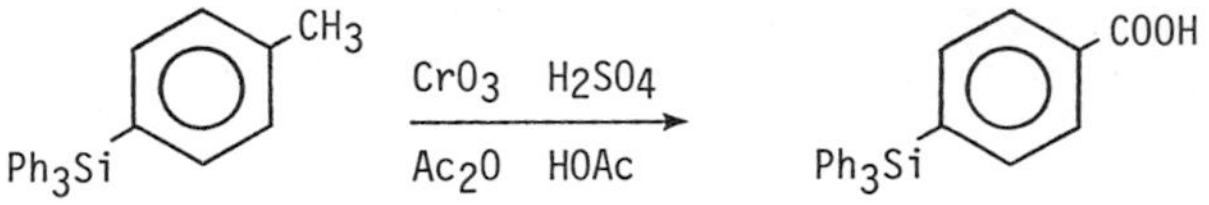

JACS (1956) 78 1689

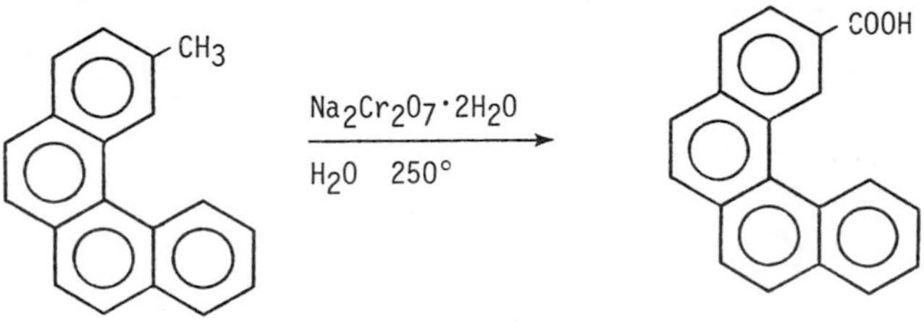

81%

JOC (1961) 26 1759

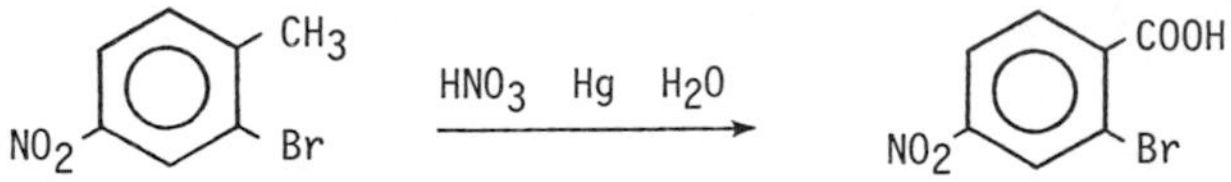

64%

Ber (1961) 94 834

76%

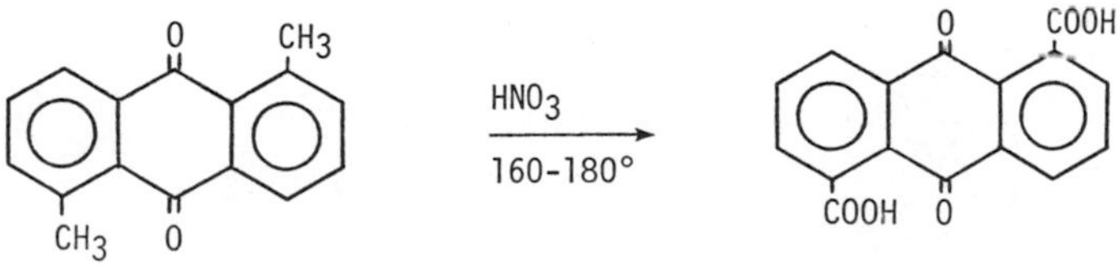

JCS (1960) 341
JOC (1960) 25 668

PhCH$_3$ $\xrightarrow{\text{Argentic picolinate}}$ PhCOOH

Tetr Lett (1967) 415

$$PhCH_3 \xrightarrow{Cl_2} PhCCl_3 \xrightarrow{TiO_2} PhCOCl \dashrightarrow PhCOOH$$

JACS (1958) <u>80</u> 3483

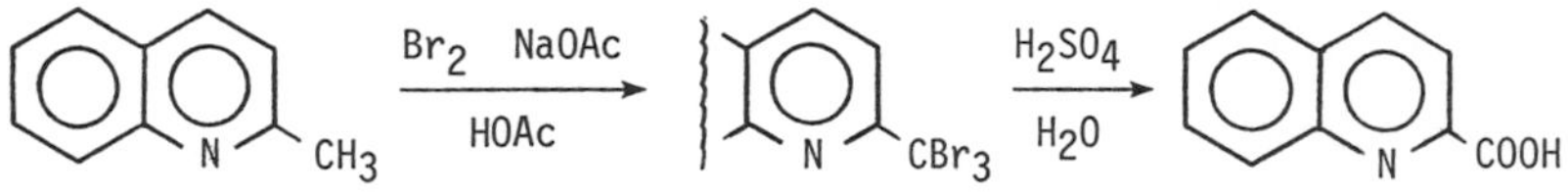

JACS (1946) <u>68</u> 1840

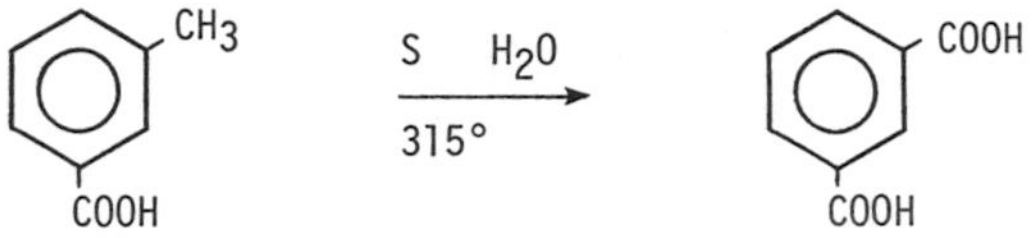

~70%

JOC (1961) <u>26</u> 2929

100%

$$PhCHMe_2 \xrightarrow[315°]{SO_2 \quad H_2O} PhCOOH$$

JOC (1961) <u>26</u> 2929

24%

$$PhCH_2Me \xrightarrow[H_2O]{H_2O_2 \quad NH_4VO_3 \quad HBr} PhCOOH$$

JCS (1961) 4082

40%

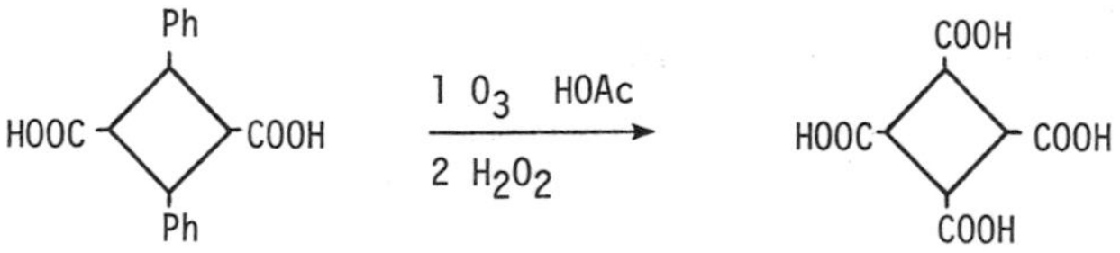

Ber (1960) <u>93</u> 2521

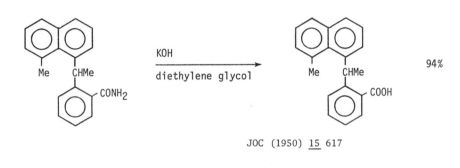

RuO₄   NaIO₄

CCl₄   H₂O

COOH

25%

Tetr Lett (1967) 4729
Chem Comm (1970) 1420

Section 21   Carboxylic Acids from Amides
           ͜ʊʊʊʊʊʊʊʊʊʊʊʊʊʊʊʊʊʊʊʊʊʊʊʊʊʊʊʊ

Me   CHMe

CONH₂

KOH

diethylene glycol

Me   CHMe

COOH

94%

JOC (1950) 15 617

CH₂CONH₂

HCl   HOAc   H₂O

CH₂COOH

90%

JACS (1941) 63 2494

Me

CONH₂

Me

H₃PO₄

145-150°

Me

COOH

Me

70%

Rec Trav Chim (1927) 46 600

N   CONH₂
COPh

Ion exch resin (acid)

Me₂CO   H₂O

N   COOH
COPh

61%

Chem Ind (1957) 736

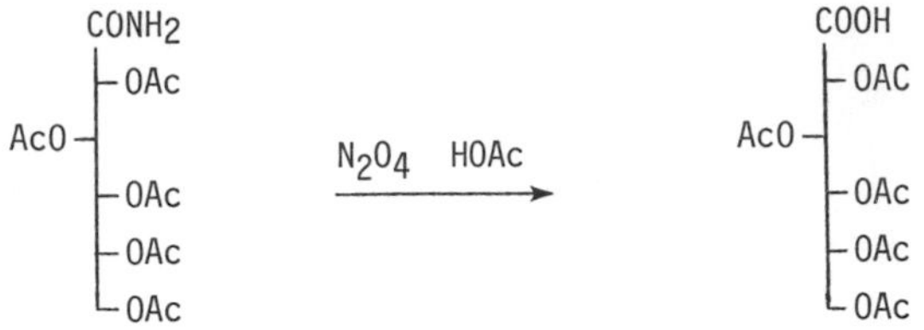

CONH$_2$
 ├─ OAc
AcO ─┤
 ├─ OAc
 ├─ OAc
 └─ OAc

$\xrightarrow{\text{N}_2\text{O}_4 \quad \text{HOAc}}$

COOH
 ├─ OAC
AcO ─┤
 ├─ OAc
 ├─ OAc
 └─ OAc

58%

JACS (1938) 60 235

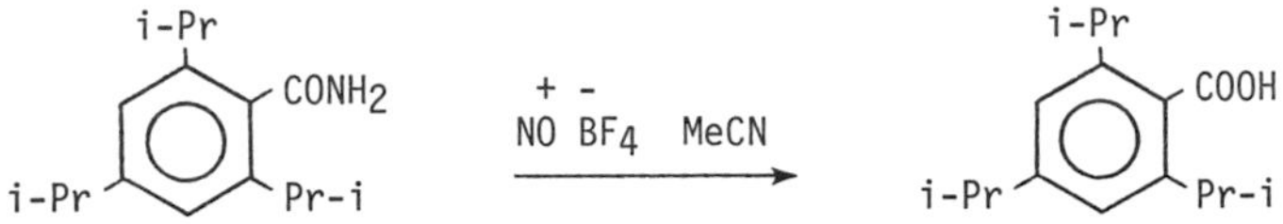

$\xrightarrow{\text{NO}^+ \text{BF}_4^- \quad \text{MeCN}}$

70%

JOC (1965) 30 2386

Bu$_3$CCONH$_2$

$\xrightarrow{\text{BuONO} \quad \text{HCl} \quad \text{HOAc}}$

Bu$_3$CCOOH   79%

JACS (1948) 70 3091
J Med Chem (1966) 9 603

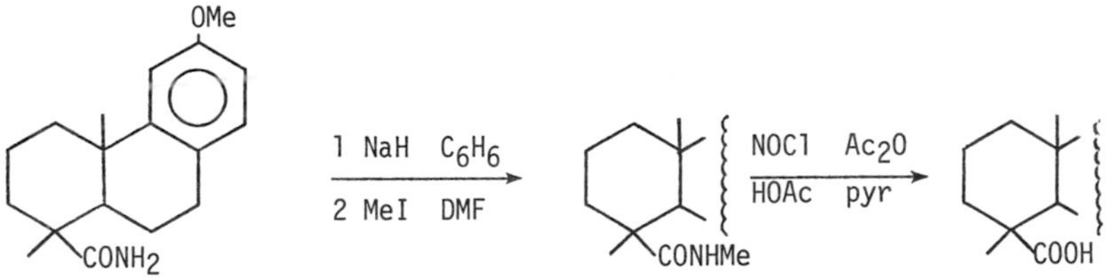

$\xrightarrow[\text{2 MeI \quad DMF}]{\text{1 NaH \quad C}_6\text{H}_6}$

$\xrightarrow[\text{HOAc \quad pyr}]{\text{NOCl \quad Ac}_2\text{O}}$ 66%

JACS (1961) 83 1492

PhCONHCHMe
       |
       Et

$\xrightarrow[\text{2 Pentane \quad 25°}]{\text{1 NaNO}_2 \quad \text{Ac}_2\text{O}}$

PhCOOH   64%

JACS (1955) 77 6011

PhCON⟨ring⟩ $\xrightarrow{\text{⟨catechol⟩}O \atop O}PCl_3$ PhCOOH   67%

Ber (1963) <u>96</u> 1387

$EtOOCCH_2NHCOCH(CH_2)_2CONHNHPh \atop \quad\quad\quad NHCOOCH_2Ph$ $\xrightarrow[H_2O]{MnO_2 \quad HOAc}$ $EtOOCCH_2NHCOCH(CH_2)_2COOH \atop \quad\quad\quad NHCOOCH_2Ph$ 82%

JOC (1963) <u>28</u> 453

$NHR' \atop RCHCONHNHPh$ $\xrightarrow[\text{cellosolve} \quad H_2O]{FeCl_3 \quad HCl}$ $NHR' \atop RCHCOOH$ 85%

JACS (1957) <u>79</u> 637   645

Carboxylic acids may also be prepared by conversion of amides into esters, followed by hydrolysis. See section 111 (Esters from Amides)

Section 22   Carboxylic Acids from Amines
∘∘∘∘∘∘∘∘∘∘∘∘∘∘∘∘∘∘∘∘∘∘∘∘∘∘∘∘∘∘

$CH_3CH_2NEt_2$ $\xrightarrow[2 \; CO_2]{1 \; i\text{-PrLi}}$ $i\text{-Pr}(CH_2)_2COOH$ 25%

JACS (1969) <u>91</u> 6362

JCS (1962) 686

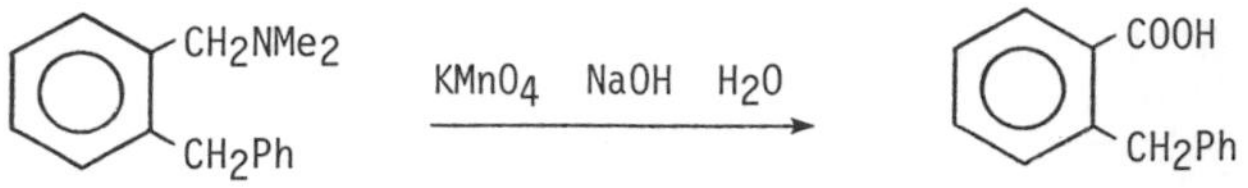

CH2NMe2
CH2Ph
KMnO4 NaOH H2O
COOH
CH2Ph

27%

JACS (1951) 73 4122

Section 23 Carboxylic Acids and Acid Halides from Esters

$C_9H_{19}CH{=}C(Me)COOMe \xrightarrow{\text{KOH EtOH}} C_9H_{19}CH{=}C(Me)COOH$

Org Synth (1963) Coll Vol 4 608

$MeOOC(CH_2)_9COOMe \xrightarrow{Ba(OH)_2\ \ MeOH} MeOOC(CH_2)_9COOH$ 60%

Org Synth (1963) Coll Vol 4 635

$(i\text{-}Pr)_3CCOOMe \xrightarrow{t\text{-}BuOK\ \ Me_2SO} (i\text{-}Pr)_3CCOOH$ 100%

Tetr Lett (1964) 2969
JCS (1965) 1290

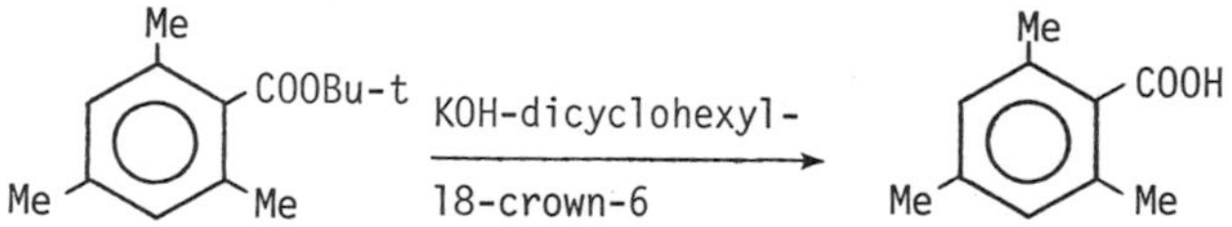

Me
COOBu-t
Me
Me
KOH-dicyclohexyl-
18-crown-6
Me
COOH
Me
Me

94%

JACS (1967) 89 7017

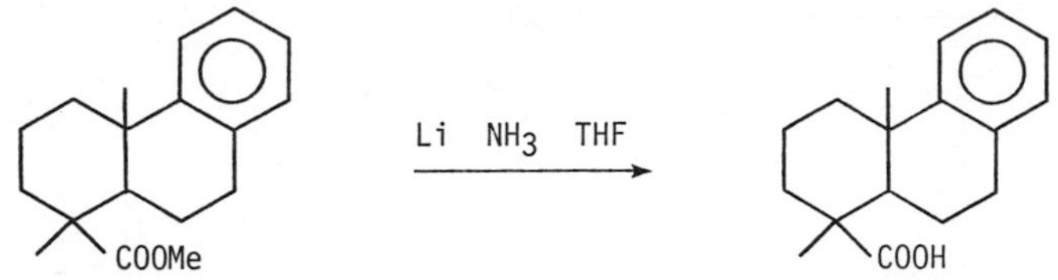

$\xrightarrow{\text{Li \quad NH}_3 \quad \text{THF}}$

59%

JACS (1958) 80 217

PhCOOMe $\xrightarrow{\text{Me}_3\text{N \quad MeOH}}$ PhCOOH

JACS (1933) 55 4079

MeCOOPh $\xrightarrow{\text{Guanidine \quad H}_2\text{O \quad EtOH}}$ MeCOOH

Bull Chem Soc Jap (1966) 39 852
JACS (1964) 86 837

BrCH$_2$CHCOOMe $\xrightarrow{\text{HBr \quad H}_2\text{O}}$ BrCH$_2$CHCOOH   72%
       | |
       Br   Br

JACS (1940) 62 3495

MeCOOC$_5$H$_{11}$ $\xrightarrow{\text{Ion exch resin (acid) \quad H}_2\text{O}}$ MeCOOH

JCS (1952) 1607

Further examples of the reaction RCOOR' → RCOOH + R'OH are included in
section 38 (Alcohols from Esters) and section 30A (Protection of
Carboxylic Acids)

PhCOOBu-t $\xrightarrow{\text{MeOH reflux 4 days}}$ PhCOOH   23%

JACS (1941) 63 3382

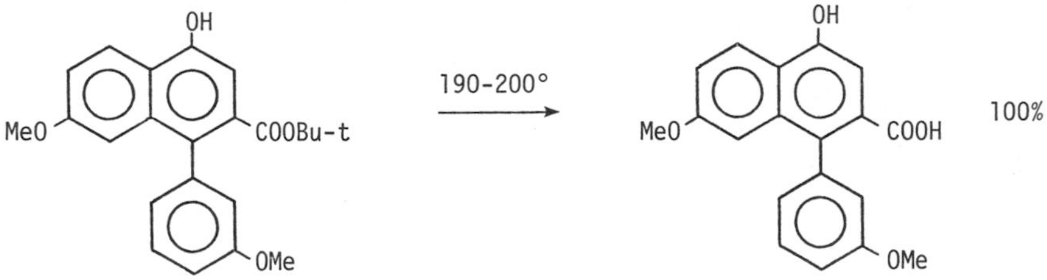

190-200°   100%

JOC (1962) 27 519

LiI DMF

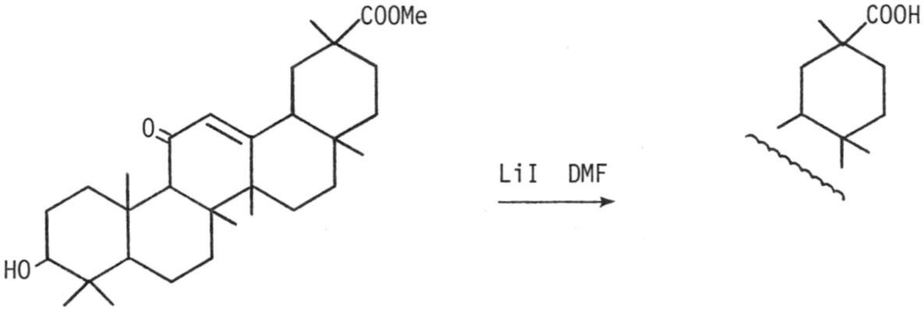

JCS (1965) 6655
For cleavage of ethyl esters see JOC (1963) 28 2184

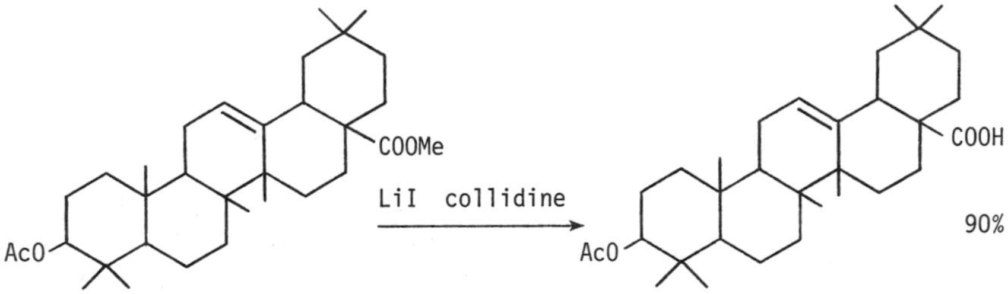

LiI collidine   90%

Helv (1960) 43 113

$(i\text{-}Pr)_3CCOOMe$ $\xrightarrow{\quad PrSLi \quad HMPA \quad}$ $(i\text{-}Pr)_3CCOOH$ 99%

Tetr Lett (1970) 4459

$PhCOOCH_2Ph$ $\xrightarrow{\quad H_2 \quad Pd \quad Et_2O \quad}$ PhCOOH

JOC (1958) 23 1700
Org React (1953) 7 263

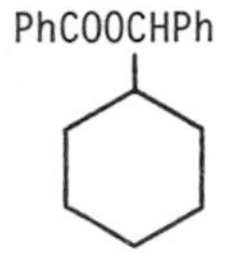

PhCOOCHPh $\xrightarrow{\quad H_2 \quad Pd \quad Et_2O \quad}$ PhCOOH

JOC (1958) 23 1700

Further examples of the cleavage of benzyl esters are included in section 30A (Protection of Carboxylic Acids)

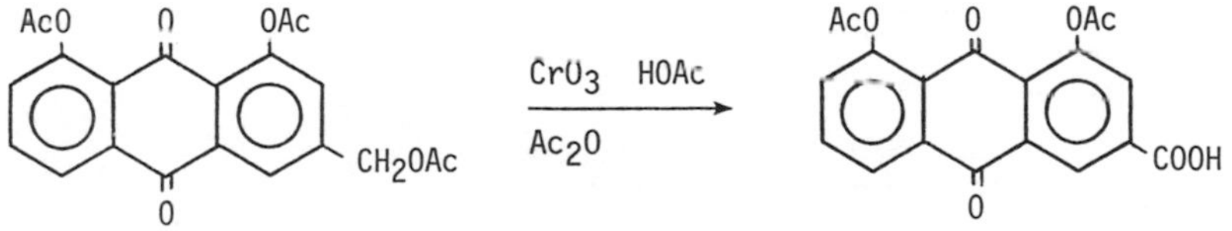

CrO3   HOAc
Ac2O

JOC (1961) 26 979

1 O3   EtOAc
2 H2O2   HOAc

RuO4   NaIO4

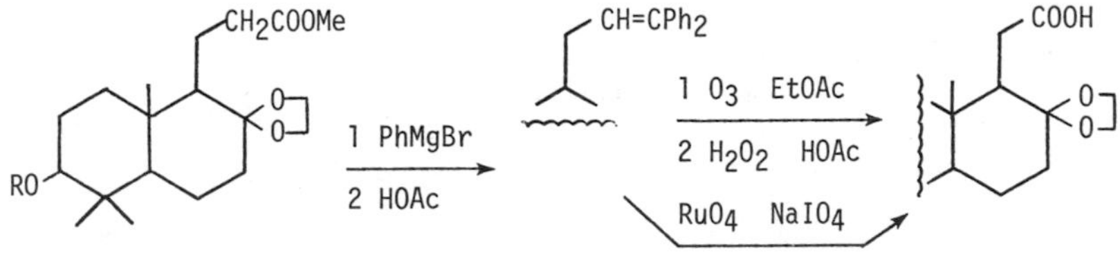

JACS (1963) 85 3419

$C_7H_{15}COOC=CH_2$ $\xrightarrow{\text{HF} \quad Et_2O}$ $C_7H_{15}COF$ 50%
     |
     Me

JOC (1969) <u>34</u> 2486

PhCOOBu   PhCOCl   91%

Ber (1963) <u>96</u> 1387

Section 24 Carboxylic Acids from Ethers

CH_3CH_2OEt $\xrightarrow[\text{2 } CO_2]{\text{1 i-PrLi} \quad Et_2O}$ $i\text{-}PrCH_2CH_2COOH$ 5%

JACS (1953) <u>75</u> 1771
(1955) <u>77</u> 2806

$CH_2=CHCH_2OPh$ $\xrightarrow[\text{2 } CO_2]{\text{1 Mg} \quad BrCH_2CH_2Br \quad THF}$ $CH_2=CHCH_2COOH$ 63%

J Organometallic Chem (1969) <u>18</u> 249

$(EtCH_2)_2O$ $\xrightarrow{Br_2 \quad H_2O}$ $EtCOOH$ 100%

JACS (1967) <u>89</u> 3550

$(PhCH_2)_2O$ $\xrightarrow{Br_2 \quad H_2O}$ $PhCOOH$ 55%

JACS (1967) <u>89</u> 3550

$Me(CH_2)_{14}CH_2OEt$ $\xrightarrow{\quad CrO_3 \quad HOAc \quad CH_2Cl_2 \quad}$ $Me(CH_2)_{14}COOH$ 55%

Chem Comm (1966) 752

Carboxylic acids may also be prepared by oxidation of ethers to esters, followed by hydrolysis. See section 114 (Esters from Ethers)

Section 25   Carboxylic Acids and Acid Halides from Halides

Cl
|
$(CH_2)_4$
|
t-Bu
$\xrightarrow[\substack{2 \ EtOOC(CH_2)_{14}CHO \\ 3 \ KOH \quad EtOH \quad H_2O}]{1 \ Mg \quad Et_2O}$
COOH
|
$(CH_2)_{14}$
|
CHOH
|
$(CH_2)_4$
|
t-Bu
$\xrightarrow[\substack{2 \ KOH \\ 3 \ H_2 \quad Pt \quad MeOH}]{1 \ PBr_3}$
COOH
|
$(CH_2)_{19}$
|
t-Bu

JACS (1950) 72 5139

Br
|
CH_2
|
CHMe
|
Et
$\xrightarrow[\substack{2 \ CdCl_2 \\ 3 \ MeOOC(CH_2)_2COCl}]{1 \ Mg \quad Et_2O}$
COOMe
|
$(CH_2)_2$
|
CO
|
CH_2
|
CHMe
|
Et
$\xrightarrow[\substack{H_2O}]{Zn \quad HCl}$
COOH
|
$(CH_2)_4$
|
CHMe
|
Et
    46%

JACS (1944) 66 46

$PhCH_2Cl$ $\xrightarrow[\substack{2 \ N_2H_4 \quad NaOH \quad MeOH}]{1 \quad \text{(cyclohexane-1,3-dione)} \quad KI \quad KOH \quad H_2O}$ $PhCH_2(CH_2)_5COOH$ 69%

Ber (1952) 85 61 1061

$C_{16}H_{33}Br$ $\xrightarrow[\text{2 Cyclohexanone}]{\text{1 Mg Et}_2\text{O}}$

cyclohexane ring with OH and $C_{16}H_{33}$ $\xrightarrow[\text{2 N}_2\text{H}_4\ \ \text{NaOH}]{\text{1 CrO}_3\ \ \text{HOAc}}$ $C_{16}H_{33}(CH_2)_5COOH$

73%

JACS (1948) 70 3352

benzene ring with Br and Me $\xrightarrow[\text{2 Cyclopentanone}]{\text{1 Mg Et}_2\text{O}}$

benzene ring with cyclopentyl-OH and Me $\xrightarrow[\substack{\text{2 N}_2\text{H}_4\ \ \text{NaOH}\\ \text{diethylene}\\ \text{glycol}}]{\text{1 CrO}_3\ \ \text{HOAc}}$ benzene ring with $(CH_2)_4COOH$ and Me

JOC (1958) 23 584

$C_9H_{19}\underset{\underset{Me}{|}}{C}HCH_2Br$ $\xrightarrow[\substack{\text{2 }\ \text{furan-CHO}\\ \text{3 HCl EtOH}\\ \text{4 N}_2\text{H}_4\ \ \text{KOH}}]{\text{1 Mg Et}_2\text{O}}$ $C_9H_{19}\underset{\underset{Me}{|}}{C}HCH_2(CH_2)_4COOH$

Z Physiol Chem (1951) 287 65

cyclohexane ring with Cl $\xrightarrow[\substack{\text{2 CH}_2\text{=CH}_2\\ \text{3 CO}_2}]{\text{1 Li Et}_2\text{O}}$ cyclohexane ring with $(CH_2)_2COOH$

47%

JACS (1969) 91 6362

$PhCH_2Cl$ $\xrightarrow[\text{KOH t-BuOH}]{\substack{\text{1 NaCH(SO}_2\text{Et})_2\\ \text{2 CH}_2\text{=CHCN}}}$ $PhCH_2\underset{\underset{SO_2Et}{|}}{\overset{\overset{SO_2Et}{|}}{C}}CH_2CH_2CN$ $\xrightarrow[\substack{\text{HOAc}\\ \text{2 Ni NaOH}\\ \text{H}_2\text{O}}]{\text{1 HCl H}_2\text{O}}$ $PhCH_2(CH_2)_3COOH$

JACS (1952) 74 1225

BuBr $\xrightarrow[\text{NaOEt}]{\text{MeCOCH}_2\text{COOEt}}$ BuCHCOOEt $\xrightarrow{\text{KOH H}_2\text{O}}$ BuCH$_2$COOH
                                    |
                                   COMe

Org Synth (1941) Coll Vol 1 248
JACS (1930) <u>52</u> 5005

1 Mg THF

2 CdCl$_2$

3 BrCH$_2$COOEt

4 NaOH

62%

JOC (1968) <u>33</u> 1675

C$_8$H$_{17}$I $\xrightarrow[\text{2 HCl H}_2\text{O dioxane}]{\text{1 CH}_3\text{COOBu-t NaNH}_2\text{ NH}_3}$ C$_8$H$_{17}$CH$_2$COOH   70%

JACS (1959) <u>81</u> 5817

PhCH$_2$Cl $\xrightarrow[\text{2 HCl H}_2\text{O dioxane}]{\text{1 EtCH}_2\text{COOBu-t NaNH}_2\text{ NH}_3}$ PhCH$_2$CHCOOH   92%
                                                                        |
                                                                        Et

JACS (1959) <u>81</u> 5817

Further examples of the alkylation of esters with halides are included
in section 113 (Esters from Esters). Examples of the alkylation of acids
with halides are included in section 17 (Carboxylic Acids from Carboxylic
Acids)

PhCH$_2$I $\xrightarrow[\text{2 KOH MeOH}]{\text{1 Ph}_3\text{P=CHCOOMe EtOAc}}$ PhCH$_2$CH$_2$COOH   75%

Tetr Lett (1960) (4) 5
Ber (1962) <u>95</u> 2921

$C_{18}H_{37}I$ $\xrightarrow{\begin{array}{l}\text{1 EtCH(COOMe)}_2 \quad \text{Na} \quad \text{BuOH}\\ \text{2 NaOH} \quad H_2O\\ \text{3 160-170°}\end{array}}$ $C_{18}H_{37}\underset{\underset{Et}{|}}{C}HCOOH$ 83%

JCS (1953) 3031
Org React (1957) 9 107

t-BuCl $\xrightarrow[\text{BF}_3\cdot\text{Et}_2\text{O}]{\text{CH}_2(\text{COOEt})_2}$ t-BuCH(COOEt)$_2$ $\xrightarrow{\text{NaOH}}$ t-BuCH$_2$COOH
                                              50%

Naturwiss (1964) 51 288

RBr $\xrightarrow{\begin{array}{c}\text{[oxazine]} \quad \text{BuLi}\end{array}}$ [oxazine-CH$_2$R] $\xrightarrow[\text{H}_2\text{O}]{\text{HBr} \quad \text{NaBr}}$ RCH$_2$COOH   85%

JACS (1969) 91 5886

t-BuCl $\xrightarrow{\text{CH}_2=\text{CCl}_2 \quad \text{BF}_3 \quad \text{H}_2\text{SO}_4}$ t-BuCH$_2$COOH   78%

Angew (1965) 77 967
(Internat Ed 4 956)

$\xrightarrow[\text{2 CO}_2]{\text{1 Mg} \quad \text{BrCH}_2\text{CH}_2\text{Br} \quad \text{Et}_2\text{O}}$ 56%

(Procedure for unreactive halides)
JOC (1959) 24 504

$\underset{\underset{Et}{|}}{\overset{\overset{Me}{|}}{Pr}C}Cl$ $\xrightarrow[\text{2 CO}_2 \quad (50 \text{ psi})]{\text{1 Mg} \quad \text{Et}_2\text{O}}$ $\underset{\underset{Et}{|}}{\overset{\overset{Me}{|}}{Pr}C}COOH$ 25%

JACS (1949) 71 1877

$$\xrightarrow[\text{2 } CO_2]{\text{1 BuLi   } Et_2O \text{ } C_6H_6}$$

JACS (1939) 61 1371

58%

i-PrI $\xrightarrow{LiC(SPr-i)_3}$ i-PrC(SPr-i)$_3$ $\xrightarrow[H_2O]{HgCl_2 \text{ } Me_2CO}$ i-PrCOOH

Angew (1967) 79 468
(Internat Ed 6 442)

RI $\xrightarrow[\substack{2 \text{ } O_2 \\ 3 \text{ } HgCl_2}]{1 \text{ } \overset{S}{\underset{S}{\langle\rangle}} \text{ BuLi}}$ RCOOH

Angew (1965) 77 1134
(Internat Ed 4 1075)

$$\xrightarrow[AlCl_3 \text{ } CS_2]{}$$

$$\xrightarrow[KOH \text{ } H_2O]{KMnO_4}$$

Helv (1935) 18 721

55%

PhBr $\xrightarrow[KOAc \text{ } H_2O]{CO \text{ (300 kg/cm}^2\text{)   } Ni(OAc)_2}$ PhCOOH

Bull Chem Soc Jap (1967) 40 2203

MeCH=CHCH$_2$Cl $\xrightarrow[(\pi\text{-allyl-PdCl})_2]{CO \text{ (500 atmos)}}$ MeCH=CHCH$_2$COCl   81%

JCS (1964) 1588

$$PhCH_2Cl \xrightarrow[\text{(Ph}_3\text{P)}_2\text{Rh(CO)Cl}]{\text{CO (100 atmos)}} PhCH_2COCl$$

Tetr Lett (1966) 4713
JACS (1968) <u>90</u> 99

$$\underset{\overset{|}{Et}}{Me_2CCl} \xrightarrow{\text{CO} \quad \text{AgClO}_4 \quad \text{PhNO}_2} \underset{\overset{|}{Et}}{Me_2CCOOH} \qquad 100\%$$

JACS (1960) <u>82</u> 1261

$$BuBr \dashrightarrow (BuS)_2 \xrightarrow[\text{di-t-butyl peroxide}]{\text{CO} \quad (\text{EtO})_3\text{P}} BuCOSBu \xrightarrow{\text{NaOH}} BuCOOH$$

JACS (1960) <u>82</u> 2181

$$BuI \xrightarrow{Me_2NLi \quad CHBr_3 \quad HMPA} BuCBr_3 \xrightarrow{\text{NaOH}} BuCOOH$$

Compt Rend (1967) <u>C</u> <u>264</u> 1609

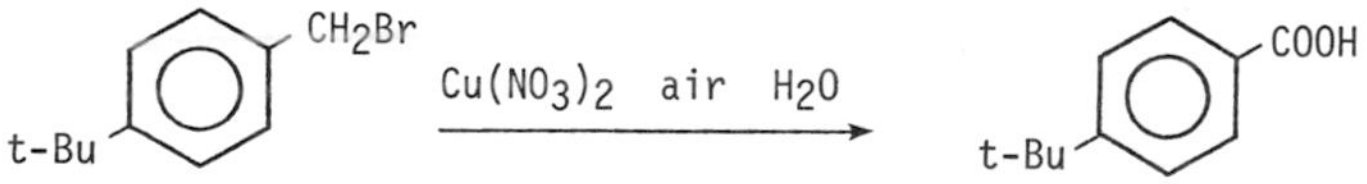

JCS (1935) 1847

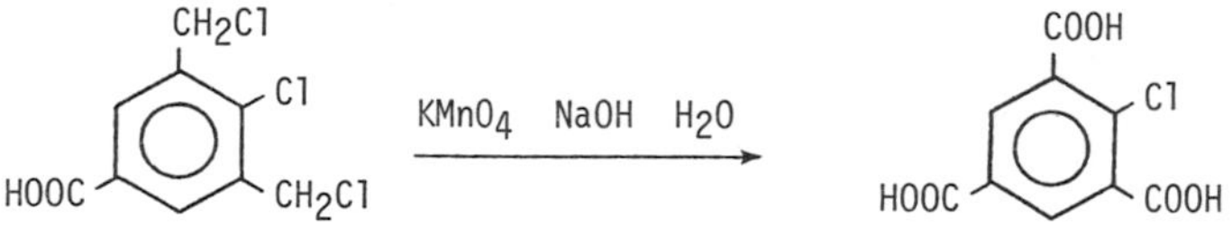

J Pharm Soc Jap (1950) <u>70</u> 538

F—⟨benzene ring⟩—⟨benzene ring⟩—CHBr₂ —(SeO₂)→ F—⟨benzene ring⟩—⟨benzene ring⟩—COOH   13%

JACS (1966) 88 3318
JCS (1962) 186

Carboxylic acids may also be prepared by conversion of halides into esters or amides followed by hydrolysis. See section 115 (Esters from Halides) and section 85 (Amides from Halides)

Section 26   Carboxylic Acids from Hydrides (RH)
ooooooooooooooooooooooooooooooooooooooo

Reactions in which a hydrogen is replaced by carboxyl or a carboxyl containing chain, e.g. RH → RCOOH or RCH₂COOH (R=alkyl, vinyl or aryl), are included in this section. For reactions in which alkyl or aryl groups are oxidized to carboxyl, e.g. RCH₃ → RCOOH, see section 20 (Carboxylic Acids from Alkyls, Methylenes and Aryls)

⟨cyclohexene⟩ —(ICH₂COOH hν)→ ⟨cyclohexene⟩—CH₂COOH

ACS Div Petr Chem (1966) 11 241
(Chem Abs 66 104510)

⟨cyclohexane⟩—Me —(CO HF SbF₅)→ ⟨cyclohexane with Me and COOH⟩

Ber (1967) 100 984

⟨cyclohexane⟩ —((COCl)₂ hν)→ ⟨cyclohexane⟩—COCl ---→ ⟨cyclohexane⟩—COOH   63%

JCS (1963) 3918

$C_8H_{17}H$ $\xrightarrow{\begin{array}{c}1\ (i\text{-}Bu)_2Hg\text{-}K\\ \hline 2\ CO_2\end{array}}$ $C_8H_{17}COOH$ 24-35%

Monatsh (1967) 98 763

$PhCH_3$ $\xrightarrow{\begin{array}{c}1\ Na\quad PhCl\\ \hline 2\ CO_2\end{array}}$ $PhCH_2COOH$ 77%

JACS (1940) 62 1514
Chem Rev (1957) 57 867

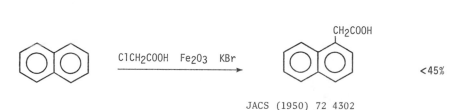

$\xrightarrow{\begin{array}{c}1\ CO(CN)_2\\ \hline 2\ H_2O\quad dioxane\end{array}}$

Chem Ind (1961) 1116

PhH $\xrightarrow{\qquad (CH_2)_nCOO \qquad AlCl_3 \qquad}$ $Ph(CH_2)_nCOOH$ 44% (n=3)
51% (n=4)

JACS (1952) 74 1591

$\xrightarrow{ClCH_2COOH\quad Fe_2O_3\quad KBr}$ <45%

JACS (1950) 72 4302
Synthesis (1970) 628

$PhCH_2Me$ $\xrightarrow{\begin{array}{c}1\ BuLi\cdot Me_2NCH_2CH_2NMe_2\\ \hline 2\ CO_2\end{array}}$ $PhCHCOOH$ +
$\underset{Me}{|}$

JOC (1970) 35 10
Chimia (1970) 24 109

MeO — (ring) $\xrightarrow[\text{Ac}_2\text{O} \quad \text{HOAc}]{\text{PdCl}_2 \quad \text{NaOAc}}$ MeO — (ring) — COOH

Tetr Lett (1969) 1019

NO_2 — (ring) $\xrightarrow[\text{Ac}_2\text{O} \quad \text{HOAc}]{\text{PdCl}_2 \quad \text{NaOAc}}$ NO_2 — (ring) — COOH

Tetr Lett (1969) 1019

PhH $\xrightarrow{\text{CO}_2 \text{ (or COCl}_2) \quad \text{AlCl}_3}$ PhCOCl ---→ PhCOOH   low yield

Chem Rev (1955) 55 229

Me — (ring) — Me $\xrightarrow[\text{CH}_2\text{Cl}_2]{\text{(benzodioxole-CCl}_2)} \quad \text{AlCl}_3$ Me — (ring) — Me, COOH   91%

Ber (1963) 96 1382

(ring)—COOMe $\xrightarrow[\text{H}_2\text{SO}_4 \quad \text{HCl}]{\text{Paraformaldehyde}}$ CH_2Cl—(ring)—COOH $\xrightarrow[\text{NaOH} \quad \text{H}_2\text{O}]{\text{KMnO}_4}$ COOH—(ring)—COOH

J Pharm Soc Jap (1950) 70 535

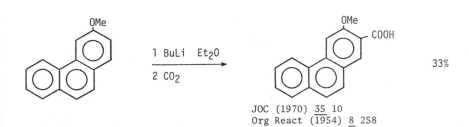

$\xrightarrow[\text{2 CO}_2]{\text{1 BuLi} \quad \text{Et}_2\text{O}}$ 33%

JOC (1970) 35 10
Org React (1954) 8 258

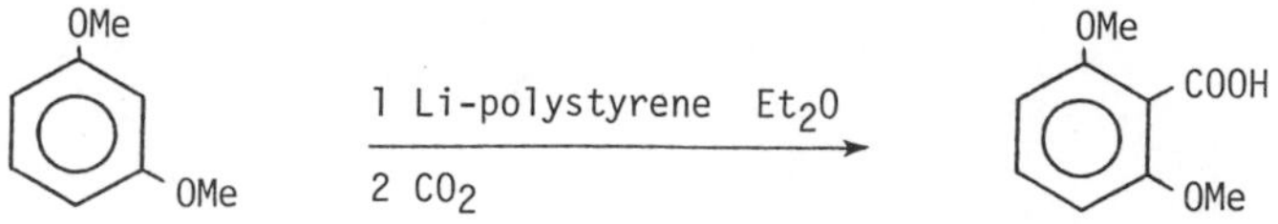

60%

Ber (1964) 97 3098

Carboxylic acids may also be prepared by conversion of hydrides (RH) into esters or amides, followed by hydrolysis. See section 116 (Esters from Hydrides) and section 86 (Amides from Hydrides)

Section 27   Carboxylic Acids from Ketones
○○○○○○○○○○○○○○○○○○○○○○○○○○○○○○○○

$$C_5H_{11}COCHMe_2 \xrightarrow[\text{t-BuOH \quad MeOH}]{\text{CH}_2\text{=CHCN \quad KOH}} C_5H_{11}COCMe_2 \underset{\underset{\text{CH}_2\text{CH}_2\text{CN}}{}}{} \xrightarrow[\substack{2 \text{ N}_2\text{H}_4 \\ \text{NaOH}}]{1 \text{ KOH}} C_5H_{11}CH_2CMe_2 \underset{\text{CH}_2\text{CH}_2\text{COOH}}{}$$

JCS (1948) 1741

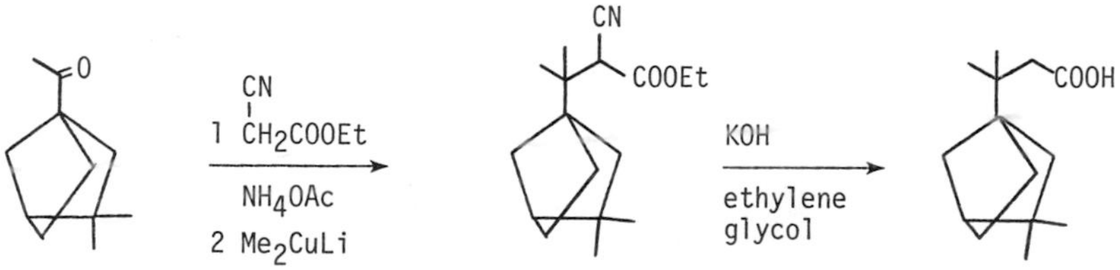

43%

Tetr Lett (1967) 2893

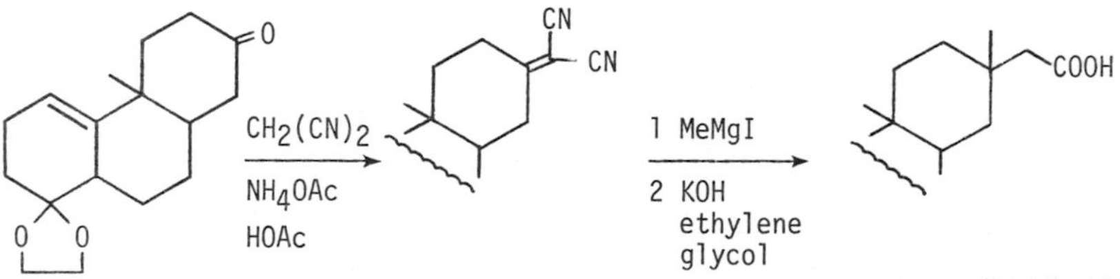

Tetr Lett (1964) 1763

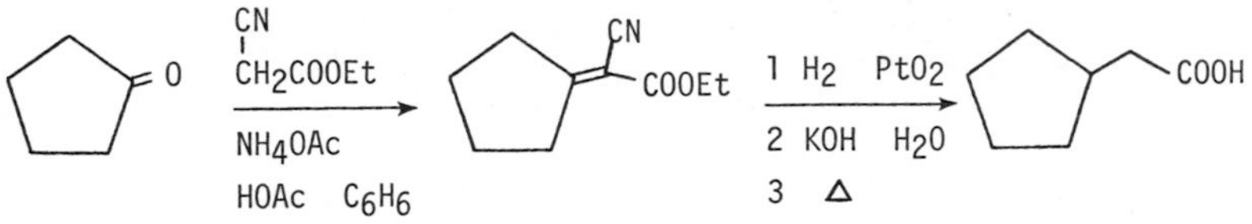

JACS (1948) 70 497
(1959) 81 5397

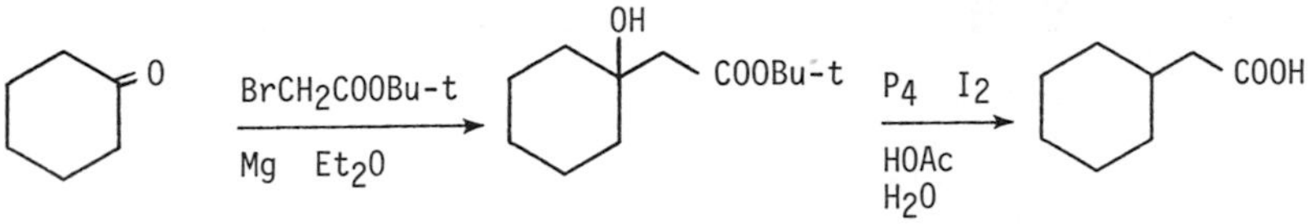

57%

JOC (1966) 31 983
JACS (1946) 68 2339

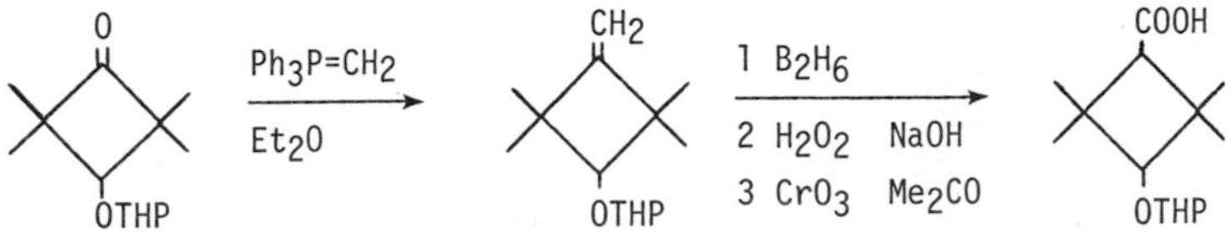

Chem Comm (1968) 206
JACS (1960) 82 2498

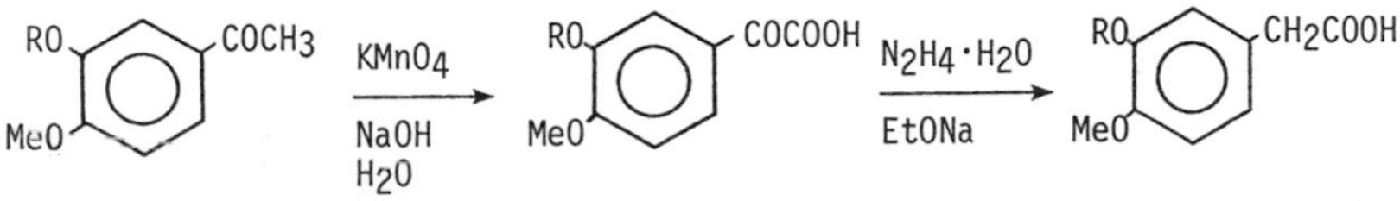

39%

Monatsh (1952) 83 883

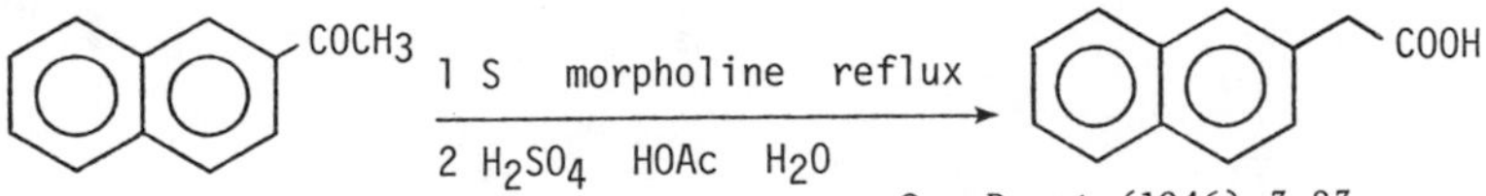

Org React (1946) 3 83

76%

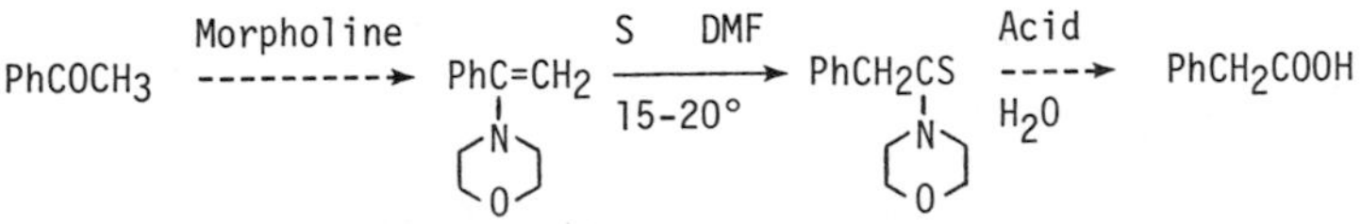

Angew (1964) 76 861
(Internat Ed 3 705)

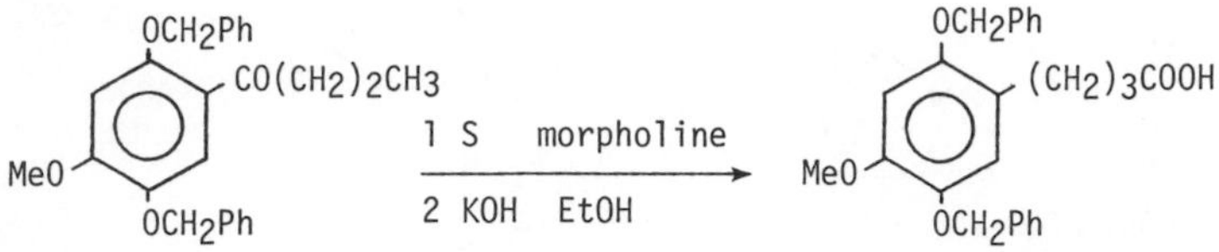

1 S   morpholine

2 KOH   EtOH

37%

Helv (1964) 47 1996

$Ph_2CHCOCH_3$ $\xrightarrow[\text{t-BuOH} \quad H_2O]{CCl_4 \quad KOH}$ Ph_2CHCH_2COOH 70%

JACS (1969) 91 7510

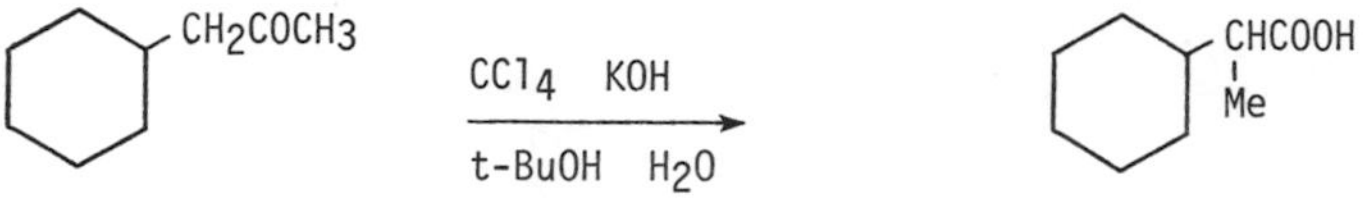

$\xrightarrow[\text{t-BuOH} \quad H_2O]{CCl_4 \quad KOH}$

60%

JACS (1969) 91 7510

KOH

Et_2O

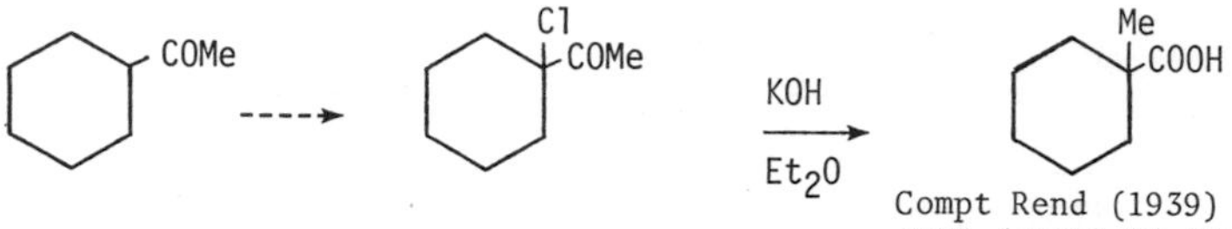

Compt Rend (1939) 208 1020
JACS (1960) 82 4307
Org React (1960) 11 261

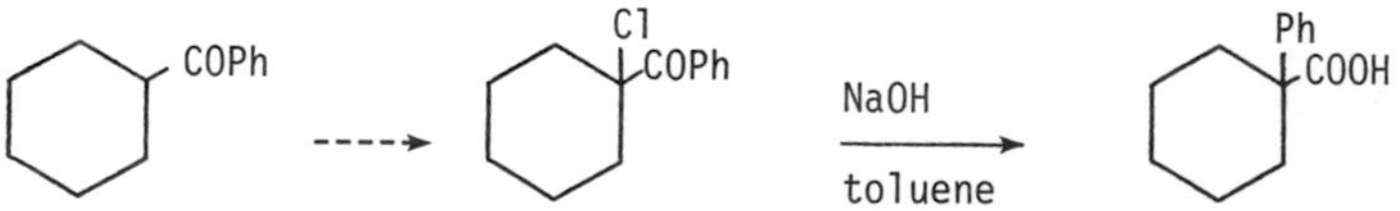

NaOH

toluene

51%

JACS (1952) 74 5352

H_2O_2 SeO_2 t-BuOH

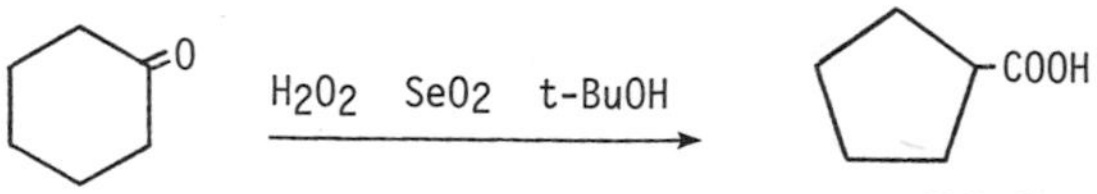

< 27%

JOC (1957) 22 1680
Compt Rend C (1967) 265 578
Annalen (1965) 681 30

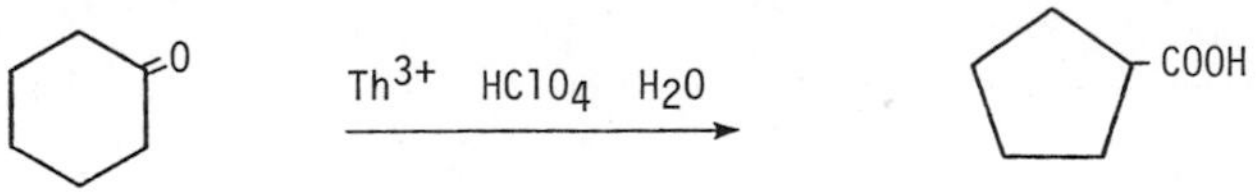
O
Th3+ HClO4 H2O
COOH

Tetr Lett (1966) 1779

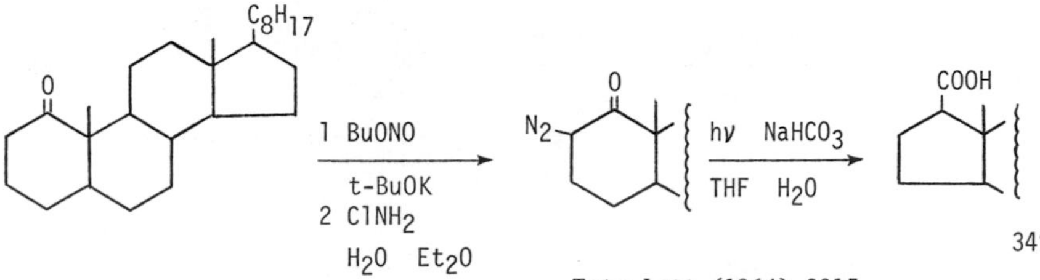
C8H17
O
1 BuONO
t-BuOK
2 ClNH2
H2O Et2O
N2
O
hν NaHCO3
THF H2O
COOH

34%

Tetr Lett (1964) 2813

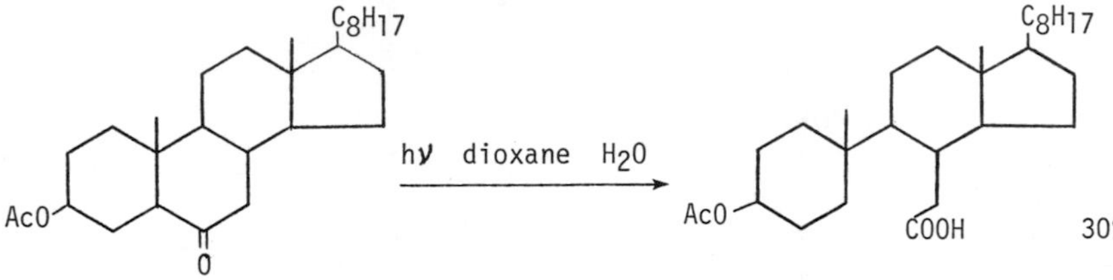
C8H17
AcO
O
hν dioxane H2O
C8H17
AcO
COOH

30%

Ber (1964) 97 958
Angew (1965) 77 229
(Internat Ed 4 211)

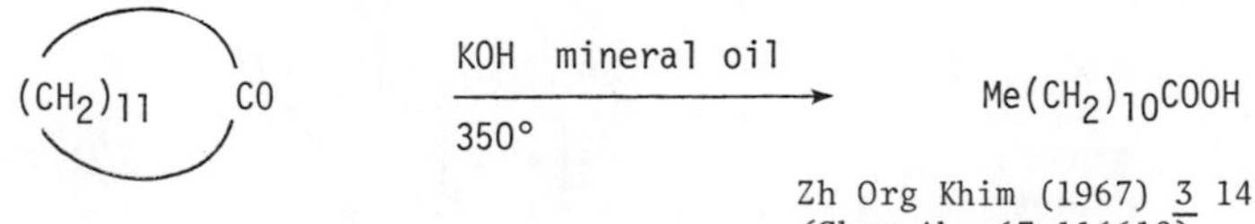
(CH2)11
CO
KOH mineral oil
350°
Me(CH2)10COOH

55%

Zh Org Khim (1967) 3 1418
(Chem Abs 67 116618)

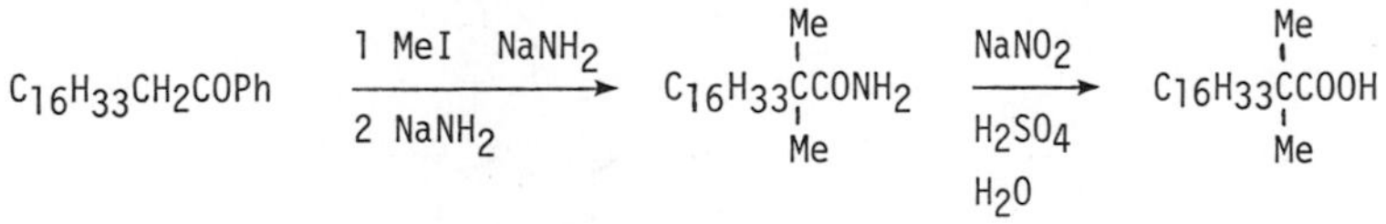
C16H33CH2COPh
1 MeI NaNH2
2 NaNH2
Me
C16H33CCONH2
Me
NaNO2
H2SO4
H2O
Me
C16H33CCOOH
Me

JCS (1942) 488

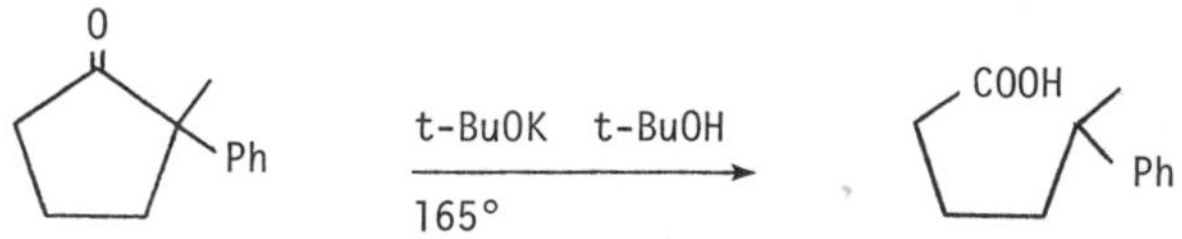
O
Ph
t-BuOK t-BuOH
165°
COOH
Ph

43%

JACS (1969) 91 1009

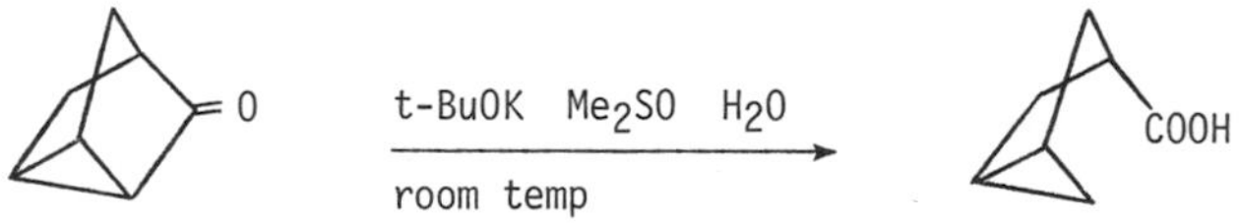
O
t-BuOK Me2SO H2O
room temp
COOH

65%

Tetr Lett (1964) 3251

Ph_2CO —(t-BuOK monoglyme H_2O; 30°)→ PhCOOH 90%

JACS (1967) 89 946

t-BuCOMe —(CCl_4 KOH; t-BuOH H_2O)→ t-BuCOOH 80%

JACS (1969) 91 7510

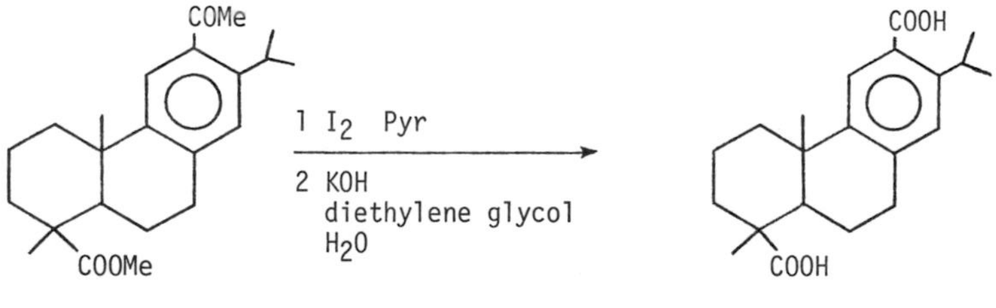
COMe
COOMe
1 I2 Pyr
2 KOH
diethylene glycol
H2O
COOH
COOH

60-70%

JACS (1951) 73 3803

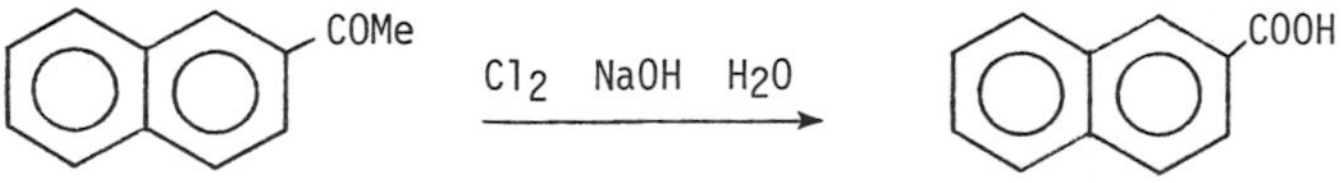
COMe
Cl2 NaOH H2O
COOH

87%

Org Synth (1943) Coll Vol 2 428

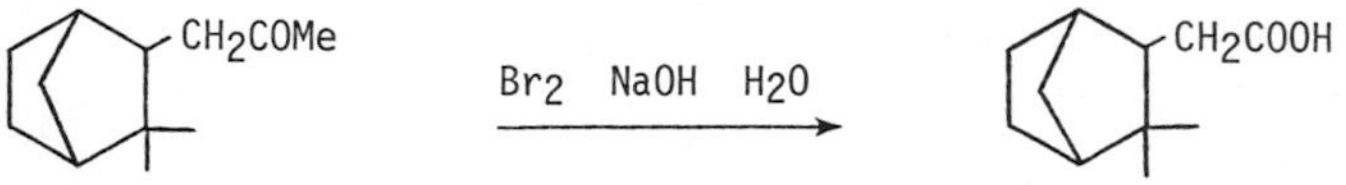

Aust J Chem (1967) <u>20</u> 2033

PhCOPr $\xrightarrow[\text{2 NaOH}]{\text{1 NaOBr H}_2\text{O}}$ PhCOOH   48%

JACS (1950) <u>72</u> 1642

Et$_3$CCOEt $\xrightarrow{\text{HNO}_3 \quad \text{H}_2\text{O}}$ Et$_3$CCOOH   71%

Bull Soc Chim Fr (1967) 2011
JACS (1942) <u>64</u> 2421

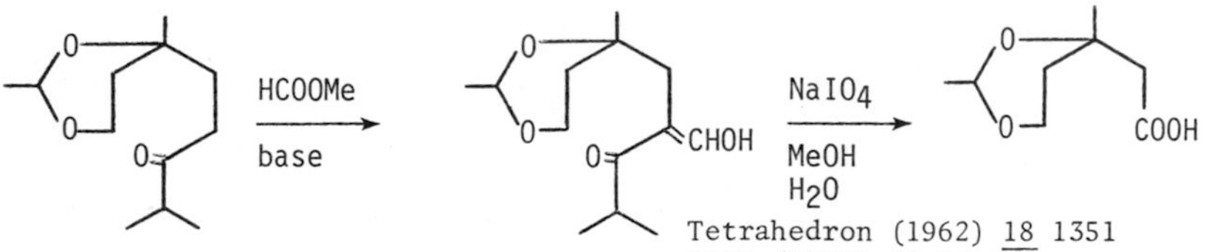

Tetrahedron (1962) <u>18</u> 1351
JACS (1956) <u>78</u> 1414

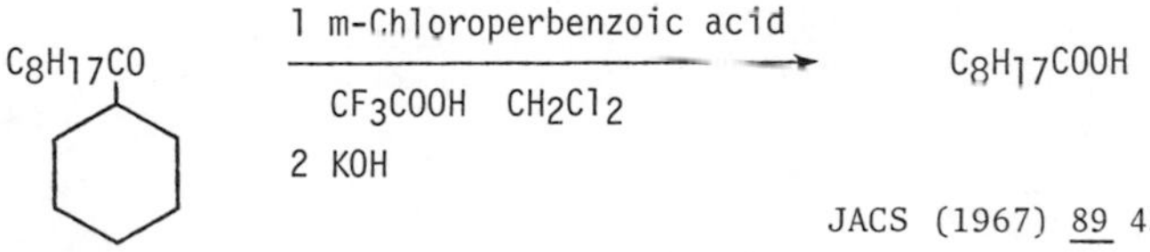

C$_8$H$_{17}$CO $\xrightarrow[\begin{array}{c}\text{CF}_3\text{COOH} \quad \text{CH}_2\text{Cl}_2 \\ \text{2 KOH}\end{array}]{\text{1 m-Chloroperbenzoic acid}}$ C$_8$H$_{17}$COOH   79%

JACS (1967) <u>89</u> 4530

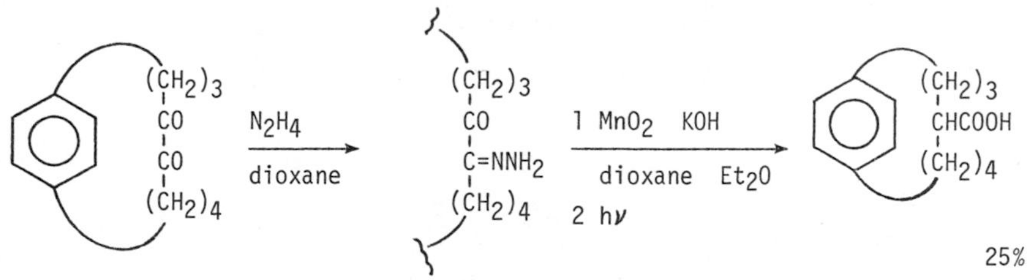

JACS (1963) <u>85</u> 1171

PhCOCOMe $\xrightarrow{\text{NaOBr \quad H}_2\text{O}}$ PhCOOH   91%

JACS (1950) <u>72</u> 1642

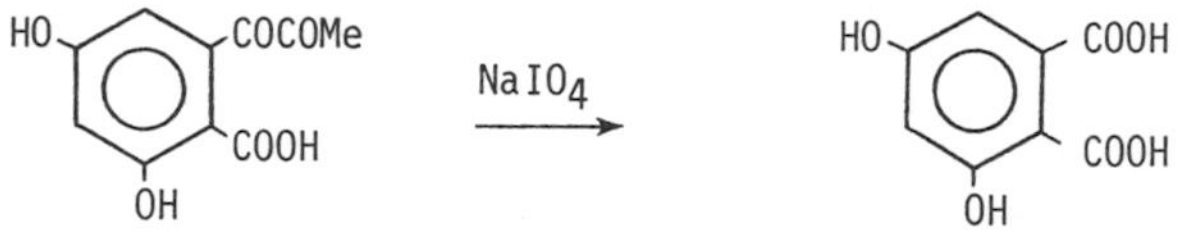

JCS (1935) 1467
Org React (1944) <u>2</u> 341

$Me(CH_2)_7COCO(CH_2)_7COOH \xrightarrow[\text{HOAc}]{\text{MeCOO}_2\text{H}} Me(CH_2)_7COOH$ 100%

Org React (1957) <u>9</u> 73

Carboxylic acids may also be prepared by conversion of ketones into esters or amides, followed by hydrolysis. See section 117 (Esters from Ketones) and section 87 (Amides from Ketones)

Section 28   Carboxylic Acids from Nitriles

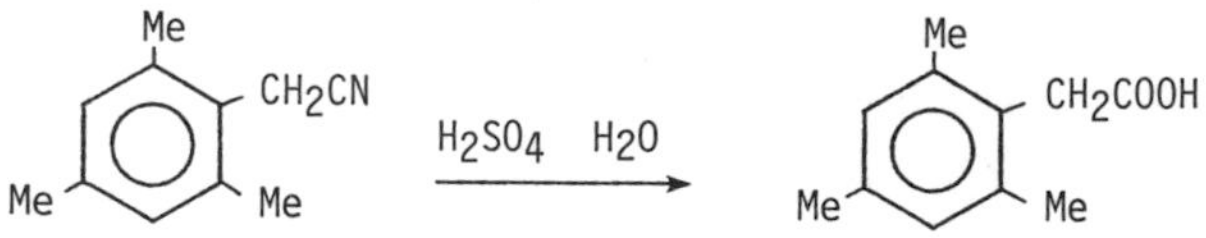

Org Synth (1955) Coll Vol 3 557

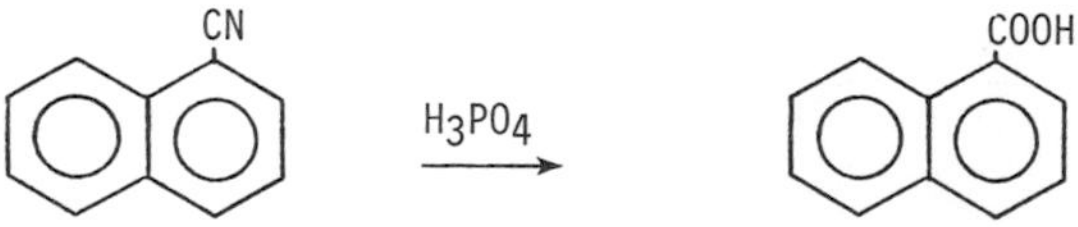

70-90%

Rec Trav Chim (1927) <u>46</u> 600

$$NO_2, CN \xrightarrow{\quad H_2SO_4 \quad NaNO_2 \quad H_2O \quad} NO_2, COOH \qquad 92\%$$

JCS (1945) 751

$$CH_2CHCN / Cl \xrightarrow{\quad HCl \quad HCOOH \quad H_2O \quad} CH_2CHCOOH / Cl \qquad 80\%$$

Proc Chem Soc (1962) 117

$$CN, CN \xrightarrow[\quad 300° \quad]{\quad \text{o-Chlorobenzoic acid} \quad} COOH, COOH \qquad 97\%$$

Annalen (1968) 716 78

$$\begin{array}{l} O \\ O \end{array}\!\!< \!\!\begin{array}{l}(CH_2)_3CN \\ (CH_2)_3CN\end{array} \xrightarrow[\quad 2 \ H_3PO_4 \quad]{\quad 1 \ H_2O_2 \quad KOH \quad H_2O \quad} O=\!\!<\!\!\begin{array}{l}(CH_2)_3COOH \\ (CH_2)_3COOH\end{array} \qquad 85\%$$

JCS (1962) 4722

$$PhCN \xrightarrow{\quad H_2S \quad NH_3 \quad EtOH \quad} PhCSNH_2 \quad ----\rightarrow \quad PhCOOH$$

Ber (1890) 23 158
JOC (1951) 16 131

$$Ph \quad CN / Cl \xrightarrow{\quad NaOH \quad EtOH \quad H_2O \quad} Ph \quad COOH / Cl \qquad 81-90\%$$

JCS C (1966) 840

93%

Chem Pharm Bull (1969) <u>17</u> 1564

Carboxylic acids may also be prepared by conversion of nitriles into esters or amides, followed by hydrolysis. See section 118 (Esters from Nitriles) and section 88 (Amides from Nitriles)

Section 29   <u>Carboxylic Acids from Olefins</u>
            ∘∘∘∘∘∘∘∘∘∘∘∘∘∘∘∘∘∘∘∘∘∘∘∘∘∘∘∘∘∘∘∘∘

1 t-BuLi   Et$_3$N

2 CO$_2$

Bu-t

COOH

30-45%

JOC (1965) <u>30</u> 917
JACS (1969) <u>91</u> 6362

$Me_2C=CH_2$ $\xrightarrow{CH_2=CCl_2\ \ BF_3\ \ H_2SO_4}$ Me_3CCH_2COOH 75%

Angew (1965) <u>77</u> 967
(Internat Ed <u>4</u> 956)

$C_6H_{13}CH=CH_2$ $\xrightarrow[\text{di-t-butyl peroxide}]{CH_3COOH}$ $C_6H_{13}(CH_2)_3COOH$ ~70%

JCS (1965) 1918

$BuCH=CH_2$ $\xrightarrow{BrCH_2COOH\ \ h\nu}$ $Bu(CH_2)_3COOH$ 80%

Chem Comm (1967) 435

$$C_6H_{13}CH=CH_2 \xrightarrow[\substack{\text{di-t-butyl} \\ \text{peroxide}}]{CH_2(COOEt)_2} C_6H_{13}CH_2CH_2CH(COOEt)_2 \xrightarrow[2 \quad \Delta]{1 \text{ Hydrolysis}} C_6H_{13}(CH_2)_3COOH$$

Chem Ind (1961) 830

$$BuCH=CH_2 \xrightarrow[\substack{\text{di-t-butyl} \\ \text{peroxide}}]{HC(COOEt)_3} BuCH_2CH_2C(COOEt)_3 \xdashrightarrow[2 \quad \Delta]{1 \text{ Hydrolysis}} Bu(CH_2)_3COOH$$

Synthesis (1970) 124

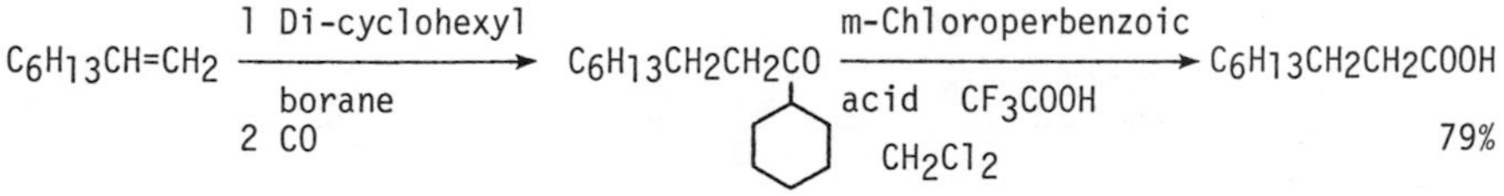

$$C_6H_{13}CH=CH_2 \xrightarrow[\substack{\text{borane} \\ 2 \text{ CO}}]{1 \text{ Di-cyclohexyl}} C_6H_{13}CH_2CH_2CO \xrightarrow[\substack{\text{acid} \quad CF_3COOH \\ CH_2Cl_2}]{\text{m-Chloroperbenzoic}} C_6H_{13}CH_2CH_2COOH$$

79%

JACS (1967) 89 4530

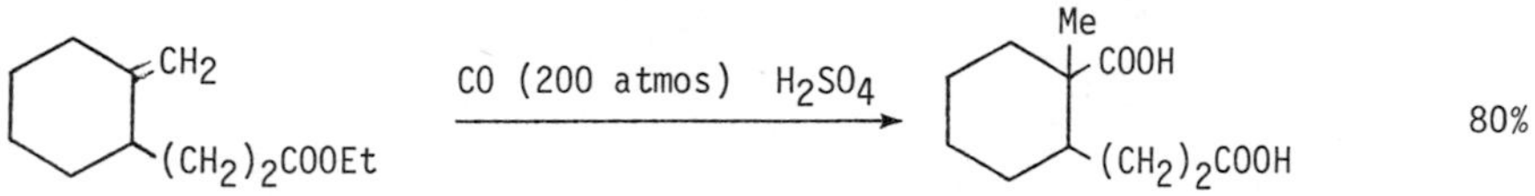

80%

JACS (1960) 82 1261

9%

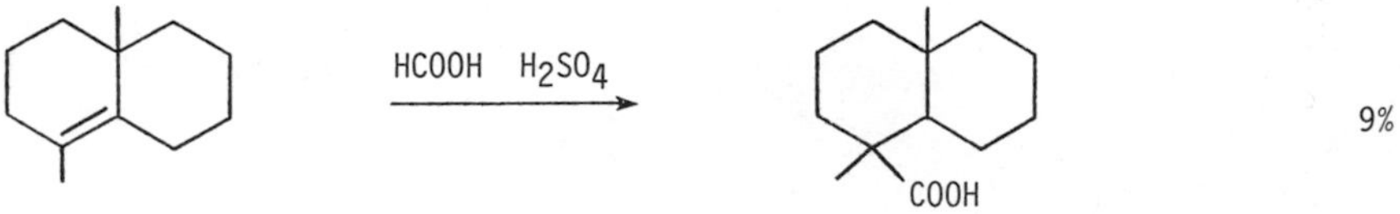

Tetrahedron (1965) 21 2641

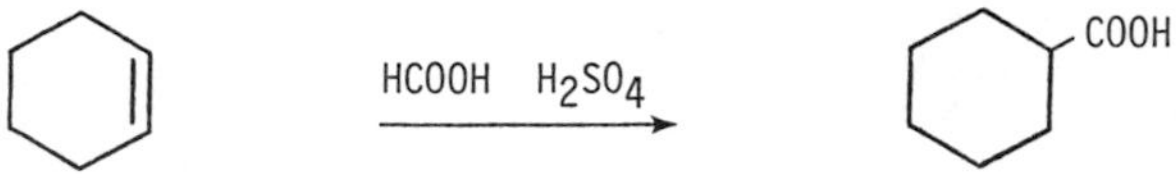

Ber (1966) 99 1149

Me Me

$\triangle$ $\xrightarrow{\text{CO (300 atmos)   H}_2\text{SO}_4}$ Me─C─Me, Et─C─COOH   60%

Ber (1964) **97** 3088

cyclohexene   - - →   bicyclic with F, Cl $\xrightarrow{\text{H}_2\text{SO}_4}$ cyclohexane─COOH   90%

Ber (1968) **101** 1291

$C_6H_{13}CH=CH_2$ $\xrightarrow[\text{HCl   Me}_2\text{CO   H}_2\text{O}]{\text{Ni(CO)}_4\quad h\nu}$ $C_6H_{13}\overset{\displaystyle |}{\underset{\displaystyle Me}{CHCOOH}}$ (+ isomers)

43%

Angew (1965) **77** 813
(Internat Ed **4** 790)

cyclohexene─CH=CH₂ $\xrightarrow[\text{2 CO}_2]{\text{1 PrMgBr   TiCl}_4\quad THF}$ cyclohexene─CH₂CH₂COOH   49%

JOC (1970) **35** 392

cycloheptane (Me, OH, CH₂CH=CH₂, Me) $\xrightarrow[\text{2 CrO}_3]{\text{1 B}_2\text{H}_6\quad THF}$ cycloheptane (Me, OH, CH₂CH₂COOH, Me)

(→ lactone)

Chem Comm (1968) 122

$\begin{matrix} CH_2 \\ \parallel \\ CH \\ | \\ (CH_2)_8 \\ | \\ COOH \end{matrix}$ $\xrightarrow[\text{2 NaOH}]{\text{1 (NH}_4)_2\text{S   (NH}_4)_2\text{S}_2\text{O}_3\quad H_2O}$ $\begin{matrix} COOH \\ | \\ CH_2 \\ | \\ (CH_2)_8 \\ | \\ COOH \end{matrix}$ 35%

JACS (1946) **68** 2033

$Me_2CH(CH_2)_3\underset{\underset{Me}{|}}{C}HCH=CH_2$ $\xrightarrow{KMnO_4 \quad NaHCO_3 \quad Me_2CO}$ $Me_2CH(CH_2)_3\underset{\underset{Me}{|}}{C}HCOOH$ 45%

JACS (1943) 65 745

$\xrightarrow[\text{t-BuOH} \quad H_2O]{KMnO_4 \quad NaIO_4 \quad K_2CO_3}$

Can J Chem (1967) 45 1439

$\begin{matrix} COOMe \\ | \\ (CH_2)_7 \\ | \\ CH \\ || \\ CH \\ | \\ (CH_2)_7 \\ | \\ Me \end{matrix}$ $\xrightarrow[\text{t-BuOH} \quad H_2O]{KMnO_4 \quad NaIO_4}$ $\begin{matrix} COOMe \\ | \\ (CH_2)_7 \\ | \\ COOH \\ \\ COOH \\ | \\ (CH_2)_7 \\ | \\ Me \end{matrix}$ 98-99%

Can J Chem (1956) 34 1413
JACS (1961) 83 4819

$\xrightarrow[Me_2CO \quad H_2O]{RuO_4 \quad NaIO_4}$

    89%

JACS (1963) 85 3419

$C_{11}H_{23}CH=CH_2$ $\xrightarrow[\text{2 Ag}_2O \quad NaOH \quad H_2O]{1 \quad O_3}$ $C_{11}H_{23}COOH$

Ber (1942) 75B 656

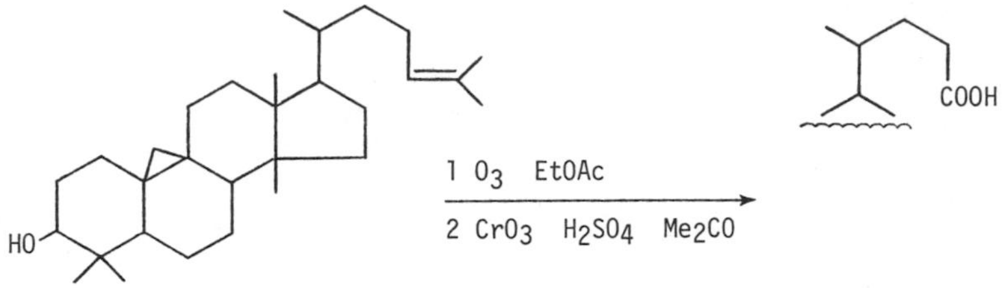

1 O_3 EtOAc

2 CrO_3 H_2SO_4 Me_2CO

80%

Tetr Lett (1969) 1733

Carboxylic acids may also be prepared by conversion of olefins into
esters or amides, followed by hydrolysis. See section 119 (Esters
from Olefins) and section 89 (Amides from Olefins)

Section 30   Carboxylic Acids from Miscellaneous Compounds

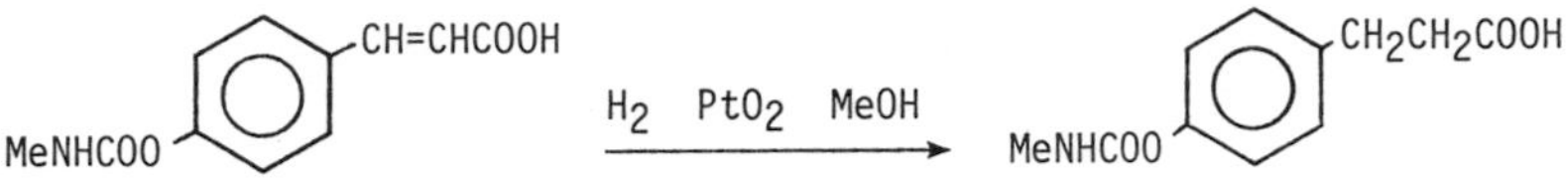

H_2 PtO_2 MeOH

70%

Chem Pharm Bull (1970) 18 243

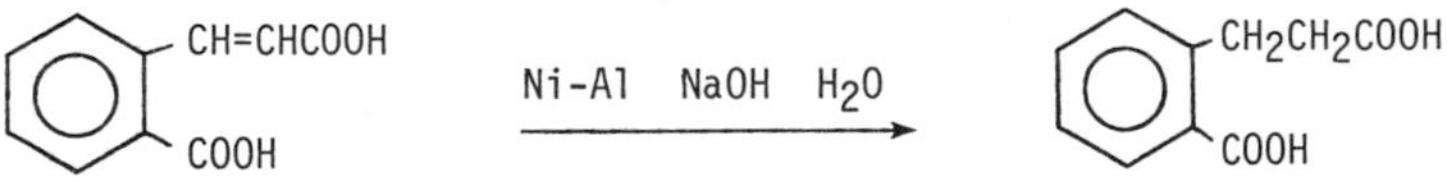

Ni-Al NaOH H_2O

92-95%

Org Synth (1963) Coll Vol 4 136

PhCH=CHCOOH $\xrightarrow{N_2H_4 \ \ Ni \ \ H_2O}$ $PhCH_2CH_2COOH$ 85%

Act Chem Scand (1961) 15 1200

PhCH=CHCOOH $\xrightarrow{\text{H}_2 \;\; (\text{Ph}_3\text{P})_3\text{RhCl}}$ PhCH$_2$CH$_2$COOH   75%

Compt Rend C (1966) 263 251
Chem Comm (1969) 1365

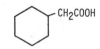

$\xrightarrow{\text{K} \;\; \text{NH}_3 \;\; \text{dioxane}}$ 95%

JACS (1969) 91 1228
Compt Rend (1969) C 268 640

PhCH=CHCOOH $\xrightarrow{\text{P}_4 \;\; \text{KI} \;\; \text{H}_3\text{PO}_4}$ PhCH$_2$CH$_2$COOH   80%

Helv (1939) 22 601

PhCHCOOH $\xrightarrow{\text{P}_4 \;\; \text{KI} \;\; \text{H}_3\text{PO}_4}$ PhCH$_2$COOH   90%
|
OH

Helv (1939) 22 601
Org Synth (1932) Coll Vol 1 224
— diphenylacetic acid

$\xrightarrow[\text{HOAc} \;\; \text{H}_2\text{O}]{\text{P}_4 \;\; \text{I}_2}$

JOC (1966) 31 983

PhCHCOOH $\xrightarrow[\text{tetralin}]{\text{H}_2 \;\; \text{Pd-BaSO}_4}$ PhCH$_2$COOH
|
OAc

Org React (1953) 7 263

$$PrCH_2NO_2 \quad \xrightarrow{H_2SO_4 \quad H_2O} \quad PrCOOH \qquad \sim 90\%$$

Ind Eng Chem (1939) <u>31</u> 118

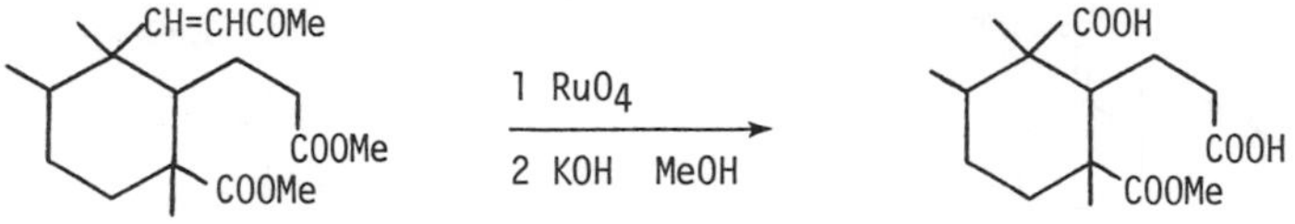

$$\xrightarrow{\substack{1 \ RuO_4 \\ 2 \ KOH \quad MeOH}}$$

Tetr Lett (1968) 2681

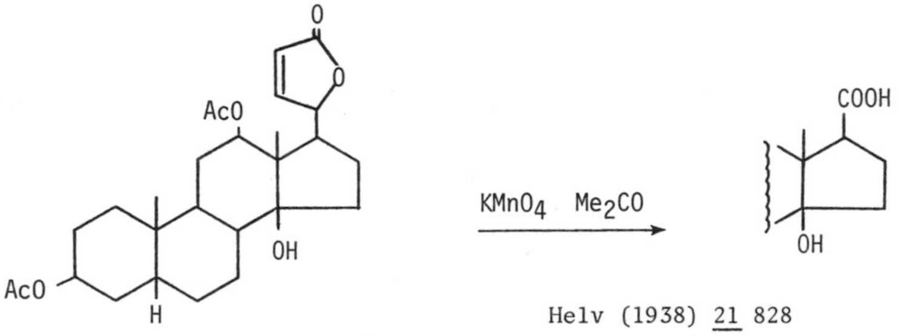

$$\xrightarrow{KMnO_4 \quad Me_2CO} \qquad <25\%$$

Helv (1938) <u>21</u> 828

$$C_8H_{17}\underset{\underset{OH}{|}}{COCH}(CH_2)_7COOH \quad \xrightarrow{KIO_4 \quad H_2SO_4 \quad EtOH} \quad C_8H_{17}COOH \qquad 92\%$$

JCS (1936) 1788
     (1935) 1467

$$PhCH\underset{\underset{OH}{|}}{COOH} \quad \xrightarrow[NaOH \quad H_2O]{Nickel \ peroxide} \quad PhCOOH \qquad 90\%$$

Chem Pharm Bull (1964) <u>12</u> 403

Section 30A   Protection of Carboxylic Acids

RCOOH   ----→   RCOOCH$_2$CCl$_3$
    ←———————————
       Zn HOAc H$_2$O

                                    JACS (1966) 88 852

RNHCH$_2$COOH $\xrightarrow{\text{NO}_2\text{-C}_6\text{H}_4\text{-SCH}_2\text{CH}_2\text{OH}}{\text{TsOH}\ \ C_6H_6}$ RNHCH$_2$COOCH$_2$CH$_2$S-C$_6$H$_4$-NO$_2$

    ←———————————
       1 H$_2$O$_2$ (NH$_4$)$_6$Mo$_7$O$_{24}$ JCS C (1969) 2495
       2 pH 10-10.5

RCOOH   ----→   RCOOCH$_2$CH$_2$SMe $\xrightarrow{\text{H}_2\text{O}_2\ \ (\text{NH}_4)_6\text{Mo}_7\text{O}_{24}}$ RCOOCH$_2$CH$_2$SO$_2$Me
    ←———————————
     1 MeI
     2 pH 10-10.5   (stable to acid)

                  pH 10-11   Tetr Lett (1968) 2525

RHN—[β-lactam-thiazolidine]—COOH $\xrightarrow{(\text{Me}_3\text{Si})_2\ \ \text{CHCl}_3}$ ⋯—COOSiMe$_3$
    ←———————————
       H$_2$O
                                    Annalen (1964) 673 166
                                    JACS (1966) 88 3390

H$_2$N—[β-lactam-cephem]—CH$_2$OAc, COOH $\xrightarrow{\text{Me}_2\text{C=CH}_2\ \ \text{H}_2\text{SO}_4}$ ⋯—COOCMe$_3$
    ←———————————
       CF$_3$COOH
                                    J Med Chem (1966) 9 444

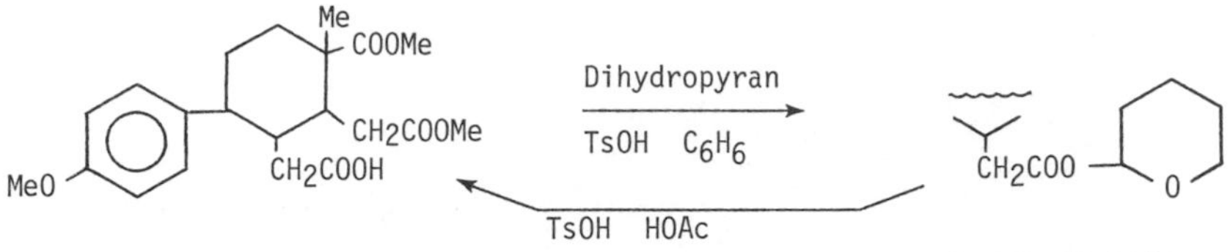

JACS (1957) 79 1995
JCS (1952) 3945

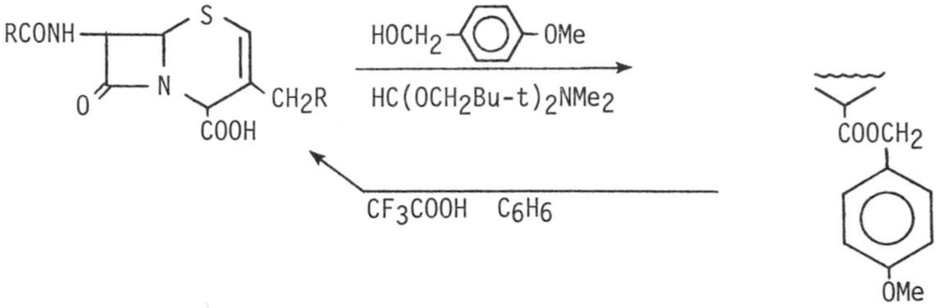

JACS (1969) 91 5674

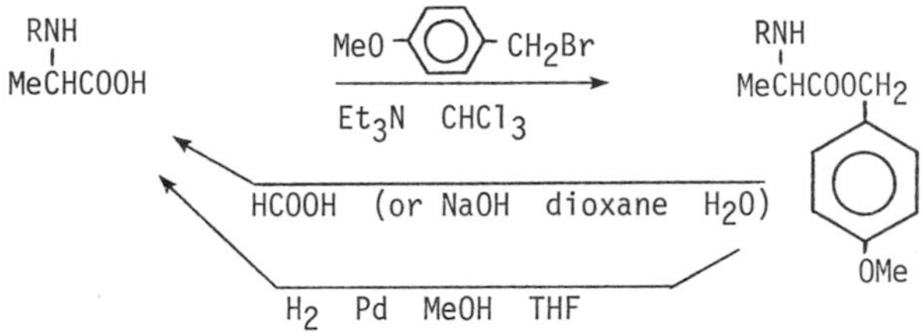

JCS C (1970) 964

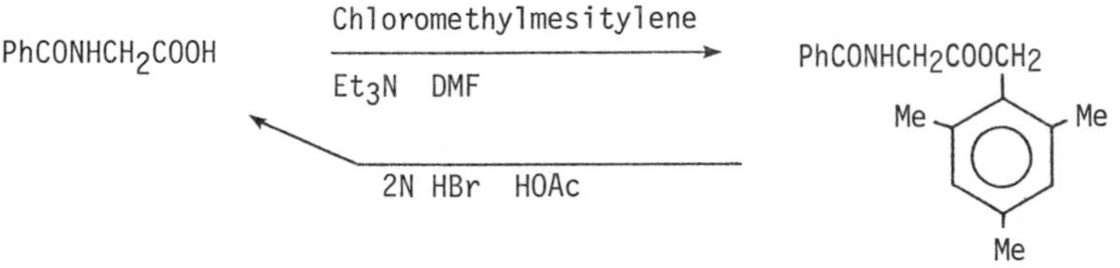

Aust J Chem (1966) 19 1067
            (1967) 20 2243

Further examples of the preparation and cleavage of benzyl esters
are included in section 107 (Esters from Carboxylic Acids) and
section 23 (Carboxylic Acids from Esters)

PhCH$_2$CH$_2$COONa $\xrightarrow[\text{NaOH \quad H}_2\text{O}]{\text{Ph}_3\text{CBr \quad C}_6\text{H}_6}$ PhCH$_2$CH$_2$COOCPh$_3$

JOC (1962) <u>27</u> 3595

RNHCH$_2$COOH $\xrightarrow[\text{HCl \quad dioxane \quad H}_2\text{O}]{\text{Ph}_3\text{CCl \quad Et}_3\text{N}}$ RNHCH$_2$COOCPh$_3$

JCS <u>C</u> (1966) 1191

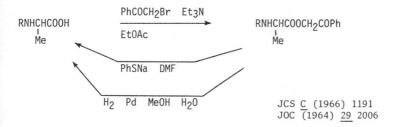

RNHCHCOOH
   |
  Me
$\xrightarrow[\text{EtOAc}]{\text{PhCOCH}_2\text{Br \quad Et}_3\text{N}}$
RNHCHCOOCH$_2$COPh
       |
      Me

PhSNa   DMF

H$_2$ Pd   MeOH   H$_2$O

JCS <u>C</u> (1966) 1191
JOC (1964) <u>29</u> 2006

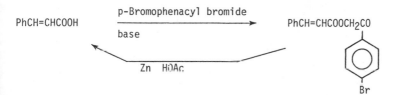

PhCH=CHCOOH $\xrightarrow[\text{base}]{\text{p-Bromophenacyl bromide}}$ PhCH=CHCOOCH$_2$CO

Zn   HOAc

Tetr Lett (1970) 343

C$_5$H$_{11}$COOH $\dashrightarrow$ C$_5$H$_{11}$COS

hν   C$_6$H$_6$

JCS (1965) 3571
JACS (1970) <u>92</u> 6333

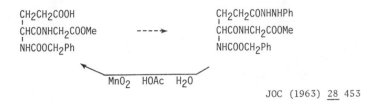

CH₂CH₂COOH
|
CHCONHCH₂COOMe   ---→
|
NHCOOCH₂Ph

CH₂CH₂CONHNHPh
|
CHCONHCH₂COOMe
|
NHCOOCH₂Ph

←——— MnO₂ HOAc H₂O

JOC (1963) 28 453

CH₂CONHCH₂COOH
|
NHCOOCH₂Ph   ---→

CH₂CONHCH₂CONHNH₂
|
NHCOOCH₂Ph

←——— NBS H₂O

JOC (1965) 30 315

RCH₂COOH   ---→   RCH₂COOMe   ---→   RCH₂CONHOH

←——— HIO₄ H₂O

JACS (1960) 82 4903

(benzoic acid)   COOH

NH₂
|
Me₂CCH₂OH
————————→

(oxazoline product with gem-dimethyl)

(Stable to RMgX
and CrO₃)

←——— HCl H₂O EtOH

JACS (1970) 92 6646

Other reactions useful for the protection of carboxylic acids are included
in section 107 (Esters from Carboxylic Acids), section 23 (Carboxylic Acids
from Esters), section 77 (Amides from Carboxylic Acids) and section 21
(Carboxylic Acids from Amides)

Chapter 3 PREPARATION OF ALCOHOLS AND PHENOLS

Section 31 Alcohols from Acetylenes

$BuC{\equiv}CH$ $\xrightarrow[\text{2 MeCOEt Et}_2O]{\text{1 EtMgBr Et}_2O}$ $\underset{\underset{OH}{|}}{BuC{\equiv}C}\overset{\overset{Et}{|}}{C}Me$ $\xrightarrow{H_2\ \ Pt}$ $BuCH_2CH_2\overset{\overset{Et}{|}}{\underset{\underset{OH}{|}}{C}}Me$

92%

JACS (1941) 63 186
Tetr Lett (1965) 1619

$PrC{\equiv}CH$ $\xrightarrow[\substack{\text{2 MeLi}\\ \text{3 EtBr}\\ \text{4 NaOH}\ H_2O}]{\text{1 }B_2H_6\ \ THF}$ $PrCH_2\underset{\underset{OH}{|}}{C}HEt$ 90%

Tetr Lett (1966) 2535
JACS (1967) 89 291

$BuC{\equiv}CH$ $\xrightarrow[\substack{\text{2 MeONa}\\ \text{3 MeI}\\ \text{4 }O_2}]{\text{1 (i-Bu)}_2AlH\ \ THF}$ $BuCH_2\underset{\underset{Me}{|}}{C}HOH$ 90%

Tetr Lett (1966) 6021

$BuC{\equiv}CH$ $\xrightarrow[\text{2 }H_2O_2\ \ NaOH]{\text{1 }B_2H_6\ \ THF}$ $BuCH_2CH_2OH$ 80%

JACS (1967) 89 291

Section 32   Alcohols and Phenols from Carboxylic Acids, Acid Chlorides and
             Anhydrides

PhCOOH $\xrightarrow[\text{H}_2\text{O}]{\text{H}_2 \text{ (205 atmos)}\quad \text{ReO}_3}$ PhCH$_2$OH   43%

JOC (1963) 28 2345

C$_9$H$_{19}$COOH $\xrightarrow{\text{H}_2 \text{ (173 atmos)}\quad \text{Rh}_2\text{O}_7}$ C$_9$H$_{19}$CH$_2$OH   100%

JOC (1959) 24 1847

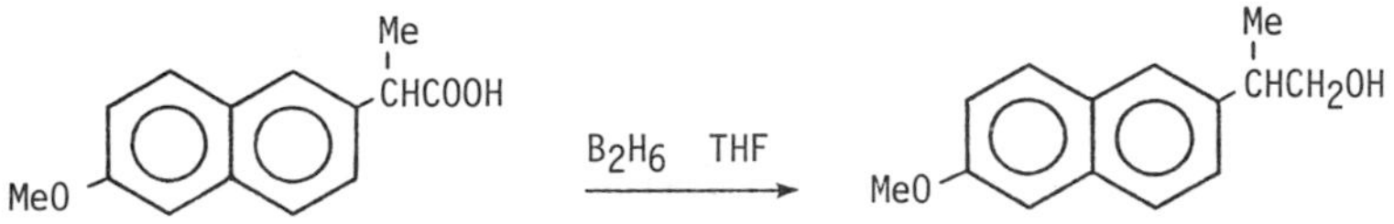

87%

J Med Chem (1970) 13 203

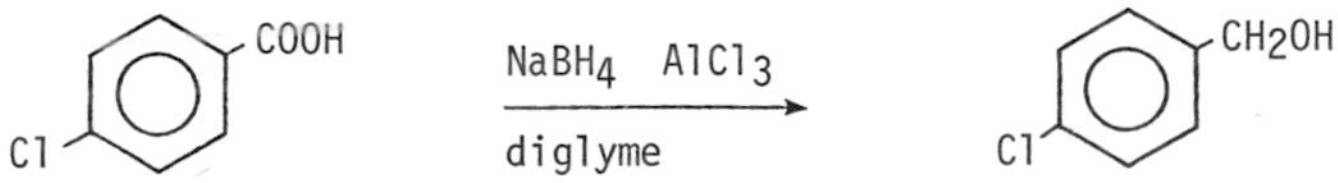

$\xrightarrow[\text{diglyme}]{\text{NaBH}_4 \quad \text{AlCl}_3}$

64%

JACS (1956) 78 2582

$\xrightarrow[\text{H}_2\text{SO}_4 \quad \text{H}_2\text{O}]{\text{Electrolysis}}$

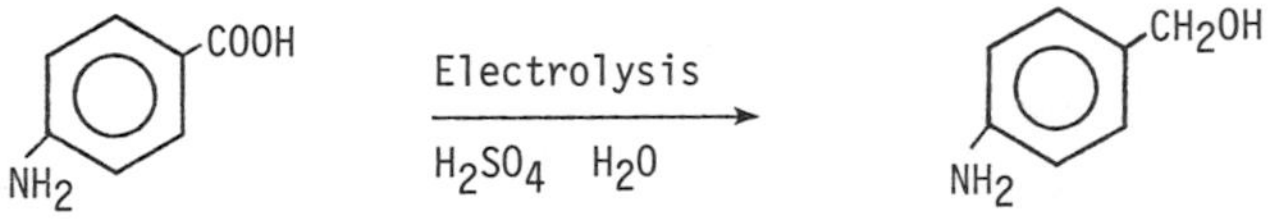

56%

JCS (1942) 98

$$Ph_3CCOOH \xrightarrow{\text{LiAlH}_4 \quad \text{THF}} Ph_3CCH_2OH$$

$$\downarrow$$

$$Ph_3CCOCl \xrightarrow{\text{LiAlH}_4 \quad \text{THF}}$$

Org React (1951) **6** 469

$$BrCH_2CH_2COCl \xrightarrow{\text{LiAlH}_4 \quad \text{AlCl}_3 \quad \text{Et}_2O} BrCH_2CH_2CH_2OH \qquad 90\%$$

JACS (1959) **81** 610

$$C_{15}H_{31}COCl \xrightarrow{\text{NaBH}_4 \quad \text{dioxane}} C_{15}H_{31}CH_2OH \qquad 87\%$$

JACS (1949) **71** 122

$$C_5H_{11}COOH \xrightarrow[\text{THF}]{\text{ClCOOEt} \quad \text{Et}_3N} C_5H_{11}CO\text{--OCOOEt} \xrightarrow[\text{H}_2O]{\text{NaBH}_4 \quad \text{THF}} C_5H_{11}CH_2OH \qquad 70\%$$

Chem Pharm Bull (1968) **16** 492

J Med Chem (1964) **7** 483

$$C_{15}H_{31}COCl \xrightarrow[\text{pyr}]{\text{PhCH}_2SH} C_{15}H_{31}COSCH_2Ph \xrightarrow{\text{Ni} \quad \text{Et}_2O} C_{15}H_{31}CH_2OH$$

Helv (1946) **29** 684

$$C_6H_{13}COCl \xrightarrow{\substack{\text{1 m-Chloroperbenzoic acid} \\ \text{pyr hexane} \\ \text{2 Hexane cyclohexane reflux} \\ \text{3 Base}}} C_6H_{13}OH \qquad 66\%$$

JOC (1965) **30** 3760

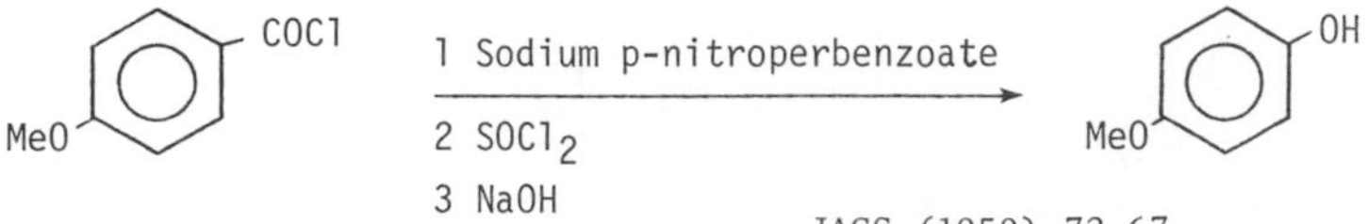

1 Sodium p-nitroperbenzoate

2 SOCl$_2$

3 NaOH

JACS (1950) **72** 67

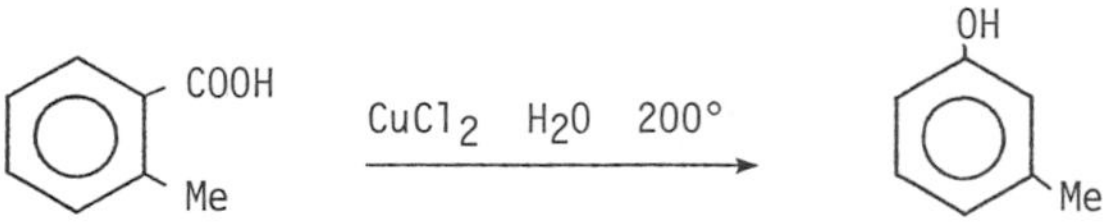

CuCl$_2$ H$_2$O 200°

JOC (1961) **26** 3144

NaN$_3$ H$_2$SO$_4$

CHCl$_3$

1 NaNO$_2$

H$_2$SO$_4$

H$_2$O

2 H$_2$SO$_4$ H$_2$O   74%

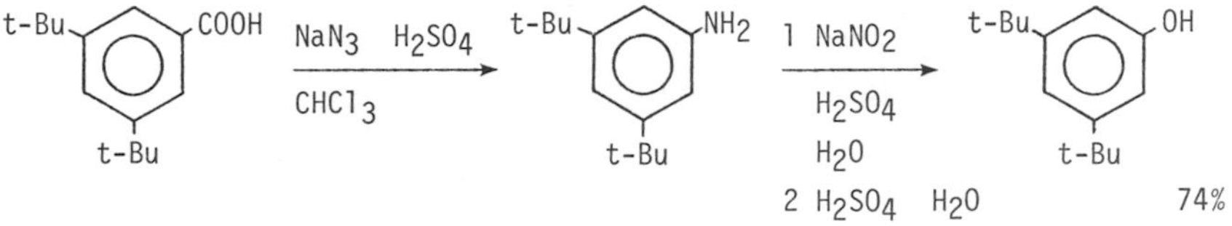

Can J Chem (1963) **41** 1653

Alcohols may also be prepared by esterification of carboxylic acids followed by reduction. See section 38 (Alcohols from Esters)

Section 33   Alcohols from Alcohols and Phenols
ooooooooooooooooooooooooooooooooo

1 TsCl pyr

2 DMF

KOH

MeOH

30%

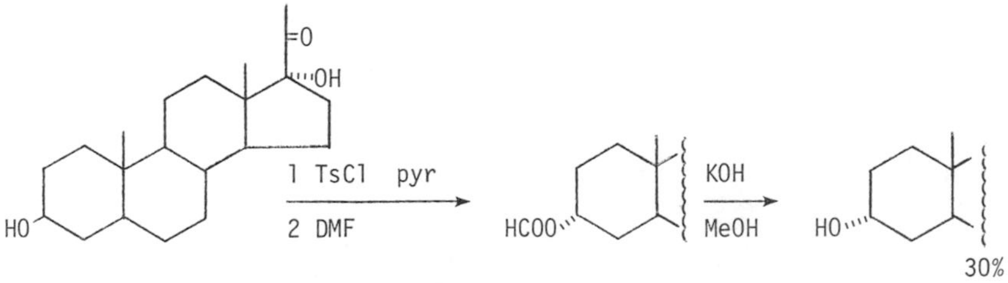

Steroids (1964) **3** 359

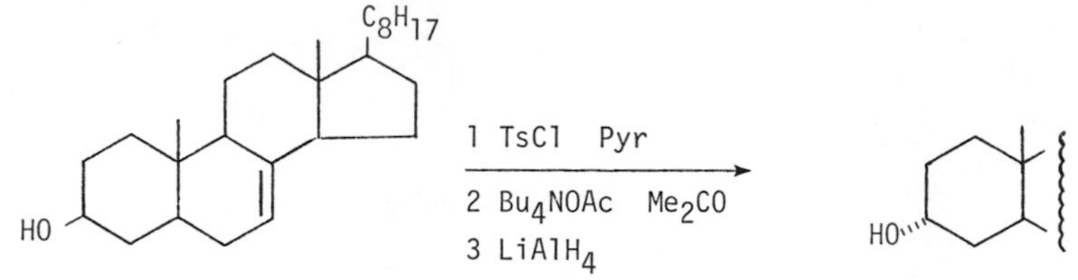
C8H17
HO
1 TsCl Pyr
2 Bu4NOAc Me2CO
3 LiAlH4
HO

92%

JCS C (1969) 1605

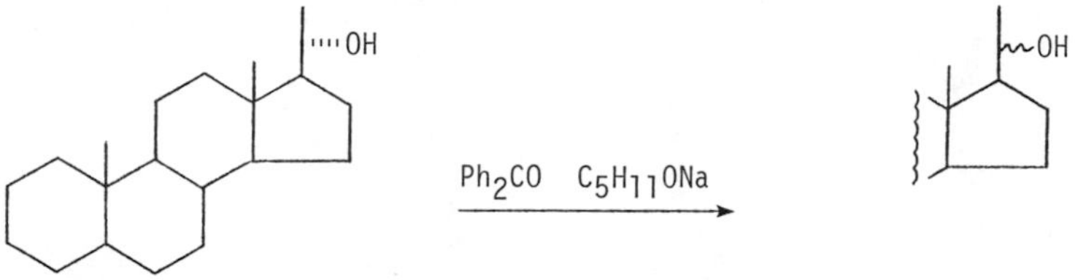
OH
Ph2CO C5H11ONa
OH

JCS C (1969) 969

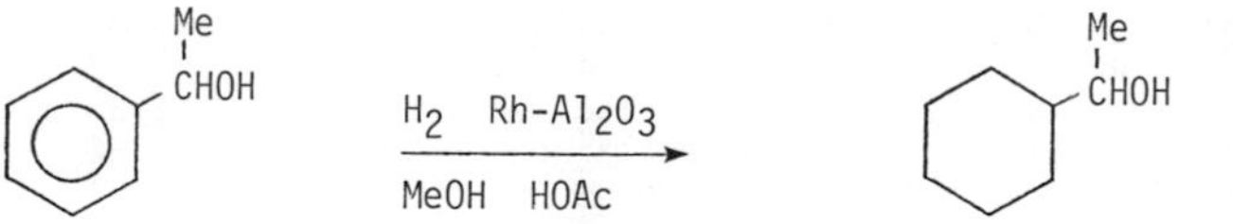
Me
CHOH
H2 Rh-Al2O3
MeOH HOAc
Me
CHOH

80%

JOC (1962) 27 2288

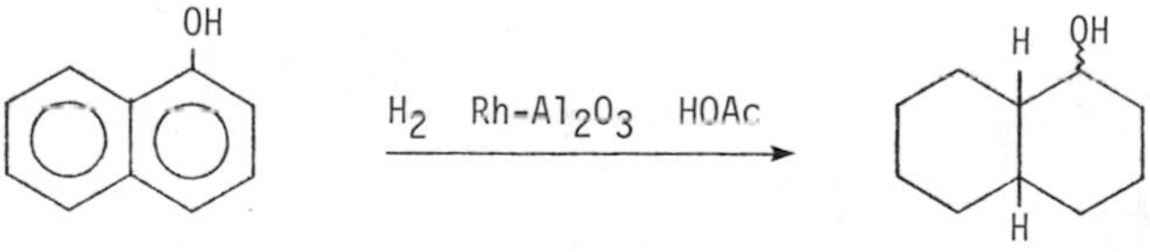
OH
H2 Rh-Al2O3 HOAc
H OH
H

90%

JOC (1964) 29 3427

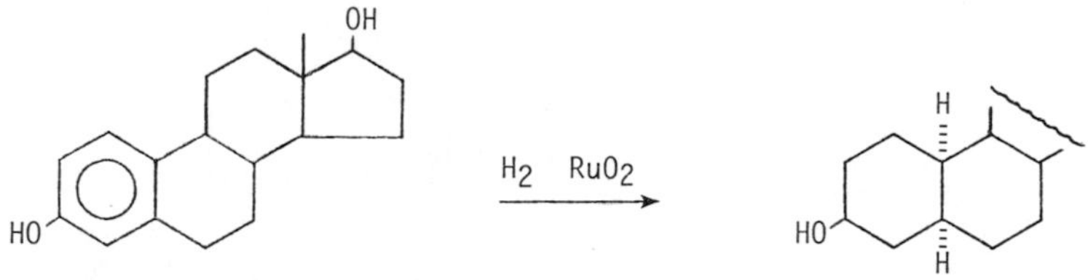
OH
HO
H2 RuO2
H
HO
H

JOC (1958) 23 1404

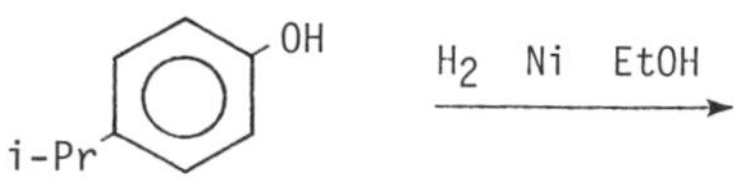

96%

JACS (1949) 71 3889

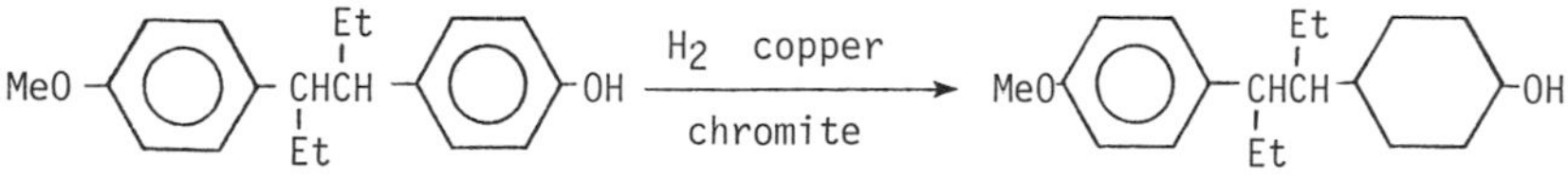

~52%

JACS (1948) 70 4127

Section 34   Alcohols and Phenols from Aldehydes

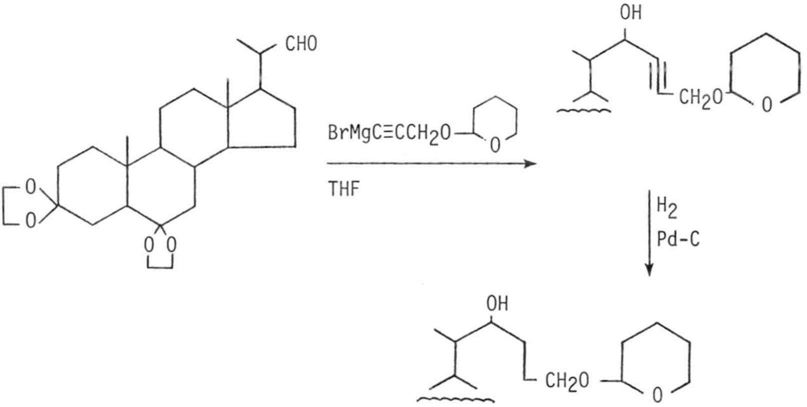

Chem Pharm Bull (1969) 17 690
Tetr Lett (1966) 3457

$$\text{Cyclohexyl-CHO} \xrightarrow{\text{CH}_2=\text{CHCH}_2\text{Br} \quad \text{Mg} \quad \text{Et}_2\text{O}} \text{Cyclohexyl-}\underset{\overset{|}{\text{OH}}}{\text{CH}}\text{CH}_2\text{CH}=\text{CH}_2$$

72%

(Procedure for unstable Grignard reagents)

JOC (1963) <u>28</u> 3269

$$\text{PhCHO} \xrightarrow{\text{PhBr} \quad \text{Li} \quad \text{THF}} \text{Ph}_2\text{CHOH}$$

96%

(One-step procedure)

Chem Comm (1970) 1160

$$\text{(2-Cl-C}_6\text{H}_4\text{)CHO} \xrightarrow{\text{MeMgBr} \quad \text{Et}_2\text{O}} \text{(2-Cl-C}_6\text{H}_4\text{)}\underset{\overset{|}{\text{CHOH}}}{\overset{\overset{\text{Me}}{|}}{}}$$

76%

JACS (1944) <u>66</u> 1295

$$\text{C}_6\text{H}_{13}\text{CHO} \xrightarrow[\text{EtOH}]{\text{H}_2 \quad \text{Pt} \quad \text{FeCl}_3} \text{C}_6\text{H}_{13}\text{CH}_2\text{OH}$$

JACS (1923) <u>45</u> 1071

$$\text{(3-OMe-C}_6\text{H}_4\text{)CHO} \xrightarrow[\text{EtOH}]{\text{H}_2 \quad \text{PtO}_2 \quad \text{FeSO}_4} \text{(3-OMe-C}_6\text{H}_4\text{)CH}_2\text{OH}$$

JACS (1940) <u>62</u> 1478

$$\text{Me}_2\text{C}=\text{CHCH}_2\text{CH}_2\underset{\overset{|}{\text{Me}}}{\text{C}}=\text{CHCHO} \xrightarrow{\text{H}_2 \quad \text{Pt} \quad \text{FeSO}_4} \text{Me}_2\text{C}=\text{CHCH}_2\text{CH}_2\underset{\overset{|}{\text{Me}}}{\text{C}}=\text{CHCH}_2\text{OH}$$

100%

JACS (1926) <u>48</u> 477
(1925) <u>47</u> 3061

$$C_6H_{13}CHO \xrightarrow{\quad H_2 \ (50 \ atmos) \quad RhCl_3(Ph_3P)_3 \quad} C_6H_{13}CH_2OH \qquad 100\%$$

Chem Comm (1965) 17

$$PrCHO \xrightarrow[RhCl_3 \cdot 3H_2O]{H_2 \quad CO \ (240 \ atmos)} PrCH_2OH \qquad 70\%$$

Ber (1966) <u>99</u> 1086

$$C_6H_{13}CHO \xrightarrow{\quad Fe \quad HOAc \quad H_2O \quad} C_6H_{13}CH_2OH \qquad 75\text{-}81\%$$

Org Synth (1932) Coll Vol 1 304
JACS (1939) <u>61</u> 2134

$$C_6H_{13}CHO \xrightarrow{\quad LiAlH_4 \quad Et_2O \quad} C_6H_{13}CH_2OH \qquad 86\%$$

Org React (1951) <u>6</u> 469
JACS (1947) <u>69</u> 1197

$$PhCHO \xrightarrow{\quad LiAl(OBu\text{-}t)_3H \quad Et_2O \quad} PhCH_2OH$$

JACS (1958) <u>80</u> 5372

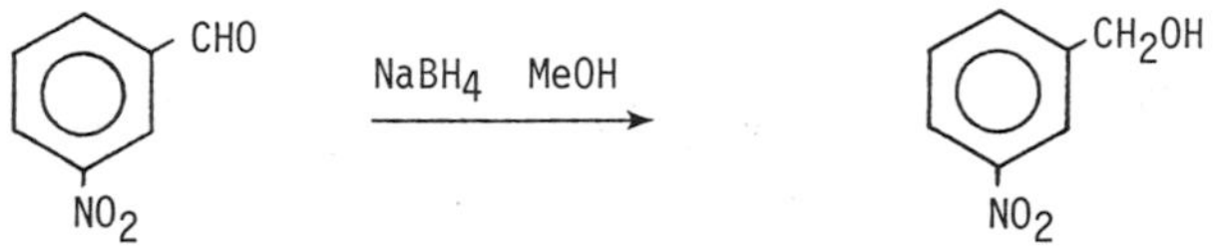

$$\xrightarrow{\quad NaBH_4 \quad MeOH \quad} \qquad 82\%$$

JACS (1949) <u>71</u> 122

PhCHO $\xrightarrow{\text{LiBH}_3\text{CN} \quad \text{MeOH}}$ PhCH$_2$OH   78%

JACS (1969) <u>91</u> 3996

PrCHO $\xrightarrow{\text{B}_2\text{H}_6 \quad \text{THF}}$ PrCH$_2$OH

JACS (1960) <u>82</u> 681

PhCH=CHCHO $\xrightarrow{\text{C}_5\text{H}_5\text{NBH}_3 \quad \text{C}_6\text{H}_6}$ PhCH=CHCH$_2$OH   84%

JOC (1958) <u>23</u> 1561

PhCHO $\xrightarrow{\text{Ph}_3\text{SnH}}$ PhCH$_2$OH   85%

JACS (1961) <u>83</u> 1246

MeCH=CHCHO $\xrightarrow{\text{Bu}_2\text{SnH}_2 \quad \text{Et}_2\text{O}}$ MeCH=CHCH$_2$OH   59%

JACS (1961) <u>83</u> 1246

PrCHO $\xrightarrow{\text{(i-PrO)}_3\text{Al} \quad \text{C}_6\text{H}_6}$ PrCH$_2$OH   36%

Org React (1944) <u>2</u> 178

$\xrightarrow{\begin{array}{c}\text{HCHO} \quad \text{NaOH}\\ \text{MeOH} \quad \text{H}_2\text{O}\end{array}}$ 85-90%

JACS (1935) <u>57</u> 905
Org React (1944) <u>2</u> 94

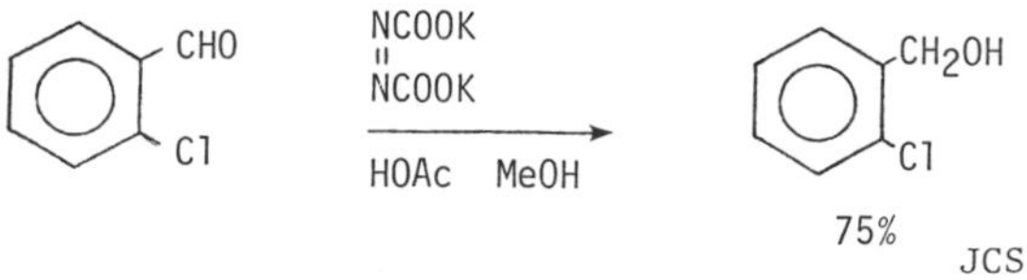

(Rate of reduction
ArC=O > AliphC=O)

75%

JCS C (1967) 1120

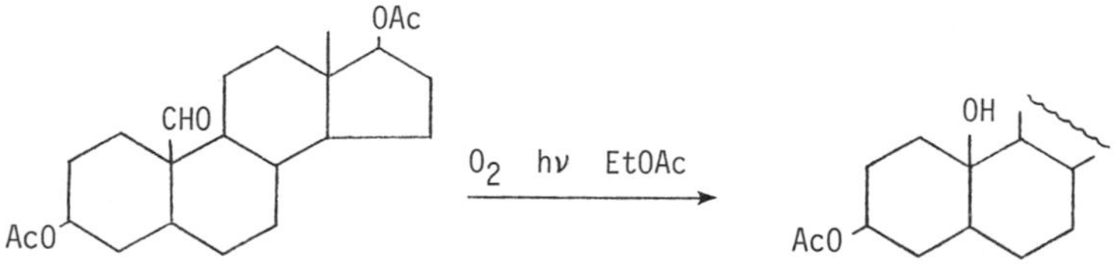

20%

Tetr Lett (1966) 2507

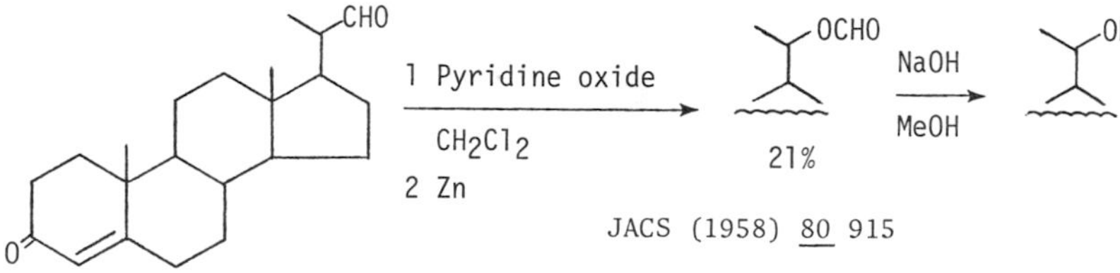

JACS (1958) 80 915

$C_5H_{11}CH_2CHO$ $\xrightarrow{H_2O_2}$ $C_5H_{11}CH_2OCHO$ $---\rightarrow$ $C_5H_{11}CH_2OH$

Org React (1957) 9 73
Ber (1941) 74B 1552

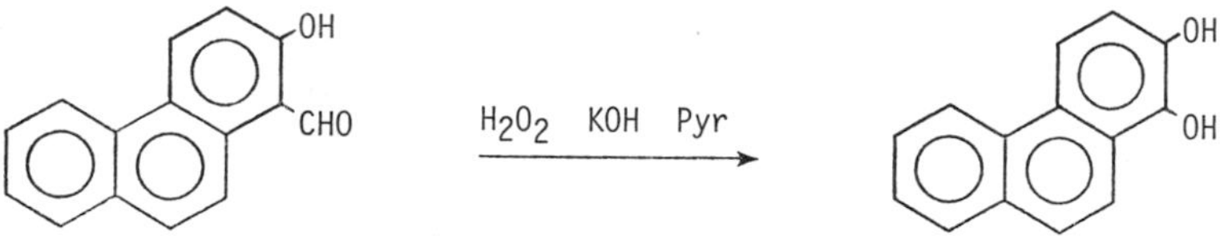

80%

Org React (1957) 9 73
Ber (1940) 73B 935

Alcohols may also be prepared from aldehydes by some of the methods
listed in section 42 (Alcohols and Phenols from Ketones)

Section 35   <u>Alcohols and Phenols from Alkyls, Methylenes and Aryls</u>
ooooooooooooooooooooooooooooooooooooooooooooooooooooooo

No examples of the reaction RR' → ROH (R'=alkyl, aryl etc.) occur in the
literature. For reactions of the type RH → ROH (R=alkyl or aryl) see
section 41 (Alcohols and Phenols from Hydrides)

Section 36   <u>Alcohols from Amides</u>
ooooooooooooooooooooooo

$\xrightarrow{\text{LiAlH}_4 \quad \text{THF}}$ 96%

Ber (1955) <u>88</u> 301

PhCON⟨⟩ $\xrightarrow{\text{LiAlH}_4 \quad \text{Et}_2\text{O}}$ PhCH$_2$OH   80%

JOC (1953) <u>18</u> 1190

$C_{11}H_{23}CONH_2$ $\xrightarrow[\text{Copper chromite \quad EtOH}]{\text{H}_2 \text{ (200-300 atmos)} \quad 250°}$ $C_{11}H_{23}CH_2OH$

JACS (1934) <u>56</u> 2419

$C_{13}H_{27}CONH_2$ $\xrightarrow{\text{Electrolysis \quad LiCl \quad MeNH}_2}$ $C_{13}H_{27}CH_2OH$ 92%

JOC (1970) <u>35</u> 1210

$PhCONHC_6H_{13}$ $\xrightarrow{\text{Electrolysis \quad Me}_4\text{NCl \quad MeOH}}$ $PhCH_2OH$ 69%

Ber (1965) <u>98</u> 3462

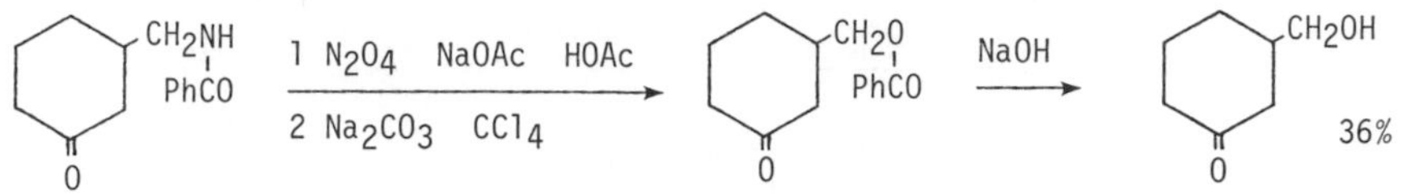

JOC (1969) 34 3834

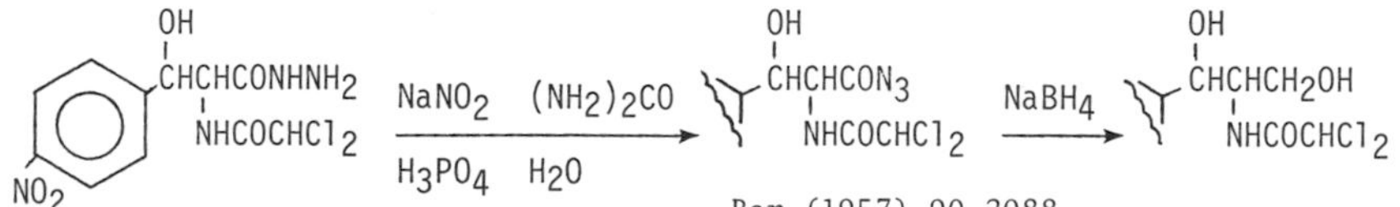

Ber (1957) 90 2088

Alcohols may also be prepared by conversion of amides into esters, followed by hydrolysis. See section 111 (Esters from Amides)

Section 37   Alcohols and Phenols from Amines
            ○○○○○○○○○○○○○○○○○○○○○○○○○○○○○○○○○○○○○○○

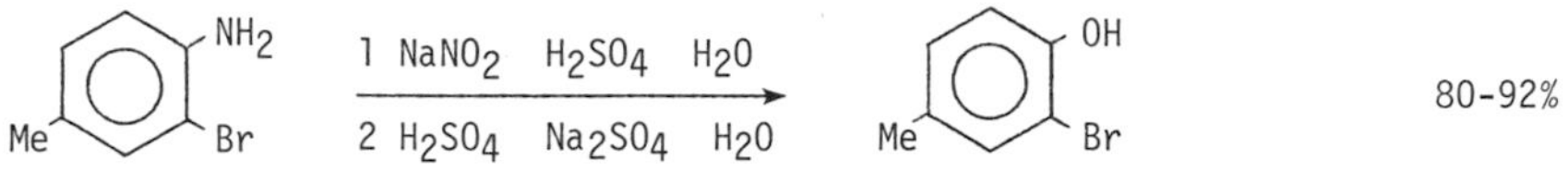

80-92%

Org Synth (1955) Coll Vol 3 130

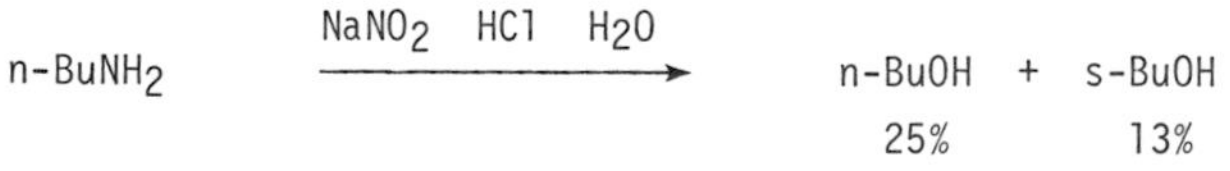

n-BuNH$_2$ $\xrightarrow{\text{NaNO}_2 \quad \text{HCl} \quad \text{H}_2\text{O}}$ n-BuOH + s-BuOH

                                    25%   13%

JACS (1932) 54 3441

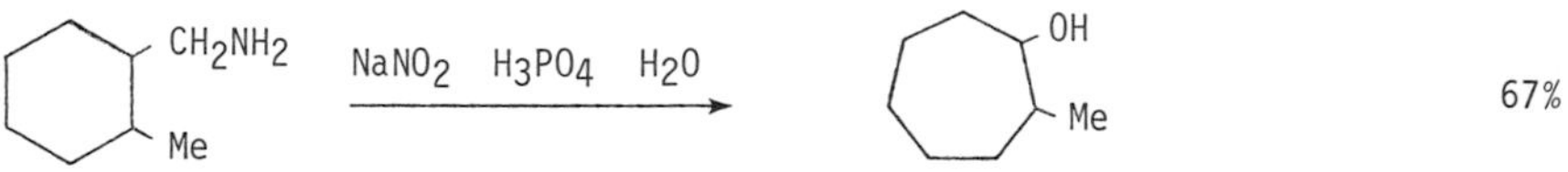

67%

JOC (1965) 30 350

NaNO$_2$ HOAc

dioxane H$_2$O

96%

JCS (1959) 345

MeO–[benzene]–CH$_2$CH$_2$NH$_2$ $\xrightarrow{Ac_2O}$ CH$_2$CH$_2$NHAc $\xrightarrow[\text{2 KOH MeOH}]{\text{1 NaNO}_2 \quad \text{HOAc Ac}_2\text{O}}$ MeO–[benzene]–CH$_2$CH$_2$OH

82%

Chem Pharm Bull (1960) 8 266

Section 111 contains further examples of the conversion of N-acyl amines
into esters from which alcohols can be prepared by hydrolysis

Section 38   Alcohols and Phenols from Esters

C$_{15}$H$_{31}$COOMe $\xrightarrow{\text{EtMgBr Et}_2\text{O}}$ C$_{15}$H$_{31}$CEt$_2$ 96%

OH

JACS (1945) 67 2239

PhCOOMe $\xrightarrow{\text{MeI Li THF}}$ PhCOH (with Me above and Me below)   (One-step procedure)   74%

Chem Comm (1970) 1160

C$_{11}$H$_{23}$COOEt $\xrightarrow{\text{Na EtOH toluene}}$ C$_{11}$H$_{23}$CH$_2$OH   65-75%

Org Synth (1943) Coll Vol 2 372

$C_7H_{15}COOEt$ $\xrightarrow[\text{H}_2\text{O} \quad \text{Et}_2\text{O}]{\text{Na} \quad \text{NaOAc} \quad \text{HOAc}}$ $C_7H_{15}CH_2OH$ >90%

Rec Trav Chim (1923) <u>42</u> 1050

$C_5H_{11}COOMe$ $\xrightarrow[\text{250}°]{\text{H}_2 \text{ (200 atmos)} \quad \text{copper chromite}}$ $C_5H_{11}CH_2OH$ 92%

JACS (1932) <u>54</u> 1145
Org React (1954) <u>8</u> 1

$PrCOOBu$ $\xrightarrow[\text{218}°]{\text{H}_2 \text{ (205 atmos)} \quad \text{ReO}_3}$ $PrCH_2OH$ 85%

JOC (1963) <u>28</u> 2345

$PhCOOEt$ $\xrightarrow{\text{Electrolysis} \quad \text{Me}_4\text{NCl} \quad \text{MeOH}}$ $PhCH_2OH$ 89%

Ber (1965) <u>98</u> 3462

$C_{15}H_{31}COOEt$ $\xrightarrow{\text{LiAlH}_4 \quad \text{Et}_2\text{O}}$ $C_{15}H_{31}CH_2OH$ 98%

JACS (1947) <u>69</u> 1197
Org React (1951) <u>6</u> 469

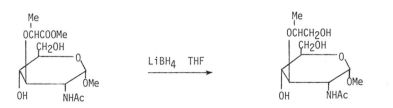

Carbohydrate Res (1967) <u>4</u> 504

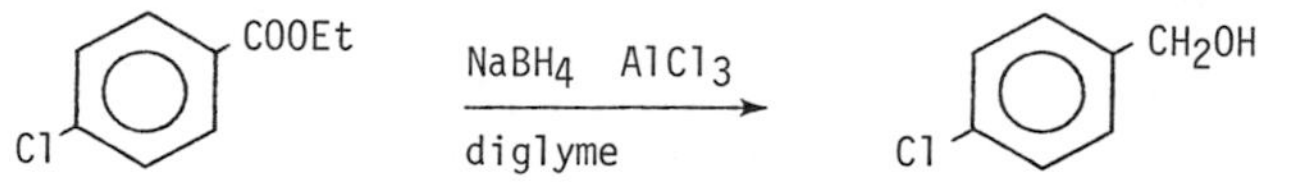

84%

JACS (1956) 78 2582

$Ph(CH_2)_2COOMe$ $\xrightarrow[\text{reflux}]{\text{NaBH}_4 \quad \text{MeOH}}$ $Ph(CH_2)_2CH_2OH$ 73%

JOC (1963) 28 3261

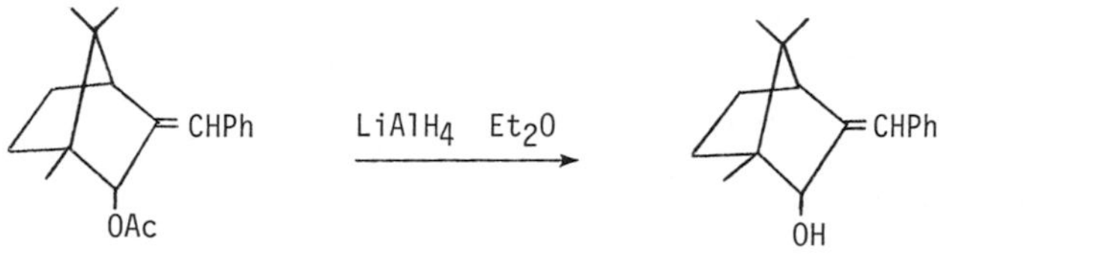

92%

Tetrahedron (1970) 26 395

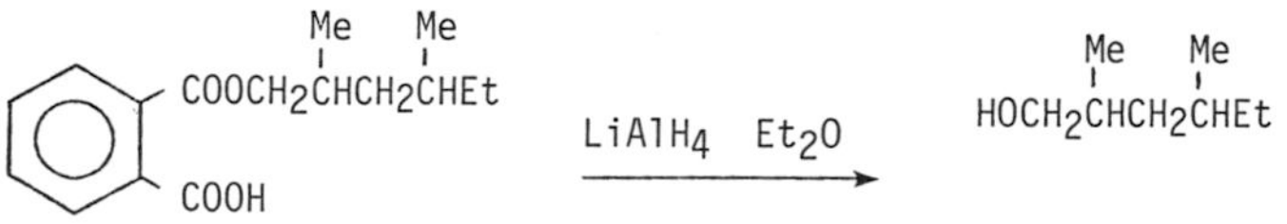

~80%

JACS (1950) 72 147

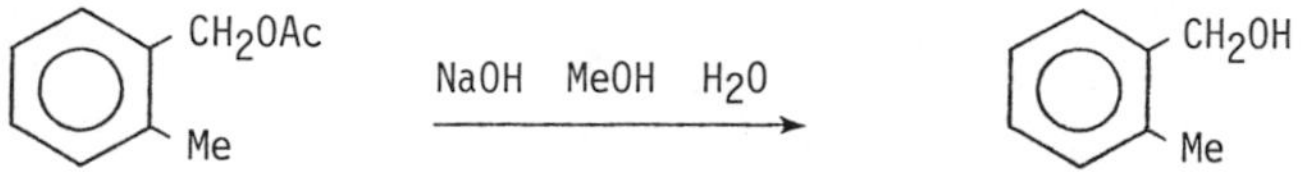

95-97%

Org Synth (1963) Coll Vol 4 582

$PhCOOEt$ $\xrightarrow[]{\text{NaOH} \quad \text{Me}_2\text{SO}}$ $EtOH$

JCS (1965) 1290

MeC=CH(CH$_2$)$_2$C=CHCH$_2$OAc $\xrightarrow{\text{K}_2\text{CO}_3 \quad \text{EtOH}}$ MeC=CH(CH$_2$)$_2$C=CHCH$_2$OH   98%
| | | |
COOEt   Me   COOEt   Me

JACS (1969) <u>91</u> 4318

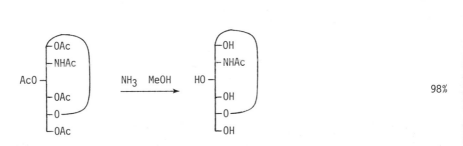

JCS (1967) 448

60%

MeO ─┬─ O ─┐   MeO ─┬─ O ─┐
     ├─ OAc │   MeOK   MeOH   ├─ OH │
AcO ─┤   │   ─────────────→   HO ─┤   │   99%
     ├─ OAc │   ├─ OH │
     └─ OAc ┘   └─ OH ┘

JACS (1948) <u>70</u> 314

     ┌─ OAc ┐   ┌─ OH ┐
     ├─ NHAc│   ├─ NHAc│
AcO ─┤   │   NH$_3$ MeOH   HO ─┤   │   98%
     ├─ OAc │   ─────────────→   ├─ OH │
     ├─ O   │   ├─ O   │
     └─ OAc ┘   └─ OH ┘

JCS (1956) 2042

HCOOBu $\xrightarrow{\text{NaH} \quad \text{diglyme}}$ BuOH   68%

Tetr Lett (1965) 1713

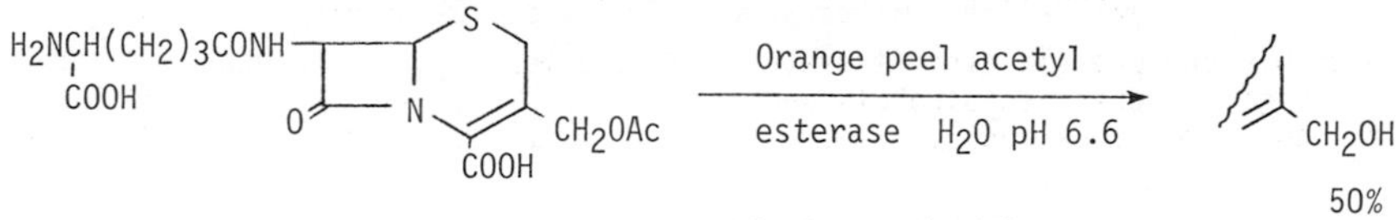

Biochem J (1961) <u>81</u> 591

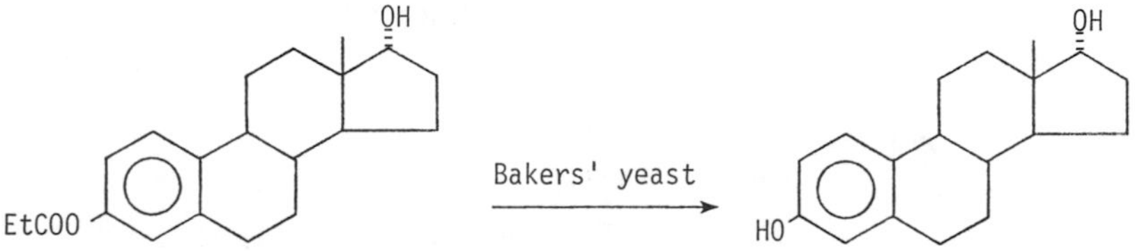

Bakers' yeast

Ber (1938) <u>71B</u> 2696

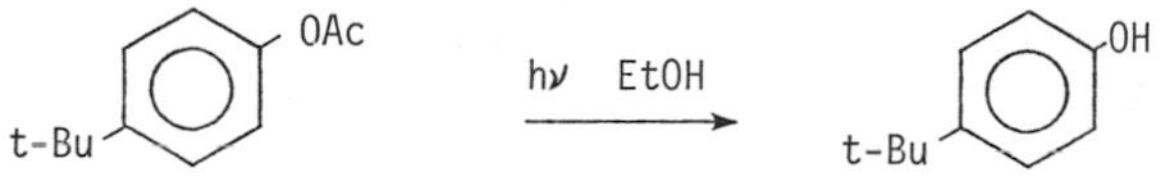

JCS (1965) 5162

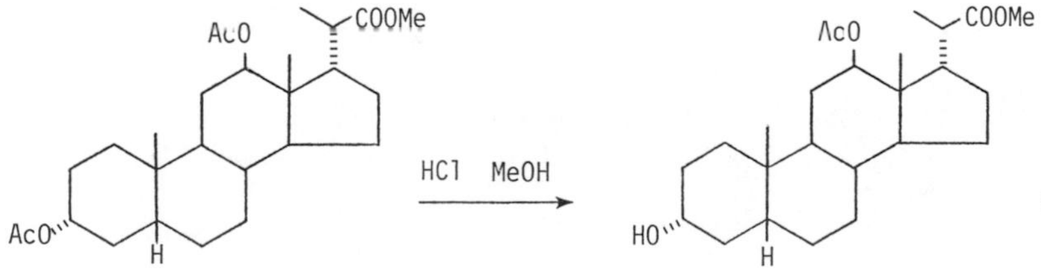

34%

HCl   MeOH

Helv (1944) <u>27</u> 713

88%

Further examples of the hydrolysis and cleavage of esters are included in
section 23 (Carboxylic Acids from Esters) and section 45A (Protection of
Alcohols and Phenols)

Section 39 Alcohols and Phenols from Ethers and Epoxides

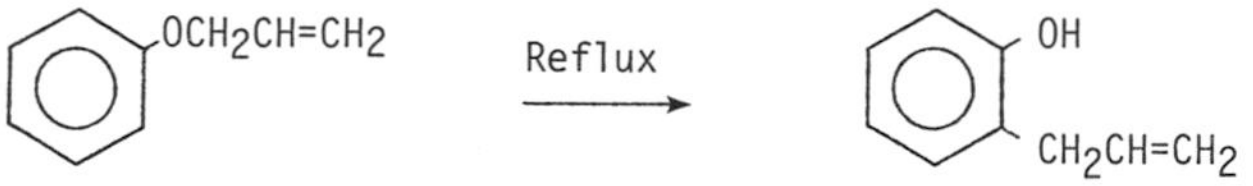

73%

Org React (1944) 2 1

$PhCH_2OMe \xrightarrow{PhLi\ \ Et_2O} PhCH(Me)OH$ 35%

Annalen (1942) 550 260
JOC (1962) 27 1933

$Me_2C{=}CHCH_2OCH_2CH{=}CMe_2 \xrightarrow{t\text{-}BuLi\ \ THF} Me_2C{=}CHCH(OH)C(Me)_2CH{=}CH_2$ 31%

Tetr Lett (1970) 353

Review: The Cleavage of Ethers Chem Rev (1954) 54 615

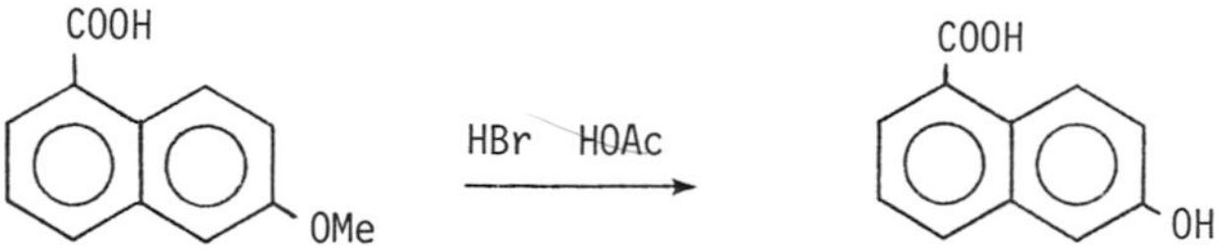

90%

JOC (1941) 6 852

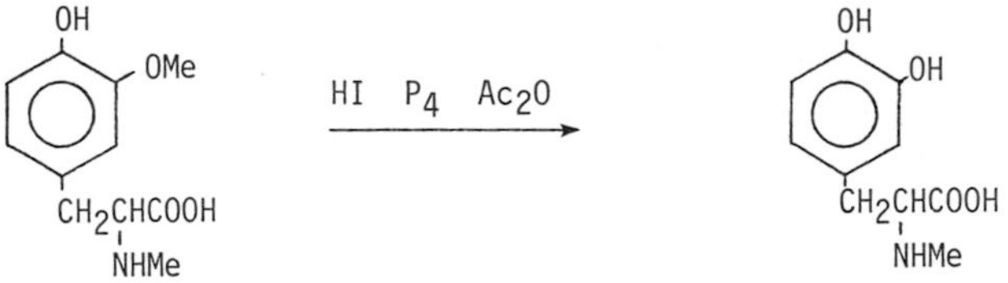

$$\xrightarrow{\text{HI P}_4\text{ Ac}_2\text{O}}$$

82%

Org Synth (1955) Coll Vol 3 586

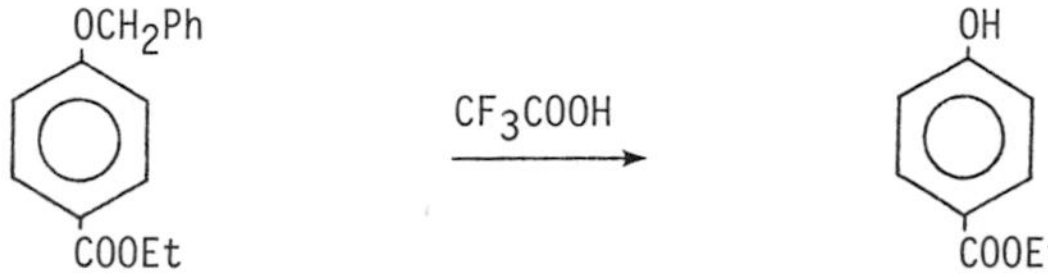

$$\xrightarrow{\text{CF}_3\text{COOH}}$$

JOC (1965) 30 2491

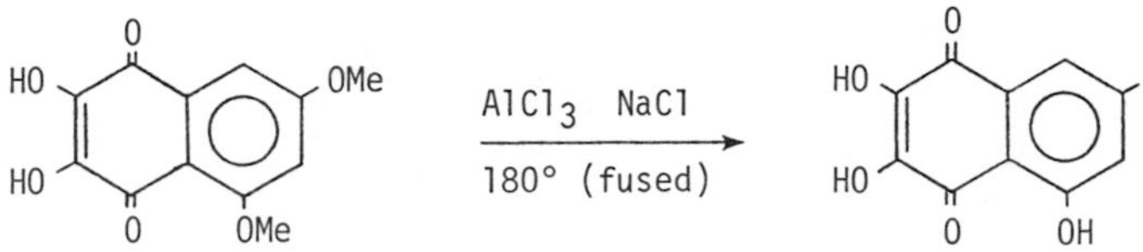

$$\xrightarrow[\text{180° (fused)}]{\text{AlCl}_3\text{ NaCl}}$$

Low yield

JCS (1961) 1008

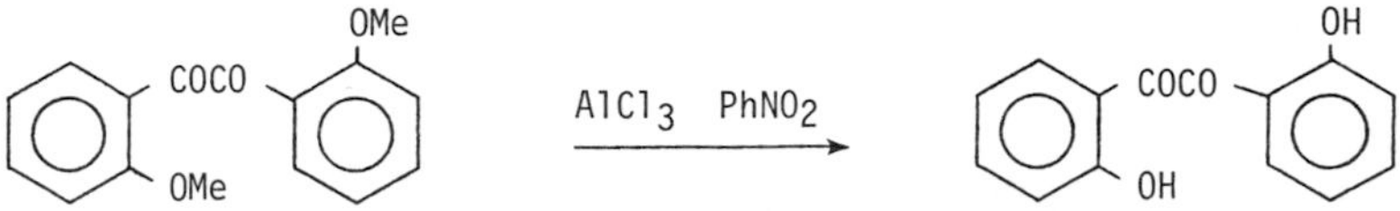

$$\xrightarrow{\text{AlCl}_3\text{ PhNO}_2}$$

50-60%

Ber (1943) 76B 900

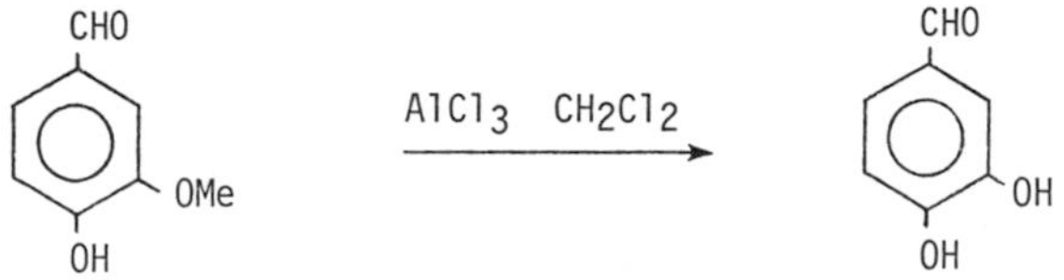

$$\xrightarrow{\text{AlCl}_3\text{ CH}_2\text{Cl}_2}$$

87%

JOC (1962) 27 2037

$$\xrightarrow{\text{AlBr}_3\text{ C}_6\text{H}_6}$$

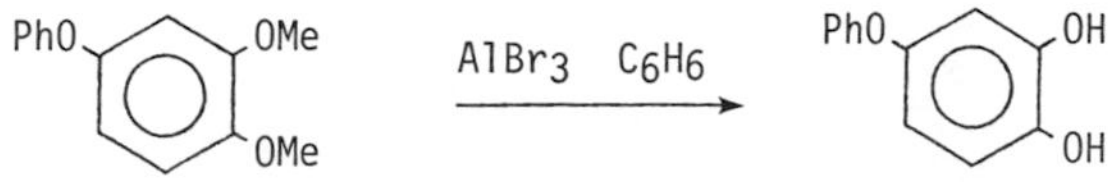

87%

Ber (1960) 93 2761

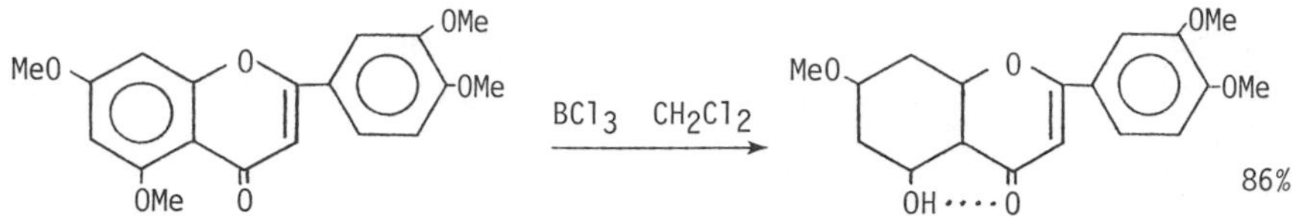

Tetr Lett (1966) 4153
JCS (1962) 1260

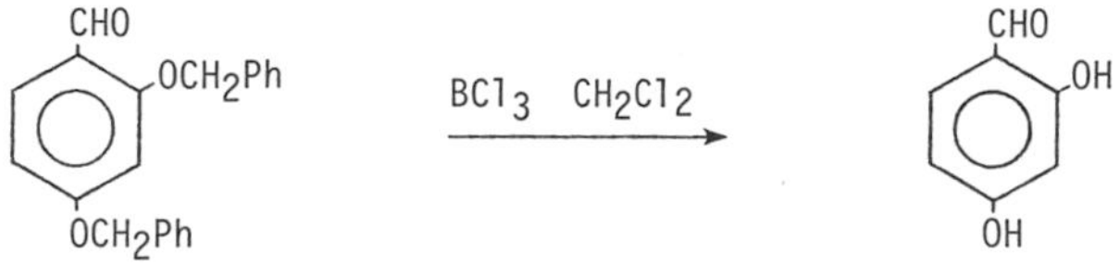

86%

Tetr Lett (1966) 4155

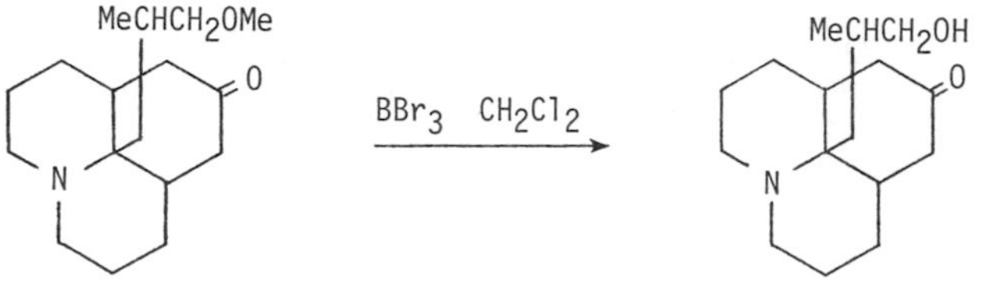

75%

JACS (1968) 90 1648

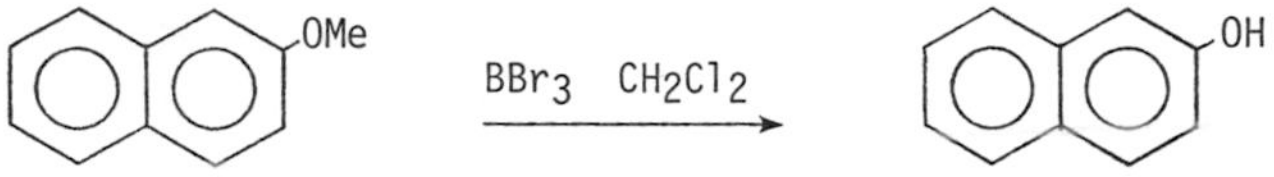

Tetrahedron (1968) 24 2289

67%

PhOEt $\xrightarrow[\text{2 } H_2O]{\text{1 } BI_3}$ PhOH

Tetr Lett (1967) 4131

$Pr_2O \xrightarrow{BI_3} (PrO)_2BI \dashrightarrow PrOH$

Tetr Lett (1967) 4131

$$Et_2O \xrightarrow{\text{C}_5\text{H}_5\text{NBH}_2\text{I} \quad \text{C}_6\text{H}_6 \quad \text{Et}_2\text{O}} C_5H_5NBH_2OEt \dashrightarrow EtOH$$

JACS (1968) <u>90</u> 6260

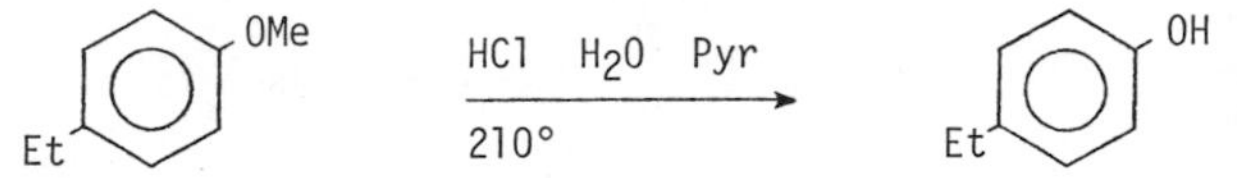

HCl H$_2$O Pyr

210°

97%

Chem Ind (1967) 1138

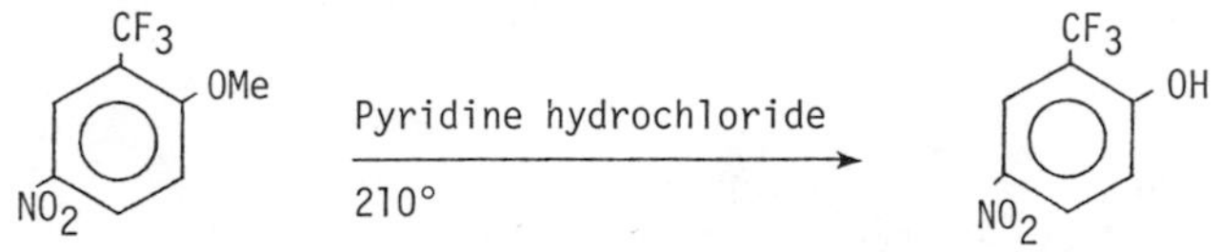

Pyridine hydrochloride

210°

75%

JOC (1962) <u>27</u> 4660

HOAc H$_2$O

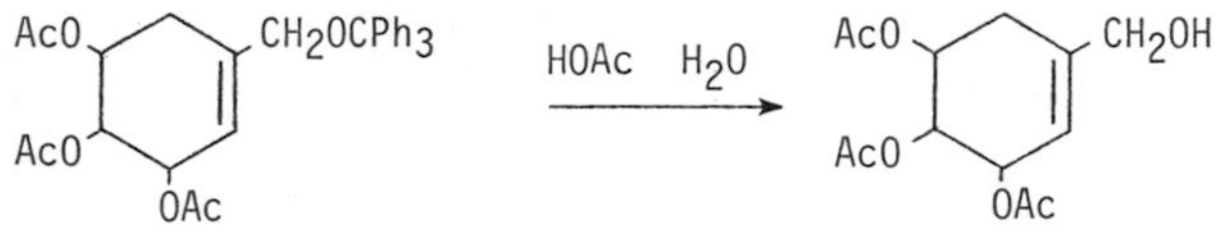

80%

Ber (1956) <u>89</u> 898

Further examples of the cleavage of triphenylmethyl ethers are included in section 45A (Protection of Alcohols and Phenols)

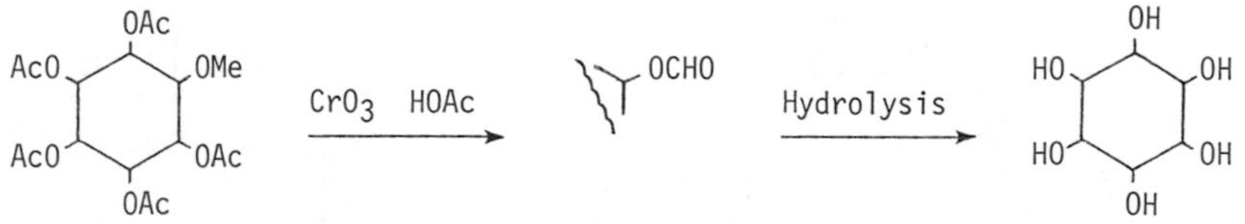

CrO$_3$ HOAc

Hydrolysis

Carbohydrate Res (1970) <u>12</u> 147

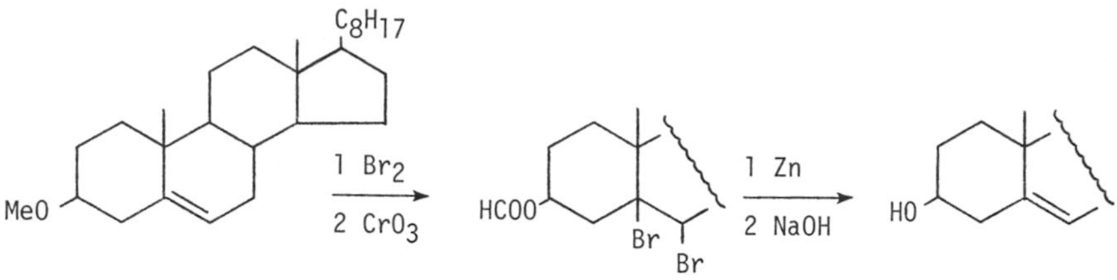

Chem Comm (1966) 752

$$BuOCH_2Pr \xrightarrow[\text{CuCl} \quad \text{C}_6\text{H}_6]{\text{PhCOO}_2\text{Bu-t}} \underset{\text{PhCOO}}{BuOCHPr} \dashrightarrow BuOH$$

50%

Acta Chem Scand (1961) 15 249
Tetr Lett (1960) (2) 4
Arkiv Kemi (1960) 16 287

$$BuOCH_2Pr \xrightarrow[\text{hν}]{\overset{\text{NCOOEt}}{\underset{\text{NCOOEt}}{\|}}} \underset{\underset{\text{HNCOOEt}}{\text{NCOOEt}}}{BuOCHPr} \xrightarrow{\text{H}_2\text{O}} BuOH$$

60%

Chem Comm (1965) 259

$$EtOCH_2Me \xrightarrow{\text{PhN=NOAc}} \underset{\text{OAc}}{EtOCHMe} \overset{\text{H}_2\text{O}}{\dashrightarrow} EtOH$$

JACS (1964) 86 3180

$$BuOCH_2Ph \xrightarrow[\quad]{\text{H}_2 \quad \text{Pd}} BuOH \qquad\qquad 100\%$$

Org React (1953) 7 263
JACS (1954) 76 3188

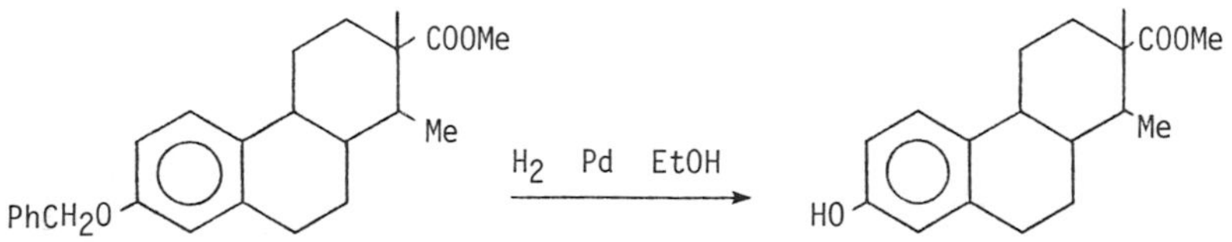

J Med Chem (1969) 12 192

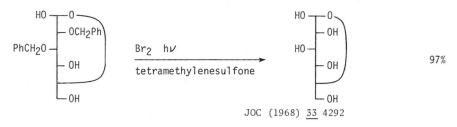

$$\text{HO}\!-\!\!\left[\begin{array}{l}\text{O}\\\text{OCH}_2\text{Ph}\\\text{OH}\\\text{OH}\end{array}\right.$$

PhCH$_2$O

$\xrightarrow[\text{tetramethylenesulfone}]{\text{Br}_2\quad h\nu}$

$$\text{HO}\!-\!\!\left[\begin{array}{l}\text{O}\\\text{OH}\\\text{OH}\\\text{OH}\end{array}\right.$$

HO

97%

JOC (1968) <u>33</u> 4292

⬡—OMe $\xrightarrow{(\text{Me}_3\text{Si})_2\text{Hg}\quad 160°}$ ⬡—OSiMe$_3$ --→   ⬡—OH

JCS <u>C</u> (1967) 2188

⬠⬠—OMe $\xrightarrow{\text{LiI}\quad \text{collidine}}$ ⬠⬠—OH   100%

Chem Comm (1969) 616
JACS (1950) <u>72</u> 3396
For cleavage with KCN see Ber (1933) <u>66B</u> 1623

⬡—OMe
Me $\xrightarrow{\text{EtS}^-\quad \text{DMF}}$ ⬡—OH
Me   94%

Tetr Lett (1970) 1327

PhOMe $\xrightarrow[185°]{\text{MeNH}_2\quad \text{H}_2\text{O}}$ PhOH

Chem Ind (1970) 1230
Aust J Chem (1970) <u>23</u> 2539

PhOR $\xrightarrow{\text{Ph}_2\text{PLi}\quad \text{THF}}$ PhOH

83% (R=Me)
88% (R=PhCH$_2$)

No reaction (R=Et, i-Pr)

JCS (1965) 4120
Tetr Lett (1967) 3661
JACS (1970) <u>92</u> 7232

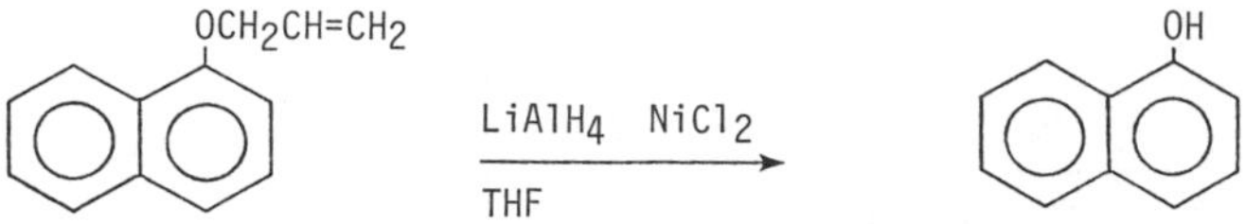

JACS (1957) 79 5463   76%

PhOMe $\xrightarrow{\text{Lithium diphenyl THF}}$ PhOH   80%

JOC (1963) 28 707

PhOR $\xrightarrow{\text{BuMgBr CoCl}_2\text{ Et}_2\text{O}}$ PhOH   86% (R=PhCH$_2$)

43% (R=Ph)

No reaction (R=Me)

JOC (1952) 17 669

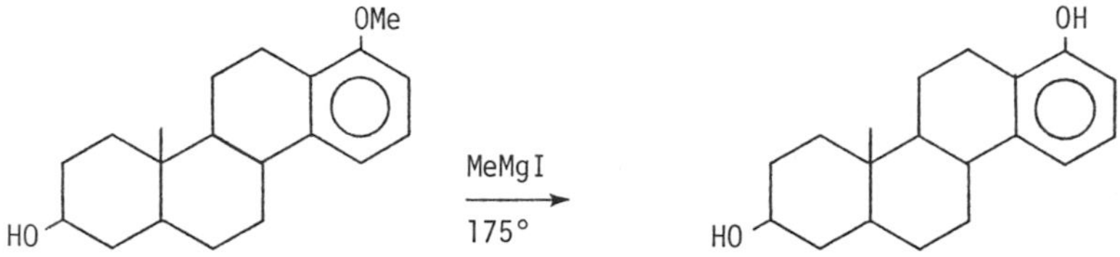

JACS (1956) 78 6322   90%

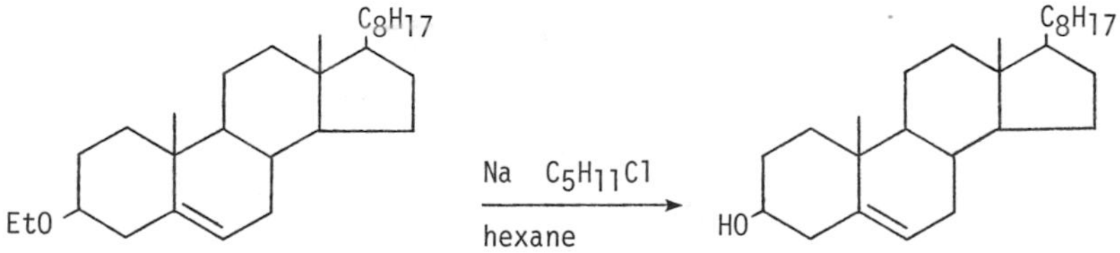

JACS (1951) 73 1263   74%

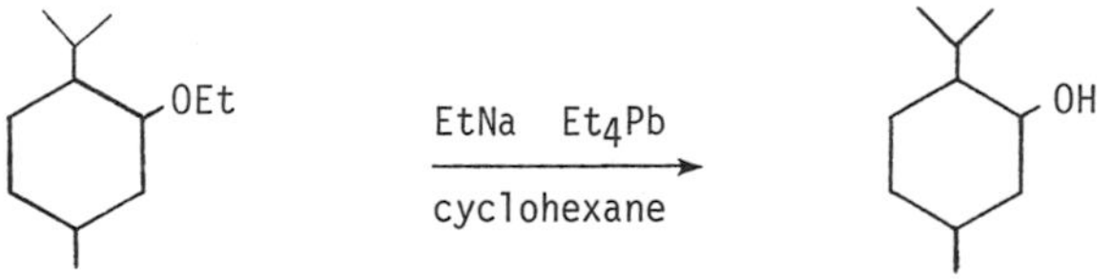

J Prakt Chem (1938) 151 61

$PhOCH_2CH=CH_2$ $\xrightarrow[\text{2 } CO_2]{\text{1 Mg   THF}}$ PhOH   86%

J Organometallic Chem (1969) <u>18</u> 249

PhOR $\xrightarrow{NaNH_2 \text{   piperidine}}$ PhOH   82% (R=Me)
                                                                              90% (R=$PhCH_2$)

Chem Ind (1957) 80
JACS (1951) <u>73</u> 1437

$PhCH_2OEt$ $\xrightarrow{\text{Li   THF}}$ EtOH

JOC (1961) <u>26</u> 3723

PhOMe $\xrightarrow{\text{K   HMPA}}$ PhOH   76-80%

Bull Soc Chim Fr (1966) 3344

PhOMe $\xrightarrow{\text{Na   Pyr}}$ PhOH   94%

Ber (1943) <u>76B</u> 156

$\xrightarrow{\text{Na   } NH_3}$ + PhOH

JACS (1937) <u>59</u> 1488

PhOMe $\xrightarrow{\text{Li   } H_2NCH_2CH_2NH_2}$ PhOH   54%

JOC (1957) <u>22</u> 891

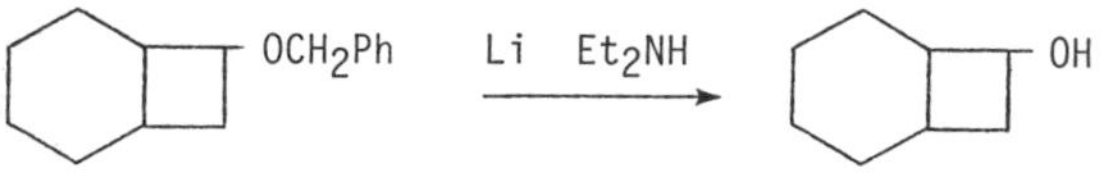

JACS (1970) <u>92</u> 553

Alcohols and phenols may also be prepared by conversion of ethers into esters, followed by hydrolysis. See section 114 (Esters from Ethers)

Further methods for the cleavage of ethers to alcohols are included in section 45A (Protection of Alcohols and Phenols)

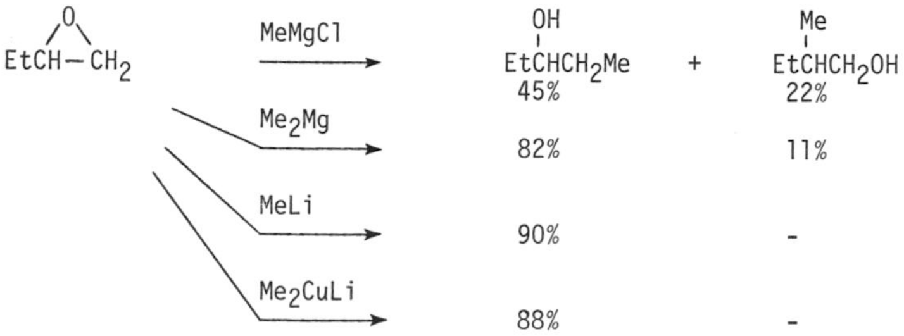

JACS (1970) <u>92</u> 4979
Chem Rev (1951) <u>49</u> 413

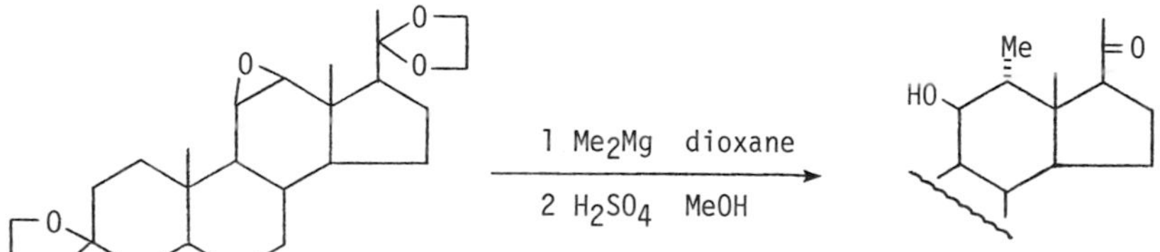

JACS (1960) <u>82</u> 3995
Bull Soc Chim Fr (1969) 4414

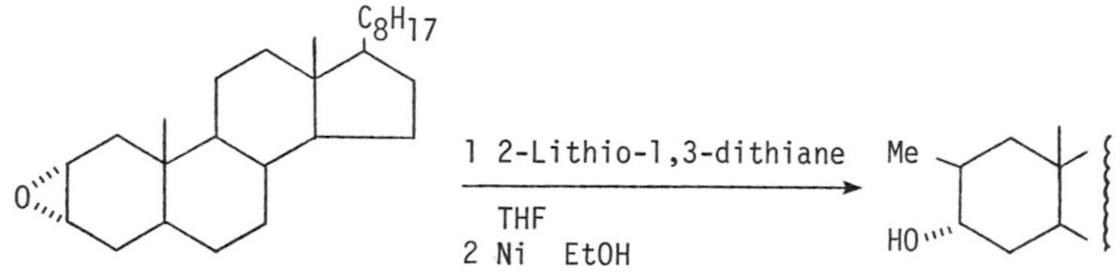

Chem Comm (1970) 141 741

$$\xrightarrow{\text{H}_2 \quad \text{Pt} \quad \text{HOAc}}$$

81%

Helv (1953) 36 1332
      (1943) 26 562

$$\overset{O}{\overset{\diagup\ \diagdown}{\text{MeCH}-\text{CH}_2}} \xrightarrow{\text{Na} \quad \text{NH}_3} \overset{\text{OH}}{\underset{|}{\text{MeCHMe}}}$$

Quart Rev (1958) 12 17

$$\xrightarrow{\text{Li} \quad \text{EtNH}_2 \quad \text{t-BuOH}}$$

JCS C (1968) 1581

$$\overset{O}{\overset{\diagup\ \diagdown}{\text{Me}_2\text{C}-\text{CH}_2}}$$

$$\xrightarrow{\text{LiAlH}_4 \quad \text{Et}_2\text{O}} \quad \overset{\text{OH}}{\underset{|}{\text{Me}_2\text{CMe}}} \qquad 26\%$$

$$\xrightarrow{\text{LiAlH}_4 \quad \text{AlCl}_3 \quad \text{Et}_2\text{O}} \quad \text{Me}_2\text{CHCH}_2\text{OH} \qquad 55\%$$

JACS (1956) 78 3226
Org React (1951) 6 469

$$\overset{O}{\overset{\diagup\ \diagdown}{\text{PhCH}-\text{CH}_2}} \xrightarrow{\text{LiAlH}_4 \quad \text{Et}_2\text{O}} \overset{\text{OH}}{\underset{|}{\text{PhCHMe}}} \qquad 94\%$$

JACS (1948) 70 3738

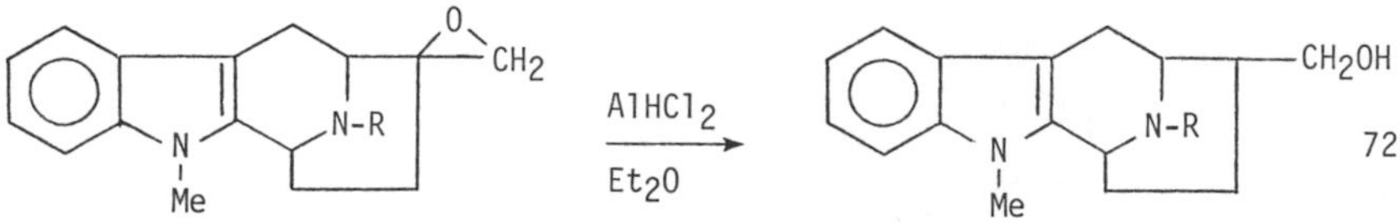

AlHCl$_2$
Et$_2$O

72%

Tetr Lett (1969) 901

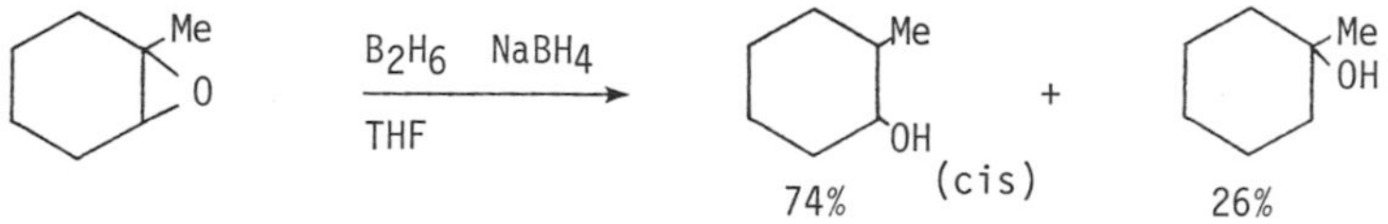

B$_2$H$_6$ NaBH$_4$
THF

74%   (cis)   26%

JACS (1968) 90 2686

B$_2$H$_6$ BF$_3$·Et$_2$O   THF

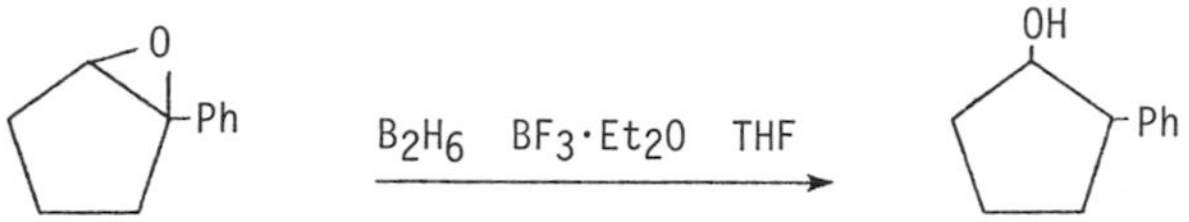

82%

Chem Comm (1968) 1549

Section 40   Alcohols and Phenols from Halides and Sulfonates

i-PrCl   ----→   i-PrLi

1 CH$_2$=CH$_2$
2 Ph$_2$CO

i-PrCH$_2$CH$_2$COH with Ph, Ph substituents

JACS (1969) 91 6362

BuBr   ----→   BuLi

CH$_2$=CHCH$_2$OH
Me$_2$NCH$_2$CH$_2$NMe$_2$

BuCHCH$_2$OH
  |
  Me

72%

Tetr Lett (1969) 325

$CH_2=CHCH_2Br$ $\xrightarrow[\text{2 (EtO)}_2CO]{\text{1 Mg Et}_2O}$ $(CH_2=CHCH_2)_3COH$ 30%

Compt Rend (1968) C 267 773
Org Synth (1943) Coll Vol 2 602

PhBr $\xrightarrow{\text{1 Mg Et}_2O}$ 2 PhCOOEt C_6H_6 / 2 Ph_2CO C_6H_6 → Ph_3COH 89-93%

Org Synth (1955) Coll Vol 3 839

$\xrightarrow[\text{2 MeCHO}]{\text{1 Mg Et}_2O}$

82-88%

Org Synth (1955) Coll Vol 3 200

BuBr $\xrightarrow[\text{2 HCOOEt}]{\text{1 Mg Et}_2O}$ Bu_2CHOH 83-85%

Org Synth (1943) Coll Vol 2 179

$C_6H_{13}F$ $\xrightarrow[\text{THF}]{\text{Mg I}_2}$ $C_6H_{13}MgF$ $\xrightarrow{\text{Ph}_2\text{CO}}$ $C_6H_{13}\overset{\text{Ph}}{\underset{\text{Ph}}{C}}OH$

95%

JACS (1970) 92 433

Mg Et$_2$O

59%

37%

JOC (1968) 33 2991
Org Synth (1932) Coll Vol 1 306

95%

Rec Trav Chim (1965) 84 1200

For Grignard reaction with HCHO gas see JACS (1933) 55 1119
and JOC (1968) 33 3408

68%

Can J Chem (1962) 40 2175
JOC (1939) 4 318

PhBr $\xrightarrow[\text{2 O}_2]{\text{1 Mg Et}_2\text{O}}$ PhOH   46%

JACS (1955) 77 6032

$\underset{\text{Me}}{C_6H_{13}CHCl}$ $\xrightarrow[\text{2 O}_2]{\text{1 Mg Et}_2\text{O}}$ $\underset{\text{Me}}{C_6H_{13}CHOOH}$ $\dashrightarrow$ $\underset{\text{Me}}{C_6H_{13}CHOH}$

91%

JACS (1955) 77 6032
(1943) 65 501

t-BuCl $\xrightarrow[\text{2 HgCl}_2]{\text{1 Mg Et}_2\text{O}}$ t-BuHgCl $\xrightarrow[\text{}]{\text{O}_3 \quad \text{CH}_2\text{Cl}_2}$ t-BuOH   50%

Tetr Lett (1970) 2679

PhCl $\xrightarrow[\substack{\text{2 (MeO)}_3\text{B} \\ \text{3 H}_2\text{O}_2 \quad \text{H}_2\text{O}}]{\text{1 Mg Et}_2\text{O}}$ PhOH   78%

JOC (1957) 22 1001

PhBr $\xrightarrow[\text{2 t-BuOOH}]{\text{1 Mg Et}_2\text{O}}$ PhOH   98%

JACS (1959) 81 4230

PhBr $\xrightarrow[\text{2 PhCOO}_2\text{Bu-t}]{\text{1 Mg Et}_2\text{O}}$ PhOBu-t $\xrightarrow{\text{Acid}}$ PhOH

Arkiv Kemi (1961) <u>17</u> 393
Org Synth (1963) <u>43</u> 55

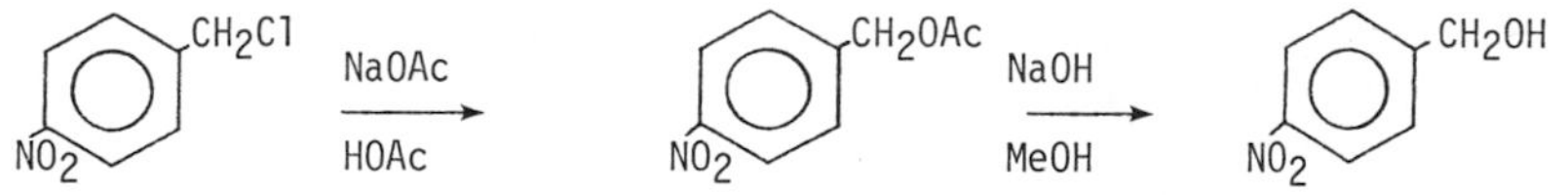

$\xrightarrow[\text{HOAc}]{\text{NaOAc}}$

$\xrightarrow[\text{MeOH}]{\text{NaOH}}$

Org Synth (1955) Coll Vol 3 650 652

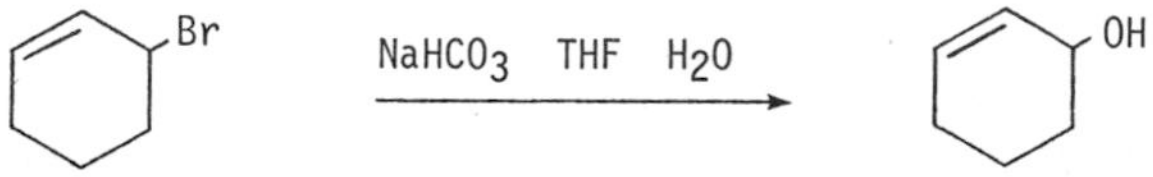

$\xrightarrow{\text{NaHCO}_3 \quad \text{THF} \quad \text{H}_2\text{O}}$

JACS (1961) <u>83</u> 198
(1946) <u>68</u> 751

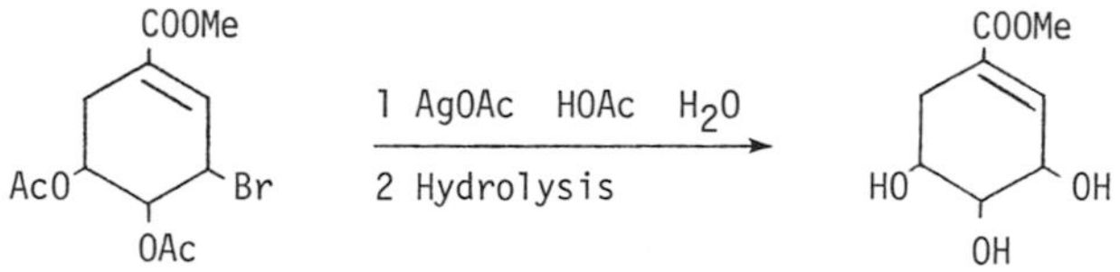

$\xrightarrow[\text{2 Hydrolysis}]{\text{1 AgOAc HOAc H}_2\text{O}}$

Ber (1964) <u>97</u> 443

$\xrightarrow[\text{methyl cellosolve}]{\text{AgNO}_3 \quad \text{H}_2\text{O}}$

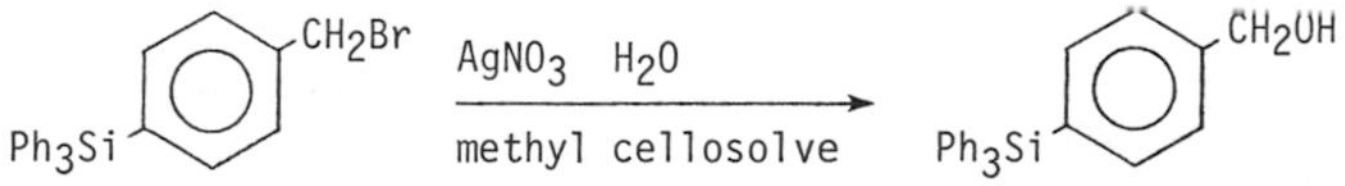

76%

JACS (1956) <u>78</u> 1689
J Biol Chem (1946) <u>164</u> 569

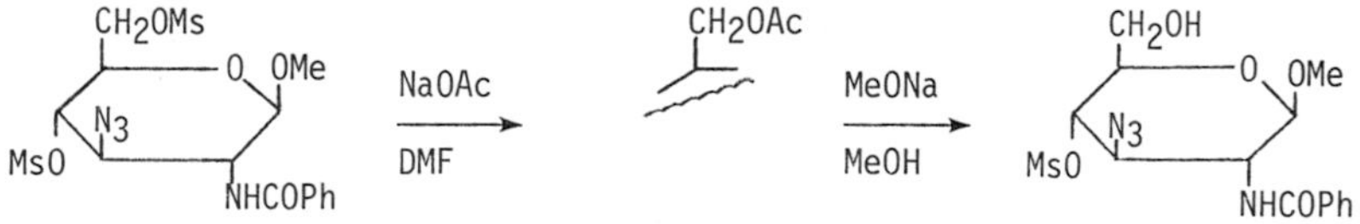

$\xrightarrow[\text{DMF}]{\text{NaOAc}}$

$\xrightarrow[\text{MeOH}]{\text{MeONa}}$

77%

Ber (1970) <u>103</u> 37

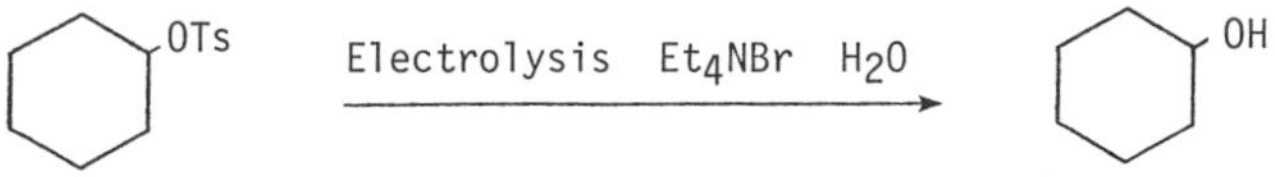

JOC (1968) 33 2716

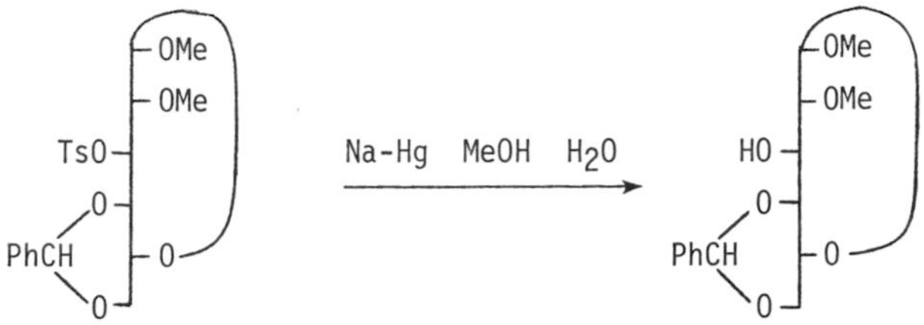

96%

Helv (1945) 28 1164

88%

PhOTs $\xrightarrow{\text{Sodium-naphthalene   THF}}$ PhOH   99%

JACS (1966) 88 1581

$C_8H_{17}OTs$ $\xrightarrow{\text{Na   NH}_3\text{   Et}_2O}$ $C_8H_{17}OH$ 56%

JOC (1956) 21 479

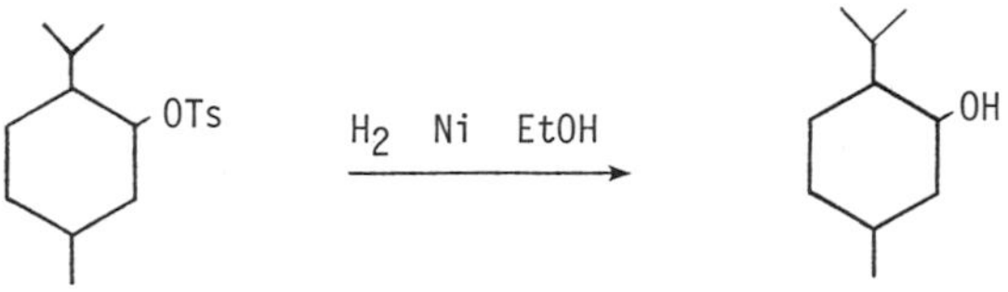

JCS (1949) S178

86%

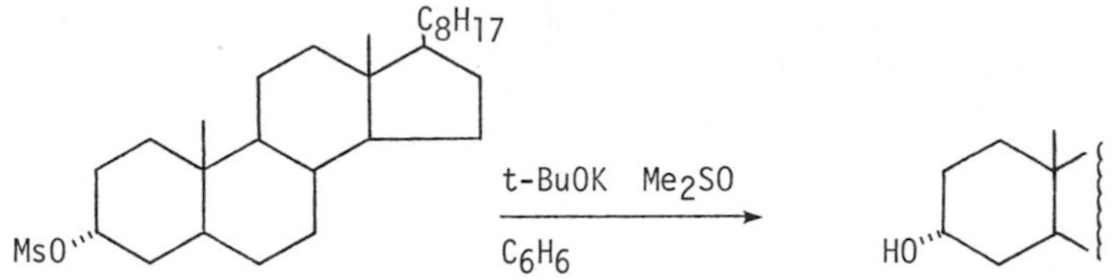

Tetr Lett (1964) 305

87%

Further examples of the cleavage of sulfonates to alcohols and phenols are included in section 45A (Protection of Alcohols and Phenols)

Section 41   Alcohols and Phenols from Hydrides (RH)

Bull Chem Soc Jap (1967) 40 2980

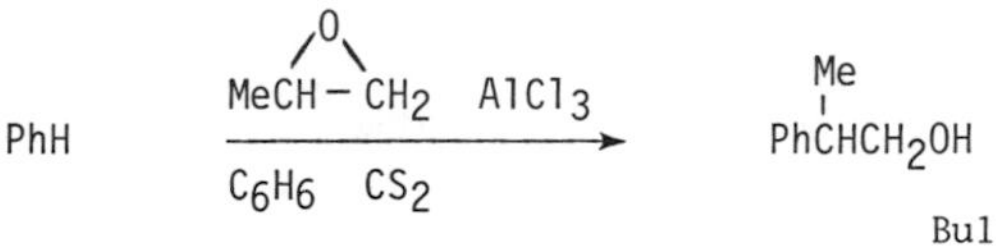

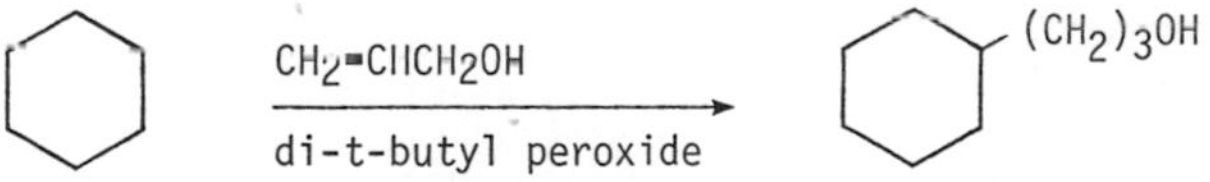

56%

Neftekhimiya (1965) 5 554
(Chem Abs 63 16223)

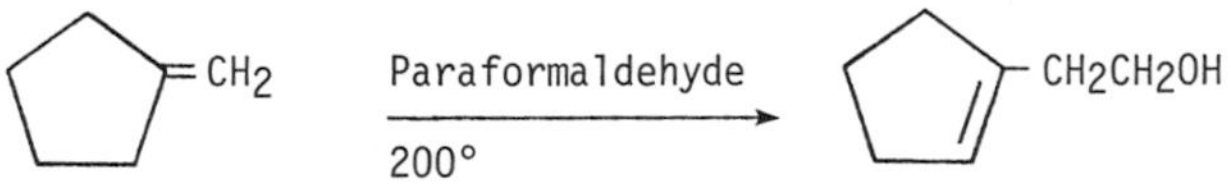

38%

JACS (1950) 72 2871
Chem Rev (1952) 51 505

38%

Aust J Chem (1967) 20 2033

C_7H_{16} $\xrightarrow[\text{2 NaOH } H_2O]{\text{1 }(CF_3COO)_4Pb}$ $C_7H_{15}OH$ (isomer mixture)   ~45%

JACS (1967) 89 3662

Air   ascorbic acid

FeSO$_4$ ethylenediamine-
tetraacetic acid   EtOH
EtOAc   pH 5.5   ~61%

Steroids (1965) 5 451

CrO$_3$ HOAc

H$_2$O   30%

Tetr Lett (1969) 1157
JCS C (1968) 2346
JACS (1948) 70 3237

Me
|
BuCH $\xrightarrow[\text{HOAc } H_2O]{\text{Na}_2\text{Cr}_2\text{O}_7 \text{ HClO}_4}$ BuCOH   6%
| |
Et   Et

JACS (1961) 83 423
Tetrahedron (1968) 24 4667

H_2O_2 V_2O_5 Me_2CO

JOC (1963) 28 2057

$Me_2C=CMe_2$

1 H_2O_2 NaOCl H_2O
2 Reduction
→   63%

                                                    Me
                                                    |
                                          $CH_2=CCMe_2$
                                                    |
                                                   OH

O_2 hν   rose bengal   MeOH
→   100%

JACS (1968) 90 975

1 O_2 hν   hematoporphyrin
2 H_2 Ni   Pyr
→   55%

Annalen (1958) 618 194 185

$PhCH_3$

Argentic picolinate
Me_2SO
→ $PhCH_2OH$

Tetr Lett (1967) 415

RH

1 $(CF_3COO)_4Pb$
2 NaOH H_2O
→   ROH   (R=Ph or $PhCH_2$)   ~45%

JACS (1967) 89 3662

CF_3COO_2H BF_3
CH_2Cl_2
→   89%

JOC (1964) 29 2397
      (1966) 31 153

1 $(CF_3COO)_3Tl$ CF_3COOH
2 $Pb(OAc)_4$ 3 Ph_3P
4 HCl   5 Base
→   62%

JACS (1970) 92 3520

PhH $\xrightarrow{\text{H}_2\text{O}_2 \quad \text{V}_2\text{O}_5 \quad \text{t-BuOH}}$ PhOH   30%

JACS (1937) 59 2342
Tetrahedron (1968) 24 3475

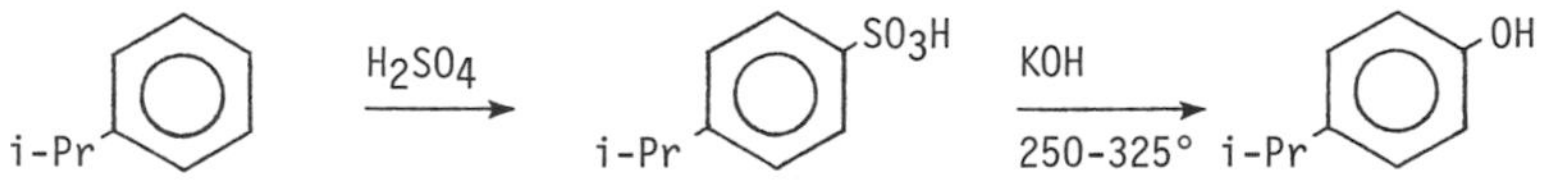

35%

JACS (1949) 71 3889
Org React (1946) 3 141

Alcohols may also be prepared by conversion of hydrides into esters,
followed by hydrolysis. See section 116 (Esters from Hydrides)

Section 42   Alcohols and Phenols from Ketones

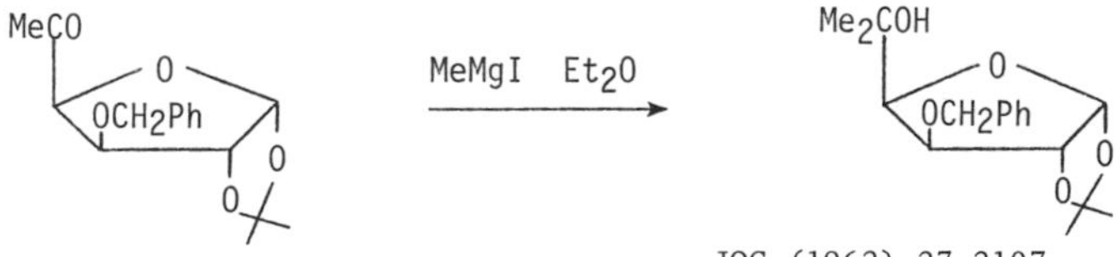

71%

JOC (1962) 27 2107

$(\text{i-Pr})_2\text{CO}$ $\xrightarrow[\text{Et}_2\text{O}]{\text{PrMgBr} \quad \text{LiClO}_4}$ $(\text{i-Pr})_2\overset{\text{COH}}{\underset{\text{Pr}}{|}}$ 70%

(Procedure for hindered ketones)

Chem Comm (1970) 470

Me_2CO $\xrightarrow{\text{PhBr} \quad \text{Li} \quad \text{THF}}$ $\underset{Ph}{Me_2COH}$ (One-step procedure)   34%

Chem Comm (1970) 1160

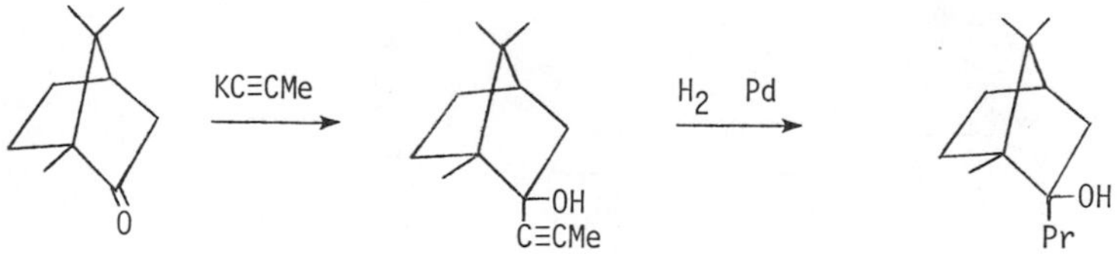

Tetr Lett (1965) 1619
JOC (1969) 34 3754

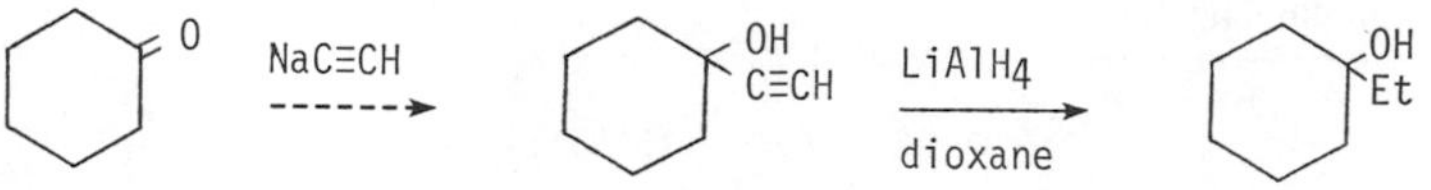

JCS (1954) 1854

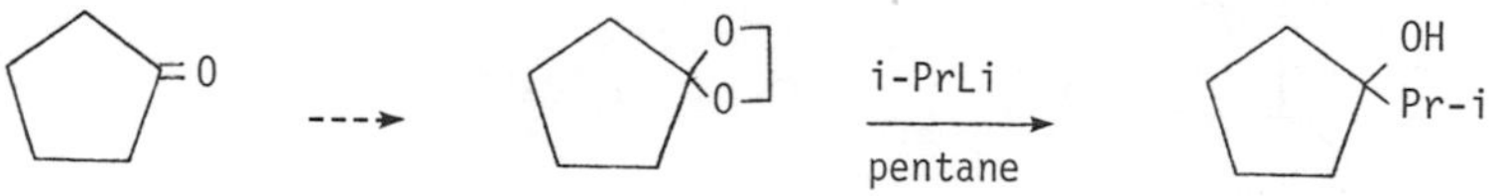

J Heterocyclic Chem (1969) 6 139

$i\text{-}PrCH_2COMe$ $\xrightarrow{\text{H}_2 \quad \text{Pd-C}}$ $\underset{}{i\text{-}PrCH_2\overset{OH}{\underset{|}{C}}HMe}$

(Pt Rh or Ru catalysts may also be used)

JOC (1959) 24 1855

For stereochemistry see Chem Rev (1957) 57 895

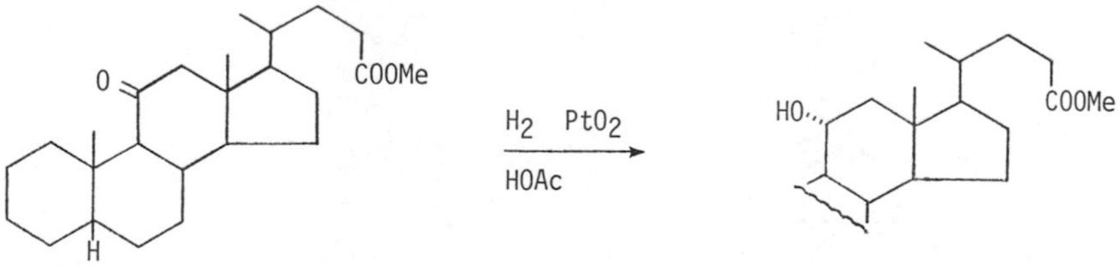

Helv (1943) 26 562
JCS (1954) 2487

$PhCOMe \xrightarrow{H_2 \quad RhH_2(PhPMe_2)_2^{+} \; ClO_4^{-}} PhCH(OH)Me$

Chem Comm (1970) 567

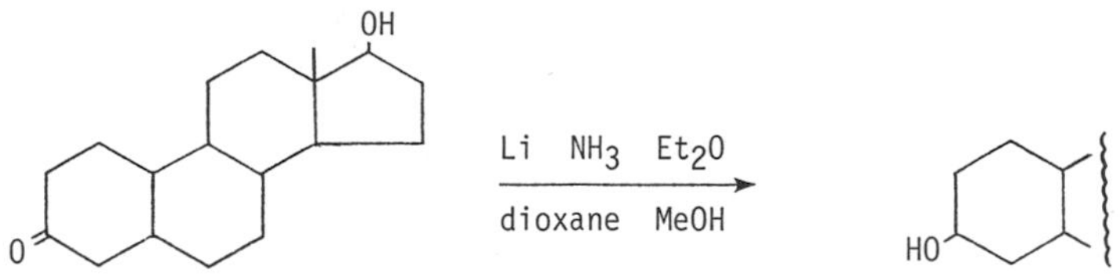
OH
O
Li NH3 Et2O
dioxane MeOH
HO

65%

JACS (1958) 80 6115
(1968) 90 6486

$Pr_2CO \xrightarrow{Li \quad NH_2CH_2CH_2NH_2} Pr_2CHOH$ 30%

JOC (1957) 22 891

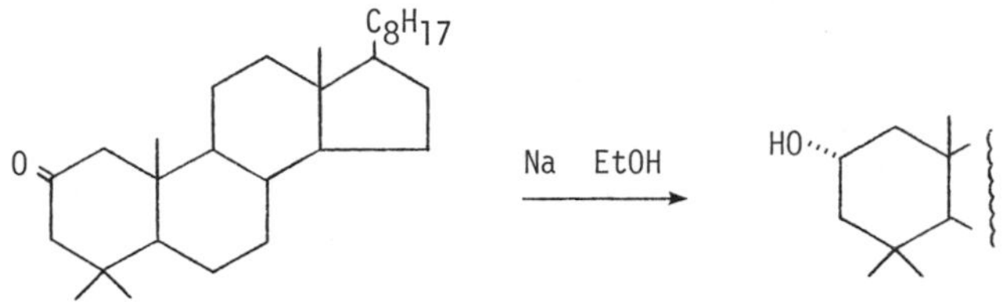
C8H17
O
Na EtOH
HO

93%

Bull Soc Chim Fr (1964) 2236
JCS C (1969) 968

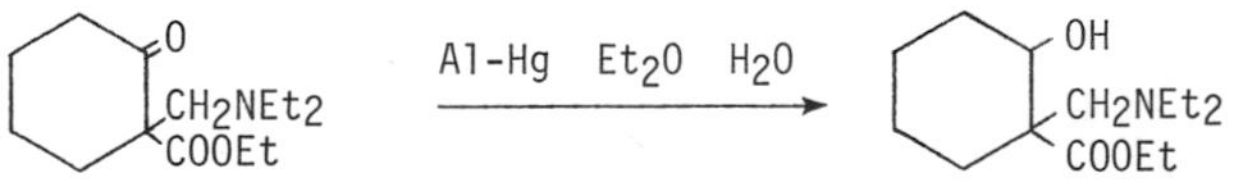
O
CH2NEt2
COOEt
Al-Hg Et2O H2O
OH
CH2NEt2
COOEt

74%

Arch Pharm (1942) 280 361

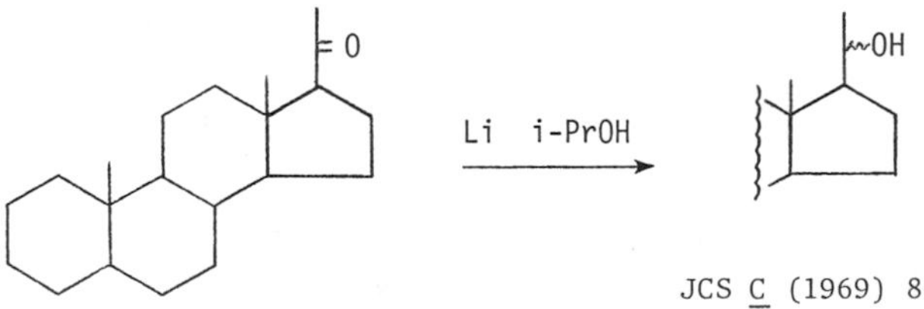
O
Li i-PrOH
OH

JCS C (1969) 804 968

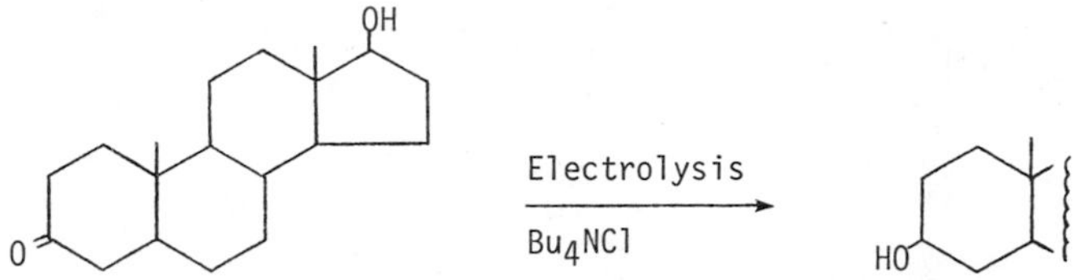
OH
O
Electrolysis
Bu4NCl
HO

100%

JOC (1961) 26 1738
(1970) 35 261

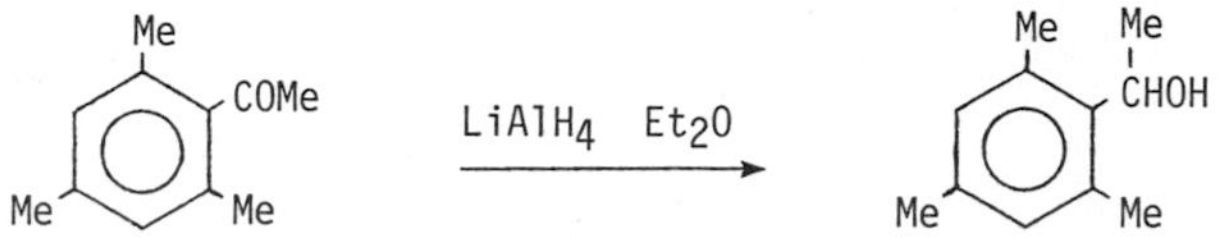
Me
COMe
Me
Me
LiAlH4 Et2O
Me
Me
CHOH
Me
Me

JACS (1947) 69 1197
Org React (1951) 6 469

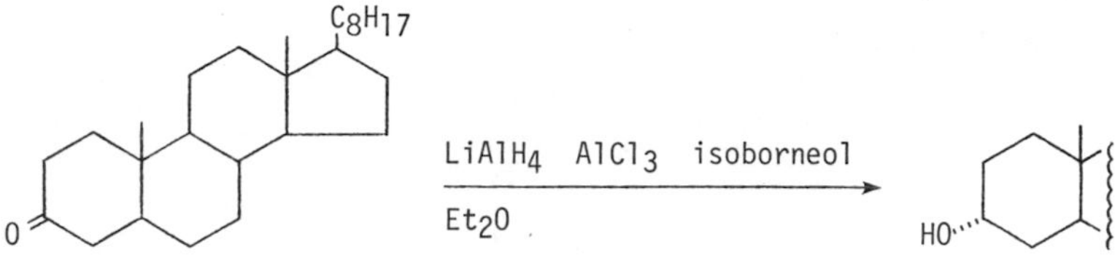
C8H17
O
LiAlH4 AlCl3 isoborneol
Et2O
HO

70%

JOC (1965) 30 3809

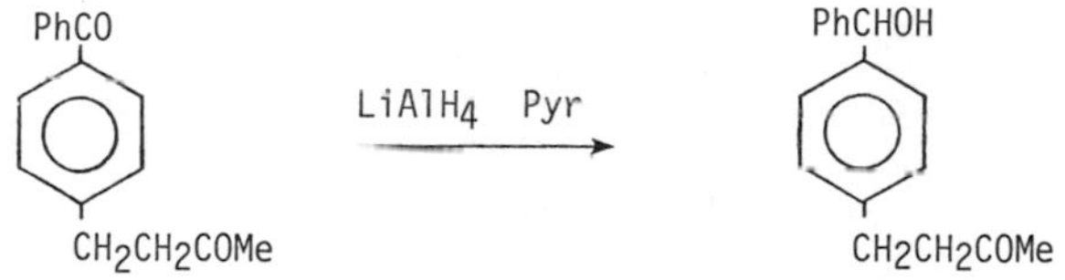
PhCO
CH2CH2COMe
LiAlH4 Pyr
PhCHOH
CH2CH2COMe

56%

JACS (1962) 84 1756

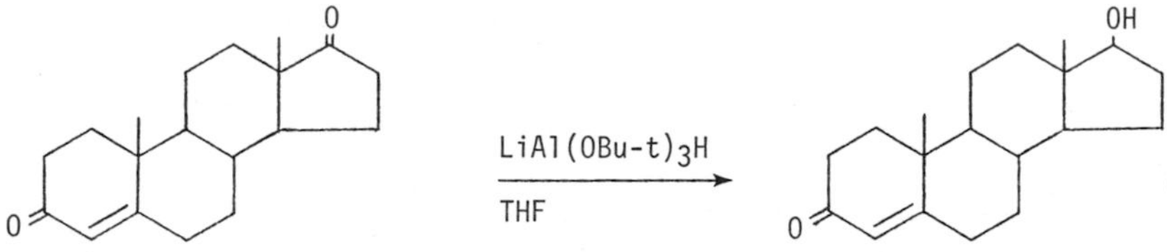
O
O
LiAl(OBu-t)3H
THF
OH
O

55%

Coll Czech (1959) 24 2284

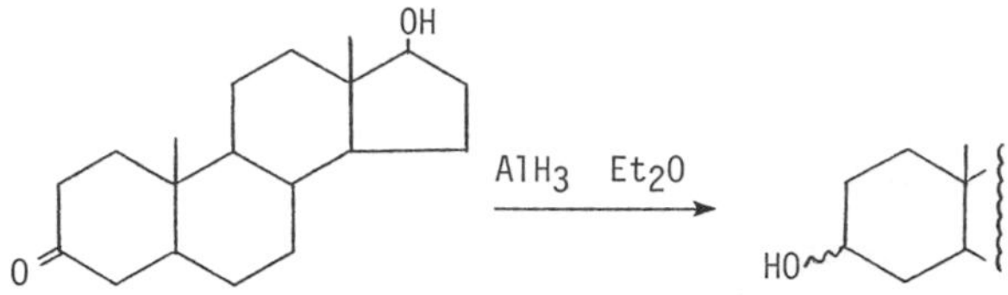
OH
AlH_3 Et_2O
O
HO

$\alpha/_\beta$ = 34/66

JCS B (1967) 581

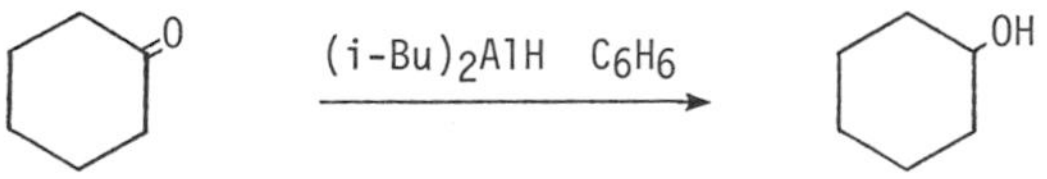
O
$(i\text{-}Bu)_2AlH$ C_6H_6
OH

83%

JOC (1959) 24 627

For reduction of unsaturated ketones see Chem Comm (1970) 213

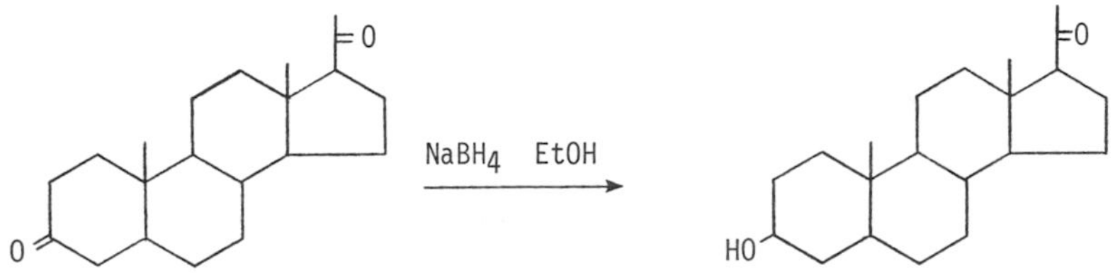
=O
$NaBH_4$ EtOH
O
HO

76%

JACS (1953) 75 1286

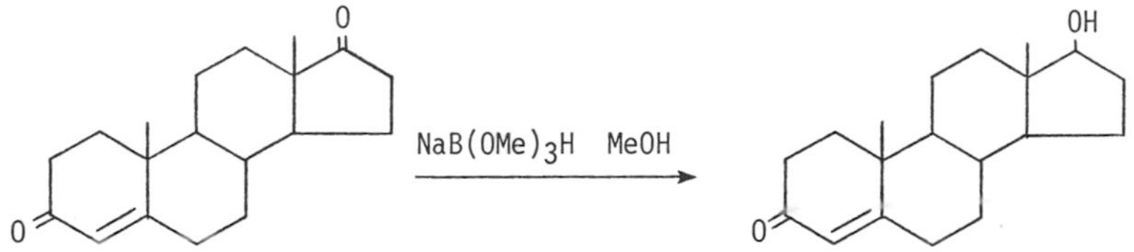
O
$NaB(OMe)_3H$ MeOH
OH
O

22%

JCS (1955) 3426

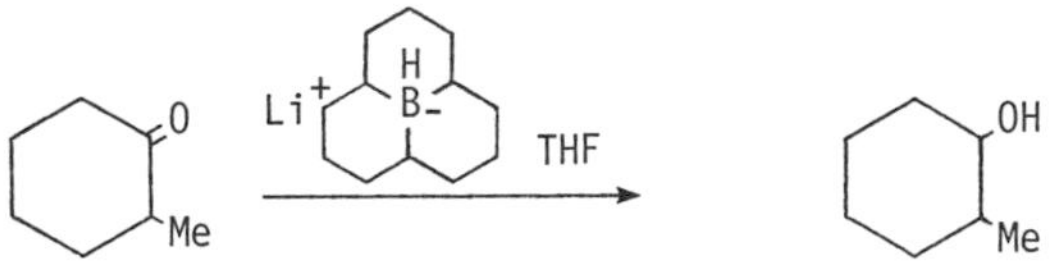
O
Me
Li^+
H
B^-
THF
OH
Me

cis/trans=99/1

JACS (1970) 92 709

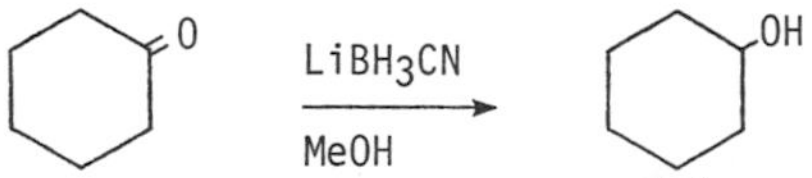
O
$LiBH_3CN$
MeOH
OH

77%

($LiBH_3CN$ is stable to pH 3)

JACS (1969) 91 3996

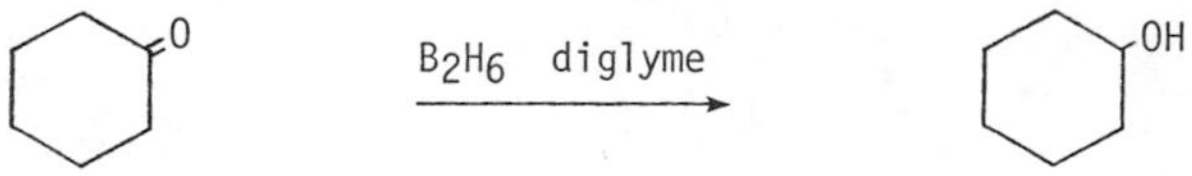

B_2H_6 diglyme

JACS (1960) $\underline{82}$ 681

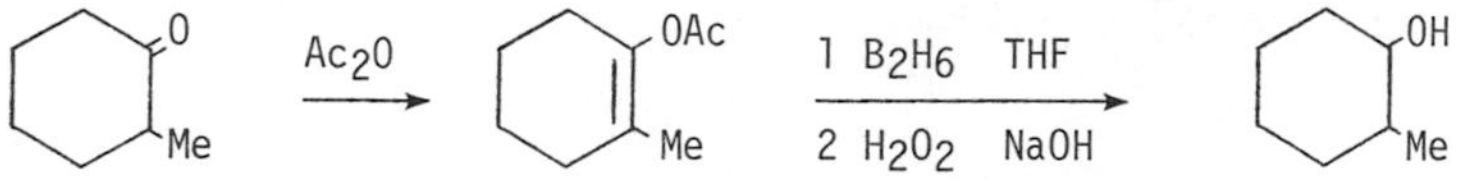

Ac_2O

1 B_2H_6 THF

2 H_2O_2 NaOH

cis

Tetr Lett (1968) 4937

$B_2H_6-\alpha$-pinene

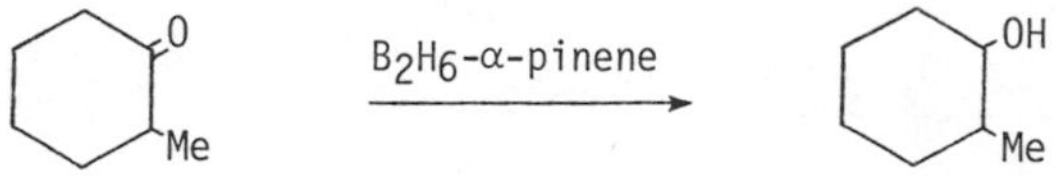

cis/trans = 92/8

JACS (1961) $\underline{83}$ 3166

Ph_2CO Pyr-B_2H_6 toluene Ph_2CHOH ~83%

JOC (1958) $\underline{23}$ 1561

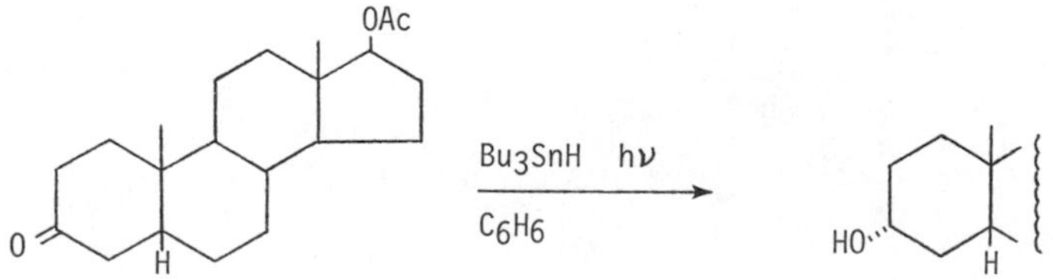

Bu_3SnH hν

C_6H_6

66%

Helv (1967) $\underline{50}$ 2259
Tetr Lett (1968) 5385

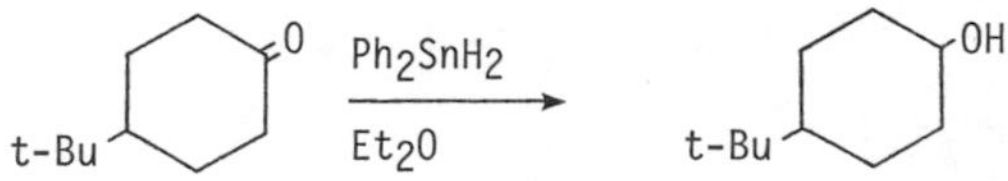

Ph_2SnH_2

Et_2O

85-93%

t-Bu   t-Bu

JACS (1958) $\underline{80}$ 3798
      (1961) $\underline{83}$ 1246

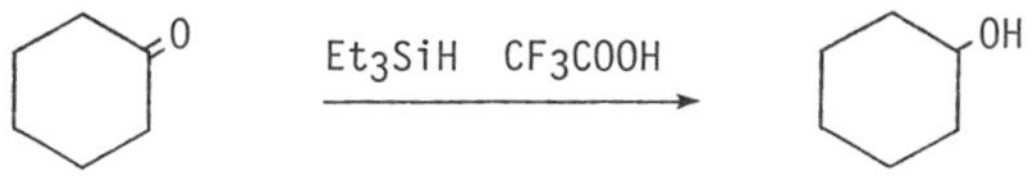
O
Et3SiH CF3COOH
OH

74%

Tetrahedron (1967) 23 2235

Ph_2CO $\xrightarrow{\text{NaH xylene}}$ Ph_2CHOH 83%

JACS (1946) 68 2647

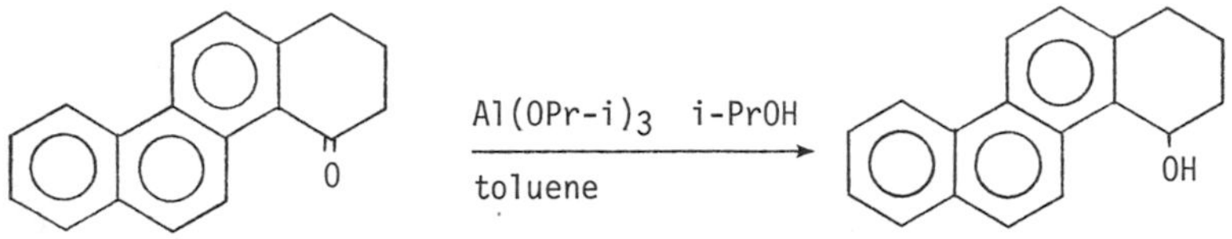
O
Al(OPr-i)3 i-PrOH
toluene
OH

76%

JOC (1939) 4 456
Org React (1944) 2 178

PhCOMe $\xrightarrow{\text{i-PrOLi i-PrOH}}$ PhCH(Me)OH >90%

JCS C (1969) 804

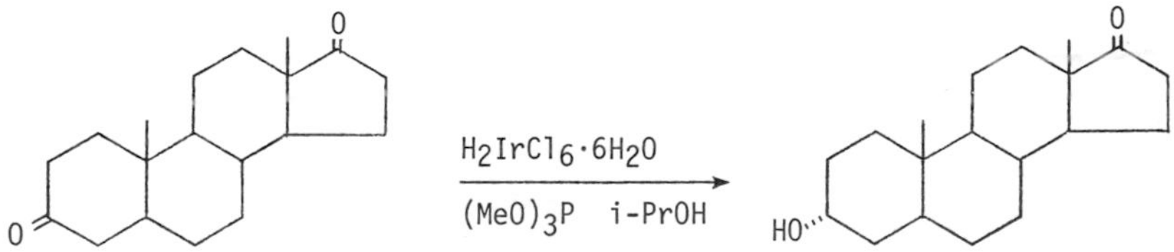
O
O
H2IrCl6·6H2O
(MeO)3P i-PrOH
O
HO

85%

JCS C (1969) 1653
(1970) 785
Chem Comm (1970) 162

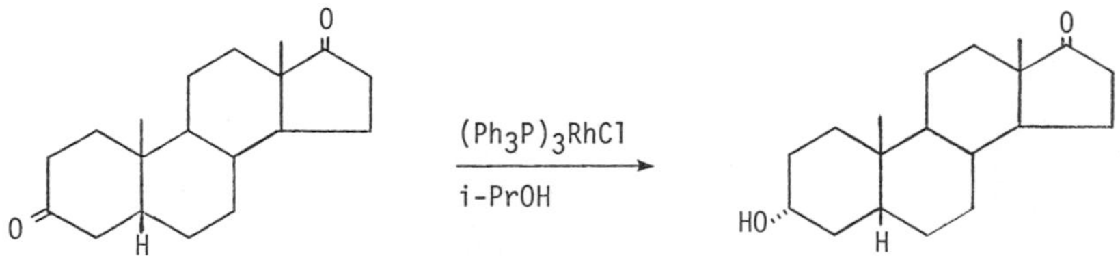
O
O
H
(Ph3P)3RhCl
i-PrOH
O
HO
H

100%

Chem Comm (1970) 162

Ph_2CO $\xrightarrow{\text{KOH   HOCH}_2\text{CH}_2\text{OH}}$ Ph_2CHOH 93%

JOC (1967) <u>32</u> 840
(1960) <u>25</u> 1707

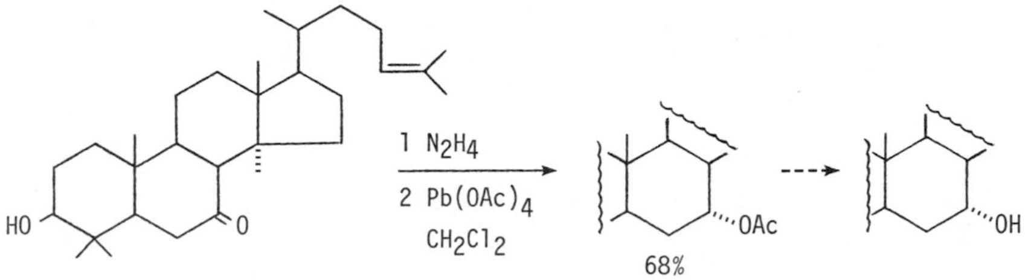

$\xrightarrow[\text{CH}_2\text{Cl}_2]{\begin{array}{l}\text{1 N}_2\text{H}_4 \\ \text{2 Pb(OAc)}_4\end{array}}$

68%

Chem Comm (1969) 450

$\xrightarrow[\text{2 EtONa   EtOH}]{\text{1 NH}_2\text{CONHNH}_2 \cdot \text{HOAc}}$

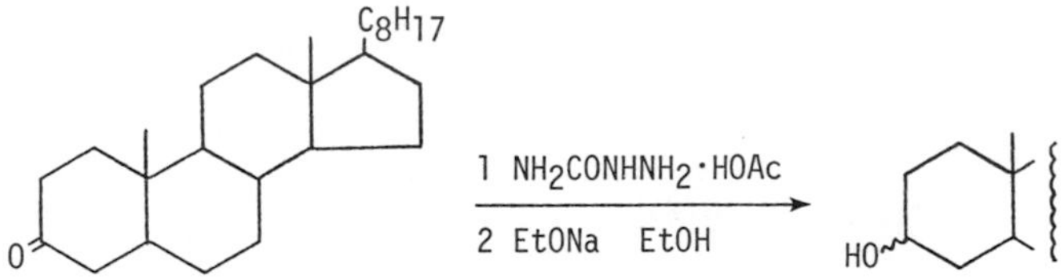

90%

JACS (1939) <u>61</u> 1992

PhCOMe $\xrightarrow{\substack{\text{NCOONa} \\ \| \\ \text{NCOONa   HOAc   MeOH}}}$ $\overset{\text{Me}}{\underset{|}{\text{PhCHOH}}}$ 31%

(Rate of reduction ArC=O > AliphC=O)

Chem Comm (1965) 71

Ph_2CO $\xrightarrow{\text{PhLi-Pyr   Et}_2\text{O}}$ Ph_2CHOH 62%

Can J Chem (1963) <u>41</u> 1961

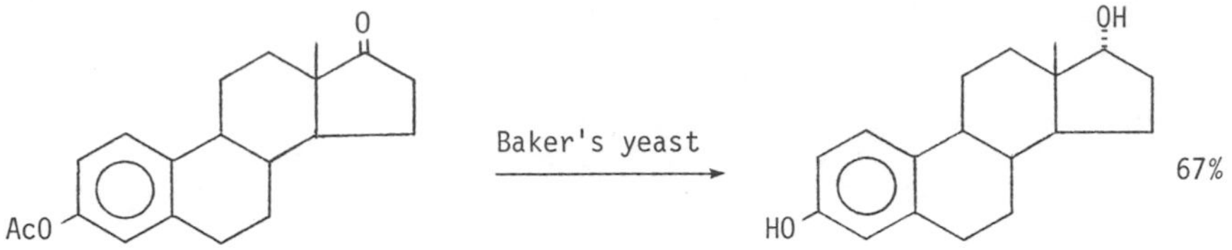

Baker's yeast

67%

Ber (1938) 71B 2696

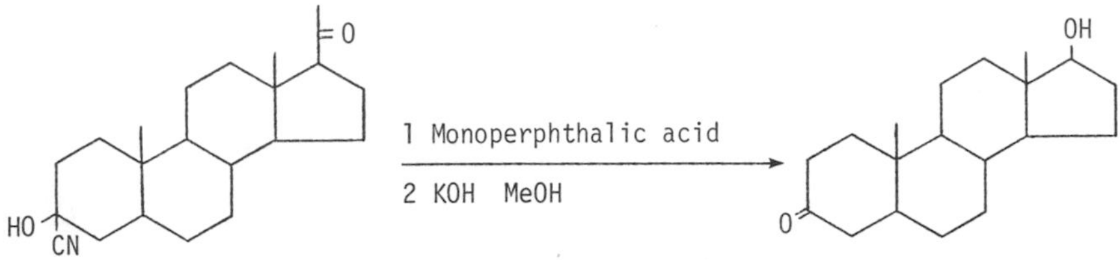

1 Monoperphthalic acid

2 KOH   MeOH

Gazz (1961) 91 1250
(Chem Abs 56 10211)

EtCOPh $\xrightarrow{\text{PhCOO}_2\text{H}\quad\text{CHCl}_3}$ EtCOOPh   - - - - →   PhOH

73%

JACS (1949) 71 14

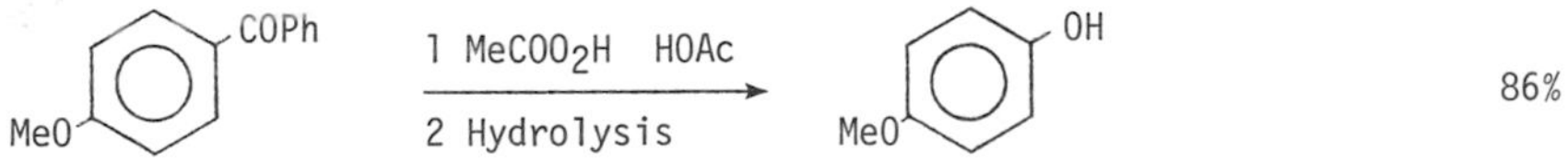

1 MeCOO$_2$H   HOAc

2 Hydrolysis

86%

JACS (1950) 72 5515

Further examples of the Baeyer-Villiger degradation of ketones to alcohols
and phenols via esters are included in section 117 (Esters from Ketones)

Some of the methods listed in section 34 (Alcohols and Phenols from
Aldehydes) may also be applied to the preparation of alcohols from
ketones

Section 43   Alcohols and Phenols from Nitriles
°°°°°°°°°°°°°°°°°°°°°°°°°°°°°°°°°°°°°°°

No examples

Section 44   Alcohols from Olefins
°°°°°°°°°°°°°°°°°°°°°°°°°°°

$C_6H_{13}CH=CH_2$ $\xrightarrow{\begin{array}{l} 1 \ NaBH_4 \ \ BF_3 \ \ diglyme \\ \hline 2 \ CO \\ 3 \ H_2O_2 \ \ NaOH \ \ H_2O \end{array}}$ $(C_6H_{13}CH_2CH_2)_3COH$ 90%

JACS (1967) <u>89</u> 2737
(1970) <u>92</u> 6648

$C_5H_{11}CH=CH_2$ $\xrightarrow{\begin{array}{l} Di\text{-}t\text{-}butyl \ peroxide \\ \hline MeCH_2OH \end{array}}$ $C_5H_{11}CH_2CH_2\underset{\underset{Me}{|}}{C}HOH$ 40-60%

Izv (1964) 894
(Chem Abs <u>61</u> 5510)

$PrCH=CH_2$ $\xrightarrow{\begin{array}{l} 1 \ B_2H_6 \ \ THF \\ \hline 2 \ MeOCHCl_2 \ \ MeLi \ \ Et_2O \\ 3 \ H_2O_2 \ \ NaOH \ \ H_2O \end{array}}$ $(PrCH_2CH_2)_2CHOH$ + $PrCH_2\underset{\underset{OH}{|}}{C}HCH_2CH_2Pr$

                                        21%   28%

Tetr Lett (1969) 2955

$C_6H_{13}CH=CH_2$ $\xrightarrow{\begin{array}{l} 1 \ B_2H_6 \ \ THF \\ \hline 2 \ CO \ \ LiBH_4 \\ 3 \ KOH \ \ EtOH \\ 4 \ H_2O_2 \end{array}}$ $C_6H_{13}(CH_2)_3OH$ 70%

JACS (1967) <u>89</u> 2740

$PhCH=CH_2$ $\xrightarrow{\begin{array}{l} HCHO \\ \hline H_2SO_4 \end{array}}$ $\xrightarrow{\begin{array}{l} H_2 \ (2,600 \ psi) \\ \hline copper \ chromite \end{array}}$ $Ph(CH_2)_3OH$ ~74%

JACS (1950) <u>72</u> 5314
Chem Rev (1952) <u>51</u> 505

$C_5H_{11}CH=CH_2$ $\xrightarrow[\text{RhCl}_3\cdot 3H_2O]{\text{CO} \quad H_2 \text{ (240 atmos)}}$ $C_5H_{11}(CH_2)_3OH$ 73%

Ber (1966) 99 1086

$\xrightarrow[\text{Co}]{\text{CO} \quad H_2 \text{ (450 atmos)}}$ 56%

JACS (1952) 74 4496

$BuCH=CH_2$ $\xrightarrow[\substack{2 \; CH_2=SMe_2 \;\; Me_2SO \\ 3 \; H_2O_2 \;\; NaOH \;\; H_2O}]{1 \; B_2H_6}$ $Bu(CH_2)_3OH$ + $Bu(CH_2)_2OH$

Chem Comm (1967) 505

$i\text{-}PrCH_2CH=CH_2$ $\xrightarrow[\substack{2 \; H_2O_2 \;\; NaOH \;\; H_2O}]{1 \; NaBH_4 \;\; BF_3\cdot Et_2O \;\; THF}$ $i\text{-}Pr(CH_2)_3OH$ 80%

Org React (1963) 13 1

$\xrightarrow[\substack{2 \; H_2O_2 \;\; NaOH \;\; H_2O}]{1 \; LiBH_4 \;\; H_2SO_4 \;\; THF}$ 22%

JOC (1963) 28 3551

$CH_2=CH(CH_2)_8COOH$ $\xrightarrow[\substack{2 \; H_2O_2}]{\substack{Me \\ 1 \; (i\text{-}PrCH)_2BH \;\; THF}}$ $HOCH_2CH_2(CH_2)_8COOH$ 82%

JACS (1961) 83 486

$Me_2C=CHMe$ $\xrightarrow[\substack{2 \; H_2O_2 \;\; NaOH \;\; H_2O}]{1 \quad \text{} \quad THF}$ $\underset{\underset{OH}{|}}{Me_2CHCHMe}$ ~95%

JACS (1968) 90 5281

$C_{10}H_{21}CH=CHC_{11}H_{23}$ $\xrightarrow[\substack{\text{bis-2-ethoxyethyl ether}\\185-190°\\ 2\ H_2O_2\ \ NaOH\ \ H_2O}]{1\ NaBH_4\ \ BF_3\cdot Et_2O}$ $C_{22}H_{45}CH_2OH$ 51%

JOC (1961) <u>26</u> 3657
JACS (1967) <u>89</u> 561 567

$PhCH=CHMe$ $\xrightarrow[2\ H_2O_2\ \ NaOH]{1\ Et_2AlCl}$ $\underset{\underset{OH}{|}}{PhCH_2CHMe}$ 75%

Tetr Lett (1970) 3471

JACS (1966) <u>88</u> 3016

$MeCO(CH_2)_2CH=CMe_2$ $\xrightarrow[2\ NaHCO_3\ \ MeOH]{1\ HCOOH}$ $\underset{\underset{Me}{|}}{\overset{\overset{Me}{|}}{MeCO(CH_2)_2CH_2COH}}$ 80-95%

Bol Inst Quim Univ Nac Aut Mex (1965) <u>17</u> 181
(Chem Abs <u>65</u> 8963)
JACS (1953) <u>75</u> 6212

$MeCO(CH_2)_2CH=CMe_2$ $\xrightarrow[2\ NaOH\ \ H_2O]{1\ H_2SO_4\ \ H_2O}$ $\underset{\underset{Me}{|}}{\overset{\overset{Me}{|}}{MeCO(CH_2)_2CH_2COH}}$ 85%

JACS (1955) <u>77</u> 1617

$BuCH=CH_2$ $\xrightarrow[\substack{2\ NaOH\ \ H_2O\\3\ NaBH_4\ \ NaOH\ \ H_2O}]{1\ Hg(OAc)_2\ \ THF\ \ H_2O}$ $\underset{\underset{OH}{|}}{BuCHMe}$ 96%

JACS (1967) <u>89</u> 1522
JOC (1970) <u>35</u> 1844

$$\xrightarrow[\text{MeOCH}_2\text{CH}_2\text{OMe}]{\text{H}_2\text{O} \quad h\nu \quad \text{xylene}}$$

50%

JACS (1967) 89 6788
Acc Chem Res (1969) 2 33

$$\text{Me}_2\text{C}=\text{CH(CH}_2)_2\overset{\overset{\text{Me}}{|}}{\text{C}}\text{HCH}_2\text{COOMe} \qquad \xrightarrow[\text{2 NaBH}_4]{\text{1 O}_3 \quad \text{MeOH}} \qquad \text{HOCH}_2(\text{CH}_2)_2\overset{\overset{\text{Me}}{|}}{\text{C}}\text{HCH}_2\text{COOMe}$$

72%

JACS (1968) 90 3525
Helv (1967) 50 2445

Section 45   Alcohols from Miscellaneous Compounds

$$\text{PhCH}=\text{CHCH}_2\text{OH} \qquad \xrightarrow{\text{H}_2 \quad \text{Pd}} \qquad \text{Ph(CH}_2)_3\text{OH}$$

55%

Annalen (1924) 439 276

$$\text{CH}_2=\text{CHCH}_2\text{OH} \qquad \xrightarrow[\text{NaOH}]{\text{ClCH}_2\text{CONHNH}_2 \cdot \text{HCl}} \qquad \text{Me(CH}_2)_2\text{OH}$$

70%

Chem Ind (1964) 839

$$\text{PhCH}=\text{CHCH}_2\text{OH} \qquad \xrightarrow{\text{LiAlH}_4 \quad \text{Et}_2\text{O}} \qquad \text{Ph(CH}_2)_3\text{OH}$$

93%

JACS (1948) 70 3484

$$\xrightarrow{\text{Ni} \quad \text{EtOH}}$$

Gazz (1963) 93 1028
(Chem Abs 60 3029)
JACS (1953) 75 1700

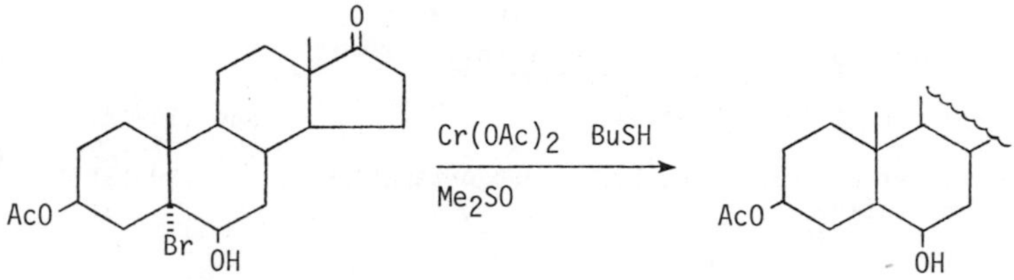
O
AcO
Br
OH
Cr(OAc)2 BuSH
Me2SO
AcO
OH

65%

JACS (1966) 88 3016

$PhCH{=}CHCHO \xrightarrow{H_2\ \ Pt} Ph(CH_2)_3OH$

JACS (1925) 47 3061

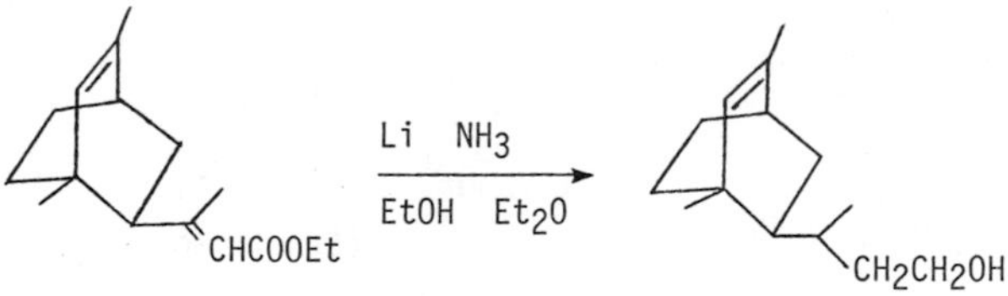
CHCOOEt
Li NH3
EtOH Et2O
CH2CH2OH

Tetr Lett (1970) 3219

$MeCH{=}CHCOOH \xrightarrow{B_2H_6} Me(CH_2)_3OH$

Hua Hsueh Hsueh Pao (1965) 31 376
(Chem Abs 64 8022)

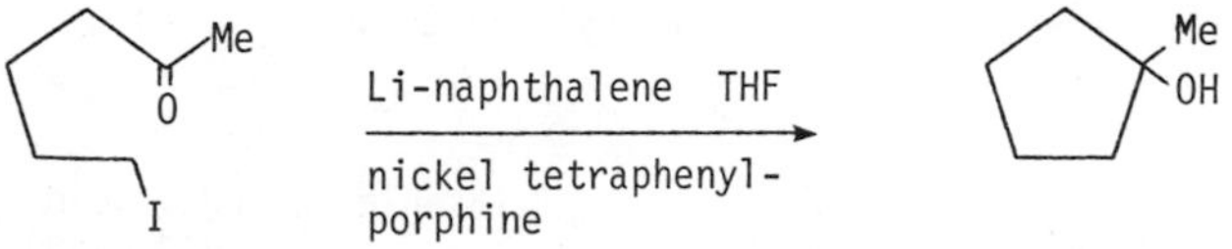
Me
O
I
Li-naphthalene THF
nickel tetraphenyl-
porphine
Me
OH

61%

JACS (1970) 92 395

Section 45A   Protection of Alcohols and Phenols

(Stable to CrO_3 and acid)

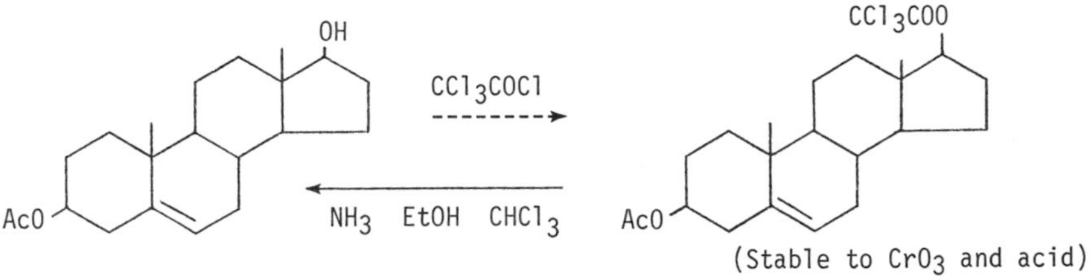

Coll Czech (1962) 27 2567
Tetr Lett (1968) 4681

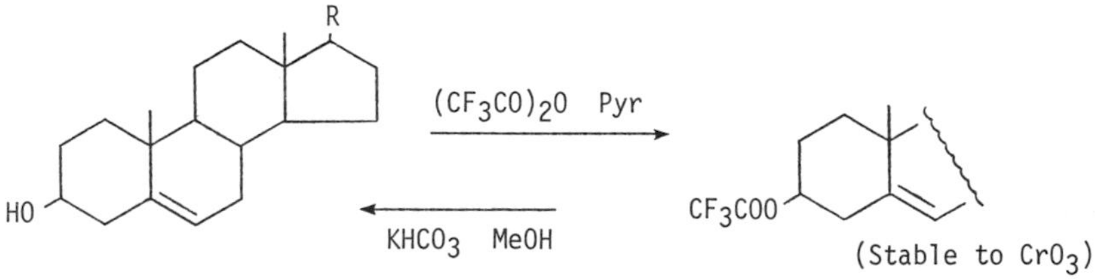

(Stable to CrO_3)

Helv (1954) 37 388

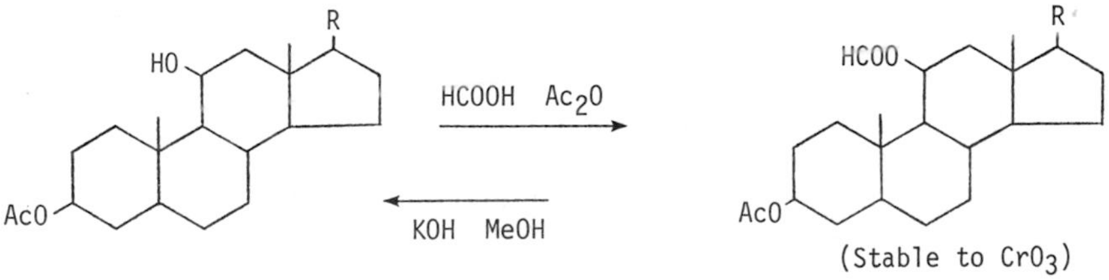

(Stable to CrO_3)

Helv (1954) 37 443

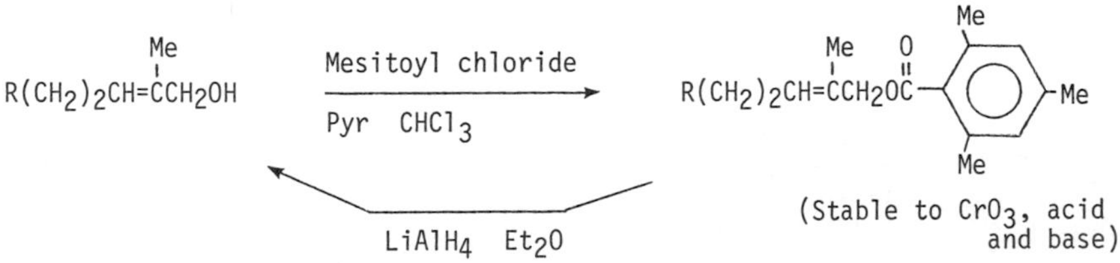

(Stable to CrO_3, acid and base)

JACS (1969) 91 4318
(1967) 89 2758

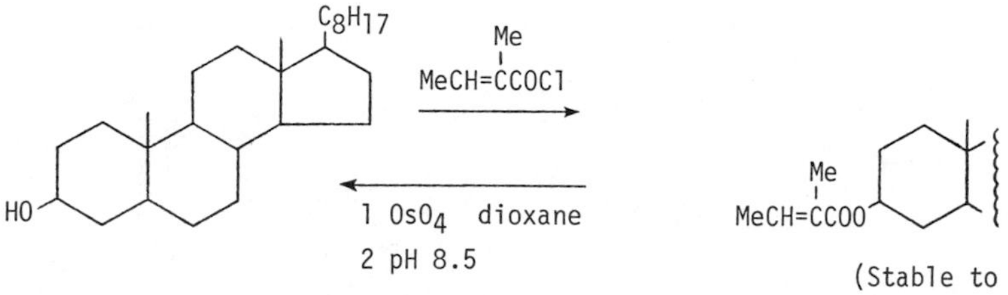

C₈H₁₇ structure

Me
MeCH=CCOCl
→

1 OsO₄ dioxane
2 pH 8.5
←

HO

Me
MeCH=CCOO

(Stable to acid)

JOC (1962) 27 3103

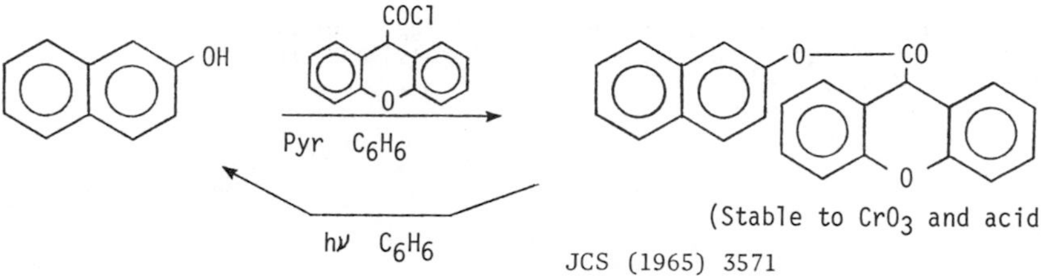

OH

COCl

Pyr C₆H₆
→

O——CO

←
hν C₆H₆

(Stable to CrO₃ and acid)

JCS (1965) 3571

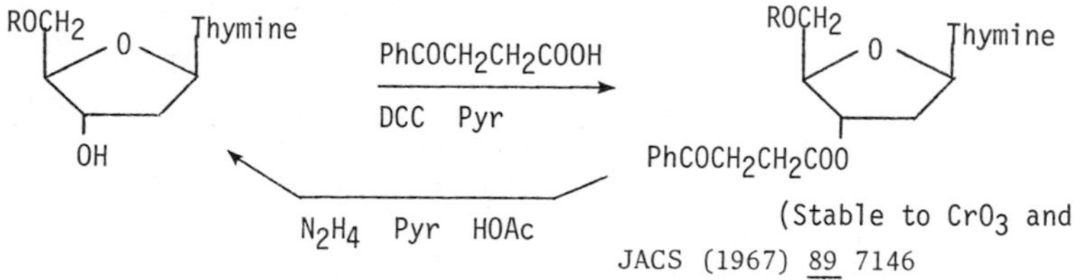

ROCH₂   O   Thymine

OH

PhCOCH₂CH₂COOH
————————————
DCC Pyr
→

←
N₂H₄ Pyr HOAc

ROCH₂   O   Thymine

PhCOCH₂CH₂COO

(Stable to CrO₃ and acid)

JACS (1967) 89 7146

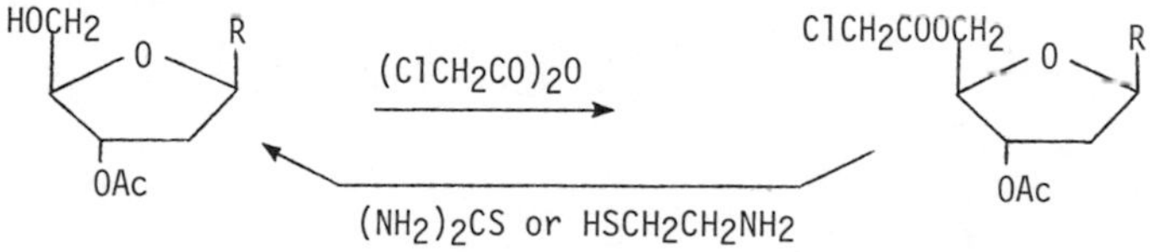

HOCH₂   O   R

OAc

(ClCH₂CO)₂O
→

←
(NH₂)₂CS or HSCH₂CH₂NH₂

ClCH₂COOCH₂   O   R

OAc

JOC (1970) 35 1940

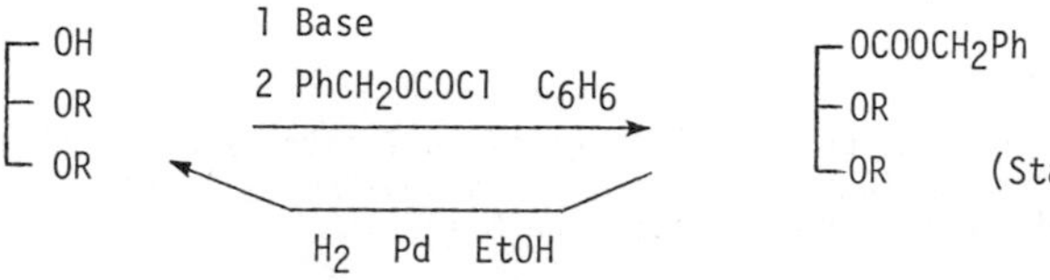

OH
OR
OR

1 Base
2 PhCH₂OCOCl C₆H₆
→

←
H₂ Pd EtOH

OCOOCH₂Ph
OR
OR   (Stable to CrO₃ and acid)

JACS (1939) 61 3328

CCl_3CH_2OCOCl Pyr

$\xleftarrow{\quad\quad}$
Zn   HOAc

CCl_3CH_2OCOO

(Stable to CrO_3 and acid)
Tetr Lett (1967) 2555
JOC (1968) 33 3589

Further examples of the preparation and cleavage of esters are included
in section 108 (Esters from Alcohols and Phenols) and section 38 (Alcohols
and Phenols from Esters)

HNO_3 Ac_2O

$\xleftarrow{\quad\quad}$
Zn   HOAc

NO_2O ONO_2

(Stable to CrO_3, acid and base)
Ber (1962) 95 1094
Chem Rev (1955) 55 485

$R[CH_2]_{10}CH_2OH$

$\xrightarrow{\quad H_3BO_3 \quad toluene \quad}$
$\xleftarrow{\quad\quad}$
H_2O

$(R[CH_2]_{10}CH_2O)_3B$

(Stable to Wittig reagents, acid and
                                    base)
Rec Trav Chim (1953) 72 411
Tetr Lett (1969) 4155

MsCl   Pyr

$\xleftarrow{\quad\quad}$
NaOH H_2O

OMs

(Stable to CrO_3 and acid)

JACS (1957) 79 717
For cleavage of mesylates with PhLi see JCS C (1968) 2283

Further examples of the preparation and cleavage of sulfonates are included
in section 138 (Halides and Sulfonates from Alcohols and Phenols) and
section 40 (Alcohols and Phenols from Halides and Sulfonates)

$$Me_2C=CH_2 \quad BF_3 \quad H_3PO_4$$

$$\xleftarrow{CF_3COOH}$$

(Stable to RMgX, LiAlH$_4$,
CrO$_3$ and base)

JCS (1963) 755

$$\xrightarrow{Ph_3CCl \quad Pyr}$$

$$\xleftarrow{HOAc \quad H_2O}$$

(Stable to RMgX, LiAlH$_4$, CrO$_3$
and base)

JOC (1951) $\underline{16}$ 349
(1950) $\underline{15}$ 264

For hydrogenolysis of trityl ethers see Rec Trav Chim (1942) $\underline{61}$ 373

$$\xrightarrow{PhCH_2Cl \quad K_2CO_3 \quad EtOH}$$

$$\xleftarrow{H_2 \quad Pd-C \quad EtOH}$$

(Stable to RMgX, LiAlH$_4$, CrO$_3$
acid and base)

J Med Chem (1967) $\underline{10}$ 262
JACS (1954) $\underline{76}$ 3188
Org React (1953) $\underline{7}$ 263

Further examples of the cleavage of benzyl ethers are included in section 39 (Alcohols and Phenols from Ethers and Epoxides)

RCH$_2$OH

1 MsCl Pyr
2 Sodium p-chlorophenoxide

$$\xrightarrow{\hspace{3cm}}$$

$$\xleftarrow[\text{2 Acid}]{\text{1 Li NH}_3 \text{ EtOH}}$$

RCH$_2$O—⟨○⟩—Cl

(Stable to RMgX, LiAlH$_4$, CrO$_3$
acid and base)

JACS (1968) $\underline{90}$ 1090

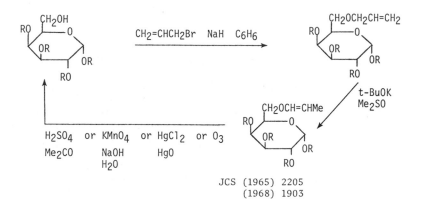

Further examples of the cleavage of ethers are included in section 39 (Alcohols and Phenols from Ethers and Epoxides)

Ber (1964) 97 2196
For trimethylsilyl ethers of 3ry alcohols see Chem Comm (1968) 466

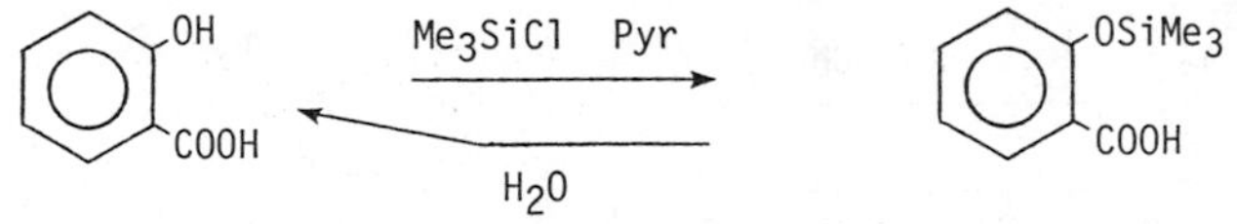

JOC (1957) 22 592

For trimethylsilyl ethers of hindered phenols see JACS (1966) 88 3390

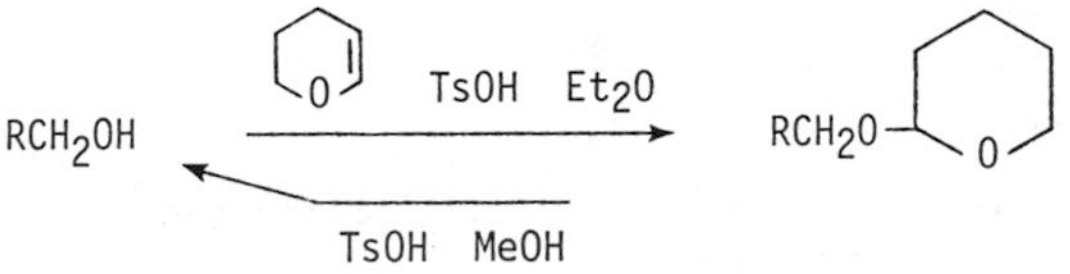

(Stable to RMgX, LiAlH$_4$, CrO$_3$ and base)

JACS (1969) 91 4318
For tetrahydropyranyl ethers of 3ry alcohols see Tetrahedron (1961) 13 241
and Steroids (1964) 4 229

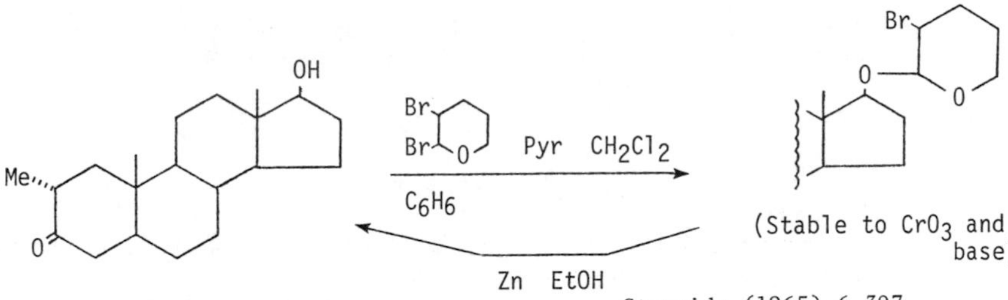

(Stable to CrO$_3$ and base)

Steroids (1965) 6 397

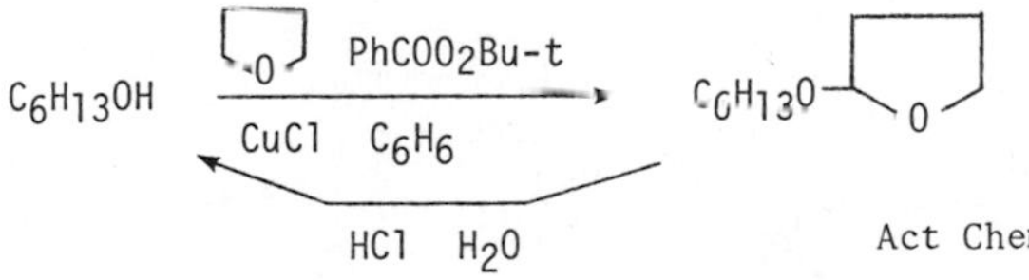

(Stable to RMgX, LiAlH$_4$, CrO$_3$ and base)

Act Chem Scand (1960) 14 1854
(1961) 15 249

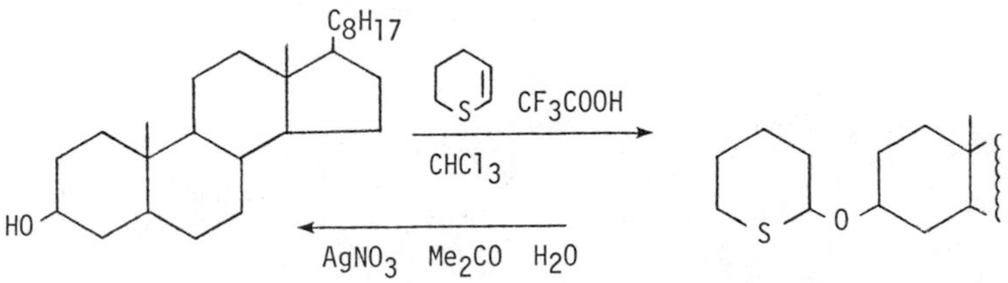

JOC (1966) 31 2333

$$\text{AcOCH}_2 \quad O \quad \text{Thymine} \quad \xrightarrow[\text{AgNO}_3 \quad \text{MeCN}]{\text{S} \quad \text{HCl} \quad \text{CHCl}_3} \quad \text{AcOCH}_2 \quad O \quad \text{Thymine}$$

JOC (1966) 31 2333

1 EtONa   EtOH

2 MeOCH$_2$Cl   CHCl$_3$

HOAc

(Stable to RMgX, LiAlH$_4$, CrO$_3$ and base)

J Med Chem (1966) 9 1
JACS (1957) 79 5792

OBu
|
PhCOOCHPr

OBu
|
t-BuOH ⟶ t-BuOCHPr   (Stable to RMgX, LiAlH$_4$, CrO$_3$ and base)

Acid

Tetrahedron (1961) 13 241

$$\text{ROCH}_2 \quad O \quad \text{Uracil} \quad \xrightarrow[\text{H}_2\text{O} \quad \text{pH 4}]{\text{CH}_2=\text{CMe} \quad \text{TsOH}} \quad \text{ROCH}_2 \quad O \quad \text{Uracil}$$

OMe
|
CH$_2$=CMe   TsOH

RO   OCMe$_2$
        |
        OMe

JACS (1967) 89 3366
Tetrahedron (1970) 26 1023

AcOCH$_2$ O   Uracil   MeO— O   TsOH   AcOCH$_2$ O   Uracil

AcO   OH   pH 2.0   AcO   O

MeO

(Stable to RMgX,
LiAlH$_4$, CrO$_3$,
and base)

JACS (1967) 89 3366
Tetrahedron (1970) 26 1023

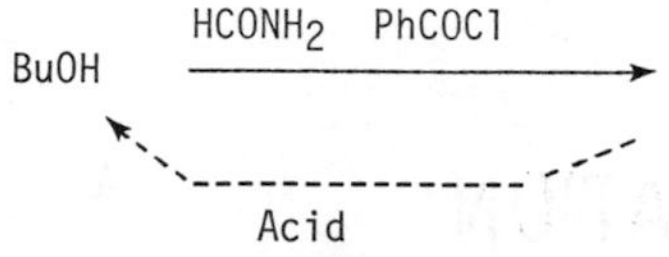
HCONH2 PhCOCl
BuOH
HC(OBu)3
Acid

(Stable to $LiAlH_4$, CrO_3 and base)

Annalen (1968) 716 207
Rec Trav Chim (1969) 88 897

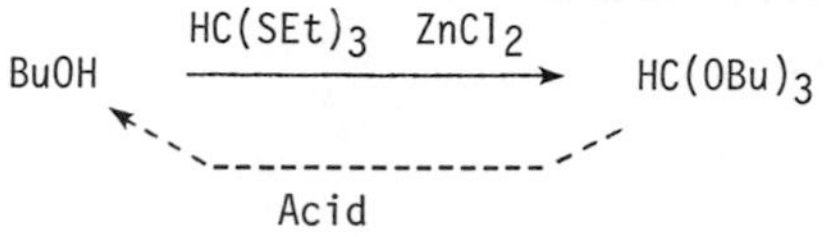
HC(SEt)3 ZnCl2
BuOH
HC(OBu)3
Acid

JACS (1948) 70 2268

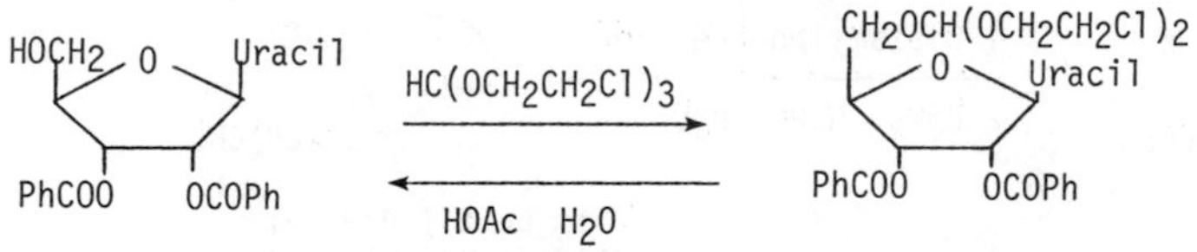
HOCH2
O
Uracil
PhCOO
OCOPh
HC(OCH2CH2Cl)3
HOAc H2O
CH2OCH(OCH2CH2Cl)2
O
Uracil
PhCOO
OCOPh

Tetr Lett (1969) 4443

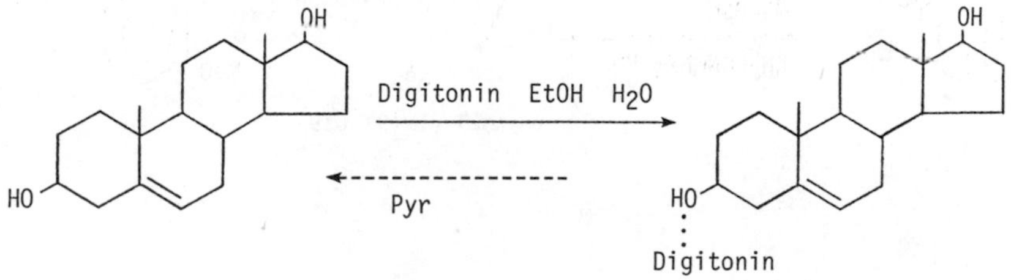
OH
HO
Digitonin EtOH H2O
Pyr
OH
HO
Digitonin

(Stable to CrO_3)

JACS (1960) 82 1257

Chapter 4 PREPARATION OF ALDEHYDES

Section 46 Aldehydes from Acetylenes

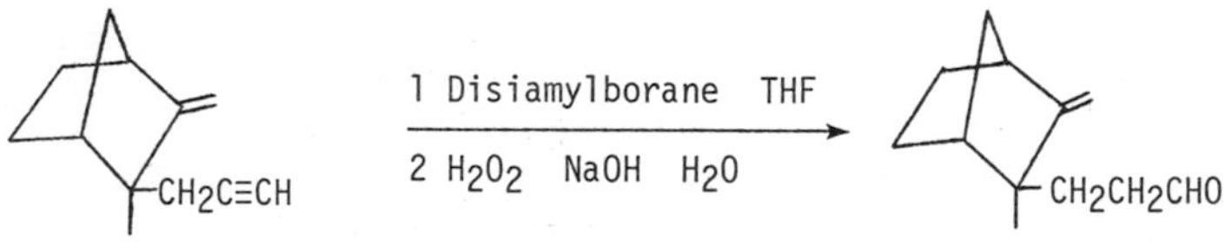

CH2C≡CH
1 Disiamylborane THF
2 H2O2 NaOH H2O
CH2CH2CHO

61%

Tetr Lett (1970) 41
JACS (1969) 91 4771
Org React (1963) 13 1

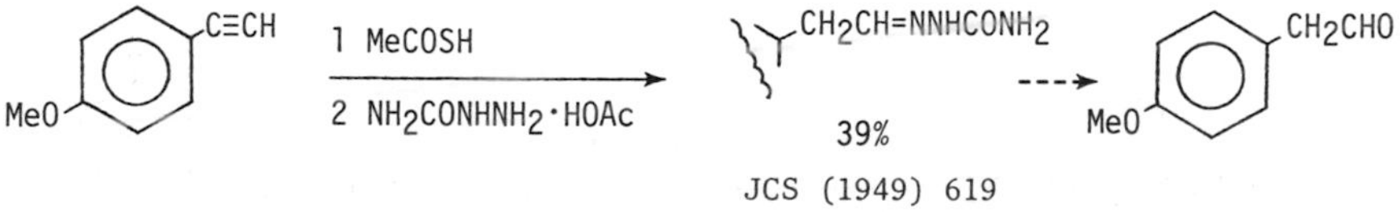

C≡CH
MeO
1 MeCOSH
2 NH2CONHNH2·HOAc
CH2CH=NNHCONH2
39%
CH2CHO
MeO

JCS (1949) 619

Section 47 Aldehydes from Carboxylic Acids, Acid Halides and Anhydrides

Review: The Synthesis of Aldehydes from Carboxylic Acids

Org React (1954) 8 218

$PrCOOH$ $\xrightarrow{\text{1 }}$ (epoxide)$CH(CH_2)_2COOH$ KOH   MeOH

electrolysis

2 HCl H_2O $\longrightarrow$ $Pr(CH_2)_2CHO$

Bull Chem Soc Jap (1965) 38 922

1 CH_2N_2 Et_2O

2 3,5-Dimethylpyrazole $h\nu$ Et_2O

$LiAlH_4$ Et_2O

55%

Annalen (1961) 642 121

$PhCH_2COCl$ $\xrightarrow[\text{2 EtSH } h\nu \text{ } C_6H_6]{\text{1 } CH_2N_2 \text{ } Et_2O}$ $PhCH_2CH_2COSEt$ $\xrightarrow[\text{2 HCl } H_2O \text{ } Et_2O]{\text{1 Ni } (PhNHCH_2)_2}$ $PhCH_2CH_2CHO$

Ber (1959) 92 528
Org React (1954) 8 218

$C_6H_{13}COOH$ $\xrightarrow[\text{2 HCOOEt}]{\text{1 (i-Pr)}_2NLi \text{ HMPA THF}}$ $C_6H_{13}CHO$ 65%

Tetr Lett (1970) 699

$BuCOOH$ $\xrightarrow{\text{Li } MeNH_2}$ $BuCHO$ 66%

JACS (1970) 92 5774

$\xrightarrow[260°]{\text{HCOOH } TiO_2}$ 36%

JCS (1943) 84
JOC (1963) 28 3029

$$C_8H_{17}COOH \quad \xrightarrow[\text{Na}_2\text{CO}_3 \quad \text{H}_2\text{O}]{\text{Na-Hg} \quad \text{H}_3\text{BO}_3 \quad \text{NaHSO}_3} \quad C_8H_{17}CHO \qquad 36\%$$

J Soc Chem Ind (1943) <u>62</u> 128

$$Ph(CH_2)_2COOH$$

1 Me$_2$CCH$_2$OH (NH$_2$)
2 MeI   MeNO$_2$
3 NaBH$_4$ MeOH
4 NaOH   H$_2$O

$Ph(CH_2)_2CH$ ⟶ (HCl / H$_2$O) $Ph(CH_2)_2CHO$

J Heterocyclic Chem (1966) <u>3</u> 531

1 N,N-Carbonyl diimidazole

THF
2 LiAl(OBu-t)$_3$H   THF

JOC (1970) <u>35</u> 458
Annalen (1962) <u>654</u> 119

1 ClCOOEt   NEt$_3$

C$_6$H$_6$
2 EtOMgCH(COOEt)$_2$
EtOH

$\xrightarrow[\text{MeOH}]{\text{NaBH}_4}$ CHO   79%

Ber (1965) <u>98</u> 3040

1 ClCOOEt   NEt$_3$

2 NaSH
3 Ni   (PhNHCH$_2$)$_2$
HOAc   THF

$\xrightarrow[\text{CH}_2\text{Cl}_2]{\text{TsOH}}$ CHO   <47%

JOC (1966) <u>31</u> 1922

H_2 Pd-BaSO$_4$

xylene

80%

JACS (1942) 64 928
      (1940) 62 49
Org React (1948) 4 362

PhCOCl $\xrightarrow{\text{Bu}_3\text{SnH}}$ PhCHO

JOC (1960) 25 284

PhCOBr $\xrightarrow{\text{h}\nu\ \ \text{Et}_2\text{O}}$ PhCHO   80%

Angew (1965) 77 169
(Internat Ed 4 146)

$\xrightarrow{\text{LiAl(OBu-t)}_3\text{H}}$ diglyme   62%

J Med Chem (1970) 13 26
JACS (1956) 78 252

t-BuCOCl $\xrightarrow[\text{Et}_3\text{N}\ \ \text{C}_6\text{H}_6]{\overset{\overset{\text{NH}}{|}}{\text{CH}_2\text{-CH}_2}}$ t-BuCON$\begin{smallmatrix}\text{CH}_2\\|\\\text{CH}_2\end{smallmatrix}$ $\xrightarrow[\text{Et}_2\text{O}]{\text{LiAlH}_4}$ t-BuCHO

JACS (1961) 83 4549

$C_{15}H_{31}COCl$ $\xrightarrow[\text{2 HOAc}]{\text{1 } CH_2N_2 \quad Et_2O}$ $C_{15}H_{31}COCH_2OAc$ $\xrightarrow[\substack{\text{i-PrOH} \\ \text{2 } Pb(OAc)_4 \\ \text{HOAc}}]{\text{1 } Al(OPr\text{-}i)_3}$ $C_{15}H_{31}CHO$

56%

Org React (1954) <u>8</u> 218

$PhCOCl$ $\xrightarrow{\text{EtSH} \quad Pyr}$ $PhCOSEt$ $\xrightarrow[H_2O]{\text{Ni} \quad EtOH}$ $PhCHO$ < 62%

JACS (1946) <u>68</u> 1455
Tetr Lett (1964) 1763
Org React (1954) <u>8</u> 218

$PhCOCl$ $\xdashrightarrow{N_2H_4}$ $PhCONHNH_2$ $\xrightarrow[\substack{\text{2 } Na_2CO_3 \\ \text{ethylene glycol}}]{\text{1 } PhSO_2Cl \quad Pyr}$ $PhCHO$ ~60%

Org React (1954) <u>8</u> 218
JOC (1961) <u>26</u> 3664

$BuCOCl$ $\xrightarrow[\text{2 } H_2SO_4 \quad H_2O]{\text{1 Quinoline} \quad HCN \quad C_6H_6}$ $BuCHO$ 62%

JACS (1941) <u>63</u> 2021
Org React (1954) <u>8</u> 218

$\left[\text{MeO} \underset{}{\bigcirc} \text{CH}_2\text{CO} \right]_2 O$ $\xrightarrow[\text{chlorobenzene}]{\text{Pyridine N-oxide}}$ $\text{MeO} \bigcirc \text{CHO}$ 69%

Tetr Lett (1965) 233

$C_6H_{13}CH_2COCl$ $\xrightarrow[\substack{C_6H_6 \\ \text{2 } HCl}]{\text{1 } BuSH \quad ZnCl_2}$ $C_6H_{13}CH=C(SBu)_2$ $\xrightarrow[\text{2 } N_2H_4 \quad MeOH]{\text{1 Perphthalic acid} \quad Et_2O}$

$C_6H_{13}CH=NNH_2$ $\xdashrightarrow{}$ $C_6H_{13}CHO$

Rec Trav Chim (1959) <u>78</u> 354

MeCHCH$_2$CH$_2$COOH

1 Bromination

2 Hydrolysis

MeCHCH$_2$$\overset{\text{OH}}{\text{C}}$HCOOH

$\xrightarrow{\text{NaIO}_4}$

MeCHCH$_2$CHO

Tetr Lett (1968) 1725

Aldehydes may also be prepared from carboxylic acids via amide or ester intermediates. See section 51 (Aldehydes from Amides) and section 53 (Aldehydes from Esters)

Section 48   Aldehydes from Alcohols

BuOH   ----→   BuOCH$_2$CH=CH$_2$ $\xrightarrow{\text{PrLi}}$ Bu(CH$_2$)$_2$CHO   29%

Tetr Lett (1969) 821

i-PrCH$_2$$\overset{\text{OH}}{\underset{\text{Me}}{\text{C}}}$CH=CH$_2$ $\xrightarrow[\text{H}_3\text{PO}_4]{\text{C}_5\text{H}_{11}\text{CH}_2\text{CH(OEt)}_2}$ i-PrCH$_2$$\overset{}{\underset{\text{Me}}{\text{C}}}$=CHCH$_2$$\overset{}{\underset{\text{C}_5\text{H}_{11}}{\text{C}}}$HCHO   60%

Helv (1967) 50 2095
Tetr Lett (1961) 493

Me$_2$$\overset{\text{OH}}{\text{C}}$CH=CH$_2$ $\xrightarrow[\text{H}_3\text{PO}_4]{\text{CH}_2\text{=CHOEt}}$ Me$_2$C=CH(CH$_2$)$_2$CHO   81%

Helv (1967) 50 2095
JACS (1961) 83 198

C$_7$H$_{15}$CH$_2$OH $\xrightarrow{\text{O}_2 \ \ \text{Pt} \ \ \text{EtOAc}}$ C$_7$H$_{15}$CHO   21%

JACS (1955) 77 190
Angew (1957) 69 600
For application to allylic alcohols see Helv (1957) 40 265

$$\text{BuCH}_2\text{OH} \xrightarrow{\text{Li} \quad \text{NH}_2\text{CH}_2\text{CH}_2\text{NH}_2} \text{BuCHO} \qquad \qquad 66\%$$

Applicable to benzylic alcohols

JOC (1962) $\underline{27}$ 2662

Benzoquinone   Al(OBu-t)$_3$

Chem Comm (1969) 799

$$\text{C}_6\text{H}_{13}\text{CH}_2\text{OH} \xrightarrow{\text{CrO}_3\text{-Pyr} \quad \text{CH}_2\text{Cl}_2} \text{C}_6\text{H}_{13}\text{CHO} \qquad \qquad 93\%$$

Applicable to allylic and benzylic alcohols

Tetr Lett (1968) 3363
JOC (1961) $\underline{26}$ 4814
(1970) $\underline{35}$ 4000

CrO$_3$ H$_2$SO$_4$

DMF   H$_2$O

80%

Helv (1968) $\underline{51}$ 772

$$\text{PhCH=CHCH}_2\text{OH} \xrightarrow[\text{Pyr·HCl} \quad \text{Pyr}]{\text{Na}_2\text{Cr}_2\text{O}_7} \text{PhCH=CHCHO} \qquad \qquad 80\%$$

Chem Ind (1969) 1594

$$\text{EtCH}_2\text{OH} \xrightarrow{\text{K}_2\text{Cr}_2\text{O}_7 \quad \text{H}_2\text{SO}_4 \quad \text{H}_2\text{O}} \text{EtCHO} \qquad \qquad 45\text{-}49\%$$

Org Synth (1943) Coll Vol 2 541
For application to allylic alcohols see Bull Soc Chim Fr (1933) $\underline{53}$ 301
and Ber (1947) $\underline{80}$ 137

$$\text{C}_6\text{H}_{11}\text{-CH}_2\text{OH} \xrightarrow[\text{t-Butyl chromate   C}_6\text{H}_6]{} \text{C}_6\text{H}_{11}\text{-CHO} \qquad 40\%$$

Applicable to allylic and benzylic alcohols

Bull Chem Soc Jap (1965) <u>38</u> 1503 893 1141

$$\text{PhCH=CHCH}_2\text{OH} \xrightarrow[\text{C}_6\text{H}_6]{\text{Nickel peroxide}} \text{PhCH=CHCHO} \qquad 86\%$$

JOC (1962) <u>27</u> 1597
JCS (19?9) <u>C</u> 2173

$$\text{Br-C}_6\text{H}_4\text{-CH}_2\text{OH} \xrightarrow[\text{HOAc   H}_2\text{O}]{\text{(NH}_4\text{)}_2\text{Ce(NO}_3\text{)}_6} \text{Br-C}_6\text{H}_4\text{-CHO} \qquad 93\%$$

JCS (1965) 5777
JOC (1967) <u>32</u> 2349   3865

$\text{MeCH(CH}_2\text{)}_2\text{CH}_2\text{OH}$

$$\xrightarrow[\text{C}_6\text{H}_6]{\text{Ag}_2\text{CO}_3\text{-celite}} \qquad \text{MeCH(CH}_2\text{)}_2\text{CHO} \qquad 95\%$$

Applicable to allylic alcohols

Compt Rend (1968) <u>C</u> <u>267</u> 900

$$\text{i-PrCH}_2\text{OH} \xrightarrow[]{\text{AgO   H}_3\text{PO}_4   \text{HOAc}} \text{i-PrCHO} \qquad 30\%$$

Applicable to allylic alcohols

Tetr Lett (1967) 4193

$$\text{BuCH}_2\text{OH} \xrightarrow[\text{Me}_2\text{SO   H}_2\text{O}]{\text{Argentic picolinate}} \text{BuCHO} \qquad 74\%$$

Applicable to benzylic alcohols

Can J Chem (1969) <u>47</u> 1649

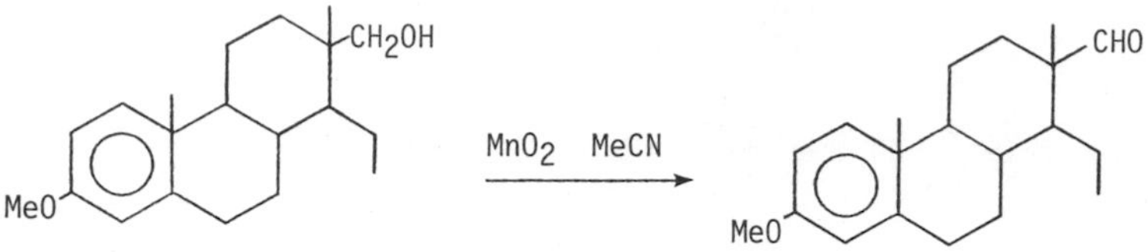

76%

Proc Chem Soc (1964) 110
Chem Comm (1966) 121

$PhCH=CHCH_2OH$ $\xrightarrow{\quad MnO_2 \quad CCl_4 \quad}$ $PhCH=CHCHO$

JCS (1952) 1094
JACS (1955) 77 4399

$PrCH_2OH$ $\xrightarrow{\quad Pb(OAc)_4 \quad Pyr \quad}$ $PrCHO$ 70%

Applicable to allylic and benzylic alcohols

Tetr Lett (1964) 3071

$C_5H_{11}CH_2OH$ $\xrightarrow{\quad INO_3\text{-}Pyr \quad CHCl_3 \quad}$ $C_5H_{11}CHO$ 20%

Applicable to benzylic alcohols

JCS C (1970) 676

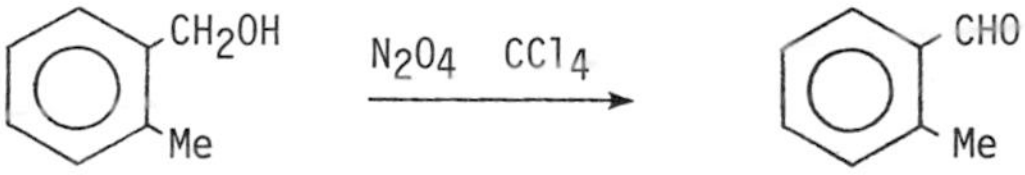

95%

JCS (1955) 1110

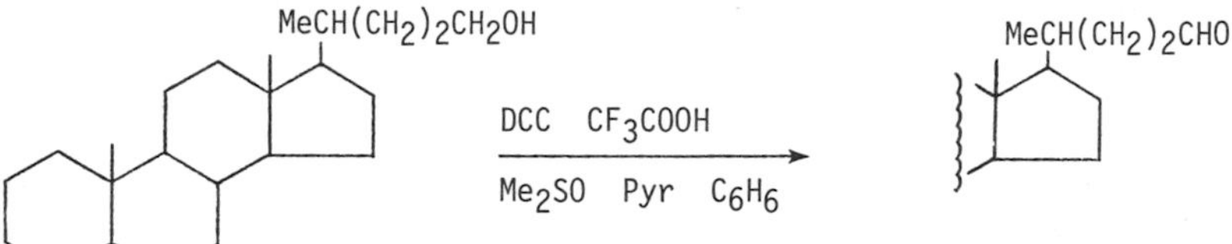

85%

Applicable to benzylic alcohols

JACS (1965) 87 5670
Chem Rev (1967) 67 247

$$\xrightarrow[\text{Et}_3\text{N}]{\text{SO}_3\cdot\text{Pyr} \quad \text{Me}_2\text{SO}}$$

70%

JACS (1967) <u>89</u> 5505

$$C_7H_{15}CH_2OH \quad \text{----}\blacktriangleright \quad C_7H_{15}CH_2OTs \xrightarrow{\text{NaHCO}_3 \quad \text{Me}_2\text{SO}} C_7H_{15}CHO \quad \sim 78\%$$

Applicable to benzylic alcohols

JACS (1959) <u>81</u> 4113

$$C_5H_{11}CH_2OH \xrightarrow{\text{COCl}_2 \quad \text{Et}_2\text{O}} C_5H_{11}CH_2OCOCl \xrightarrow[\text{2 Et}_3\text{N}]{\text{1 Me}_2\text{SO}} C_5H_{11}CHO \quad 68\%$$

JCS (1964) 1855

$$PhCH_2OH \xrightarrow{\text{KOCl} \quad \text{MeOH} \quad \text{H}_2\text{O}} PhCHO \quad 77\%$$

JOC (1961) <u>26</u> 1046

$$\xrightarrow[\text{Pyr}]{\text{N-Chlorosuccinimide}}$$

71%

Bull Soc Chim Belg (1951) <u>60</u> 54
Compt Rend (1954) <u>238</u> 2538
Chem Rev (1963) <u>63</u> 21

$$i\text{-PrCH}_2\text{OH} \xrightarrow{\text{K}_2\text{S}_2\text{O}_8 \quad \text{AgNO}_3 \quad \text{H}_2\text{O}} i\text{-PrCHO} \quad 31\%$$

JACS (1954) <u>76</u> 6345
For application to benzylic alcohols see JCS (1960) 1332

$$C_{11}H_{23}CH_2OH \xrightarrow[\text{NCOOEt}]{\overset{\text{NCOOEt}}{\underset{\|}{\text{NCOOEt}}} \quad \text{toluene}} C_{11}H_{23}CHO \qquad 20\%$$

Applicable to benzylic alcohols

JOC (1967) 32 727

$$PhCH_2OH \xrightarrow[\text{C}_6\text{H}_6]{\text{4-Phenyl-1,2,4-triazoline-3,5-dione}} PhCHO \qquad 78\%$$

Chem Comm (1966) 744
JCS C (1969) 1474

2,3-Dichloro-5,6-
dicyanoquinone (DDQ)

Applicable only to allylic and benzylic alcohols

J Med Chem (1962) 5 409
Chem Rev (1967) 67 153

$$PhCH=CHCH_2OH \xrightarrow[\text{CCl}_4]{\text{Tetrachloro-1,2-benzoquinone}} PhCH=CHCHO \qquad 100\%$$

Applicable only to allylic and benzylic alcohols

JCS (1956) 3070

1 MsCl   DMF

2 MeONa
3 NaN₃   DMF

1 Acetylation

2 hν   MeOH

Carbohydrate Res (1968) 8 366

$$\begin{array}{c} (CH_2)_7COOH \\ | \\ CHOH \\ | \\ C_8H_{17}CHOH \end{array} \xrightarrow[\text{EtOH}\;\;\text{H}_2\text{O}]{\text{KIO}_4\;\;\text{H}_2\text{SO}_4} \begin{array}{c} (CH_2)_7COOH \\ | \\ CHO \end{array} \qquad 76\%$$

Org React (1944) 2 341

$BuOCH_2CHCH_2OH$ $\xrightarrow{\text{Pb(OAc)}_4 \quad C_6H_6}$ $BuOCH_2CHO$ 57%
       $\overset{|}{HO}$

                                        JACS (1945) 67 39
                                        JOC (1970) 35 249

$Ph(CH_2)_3CHCH(CH_2)_3Ph$ $\xrightarrow[CH_2Cl_2]{I_2 \quad HgO}$ $Ph(CH_2)_3CHO$ 99%
         $\overset{|}{HO}$ $\overset{|}{OH}$

                                        JCS C (1969) 383

$PhCHCH_2OH$ $\xrightarrow{K_2S_2O_8 \quad AgNO_3 \quad H_2O}$ $PhCHO$ 61%
 $\overset{|}{OH}$

                                        JACS (1954) 76 6345

$\xrightarrow{O_2 \quad Co(OAc)_2 \quad PhCN}$

                                        Tetr Lett (1968) 5689

$MeCHCHMe$ $\xrightarrow{XeO_3 \quad H_2O}$ $MeCHO$
 $\overset{|}{HO}$ $\overset{|}{OH}$

                                        JACS (1964) 86 2078

Section 49   Aldehydes from Aldehydes

PhCHO $\xrightarrow[\text{2 HClO}_4 \quad \text{H}_2\text{O} \quad \text{Et}_2\text{O}]{\text{1 Ph}_3\text{P=CHO}-\langle\text{C}_6\text{H}_4\rangle-\text{Me}}$ PhCH$_2$CHO   55%

Ber (1962) <u>95</u> 2514

$$\begin{array}{c}\text{CHO}\\ \text{HO}-|\\ |-\text{OH}\\ |-\text{OH}\\ \llcorner\text{OH}\end{array} \xrightarrow[\text{2 HgCl}_2 \quad \text{HgO} \quad \text{H}_2\text{O}]{\text{1 Ph}_3\text{P=CHSPh} \quad \text{Me}_2\text{SO}} \begin{array}{c}\text{CH}_2\text{CHO}\\ \text{HO}-|\\ |-\text{OH}\\ |-\text{OH}\\ \llcorner\text{OH}\end{array}$$ 36%

Tetr Lett (1969) 3665

$$\begin{array}{c}\text{CHO}\\ \llcorner\text{O}\diagup\\ \text{O}=\diagdown\text{O}\diagup\\ \llcorner\text{O}\diagdown\end{array} \xrightarrow[\text{2 HClO}_4 \quad \text{Et}_2\text{O}]{\text{1 Ph}_3\text{P=CHOMe} \quad \text{Et}_2\text{O}} \begin{array}{c}\text{CH}_2\text{CHO}\\ \llcorner\text{O}\diagup\\ \text{O}=\diagdown\text{O}\diagup\\ \llcorner\text{O}\diagdown\end{array}$$ 28%

Carbohydrate Res (1969) <u>10</u> 184

C$_6$H$_{13}$CHO $\xrightarrow[\substack{\text{Mg-Hg} \quad \text{Et}_2\text{O} \\ \text{2 EtONa}}]{\text{1 Cl}_2\text{CHCOOEt}}$ C$_6$H$_{13}$CHCHCOOEt $\xrightarrow[\text{2 } \Delta]{\text{1 NaOH}}$ C$_6$H$_{13}$CH$_2$CHO

(with $\overset{\backslash}{O}\diagup$ bridging)

Compt Rend (1937) <u>204</u> 272
Org React (1949) <u>5</u> 413

C$_6$H$_{13}$CHO $\xrightarrow[\text{HgCl}_2 \quad \text{THF}]{\text{ClCH}_2\text{OEt} \quad \text{Mg}}$ C$_6$H$_{13}$CHCH$_2$OEt $\xrightarrow{\text{HCOOH}}$ C$_6$H$_{13}$CH$_2$CHO   10%

(OH below)

Bull Soc Chim Fr (1959) 459

$C_5H_{11}CH_2CHO$ $\xrightarrow{\begin{array}{l}1 \ t\text{-}BuNH_2 \\ 2 \ EtMgBr \\ 3 \ BuI\end{array}}$ $C_5H_{11}\underset{Bu}{CH}CH=NBu\text{-}t$ $\xrightarrow[H_2O]{Acid}$ $C_5H_{11}\underset{Bu}{CH}CHO$ 50%

JACS (1963) <u>85</u> 2178

$PrCH_2CHO$ $\xrightarrow{BuNHCH_2Pr\text{-}i}$ $PrCH=\underset{Bu}{CH}NCH_2Pr\text{-}i$ $\xrightarrow{\begin{array}{l}1 \ EtI \\ 2 \ Acid \ H_2O\end{array}}$ $Pr\underset{Et}{CH}CHO$ 78%

Chem Comm (1967) 510

$\underset{(CH_2)_2COOMe}{EtCHCHO}$ $\xrightarrow{Pyrrolidine}$ $\underset{(CH_2)_2COOMe}{EtC=CHN}$ $\xrightarrow{\begin{array}{l}1 \ CH_2=CHCH_2Br \\ 2 \ Hydrolysis\end{array}}$ $\overset{CH_2CH=CH_2}{\underset{(CH_2)_2COOMe}{EtCCHO}}$

Chem Comm (1965) 197
JACS (1963) <u>85</u> 207
Angew (1960) <u>72</u> 169

$PhCH_2CHO$ $\xrightarrow{\begin{array}{l}1 \ PhNHNH_2 \\ 2 \ KNH_2 \\ PhCH_2Cl \quad NH_3\end{array}}$ $\underset{CH_2Ph}{PhCH}CH=NNHPh$ $\dashrightarrow$ $\underset{CH_2Ph}{PhCH}CHO$

JACS (1967) <u>89</u> 463

$\underset{Me}{PhCH}CHO$ $\xrightarrow{\begin{array}{l}1 \ Br_2 \\ 2 \ CN^-\end{array}}$ $Ph\overset{O}{C}\text{-}\underset{Me}{CH}CN$ $\xrightarrow{\begin{array}{l}1 \ AlMe_3 \\ 2 \ \triangle\end{array}}$ $\overset{Me}{\underset{Me}{PhCCHO}}$

Tetr Lett (1970) 2947

$MeCH(CH_2)_2CHO$ $\overset{HO \ \ OH}{MeCHCH_2CHCHOEt}$ $MeCHCH_2CHO$

$\xrightarrow{\begin{array}{l}1 \ IBr \quad HBr \\ HOAc \\ 2 \ KOH \quad EtOH\end{array}}$ $\xrightarrow[\begin{array}{c}HOAc \\ Me_2CO\end{array}]{NaIO_4}$

Chem Comm (1968) 851
Tetr Lett (1970) 5229

MeCH$_2$CHO ⟍

                    O$_2$ Cu(OAc)$_2$ Pyr Et$_3$N

Me(CH$_2$)$_5$CHO ⟋   ⟶   MeCHO

Rec Trav Chim (1966) <u>85</u> 437

Some of the methods included in section 57 (Aldehydes from Ketones) may
also be applied to the preparation of aldehydes from aldehydes

Section 50   Aldehydes from Alkyls

Reactions in which methyl groups are converted into aldehyde (RMe →RCHO)
are included in this section. For reactions in which a hydrogen is
replaced by aldehyde, RH → RCHO (R=alkyl, vinyl or aryl), see section 56
(Aldehydes from Hydrides)

1 CrO$_2$Cl$_2$ CS$_2$

2 H$_2$O

50%

JCS (1949) S230

1 CrO$_2$Cl$_2$ Me$_2$C=CHMe

CCl$_4$

2 H$_2$O

25%

JACS (1951) <u>73</u> 221

CrO$_3$ Ac$_2$O

H$_2$SO$_4$ HOAc

H$_2$SO$_4$

EtOH
H$_2$O

JOC (1968) <u>33</u> 4176
Org Synth (1943) Coll Vol 2 441

PhCH$_3$

Na$_2$S$_2$O$_8$ AgNO$_3$ H$_2$O

⟶   PhCHO

50%

JCS (1960) 1332

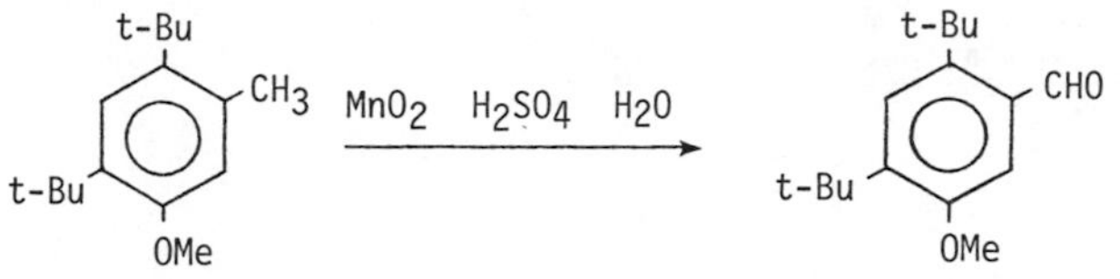

23%

JOC (1951) <u>16</u> 586

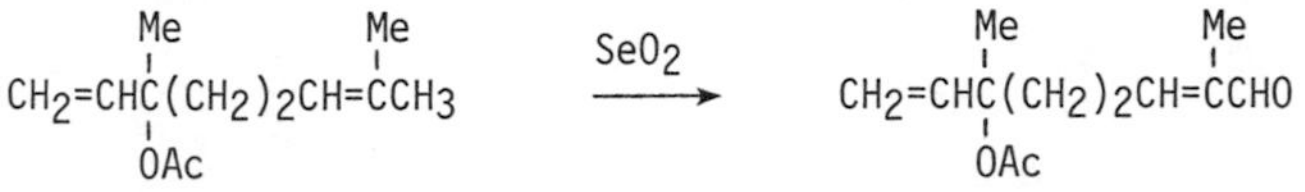

Chem Comm (1968) 1277
Org React (1949) <u>5</u> 331

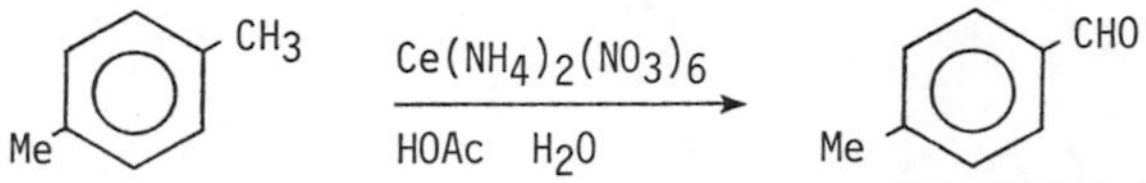

73%

JOC (1966) <u>31</u> 2033
JCS <u>C</u> (1970) 1630

PhCH₃ $\xrightarrow{\text{Argentic picolinate \quad Me}_2\text{SO}}$ PhCHO

Tetr Lett (1967) 415

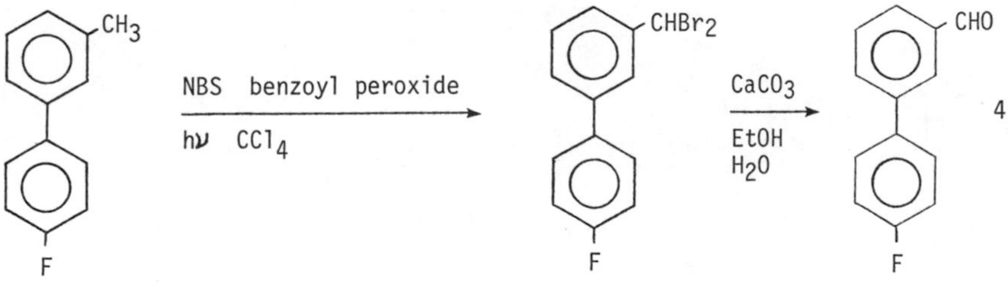

41%

JACS (1966) <u>88</u> 3318
       (1956) <u>78</u> 1689
Org React (1954) <u>8</u> 197

Section 51   Aldehydes from Amides

$C_{15}H_{31}CONMe_2$ $\xrightarrow{\text{NaAlH}_4 \quad \text{THF}}$ $C_{15}H_{31}CHO$ 78%

Tetrahedron (1969) 25 5555

$\xrightarrow{\text{LiAlH}_4 \quad \text{THF} \quad \text{Et}_2\text{O}}$ ~70%

Ber (1951) 84 625

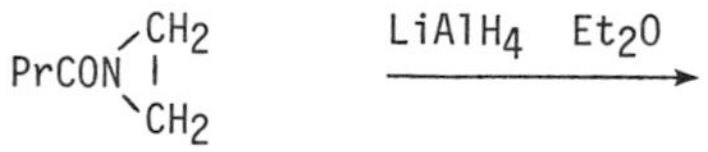

$\xrightarrow{\text{LiAlH}_4 \quad \text{Et}_2\text{O}}$ PrCHO   88%

JACS (1961) 83 4549

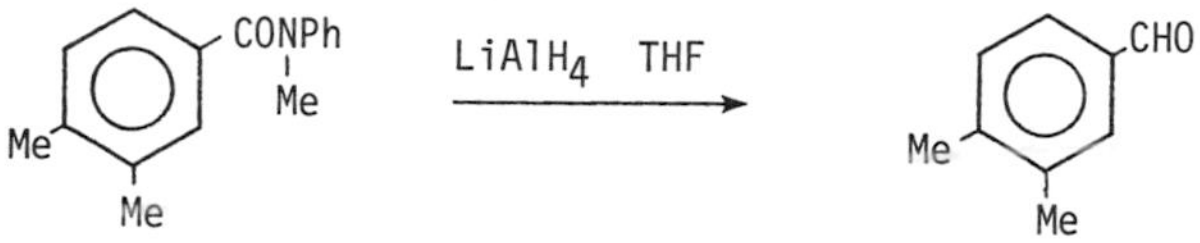

$\xrightarrow{\text{LiAlH}_4 \quad \text{THF}}$ 57%

Ber (1955) 88 301
Org React (1954) 8 218

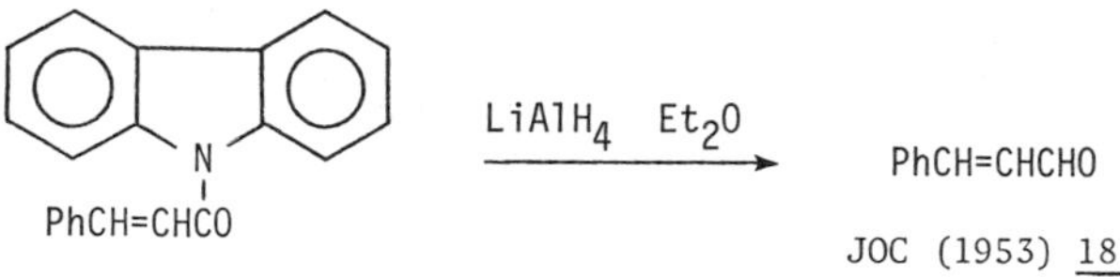

$\xrightarrow{\text{LiAlH}_4 \quad \text{Et}_2\text{O}}$ PhCH=CHCHO   45%

JOC (1953) 18 1190

$BuCONMe_2$ $\xrightarrow{\text{LiAl(OEt)}_3\text{H} \quad \text{Et}_2\text{O}}$ BuCHO   92%

JACS (1964) 86 1089

$C_7H_{15}CONPh$ $\xrightarrow{\text{(i-Bu)}_2\text{AlH} \quad \text{Et}_2\text{O}}$ $C_7H_{15}CHO$ 56%
   Me

                                      Izv (1959) 2146
                                      (Chem Abs $\underline{54}$ 10932)

$C_5H_{11}CONHPh$ $\xrightarrow[\text{2 HOAc}]{\text{1 Na} \quad \text{NH}_3}$ $C_5H_{11}CHO$ 53%

                                        Aust J Chem (1955) $\underline{8}$ 512

$C_5H_{11}CONHMe$ $\xrightarrow[\text{MeNH}_2 \quad \text{EtOH}]{\text{Electrolysis} \quad \text{LiCl}}$ $C_5H_{11}CHO$ 50%

                                        JOC (1970) $\underline{35}$ 1210

$PhCONH_2$ $\xrightarrow{\text{P}_2\text{S}_5 \quad \text{Pyr}}$ $PhCSNH_2$ $\xrightarrow{\text{Ni} \quad \text{EtOH}}$ $PhCHO$

                                        JACS (1952) $\underline{74}$ 3936
                                        Org React (1962) $\underline{12}$ 356

                                       Cl

$PhCH{=}CHCONHPh$ $\xrightarrow[\text{toluene}]{\text{PCl}_5}$ $PhCH{=}CHC{=}NPh$ $\xrightarrow[\text{Et}_2\text{O}]{\text{SnCl}_2 \quad \text{HCl}}$ $PhCH{=}CHCHO$ 92%

                                        Org React (1954) $\underline{8}$ 218

$\xrightarrow[\substack{\text{2 Na}_2\text{CO}_3 \\ \text{ethylene glycol}}]{\text{1 TsCl} \quad \text{Pyr}}$ 60%

                                        JOC (1961) $\underline{26}$ 3664
                                        Org React (1954) $\underline{8}$ 218

$\xrightarrow{\text{NaIO}_4 \quad \text{NH}_3 \quad \text{H}_2\text{O}}$ 60-70%

                                        JACS (1952) $\underline{74}$ 5796

PhCH$_2$NHAc $\xrightarrow{\text{Argentic picolinate Me}_2\text{SO}}$ PhCHO

Tetr Lett (1967) 415

EtNHAc $\xrightarrow{\text{K}_2\text{S}_2\text{O}_8 \quad \text{K}_2\text{HPO}_4 \quad \text{H}_2\text{O}}$ MeCHO   48%

JOC (1964) 29 3632

Section 52   Aldehydes from Amines

$\xrightarrow[\substack{2 \text{ HCH=NOH NaOAc CuSO}_4 \\ \text{Na}_2\text{SO}_3 \\ 3 \text{ HCl H}_2\text{O}}]{1 \text{ NaNO}_2 \quad \text{NaOAc HCl H}_2\text{O}}$ 41%

JCS (1954) 1297

PhCH$_2$NH$_2$ $\xrightarrow[2 \text{ HCl H}_2\text{O}]{1 \text{ 2,4-Dinitrobenzaldehyde Pyr}}$ PhCHO   63%

JCS (1954) 209

C$_{11}$H$_{23}$CH$_2$NH$_2$ $\xrightarrow[\substack{2 \text{ Diazabicyclononene Me}_2\text{SO THF} \\ 3 \text{ (COOH)}_2 \text{ MeOH THF}}]{1 \text{ 3,5-Dinitromesitylglyoxal C}_6\text{H}_6}$ C$_{11}$H$_{23}$CHO   34%

JACS (1969) 91 1429

C$_5$H$_{11}$CH$_2$NH$_2$ $\xrightarrow[\text{HOAc H}_2\text{O}]{\text{Hexamine HCHO}}$ C$_5$H$_{11}$CHO   42%

JCS (1953) 1737
Org React (1954) 8 197

$$\text{Me-C}_6\text{H}_4\text{-CH}_2\text{NH}_2 \xrightarrow[\text{Me}_2\text{SO}]{\text{NaNO}_2 \quad \text{CF}_3\text{COOH}} \text{Me-C}_6\text{H}_4\text{-CHO}$$

60%

Angew (1965) <u>77</u> 811
(Internat Ed <u>4</u> 787)

$$\text{PhCH}_2\text{NEt}_2 \xrightarrow[\text{HOAc} \quad \text{H}_2\text{O}]{\text{NaNO}_2 \quad \text{NaOAc}} \text{PhCHO}$$

JACS (1967) <u>89</u> 1147

$$\text{MeCH}_2\text{NH}_2 \xrightarrow{\text{OF}_2 \quad \text{freon 11}} \text{MeCH=NOH} \dashrightarrow \text{MeCHO}$$

44%

JACS (1964) <u>86</u> 1392

$$(\text{PrCH}_2)_3\text{N} \xrightarrow{\text{O}_3 \quad \text{H}_2\text{O}} \text{PrCHO}$$

47%

JCS (1964) 711

$$(\text{PrCH}_2)_3\text{N} \xrightarrow{\text{Br}_2 \quad \text{H}_2\text{O} \quad \text{pH 5}} \text{PrCHO}$$

84%

JACS (1968) <u>90</u> 3502

$$(\text{EtCH}_2)_3\text{N} \xrightarrow{\text{NBS} \quad \text{dioxane} \quad \text{H}_2\text{O}} \text{EtCHO}$$

68%

JCS (1957) 4905
Chem Rev (1963) <u>63</u> 21

$$\text{PhCH}_2\text{CH}_2\text{NH}_2 \xrightarrow[\substack{\text{NaHCO}_3 \quad \text{Et}_2\text{O} \\ \text{2 EtONa EtOH} \\ \text{3 H}_2\text{SO}_4 \quad \text{H}_2\text{O}}]{\text{1 t-Butyl hypochlorite}} \text{PhCH}_2\text{CHO}$$

39%

JACS (1954) <u>76</u> 5554

PrCH₂NH₂ $\xrightarrow[\text{H}_2\text{O} \quad \text{EtOH}]{\text{H}_2\text{O}_2 \quad \text{Na}_2\text{WO}_4}$ PrCH=NOH   ----→   PrCHO
                                              57%

Ber (1960) 93 132

$$\left.\begin{array}{l} \text{PrCH}_2\text{NH}_2 \\ (\text{PrCH}_2)_2\text{NH} \\ (\text{PrCH}_2)_3\text{N} \end{array}\right\} \xrightarrow[\text{t-BuOH} \quad \text{H}_2\text{O}]{\text{KMnO}_4 \quad \text{MSO}_4}$$ PrCHO

(M=Ca or Zn)

JOC (1967) 32 3129

PhCH₂NH₂ $\xrightarrow{\text{MnO}_2}$ PhCHO   34%

JACS (1955) 77 4399

PrCH₂NH₂ $\xrightarrow{\text{Argentic picolinate} \quad \text{H}_2\text{O}}$ PrCHO   14%

JCS (1965) 4962
Tetr Lett (1967) 415

Aldehydes may also be prepared by oxidation of N-acyl amines. See
section 51 (Aldehydes from Amides)

Section 53   Aldehydes from Esters
            ○○○○○○○○○○○○○○○○○○○○○○○

PrCOOMe $\xrightarrow{\text{NaAlH}_4 \quad \text{THF}}$ PrCHO   81%

Tetr Lett (1963) 2087

$\xrightarrow{\text{LiAl(OBu-t)}_3\text{H} \quad \text{THF}}$ 70%

JOC (1966) 31 283

i-PrCOOEt $\xrightarrow[\text{hexane Et}_2\text{O}]{\text{(i-Bu)}_2\text{AlH toluene}}$ i-PrCHO   80%

Tetr Lett (1962) 619
(1969) 1779

$\xrightarrow[\text{2 PhSO}_2\text{Cl Pyr}]{\text{1 N}_2\text{H}_4\text{ H}_2\text{O}}$

$\xrightarrow[\substack{\text{ethylene} \\ \text{glycol}}]{\text{Na}_2\text{CO}_3}$

87%

JCS (1936) 584
JOC (1961) 26 3664
Org React (1954) 8 218

$\xrightarrow[\substack{\text{Zn C}_6\text{H}_6 \\ \text{2 KBH}_4\text{ MeOH}}]{\substack{\text{Br} \\ \text{1 Me}_2\text{CCOOMe}}}$

$\xrightarrow{280°}$

87%

Bull Soc Chim Fr(1965) 1864

Et(CH=CHCH$_2$)$_3$(CH$_2$)$_6$COOMe $\xrightarrow{\text{Na xylene}}$ Et(CH=CHCH$_2$)$_3$(CH$_2$)$_6$CHOH
Et(CH=CHCH$_2$)$_3$(CH$_2$)$_6$CO

$\downarrow$ 1 LiAlH$_4$
2 Pb(OAc)$_4$

Et(CH=CHCH$_2$)$_3$(CH$_2$)$_6$CHO   68%

J Am Oil Chem Soc (1960) 37 425
(Chem Abs 54 23373)

PrCH(COOEt)$_2$ $\xrightarrow{\text{N}_2\text{H}_4}$ PrCH(CONHNH$_2$)$_2$ $\xrightarrow[\text{2 H}_2\text{SO}_4\text{ H}_2\text{O}]{\text{1 HNO}_2\text{ H}_2\text{O}}$ PrCHO   46%

Org React (1946) 3 337

Section 54   Aldehydes from Ethers and Epoxides

$(PrCH_2)_2O$ $\xrightarrow[\text{CuCl}]{C_6H_{13}OH \quad \text{t-butyl perbenzoate}}$ $PrCH(OC_6H_{13})_2$ ---→ $PrCHO$

                                               70%

Angew Chem (1961) 73 65
Act Chem Scand (1961) 15 249
Tetrahedron (1961) 13 241

$PrCH_2OBu$ $\xrightarrow{\begin{array}{c}NCOOEt \\ \| \\ NCOOEt \quad h\nu\end{array}}$ $\begin{array}{c}PrCHOBu \\ | \\ NCOOEt \\ | \\ HNCOOEt\end{array}$ $\xrightarrow{H_2O}$ $PrCHO$

                                    60%   Chem Comm (1965) 259

$PrCH_2OBu$ $\xrightarrow{Pb(OAc)_4 \quad h\nu}$ $\begin{array}{c}PrCHOBu \\ | \\ OAc\end{array}$ $\xrightarrow{Acid}$ $PrCHO$

                                          Annalen (1970) 735 47

$MeCH_2OEt$ $\xrightarrow[\text{2 Acid}]{\begin{array}{c}Ac \\ | \\ 1 \; PhNNO\end{array}}$ $MeCHO$ 50%

                                   JACS (1964) 86 3180

$MeOCH_2CH_2OMe$ $\xrightarrow[\text{NaOAc} \quad H_2O]{\text{p-Chlorobenzendiazonium chloride}}$ $MeOCH_2CHO$

                                          Angew (1958) 70 211

$MeCH_2OEt$ $\xrightarrow{PhICl_2 \quad h\nu}$ $\begin{array}{c}MeCHOEt \\ | \\ Cl\end{array}$ ----→ $MeCHO$

                            79%

                                          Annalen (1969) 728 12

$PhCH_2OMe$ $\xrightarrow{\quad Br_2 \quad HOAc \quad H_2O \quad}$ $PhCHO$ 83%

JACS (1967) <u>89</u> 3550

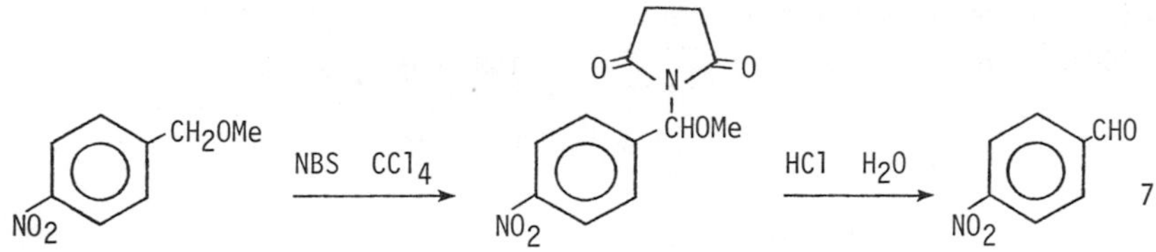

JOC (1958) <u>23</u> 1490
Angew (1959) <u>71</u> 349

$PhCH=CHCH_2OMe$ $\xrightarrow{\quad SeO_2 \quad HOAc \quad dioxane \quad}$ $PhCH=CHCHO$ 66%

Tetr Lett (1970) 2885

$PrCH_2OMe$ $\xrightarrow[\text{2 Acid}]{\text{1 } O_2 \quad Ph_2CO \quad h\nu}$ $PrCHO$ 55%

Ber (1963) <u>96</u> 509

$MeCH_2OCPh_3$ $\xrightarrow{\quad 262° \quad}$ $MeCHO$

JACS (1930) <u>52</u> 753
(1924) <u>46</u> 2580

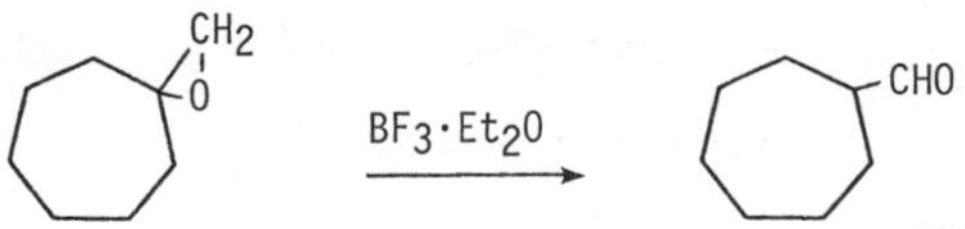

91%

JACS (1962) <u>84</u> 867
Chem Rev (1959) <u>59</u> 737

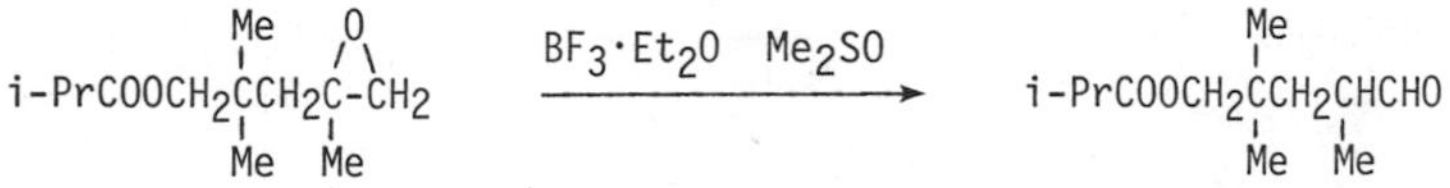

64%

JOC (1965) <u>30</u> 3480

$$C_6H_{13}\overset{O}{\overset{\triangle}{CH}}CH_2 \xrightarrow{\quad PrI \quad Me_2SO \quad} C_6H_{13}CH_2CHO \qquad\qquad 80\%$$

Chem Comm (1968) 227

$$\binom{(CH_2CH_2CH=\overset{Me}{C})_2CH_2CH_2\overset{O}{\overset{\triangle}{CH}}CHCMe_2}{(CH=\overset{}{C}CH_2CH_2)_2CH=CMe_2} \xrightarrow[Et_2O]{HIO_4} \binom{(CH_2CH_2CH=\overset{Me}{C})_2CH_2CH_2CHO}{(CH=\overset{}{C}CH_2CH_2)_2CH=CMe_2}$$
                                                                    Me

Chem Comm (1968) 1067

Section 55   Aldehydes from Halides and Sulfonates

JACS (1939) 61 2134   59%

$$i\text{-}PrBr \xrightarrow[2 \text{ MeOCH}_2\text{COMe}]{1 \text{ Mg} \quad Et_2O} i\text{-}Pr\overset{OH}{\underset{Me}{\overset{|}{C}}}CH_2OMe \xrightarrow{(COOH)_2} i\text{-}Pr\overset{}{\underset{Me}{CH}}CHO$$

JACS (1946) 68 2339

$$BuBr \dashrightarrow BuOCH_2CH=CH_2 \xrightarrow[pentane]{PrLi \quad THF} Bu(CH_2)_2CHO \qquad 29\%$$

Tetr Lett (1969) 821

$$CH_2=CH\overset{}{\underset{CH_2}{C}}(CH_2)_2CH=\overset{}{\underset{Me}{C}}CH_2Br \xrightarrow[\substack{2 \ (COOH)_2 \ H_2O}]{1 \ CH_3CH=NBu\text{-}t \\ (i\text{-}Pr)_2NLi \ Et_2O} CH_2=CH\overset{}{\underset{CH_2}{C}}(CH_2)_2CH=\overset{}{\underset{Me}{C}}(CH_2)_2CHO$$
                                                                                                                                              61%

Helv (1967) 50 2440

PrI $\xrightarrow[\substack{2 \ NaBH_4 \ \ THF \ \ EtOH \\ 3 \ (COOH)_2 \ \ H_2O}]{1 \ \text{(oxazoline)} \ \ BuLi \ \ THF}$ PrCH$_2$CHO   65%

JACS (1969) <u>91</u> 763 5886
       (1970) <u>92</u> 6676

$\xrightarrow[\substack{2 \ NaBH_4}]{1 \ \text{(oxazoline)} \ \ BuLi}$ —CHO   36%

JACS (1969) <u>91</u> 765

PhBr $\xrightarrow[\substack{2 \ CH_2=CHCH_2Br}]{1 \ Mg \ \ Et_2O}$ PhCH$_2$CH=CH$_2$ $\xrightarrow[\substack{2 \ NaOH \\ 3 \ Pb(OAc)_4}]{1 \ PhCOOAg \ \ I_2}$ PhCH$_2$CHO

Helv (1934) <u>17</u> 351

$\xrightarrow[\substack{2 \ HC(OEt)_3 \\ 3 \ HCl \ \ H_2O}]{1 \ Mg \ \ Et_2O}$ 40-42%

Org Synth (1955) Coll Vol 3 701
            (1943) Coll Vol 2 323
Ber (1970) <u>103</u> 643

$\xrightarrow[\substack{2 \ HCOOEt}]{1 \ Mg \ \ Et_2O}$ 40%

Annalen (1912) <u>393</u> 215

$\xrightarrow[\substack{2 \ HCONMe_2}]{1 \ Mg \ \ THF}$ 80%

Chimia (1964) <u>18</u> 141
JCS (1956) 4691

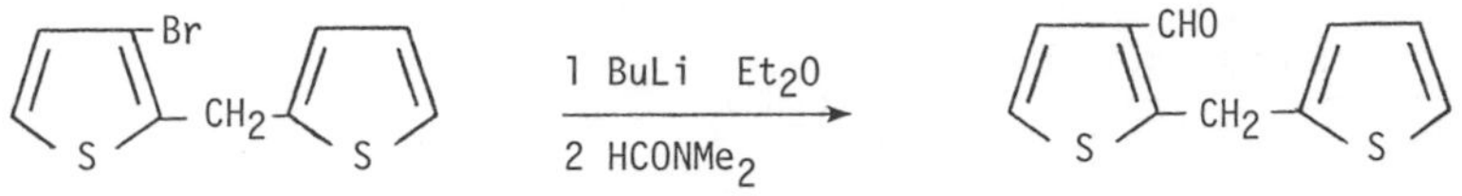
Br
CH2
S
S
1 BuLi Et2O
2 HCONMe2
CHO
CH2
S
S

JOC (1970) 35 711

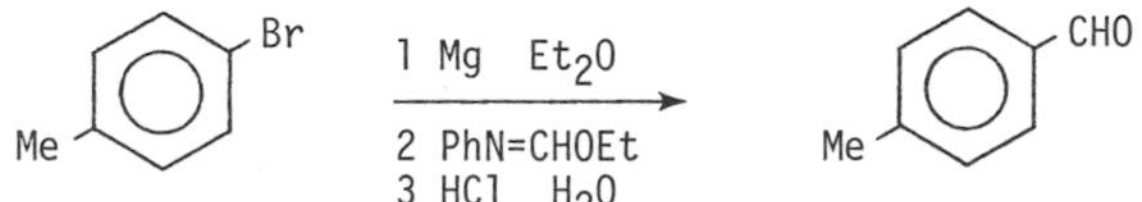
Br
Me
1 Mg Et2O
2 PhN=CHOEt
3 HCl H2O
CHO
Me

82%

JOC (1941) 6 489

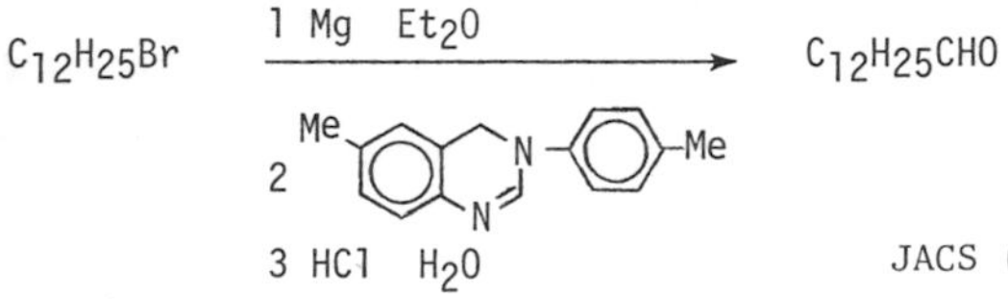
C12H25Br
1 Mg Et2O
C12H25CHO
2
Me
N
N
Me
3 HCl H2O

73%

JACS (1955) 77 5118

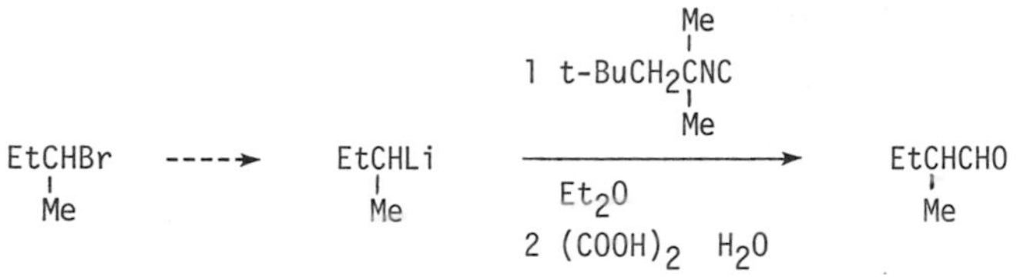
EtCHBr
Me
EtCHLi
Me
Me
1 t-BuCH2CNC
Me
Et2O
2 (COOH)2 H2O
EtCHCHO
Me

92%

JACS (1969) 91 7778
(1970) 92 6675

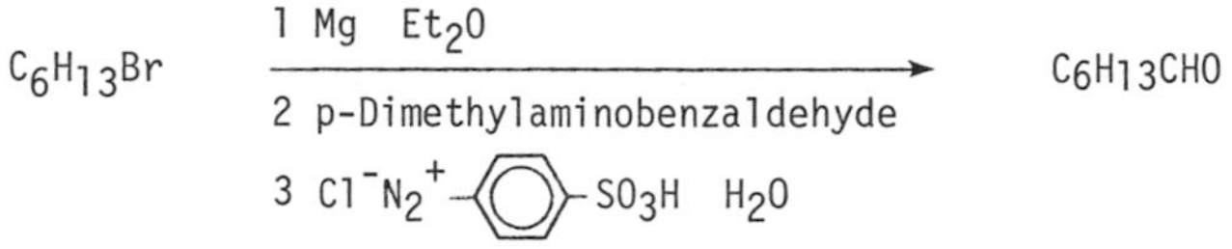
C6H13Br
1 Mg Et2O
2 p-Dimethylaminobenzaldehyde
3 Cl⁻N2⁺ SO3H H2O
C6H13CHO

75%

JOC (1960) 25 1691
(1962) 27 279

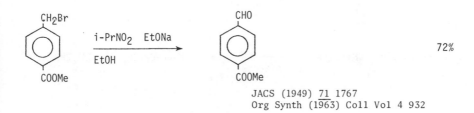

$$\text{(Benzene ring with Br and Me)} \xrightarrow[\text{2 } CS_2]{\text{1 Mg   Et}_2O} \text{(Benzene ring with CSSH and Me)} \xrightarrow[\substack{\text{Pyr} \\ \text{2 Hydrolysis}}]{\text{1 } NH_2CONHNH_2 \cdot HCl} \text{(Benzene ring with CHO and Me)} \quad 60\%$$

JOC (1941) 6 489

$$(EtO)_2CHCH_2Br \xrightarrow[\text{2 } HgCl_2]{\substack{\text{1 } \langle S \; S \rangle \quad BuLi \quad THF}} (EtO)_2CHCH_2CHO \qquad <77\%$$

Angew (1965) 77 1134
(Internat Ed 4 1075)

$$C_9H_{19}Br \xrightarrow[\text{2 HOAc}]{\text{1 } Na_2Fe(CO)_4 \quad Ph_3P \quad THF} C_9H_{19}CHO \qquad 77\%$$

JACS (1970) 92 6080

$$PhCH_2Br \xrightarrow[\text{EtONa}]{CH_2(COOEt)_2} PhCH_2CH(COOEt)_2 \xrightarrow[\substack{\text{2 } HNO_2 \\ \text{3 HCl}}]{\text{1 } N_2H_4} PhCH_2CHO$$

Org React (1946) 3 337

$$\text{(Benzene ring with } CH_2Br \text{ and } COOMe) \xrightarrow[\text{EtOH}]{i\text{-}PrNO_2 \quad EtONa} \text{(Benzene ring with } CHO \text{ and } COOMe) \qquad 72\%$$

JACS (1949) 71 1767
Org Synth (1963) Coll Vol 4 932

$$\underset{\underset{Me}{|}}{EtC}=CH(CH_2)_2\underset{\underset{Me}{|}}{C}=CHCH_2Br \xrightarrow[\text{KOH   i-PrOH   } H_2O]{\text{nitrocyclohexane}} \underset{\underset{Me}{|}}{EtC}=CH(CH_2)_2\underset{\underset{Me}{|}}{C}=CHCHO$$

Zh Obshch Khim (1959) 29 3965
(Chem Abs 54 21170)

AgNO$_3$

dioxane   H$_2$O

CH$_2$Br

KOH

dioxane   H$_2$O

CH$_2$ONO$_2$

CHO

53%

JACS (1961) $\underline{83}$ 193

C$_7$H$_{15}$CH$_2$X

$\overset{+-}{Me_3NO}$ CHCl$_3$

C$_7$H$_{15}$CHO

50% (X=I)
55% (X=OTs)

Ber (1961) $\underline{94}$ 1360

C$_6$H$_{13}$CH$_2$Br

AgOTs

MeCN

C$_6$H$_{13}$CH$_2$OTs

NaHCO$_3$

Me$_2$SO

C$_6$H$_{13}$CHO

74%

JACS (1959) $\underline{81}$ 4113

CH$_2$OTs

Br

NaHCO$_3$

Me$_2$SO

CHO

Br

65%

JACS (1959) $\underline{81}$ 4113
Chem Rev (1967) $\underline{67}$ 247

Me$_2$C=CH(CH$_2$CH$_2$C=CH)$_2$CH$_2$Br
             Me

1 Pyr

2 p-Nitroso-
dimethylaniline

Me$_2$C=CH(CH$_2$CH$_2$C=CH)$_2$CHO
             Me

Helv (1941) $\underline{24}$ 1039

C$_6$H$_{13}$CH$_2$I

1 Hexamine   CHCl$_3$

2 HOAc

C$_6$H$_{13}$CHO

41%

Org React (1954) $\underline{8}$ 197

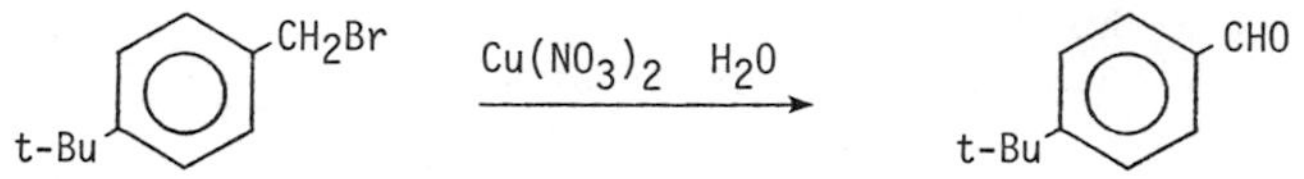

$t\text{-Bu}$—（）—CH_2Br $\xrightarrow{\text{Cu(NO}_3)_2 \ \ H_2O}$ $t\text{-Bu}$—（）—CHO

JCS (1935) 1847

$PhCH_2Cl$ $\xrightarrow[\substack{2 \ O_2 \\ 3 \ NaOH}]{1 \ Mg \ \ Et_2O}$ $PhCHO$

JACS (1955) 77 6032

$C_7H_{15}CH_2I$ $\xrightarrow[\text{butyl carbitol}]{NaN_3}$ $C_7H_{15}CH_2N_3$ $\xrightarrow[\substack{2 \ \text{Acid}}]{1 \ h\nu}$ $C_7H_{15}CHO$ 50%

Tetrahedron (1965) 21 2877

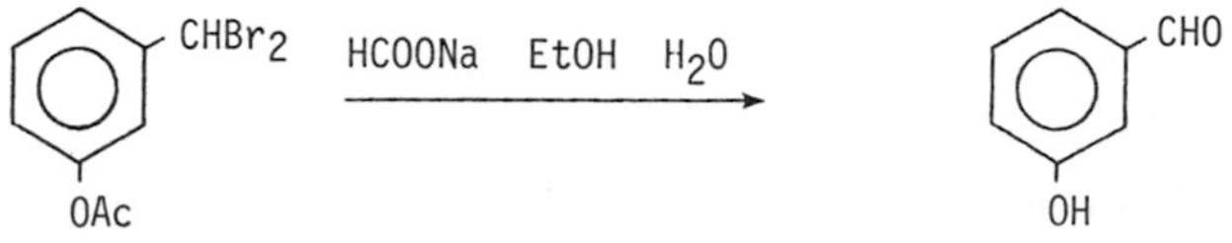

$\xrightarrow{\text{HCOONa} \ \ \text{EtOH} \ \ H_2O}$ 70%

JCS (1955) 1628

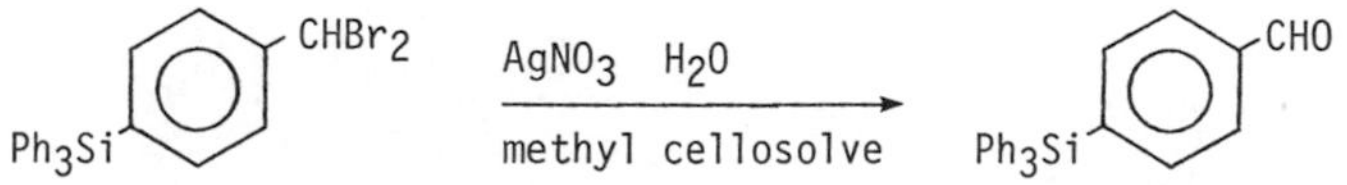

Ph_3Si—（）—$CHBr_2$ $\xrightarrow[\text{methyl cellosolve}]{AgNO_3 \ \ H_2O}$ Ph_3Si—（）—CHO 67%

JACS (1956) 78 1689

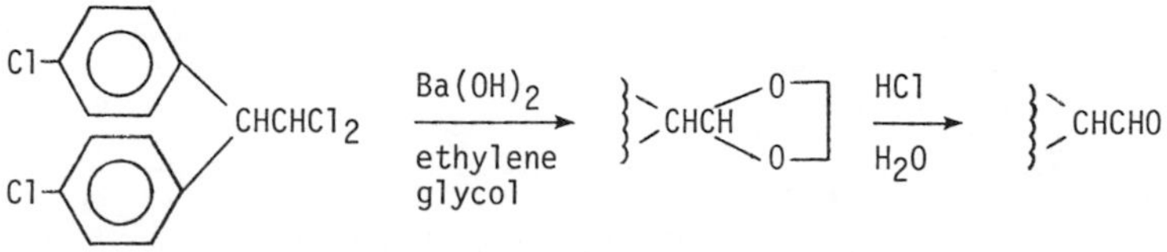

$\xrightarrow[\substack{\text{ethylene} \\ \text{glycol}}]{Ba(OH)_2}$ }CHCH\<O—O\> $\xrightarrow[H_2O]{HCl}$ }CHCHO   32%

Ber (1959) 92 900

Me_2CCH_2Br $\xrightarrow{\text{H}_2\text{O}}$ Me_2CHCHO 75%
|
Br   JACS (1933) <u>55</u> 1136

Section 56   <u>Aldehydes from Hydrides (RH)</u>

Reactions in which a hydrogen is replaced by CHO, e.g. RH $\longrightarrow$ RCHO
(R=vinyl, ethinyl or aryl), are included in this section. For reactions
in which alkyl groups are oxidized to aldehyde, e.g. $RCH_3 \longrightarrow$ RCHO, see
section 50 (Aldehydes from Alkyls)

$\xrightarrow{POCl_3 \quad HCONMe_2}$ 41%

Z Chem (1967) <u>7</u> 346
Tetrahedron (1969) <u>25</u> 4535

$\xrightarrow{POCl_3 \quad HCONMe_2}$

Ber (1964) <u>97</u> 1252
JCS <u>C</u> (1969) 913

$\xrightarrow{Ph_3P \cdot Br_2 \quad HCONMe_2}$ 45%

Annalen (1968) <u>718</u> 24

$\xrightarrow[POCl_3 \quad HCONPh]{\overset{\text{Me}}{|}}$ 71%

Org Synth (1963) Coll Vol 4 915
           (1955) Coll Vol 3 98

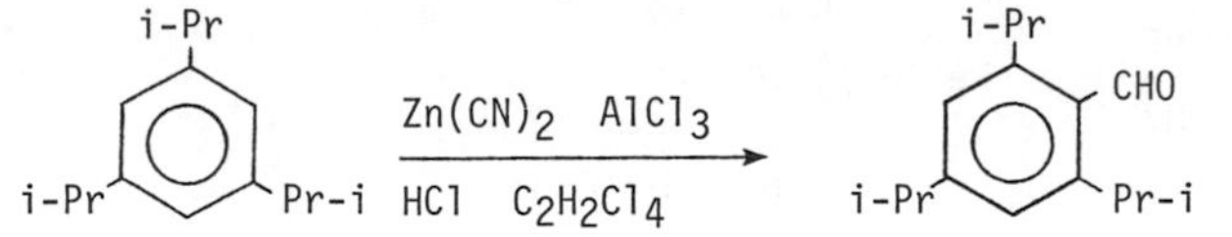

i-Pr, i-Pr, Pr-i (1,3,5-triisopropylbenzene)

$Zn(CN)_2$ $AlCl_3$
HCl $C_2H_2Cl_4$

→ i-Pr, CHO, i-Pr, Pr-i

66%

JACS (1942) 64 30

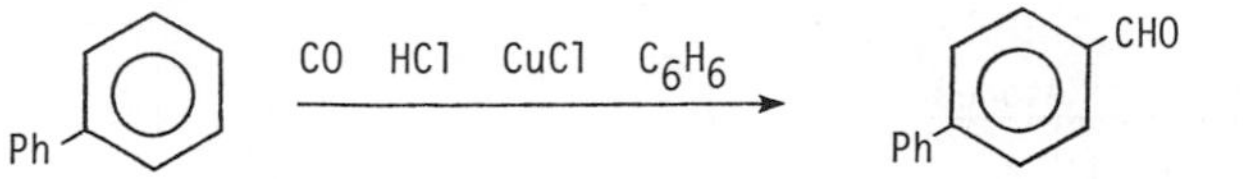

Ph (biphenyl)

CO HCl $CuCl$ C_6H_6

→ Ph, CHO

73%

Org React (1949) 5 290
Org Synth (1943) Coll Vol 2 583

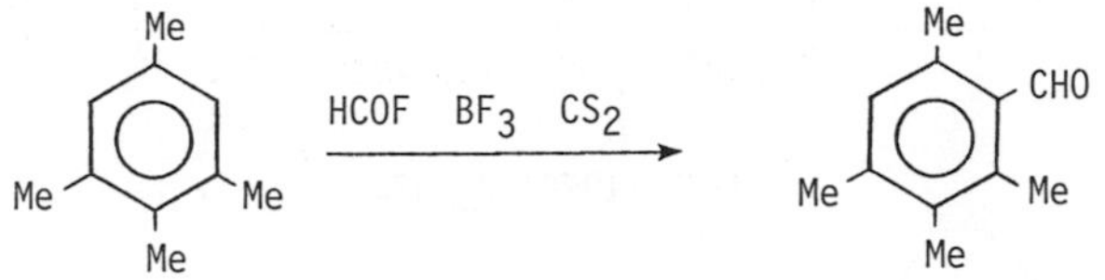

Me, Me, Me, Me (durene)

$HCOF$ BF_3 CS_2

→ Me, CHO, Me, Me, Me

72%

JACS (1960) 82 2380

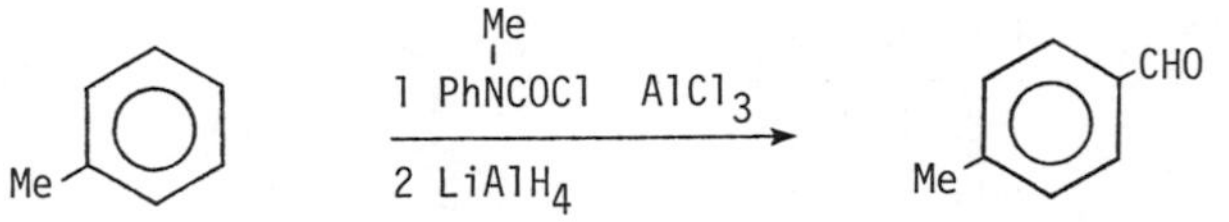

Me (toluene)

1 PhNCOCl $AlCl_3$ (Me on N)
2 $LiAlH_4$

→ Me, CHO

Ber (1955) 88 301

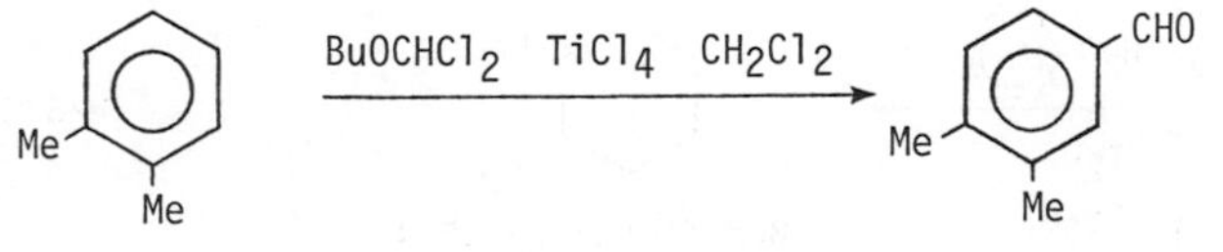

Me, Me (o-xylene)

$BuOCHCl_2$ $TiCl_4$ CH_2Cl_2

→ Me, Me, CHO

68%

Ber (1960) 93 88

OMe, OMe (1,3-dimethoxybenzene)

1 PhLi Et_2O
2 HCONPh (Me on N)

→ OMe, CHO, OMe

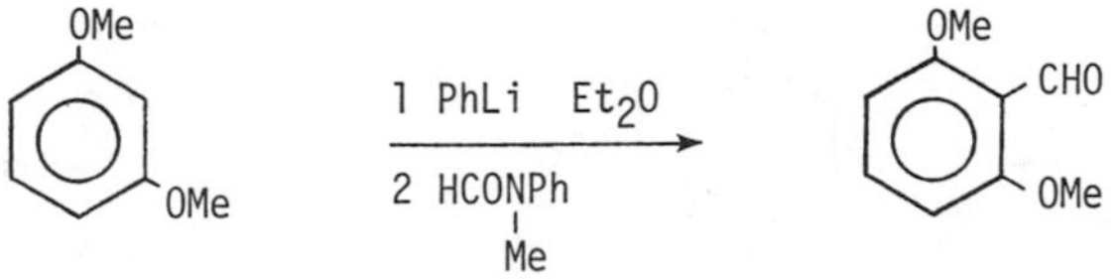

Org React (1954) 8 258

$$Me(CH_2)_3C\equiv CH \xrightarrow[\text{2 HCONMe}_2]{\text{1 EtMgBr}} Me(CH_2)_3C\equiv CCHO$$

JCS (1958) 1054

Section 57   Aldehydes from Ketones

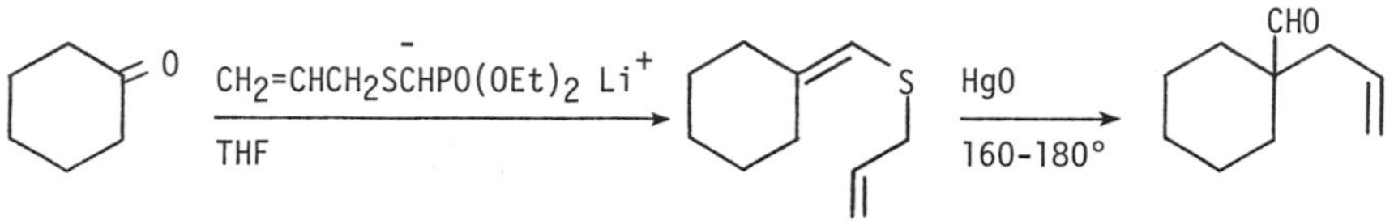

JACS (1970) 92 5522

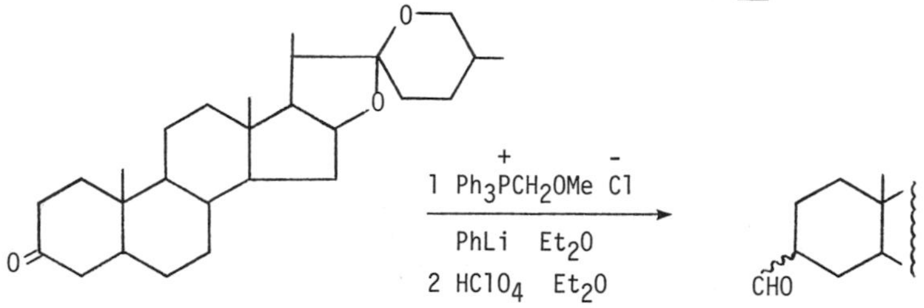

~85%

JACS (1958) 80 6150
Org React (1965) 14 270

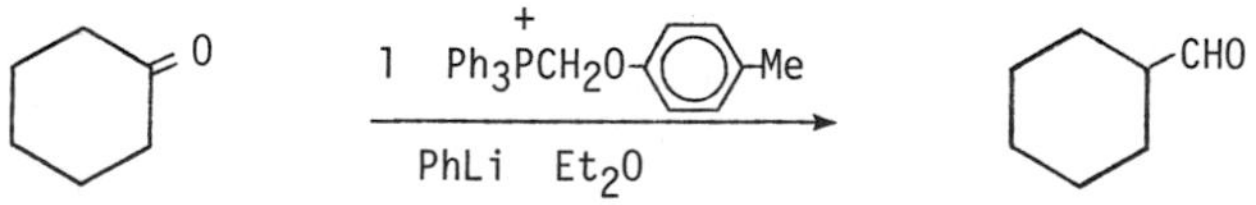

~82%

Ber (1962) 95 2514

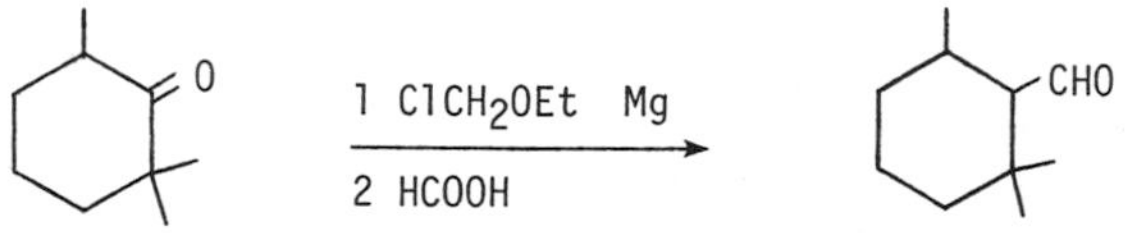

51%

JACS (1968) 90 3282
Compt Rend (1963) 256 1996

PhCOMe $\xrightarrow[\text{NaNH}_2]{\text{ClCH}_2\text{COOEt}}$ PhC−CHCOOEt (epoxide), Me $\xrightarrow[\text{2 } 100°]{\text{1 NaOH}}$ PhCHCHO, Me   42%

Org React (1949) 5 413
JOC (1970) 35 1600
Chem Rev (1955) 55 283

$\xrightarrow[\text{t-BuOK}]{\text{ClCH}_2\text{COOBu-t}}$ (epoxide)CHCOOBu-t $\xrightarrow{350-360°}$ (cyclohexane)CHO   39%

JACS (1963) 85 955

Et$_2$CO $\xrightarrow[\text{C}_6\text{H}_6]{\text{BrCH}_2\text{COOEt} \quad \text{Zn}}$ Et$_2$CCH$_2$COOEt, OH $\xrightarrow[\begin{array}{l}\text{2 NaNO}_2 \quad \text{HCl}\\ \text{3 KOH}\end{array}]{\text{1 N}_2\text{H}_4}$ Et$_2$CHCHO   79%

JACS (1951) 73 4199

$\xrightarrow[\text{Me}_2\text{SO}]{\text{Me}_3\text{SO}^+ \text{I}^- \quad \text{NaH}}$ (epoxide)CH$_2$ $\xrightarrow{\text{BF}_3\cdot\text{Et}_2\text{O}}$ (cycloheptane)CHO   65%

JACS (1962) 84 867

(CH$_2$)$_2$COMe $\xrightarrow[\text{2 HOAc}]{\text{1 LiCH}_2\text{CH=N}\langle\text{piperidine}\rangle}$ (CH$_2$)$_2$C=CHCHO (Me) $\xrightarrow[\begin{array}{l}\text{SeO}_2\\ \text{2 Al}_2\text{O}_3\end{array}]{\text{1 H}_2\text{O}_2}$ (CH$_2$)$_2$CHCHO (Me)

Tetr Lett (1970) 381

COMe $\xrightarrow[\text{Pyr} \quad \text{H}_2\text{O}]{\text{KMnO}_4 \quad \text{KOH}}$ COCOOH $\xrightarrow[\text{toluidine}]{\text{N,N-Dimethyl-}}$ CHO

JACS (1965) 87 1247

$$PhCOCH_2Me \xrightarrow{\quad BuN_3 \quad H_2SO_4 \quad toluene \quad} PhCHO \quad + \quad MeCHO$$

62%   50%

JACS (1959) <u>81</u> 3369

Some of the methods included in section 49 (Aldehydes from Aldehydes) may also be applied to the preparation of aldehydes from ketones

Section 58   Aldehydes from Nitriles

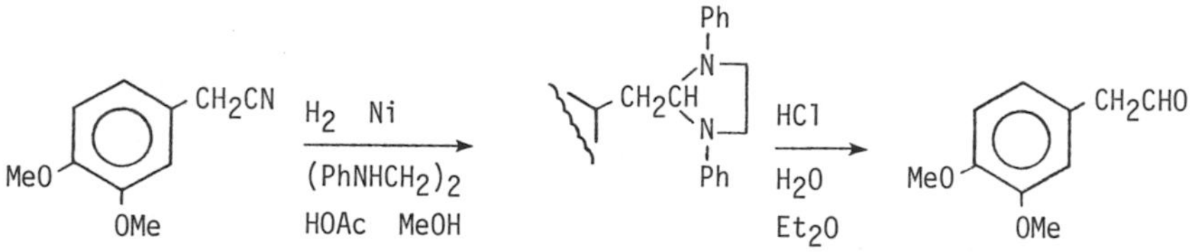

Ber (1957) <u>90</u> 617

$$PhCH_2CN \xrightarrow[\text{NaOAc \quad MeOH \quad H_2O}]{\text{1 H}_2 \quad Ni \quad NH_2NHCONH_2 \cdot HCl} PhCH_2CH=NNHCONH_2 \dashrightarrow PhCH_2CHO$$

Angew (1955) <u>67</u> 156
JOC (1962) <u>27</u> 850

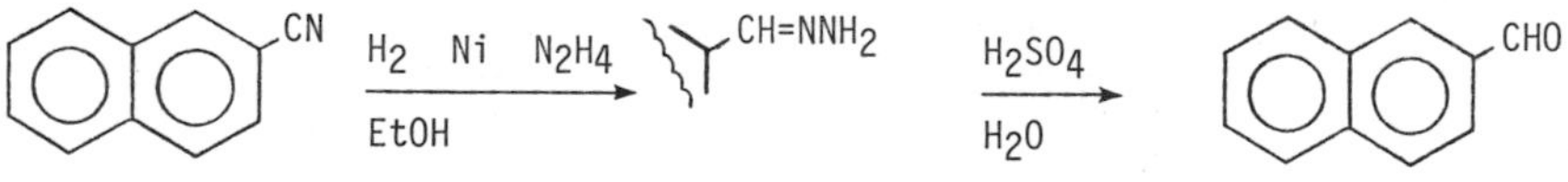

Gazz (1955) <u>85</u> 1705
(Chem Abs <u>50</u> 10037)

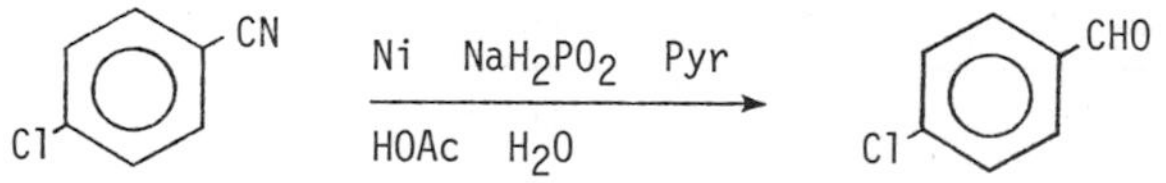

80%

JCS (1962) 3961

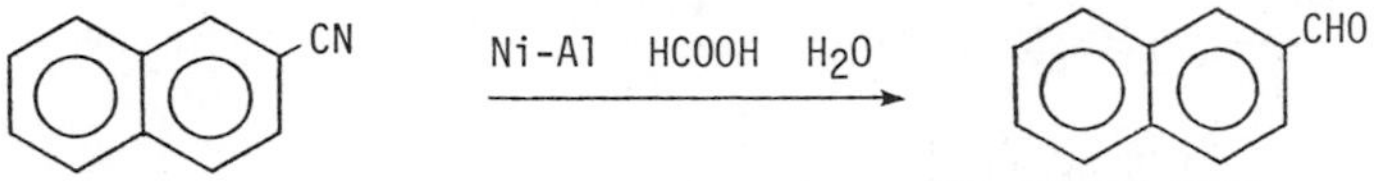

$$\text{Ni-Al} \quad \text{HCOOH} \quad H_2O$$

95%

JCS (1965) 5775
(1964) 5880

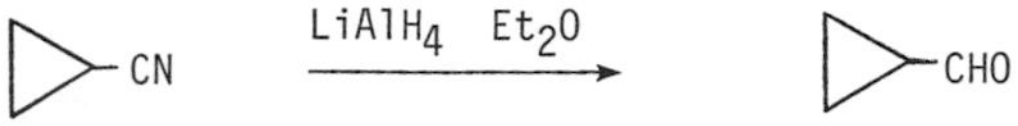

$$\text{LiAlH}_4 \quad Et_2O$$

48%

JACS (1951) 73 4047
Org React (1954) 8 252

i-PrCN

$$\text{LiAl(OEt)}_3\text{H} \quad Et_2O$$

i-PrCHO   81%

JACS (1964) 86 1085

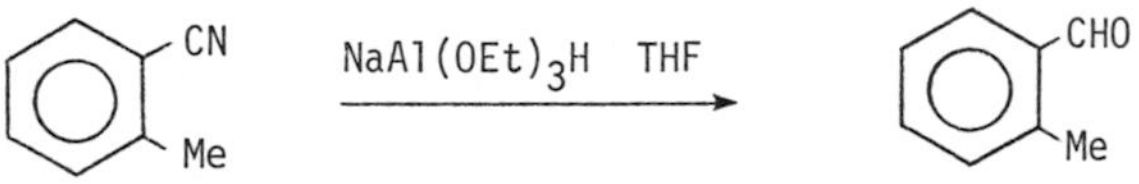

$$\text{NaAl(OEt)}_3\text{H} \quad \text{THF}$$

88%

Annalen (1957) 607 24

PhCN

$$\text{(i-Bu)}_2\text{AlH} \quad C_6H_6$$

PhCHO   90%

JOC (1959) 24 627

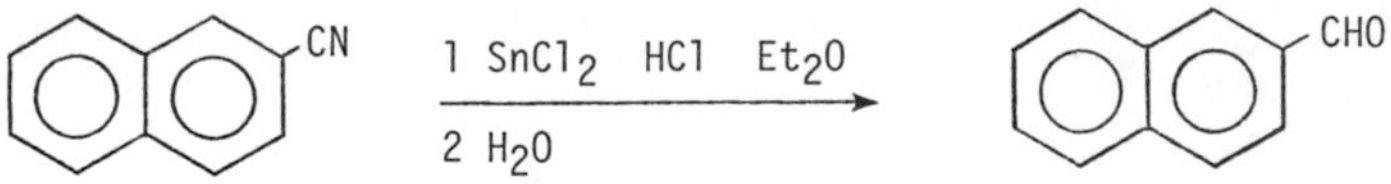

$$1 \quad \text{SnCl}_2 \quad \text{HCl} \quad Et_2O$$
$$2 \quad H_2O$$

73-80%

Org Synth (1955) Coll Vol 3 626

$C_{15}H_{31}CN$

$$1 \quad \text{SnCl}_2 \quad \text{HCl} \quad Et_2O$$
$$2 \quad H_2O$$

$C_{15}H_{31}CHO$

JCS (1925) 127 1874

$$MeSCH_2CH_2CN \xrightarrow{\text{MeOH \quad HCl}} MeSCH_2CH_2C(OMe)_3 \xrightarrow[\substack{C_6H_6 \\ \text{2 Acid}}]{\text{1 LiAlH}_4 \quad Et_2O} MeSCH_2CH_2CHO$$

JACS (1951) 73 5005

$$PhCH_2CN \dashrightarrow PhCH_2 \xrightarrow[\text{2 Base}]{\text{1 NaBH}_4 \quad THF \quad EtOH} PhCH_2CHO$$

Chem Comm (1967) 1163

$$PhCN \xrightarrow[\text{Me}]{\substack{Me \\ BrCCOOEt \\ Me}} \text{Zn} \quad C_6H_6 \quad \underset{\substack{| \\ Me}}{PhCOCCOOEt} \xrightarrow[\text{2 260°}]{\text{1 H}_2 \quad Ni \quad EtOH} PhCHO$$

Compt Rend (1953) 236 826

$$PhCN \xdashrightarrow{\text{H}_2S \quad NH_3} PhCSNH_2 \xrightarrow{\text{Ni \quad EtOH}} PhCHO$$

JACS (1952) 74 3936

Section 59   Aldehydes from Olefins

$$EtCH=CH_2 \xrightarrow[\substack{\text{2 MeCH=CHCHO} \\ \text{air \quad i-PrOH}}]{\text{1 B}_2H_6 \quad THF} \underset{Me}{Et(CH_2)_2CHCH_2CHO} \qquad 90\%$$

JACS (1970) 92 714

$$\begin{array}{c} CH_2 \\ \| \\ CH \\ | \\ (CH_2)_8 \\ | \\ CN \end{array} \xrightarrow[\substack{\text{2 CO \quad LiAl(OBu-t)}_3H \\ \text{3 H}_2O_2 \quad NaH_2PO_4 \quad K_2HPO_4 \quad H_2O}]{1 \qquad THF} \begin{array}{c} CHO \\ | \\ (CH_2)_2 \\ | \\ (CH_2)_8 \\ | \\ CN \end{array}$$

JACS (1969) 91 4606

$Me_2C=CH_2$ $\xrightarrow[\text{2 CO LiAl(OMe)}_3\text{H}]{\text{1 B}_2\text{H}_6 \quad \text{THF}}$ Me_2CHCH_2CHO 91%

JACS (1968) <u>90</u> 499 818

$C_6H_{13}CH=CH_2$ $\xrightarrow[\text{(PhO)}_3\text{P} \quad \text{toluene}]{\text{CO} \quad \text{H}_2 \quad \text{(100 psi)} \quad \text{Rh-C}}$ $C_6H_{13}(CH_2)_2CHO$ 89%

JOC (1969) <u>34</u> 327
Chem Comm (1967) 305
                (1965) 17

$\underset{\underset{Me}{|}}{PrCHCH=CH_2}$ $\xrightarrow[\text{Co}_2\text{(CO)}_8 \quad \text{C}_6\text{H}_6]{\text{CO} \quad \text{H}_2 \quad \text{(200 atmos)}}$ $\underset{\underset{Me}{|}}{PrCH(CH_2)_2CHO}$ ~62%

JACS (1968) <u>90</u> 6847

45%
Tetr Lett (1969) 43
JOC (1968) <u>33</u> 805

$C_6H_{13}CH=CH_2$ $C_6H_{13}(CH_2)_2CHO$ ~20%

Tetr Lett (1968) 4339

86%

Acc Chem Res (1970) <u>3</u> 338
Tetr Lett (1970) 5275

$(\text{t-BuCH}_2)_2\text{C=CH}_2$ $\xrightarrow[\text{2 Zn}]{\text{1 CrOCl}_2 \quad \text{CH}_2\text{Cl}_2}$ $(\text{t-BuCH}_2)_2\text{CHCHO}$ 80%

JOC (1968) <u>33</u> 3970

1 Hg(OAc)$_2$ Me$_2$CO

2 NaBr

3 Br$_2$ CCl$_4$

i-PrMgBr

Et$_2$O C$_6$H$_6$

JOC (1968) <u>33</u> 3953

$\text{PhCH}_2\text{CH=CH}_2$ $\xrightarrow{\text{DDQ} \quad \text{C}_6\text{H}_6}$ PhCH=CHCHO 50%

Chem Comm (1969) 773

$\text{C}_8\text{H}_{17}\text{CH=CH(CH}_2)_7\text{COOMe}$ $\xrightarrow[\substack{\text{2 H}_2 \ \text{Pd-C} \ \text{MeOH} \ \text{EtOAc} \\ \text{(or Zn HOAc)}}]{\text{1 O}_3 \quad \text{MeOH}}$ $\text{C}_8\text{H}_{17}\text{CHO}$ 78%

	JOC (1960) <u>25</u> 618
	JACS (1953) <u>75</u> 4952
For reduction of ozonides with tetracyanoethylene see	Ber (1963) <u>96</u> 1564
" " " " " (MeO)$_3$P	" JOC (1960) <u>25</u> 1031
" " " " " (Me$_2$N)$_3$P	" Helv (1967) <u>50</u> 2387
" " " " " Ph$_3$P	" JOC (1968) <u>33</u> 787
" " " " " Me$_2$S	" Helv (1967) <u>50</u> 2445
" " " " " dinitrophenylhydrazine	see JACS (1956) <u>78</u> 5601
" " " " " Na$_2$SO$_3$	see JOC (1961) <u>26</u> 4912
" selective ozonolysis	" JACS (1958) <u>80</u> 915
	(1969) <u>91</u> 4318
	(1958) <u>80</u> 915

PhCH=CH_2 $\xrightarrow[\text{CCl}_4]{\text{Bistriphenylsilyl chromate}}$ PhCHO

JOC (1970) <u>35</u> 774

CH=CHMe → Na₂Cr₂O₇ / Sulfanilic acid H₂O → CHO

$$\text{CH=CHMe} \xrightarrow[\text{Sulfanilic acid \quad H}_2\text{O}]{\text{Na}_2\text{Cr}_2\text{O}_7} \text{CHO}$$

86%

J Soc Chem Ind (1943) 62 90
JCS (1937) 369
Org React (1959) 10 1

$$\text{C}_6\text{H}_{13}\text{CH=CH}_2 \xrightarrow{\text{RuO}_4 \quad \text{CCl}_4} \text{C}_6\text{H}_{13}\text{CHO}$$

12%

JACS (1958) 80 6682

$$\xrightarrow[\substack{\text{2 Pb(OAc)}_4 \\ \text{Me}_2\text{CO}}]{\text{1 OsO}_4 \quad \text{Pyr}}$$

43%

JACS (1968) 90 3245

$$\text{C}_{10}\text{H}_{21}\text{CH=CH}_2 \xrightarrow[\text{dioxane}]{\text{OsO}_4 \quad \text{NaIO}_4} \text{C}_{10}\text{H}_{21}\text{CHO}$$

68%

JOC (1956) 21 478
Tetrahedron (1968) 24 3095

$$\begin{array}{c}\text{CH}_2 \\ \| \\ \text{CH} \\ | \\ \text{(CH}_2)_8 \\ | \\ \text{COOMe}\end{array} \xrightarrow[\text{2 MeONa \quad MeOH}]{\text{1 H}_2\text{O}_2 \quad \text{HCOOH}} \begin{array}{c}\text{CH}_2\text{OH} \\ | \\ \text{CHOH} \\ | \\ \text{(CH}_2)_8 \\ | \\ \text{COOMe}\end{array} \xrightarrow[]{\text{Pb(OAc)}_4 \quad \text{C}_6\text{H}_6} \begin{array}{c}\text{CHO} \\ | \\ \text{(CH}_2)_8 \\ | \\ \text{COOMe}\end{array}$$

51%

J Med Chem (1969) 12 911

$$\text{CH=CHMe} \xrightarrow[\text{H}_2\text{O}]{\text{Electrolysis \quad Na}_2\text{SO}_4} \text{CHO}$$

52%

Helv (1925) 8 332

Can J Chem (1968) <u>46</u> 75

Section 60   Aldehydes from Miscellaneous Compounds

$$PrCH=CCHO \atop \quad\ \ |\atop \qquad Et \xrightarrow[\text{2 HCl } H_2O]{1\ H_2\ \ Ni\ \ EtCHCH_2OH \atop \qquad\quad |\ NH_2}}$$

$$PrCH_2CHCHO \atop \qquad\quad | \atop \qquad\qquad Et$$

JACS (1955) <u>77</u> 359

$$MeCH=CHCHO \xrightarrow{H_2\ \ (Ph_3P)_3RhCl\ \ C_6H_6} Me(CH_2)_2CHO$$

JCS <u>C</u> (1967) 270

Chem Comm (1968) 97

$$Me_2C=CH(CH_2)_2C=CHCHO \atop \qquad\qquad\qquad\quad |\atop \qquad\qquad\qquad\quad Me \xrightarrow[\text{2 Hydrolysis}]{1\ PhCOO_2H} Me_2C-CH(CH_2)_2CHCHO \atop \qquad\qquad\qquad\qquad |\atop \qquad\qquad\qquad\qquad Me$$

Org React (1957) <u>9</u> 73
Tetr Lett (1970) 381

$$PhCH=CHCN \xrightarrow[\text{MeOH } H_2O]{Co_2(CO)_8\ \ HCl} Ph(CH_2)_2CHO \qquad\qquad 50\%$$

Chem Comm (1967) 1140

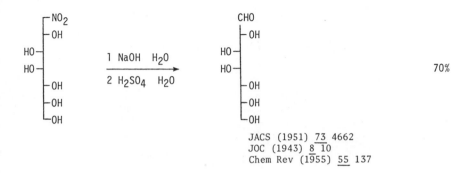

$PhCH=CHCONH_2$ $\xrightarrow{\begin{array}{c} 1\ NaOCl\quad MeOH \\ \hline 2\ Acid\quad H_2O \end{array}}$ $PhCH_2CHO$ ~70%

Org React (1946) <u>3</u> 267

1 NaOH H_2O
—————————
2 H_2SO_4 H_2O

70%

JACS (1951) <u>73</u> 4662
JOC (1943) <u>8</u> 10
Chem Rev (1955) <u>55</u> 137

$PrCH_2NO_2$ $\xrightarrow{\begin{array}{c} KMnO_4\quad KOH \\ \hline MgSO_4\quad H_2O \end{array}}$ $PrCHO$ 97%

JOC (1962) <u>27</u> 3699

$C_{10}H_{21}\underset{\underset{OH}{|}}{C}HCOOH$ $\xrightarrow{190-200°}$ $C_{10}H_{21}CHO$ 96%

J Soc Chem Ind (1943) <u>62</u> 128

$Ph\underset{\underset{OEt}{|}}{C}HCOOH$ $\xrightarrow{Electrolysis\quad MeOH}$ $Ph\underset{\underset{OEt}{|}}{C}HOMe$ $\xrightarrow{HCl\quad H_2O}$ $PhCHO$

71%

JOC (1962) <u>27</u> 281

Section 60A   Protection of Aldehydes

$$i\text{-PrCHO} \xrightarrow[\text{Acid}]{\text{HC(OEt)}_3 \quad NH_4Cl \quad EtOH} i\text{-PrCH(OEt)}_2$$

(Stable to RMgX, LiAlH$_4$, CrO$_3$
and base)

Bull Soc Chim Fr (1965) 1007

$$\xrightarrow[\text{PhSO}_2\text{NHOH} \quad H_2O \quad \text{dioxane}]{\text{PhSO}_2\text{NHOH} \quad MeOH}$$

JOC (1970) 35 1962

$$\xrightarrow[\text{Acid}]{\text{Me}_2\text{SO}_4 \quad NaOH \quad MeOH \quad H_2O}$$

Ber (1958) 91 410

$$\text{Me}_2\text{C=CH(CH}_2)_2\underset{\underset{\text{Me}}{|}}{\text{CHCH}_2\text{CHO}} \xrightarrow[\text{Acid}]{\text{HOCH}_2\text{CH}_2\text{OH} \quad \text{TsOH} \quad C_6H_6} \text{Me}_2\text{C=CH(CH}_2)_2\underset{\underset{\text{Me}}{|}}{\text{CHCH}_2\text{CH}}\begin{smallmatrix}O\\ \\O\end{smallmatrix}$$

(Stable to RMgX, LiAlH$_4$,
CrO$_3$ and base)

Tetrahedron (1959) 6 217

$$\text{PhOCH}_2\text{CH}_2\text{CHO} \xrightarrow[\text{Acid}]{\text{2-Methyl-2,4-pentanediol} \quad \text{TsOH}} \text{PhOCH}_2\text{CH}_2\text{CH}$$

JOC (1963) 28 594

$C_6H_{13}CHO$

$$\xrightarrow[\text{HgCl}_2 \quad \text{dioxane} \quad \text{EtOH}]{\substack{\text{HSCH}_2 \diagdown \diagup \text{Me} \\ \text{HSCH}_2 \diagup \diagdown \text{Me} \quad \text{HCl} \quad \text{dioxane}}}$$

$C_6H_{13}CH \diagup^{SCH_2} \diagdown_{SCH_2}$ Me / Me

(Stable to $LiAlH_4$ and base)

JCS C (1966) 1005

$$\begin{array}{c} Me \\ | \\ PhCCHO \\ | \\ OH \end{array} \xrightleftharpoons[\text{I}_2 \quad \text{NaHCO}_3 \quad \text{dioxane} \quad \text{H}_2\text{O}]{\text{MeSH}} \begin{array}{c} Me \\ | \\ PhCCH(SMe)_2 \\ | \\ OH \end{array}$$

JOC (1969) 34 3618

For cleavage of thioketals with $HgCl_2$ see Ber (1958) 91 1043

$$PrCHO \xrightleftharpoons[\text{HCl} \quad \text{H}_2\text{O}]{(PhNHCH_2)_2} PrCH \diagup^{N-}_{N-} \diagdown Ph / Ph$$

Ber (1953) 86 1463

$$\xrightleftharpoons[\text{Acid}]{\text{Piperidine} \quad \text{TsOH} \quad C_6H_6}$$

(Stable to $LiAlH_4$ and base)

JACS (1955) 77 1216

$$Me_2CHCHO \xrightleftharpoons[\text{Acid}]{\substack{1 \quad HC(OEt)_3 \quad NH_4Cl \quad EtOH \\ 2 \quad H_3PO_4}} Me_2C=CHOEt$$

(Stable to RMgX, $LiAlH_4$, CrO_3 and base)

Bull Soc Chim Fr (1965) 1007

$$Me_2CHCHO \xleftarrow{} \xrightarrow[\text{2 PhNMe}_2]{\text{1 MeOH   HCl}} Me_2C=CHOMe$$

Acid

Ber (1956) **89** 1468

$$\xrightarrow[\text{H}_2\text{SO}_4 \quad \text{H}_2\text{O} \quad \text{pet ether}]{\text{NH}_2\text{CONHNH}_2 \cdot \text{HOAc   MeOH   H}_2\text{O}}$$

JCS (1950) 3361

For cleavage of semicarbazones with HNO_2 see JACS (1934) **56** 1794

$$\begin{array}{c} CHO \\ | \\ (CHOAc)_4 \\ | \\ CH_2OAc \end{array} \xrightarrow[\text{NaNO}_2 \quad \text{HCl} \quad \text{EtOH}]{\text{NH}_2\text{OH}} \begin{array}{c} CH=NOH \\ | \\ (CHOAc)_4 \\ | \\ CH_2OAc \end{array}$$

JACS (1934) **56** 1794

For cleavage of oximes with $(NH_4)_2Ce(NO_3)_6$ see Can J Chem (1969) **47** 145

"   "   "   "   " $NaHSO_3$ " JOC (1966) **31** 3446

"   "   "   "   " $Fe(CO)_5$ " JOC (1967) **32** 2938

$$C_7H_{15}CHO \xrightarrow[\text{KHCO}_3 \quad \text{ethylene glycol} \quad \text{H}_2\text{O}]{\text{2,4-Dinitrophenylhydrazine   HCl}} C_7H_{15}CH=NNH$$

Aust J Chem (1968) **21** 271

$$RCH=CHCHO \xrightarrow[\text{H}_2\text{SO}_4 \quad \text{EtOH} \quad \text{H}_2\text{O}]{\text{Ac}_2\text{O} \quad \text{H}_3\text{PO}_4} RCH=CHCH(OAc)_2$$

JCS (1955) 1384

Some of the methods included in section 180A (Protection of Ketones) may also be applied to the protection of aldehydes

Chapter 5   PREPARATION
OF
ALKYLS
METHYLENES
AND ARYLS

This chapter lists the conversion of functional groups into Me, Et···, CH_2, Ph etc.

Section 61   Alkyls and Methylenes from Acetylenes

$BuC{\equiv}CH$ $\xrightarrow{\begin{array}{l}1\ (i\text{-Bu})_2AlH\quad THF\\[4pt]2\ MeONa\quad THF\\3\ MeI\\4\ Hydrolysis\end{array}}$ $Bu(CH_2)_2Me$ 74%

Tetr Lett (1966) 6021

$\left(\begin{array}{c}C{\equiv}C\\(CH_2)_8\end{array}\right)$ $\xrightarrow{\ H_2\quad PtO_2\quad HOAc\ }$ $\left(\begin{array}{c}CH_2CH_2\\(CH_2)_8\end{array}\right)$ 71%

JACS (1952) <u>74</u> 3636

For hydrogenation with Ni   catalyst see JCS (1952) 5032

  "   "   " Ru   "   " JOC (1959) <u>24</u> 708

  "   "   " Re_2S_7 "   " JACS (1954) <u>76</u> 1519

and JACS (1959) <u>81</u> 3587

BuC≡CH $\xrightarrow[\text{Ph}_3\text{P} \quad \text{MeOH}]{\text{H}_2 \quad \text{Rh}_2(\text{COOMe})_4}$ BuCH$_2$Me

Chem Comm (1969) 825
                (1965) 131

C$_5$H$_{11}$C≡CH $\xrightarrow{\text{Li} \quad \text{NH}_2\text{CH}_2\text{CH}_2\text{NH}_2}$ C$_5$H$_{11}$CH$_2$Me

JOC (1957) <u>22</u> 891

EtC≡CEt $\xrightarrow{\text{Na} \quad \text{HMPA} \quad \text{t-BuOH}}$ EtCH$_2$CH$_2$Et   79%

JOC (1970) <u>35</u> 3565

PhC≡CPh $\xrightarrow{\text{N}_2\text{H}_4 \quad \text{air} \quad \text{Cu}^{2+} \quad \text{MeOH}}$ PhCH$_2$CH$_2$Ph   80%

Tetr Lett (1961) 347

Section 62   Alkyls from Carboxylic Acids

Reactions in which carboxyl groups are converted into alkyl e.g.
RCOOH ⟶ RCH$_3$, are included in this section. For the conversion
RCOOH ⟶ RH see section 152 (Hydrides from Carboxylic Acids)

Chem Comm (1967) 96
Advances in Org Chem (1960) <u>1</u> 1

COOH
|
$(CH_2)_4$ $\xrightarrow{\text{Electrolysis   MeCOOH   MeOH}}$ Me
| |
COOMe $(CH_2)_4$ 40%
                                                                      |
                                                                      COOMe

JCS (1950) 3326

$C_{17}H_{35}COOH$ - - - - → $C_{17}H_{35}CONHNHTs$ $\xrightarrow{\text{LiAlH}_4}$ $C_{17}H_{35}Me$ 50-60%

Tetrahedron (1966) 22 487
Chim Ind (Milan) (1965) 47 62
(Chem Abs 62 10388)

Helv (1954) 37 1689
     (1946) 29 360

1 SiHCl$_3$ MeCN

2 Pr$_3$N

3 KOH MeOH H$_2$O

82%

JACS (1970) 92 3232

$C_{17}H_{35}COOH$ $\xrightarrow{\text{HI   P}_4\ \ 200\text{-}250°}$ $C_{17}H_{35}Me$

Ber (1882) 15 1687

RCOOH $\xrightarrow{\text{H}_2\ (120\ \text{atmos})\ \ MoS_2\ \ 350°}$ RMe

Chem Listy (1956) 50 569
(Chem Abs 50 13854)

Section 63   Alkyls and Aryls from Alcohols

Reactions in which hydroxyl groups are replaced by alkyl or aryl e.g.
ROH ⟶ RPh, are included in this section. For the conversion ROH ⟶ RH
see section 153 (Hydrides from Alcohols and Phenols)

$$\text{PhCH}_2\text{OH} \quad \xrightarrow[\text{2 TiCl}_4 \quad K \quad C_6H_6]{\text{1 NaH}} \quad \text{PhCH}_2\text{CH}_2\text{Ph} \qquad\qquad 51\%$$

JACS (1965) <u>87</u> 3277

JACS (1968) <u>90</u> 3284
Chem Comm (1969) 53

$$\text{PrOH} \quad \xrightarrow{\text{AlCl}_3 \quad C_6H_6 \quad H_2O} \quad \text{PrPh}$$

JOC (1940) <u>5</u> 253
JACS (1942) <u>64</u> 1576

63%

Org React (1946) <u>3</u> 1

Section 64   Alkyls from Aldehydes

Reactions which convert the group CHO into CH_3 are included in this
section. For the conversion RCHO $\longrightarrow$ RH see section 154 (Hydrides from
Aldehydes)

$C_6H_{13}CHO$ $\xrightarrow{\quad H_2 \quad WS_2 \quad 320\text{-}330° \quad}$ $C_6H_{13}Me$ 38%

                            Chem Listy (1956) 50 565
                            (Chem Abs 50 13831)

$\xrightarrow[\text{$H_2O$ \quad HOAc}]{\text{H_2 \quad $PdCl_2$}}$

JCS (1954) 4676
JACS (1956) 78 3769

  86%

$\xrightarrow{\text{$NaBH_4$ \quad Pd-C \quad i-PrOH}}$

Can J Chem (1964) 42 514

  89%

$\xrightarrow{\text{$LiAlH_4$ \quad $AlCl_3$ \quad Et_2O}}$

JCS (1957) 3755
Tetr Lett (1967) 1849

  78%

$\xrightarrow[\text{$H_2SO_4$ \quad dioxane}]{\text{Electrolysis}}$

JACS (1967) 89 4789

  90%

$C_6H_{13}CHO$ $\xrightarrow{\quad Zn \quad HCl \quad H_2O \quad}$ $C_6H_{13}Me$ 72%

                           Org React (1942) 1 155

$$\text{1 } N_2H_4 \cdot HCl \quad N_2H_4$$

ethylene glycol

2 KOH

48%

Tetrahedron (1969) 25 1335
Org React (1948) 4 378

$$C_{11}H_{23}CHO \xrightarrow[\text{2 NaBH}_4 \text{ MeOH}]{\text{1 TsNHNH}_2} C_{11}H_{23}Me$$

Chem Ind (1964) 153 1689
Tetrahedron (1963) 19 1127

$$\xrightarrow[\text{2 } I_2 \text{ Et}_3\text{N THF}]{\text{1 } N_2H_4 \text{ Et}_3\text{N}} \quad CH_2I_2 \quad \xrightarrow[\text{HOAc}]{Zn}$$

JCS (1962) 470

$$C_5H_{11}CHO \xrightarrow{\text{EtSH ZnCl}_2 \text{ Na}_2\text{SO}_4} C_5H_{11}CH(SEt)_2 \xrightarrow{\text{Ni EtOH}} C_5H_{11}Me \quad 40\%$$

JACS (1944) 66 909
JOC (1968) 33 3551
Org React (1962) 12 356

Further examples of the conversion C=O $\longrightarrow$ CH$_2$ are included in section 72 (Alkyls and Methylenes from Ketones)

Section 65   Alkyls and Aryls from Alkyls and Aryls

The conversion RR' $\longrightarrow$ ArR (R,R'=alkyl), the epimerization, isomerization and transposition of alkyl groups and the hydrogenation of aryl groups are included in this section. For the conversions RR' $\longrightarrow$ RH and ArR $\longrightarrow$ ArH (R,R'=alkyl) see section 155 (Hydrides from Alkyls)

$$t\text{-BuCH}_2\text{Pr-i} \xrightarrow{\text{PhH AlCl}_3} t\text{-BuPh}$$

JACS (1935) 57 2415

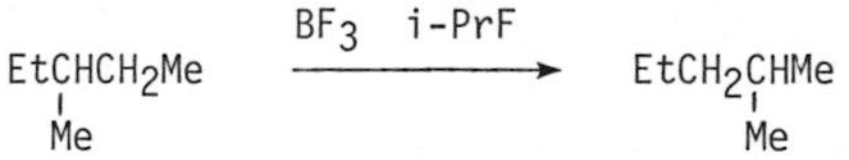
BF3 i-PrF
EtCHCH2Me
Me
EtCH2CHMe
Me

(+ higher mol wt compds)

JACS (1951) <u>73</u> 5013
JOC (1964) <u>29</u> 94

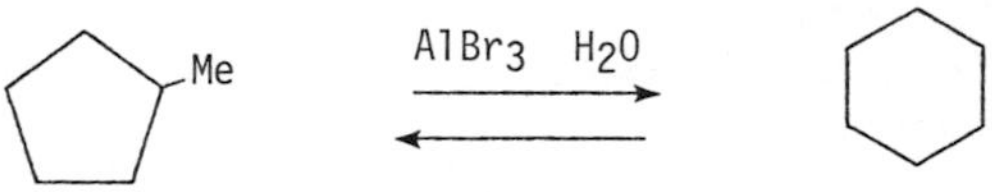
Me
AlBr3 H2O

JACS (1948) <u>70</u> 2773

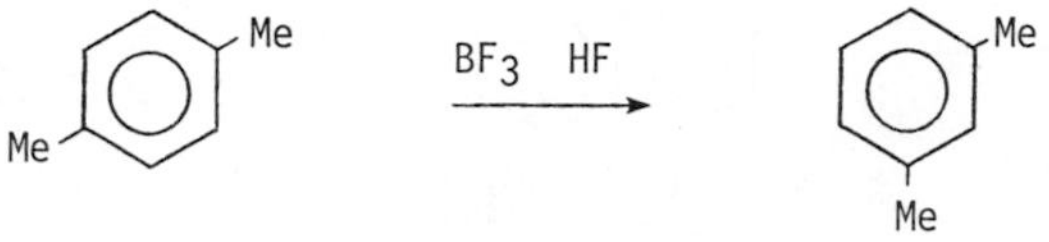
Me
Me
BF3 HF
Me
Me

100%

JACS (1952) <u>74</u> 6246
JCS (1952) <u>100</u>

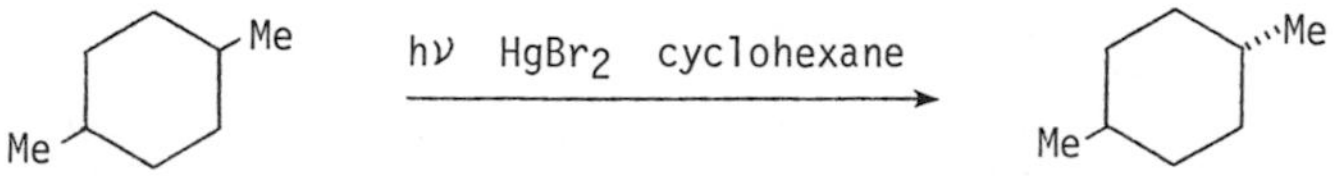
Me
Me
hν HgBr2 cyclohexane
Me
Me

JACS (1970) <u>92</u> 1094

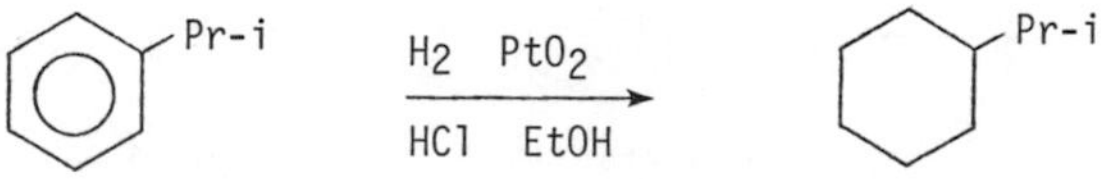
Pr-i
H2 PtO2
HCl EtOH
Pr-i

JACS (1936) <u>58</u> 1594

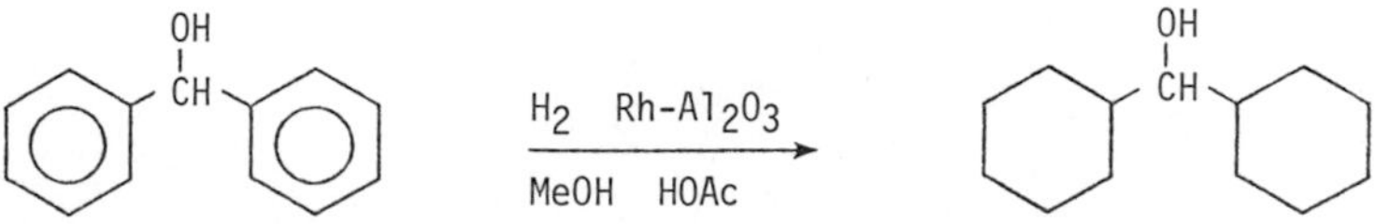
OH
CH
H2 Rh-Al2O3
MeOH HOAc
OH
CH

80%

JOC (1962) <u>27</u> 2288

Section 66   Alkyls and Aryls from Amides

The conversions ArCONR'R" → ArMe and RCONHAr → ArAr' are included
in this section. For the conversion RCONR'R" → RH see section 156
(Hydrides from Amides)

PhCON⟨cyclohexyl⟩ $\xrightarrow[\text{copper chromite   dioxane}]{\text{H}_2 \text{ (200-300 atmos)}}$ PhMe   79%

JACS (1934) 56 2419

NHAc (naphthalene) $\xrightarrow[\text{2 C}_6\text{H}_6]{\text{1 HNO}_2 \text{ HOAc}}$ Ph (naphthalene)

JCS (1940) 361 369
Org React (1944) 2 224

Section 67   Alkyls and Aryls from Amines

Reactions in which an amine group is replaced by alkyl or aryl are included
in this section. For the conversion RNH$_2$ → RH see section 157 (Hydrides
from Amines)

Br⟨C$_6$H$_4$⟩NH$_2$ $\xrightarrow[\text{2 NaOH   C}_6\text{H}_6 \text{ H}_2\text{O}]{\text{1 NaNO}_2 \text{ HCl   H}_2\text{O}}$ Br⟨C$_6$H$_4$⟩Ph   34%

Org Synth (1932) Coll Vol 1 113
Org React (1944) 2 224
JCS (1962) 4257

⟨bicyclic⟩—NH$_2$ $\xrightarrow{\text{NOCl   PhCl}}$ ⟨bicyclic⟩⟨C$_6$H$_4$⟩Cl   15%

JACS (1969) 91 5073

PhCHNHCH₂Ph $\xrightarrow[\text{2 LiAlH}_4 \quad \text{Et}_2\text{O}]{\text{1 NaNO}_2 \quad \text{HCl} \quad \text{H}_2\text{O}}$ PhCHNCH₂Ph $\xrightarrow[\text{EtOH}]{\text{HgO}}$ PhCHCH₂Ph

|
Me

NH₂
|
Me

|
Me

Chem Pharm Bull (1970) <u>18</u> 1137

Section 68   Alkyls and Aryls from Esters

This section lists the conversion of the group COOR to alkyl or aryl.
For the conversion RCOOR → RH see section 158 (Hydrides from Esters)

$C_{15}H_{31}COOMe$ $\xrightarrow[\text{Et}_2\text{O}]{\text{EtMgBr}}$ $C_{15}H_{31}\underset{\underset{\text{Et}}{|}}{\overset{\overset{\text{Et}}{|}}{C}}OH$ $\xrightarrow[\text{2 H}_2 \quad \text{Pt} \quad \text{EtOH}]{\text{1 HCOOH}}$ $C_{15}H_{31}\underset{\text{Et}}{\overset{}{C}}HEt$ 86%

JACS (1945) <u>67</u> 2239

$\xrightarrow{\text{BF}_3 \quad \text{C}_6\text{H}_6}$

37%

JACS (1937) <u>59</u> 1204
Org React (1946) <u>3</u> 1

$\xrightarrow{\text{EtMgBr} \quad \text{Et}_2\text{O}}$ MeCH=CCH₂Et

Me
|

JCS (1965) 702

$\xrightarrow[260\text{-}300°]{\text{H}_2 \quad \text{MoS}_2 \quad \text{cyclohexane}}$

+

Coll Czech (1958) <u>23</u> 1322

Section 69   Alkyls and Aryls from Ethers

The conversion ROR ⟶ RR' (R'=alkyl or aryl) is included in this section.
For the hydrogenolysis of ethers (ROR ⟶ RH) see section 159 (Hydrides
from Ethers)

Et_2O $\xrightarrow{\text{BF}_3 \quad \text{C}_6\text{H}_6}$ EtPh   (+ diethylbenzenes)   25%

JACS (1938) 60 125
Org React (1946) 3 1

33%

J Prakt Chem (1938) 151 61

Section 70   Alkyls and Aryls from Halides and Sulfonates

The replacement of halo and tosyloxy groups by alkyl or aryl groups is
included in this section.   For the conversion RX ⟶ RH (X=halo or OTs)
see section 160 (Hydrides from Halides and Tosylates)

JCS (1950) 711
Chem Rev (1946) 38 139

PhI $\xrightarrow{\text{PhCu} \quad \text{Pyr}}$ PhPh   60%

Tetr Lett (1968) 3307
JCS (1950) 7'1

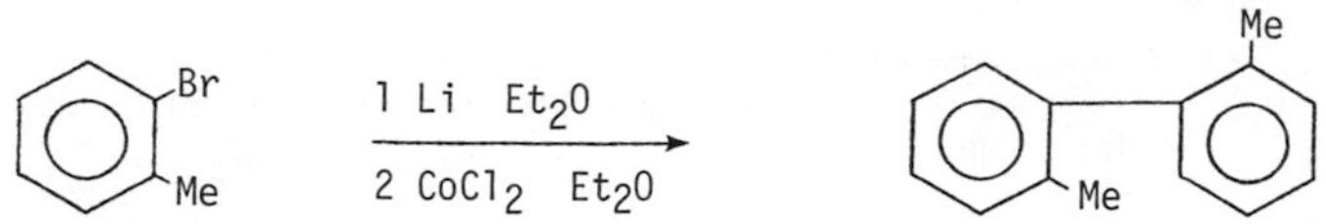
Br
Me
1 Li Et2O
2 CoCl2 Et2O
Me
Me

65%

Bull Soc Chim Fr (1964) 1331

$$\mathrm{PhCl} \xrightarrow{\text{PhLi piperidine } Et_2O} \mathrm{PhPh}$$

61%

Angew (1957) 69 267

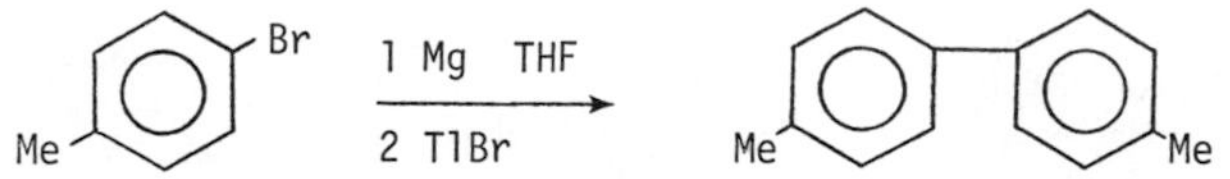
Br
Me
1 Mg THF
2 TlBr
Me
Me

91%

JACS (1968) 90 2423
Tetrahedron (1970) 26 4041

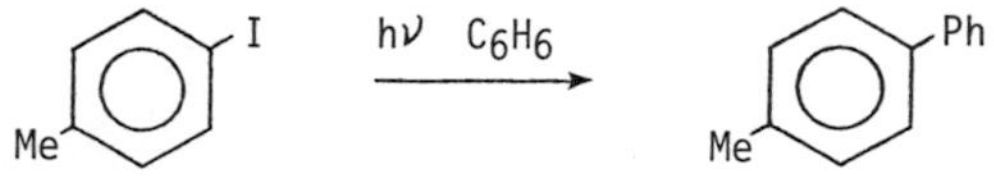
I
Me
hν C6H6
Ph
Me

75%

JOC (1965) 30 2493

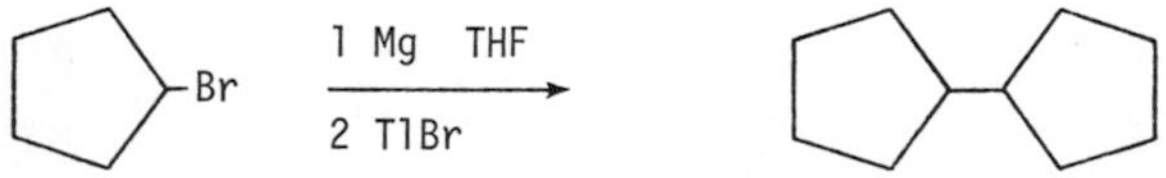
Br
1 Mg THF
2 TlBr

56%

JACS (1968) 90 2423

$$C_6H_{13}\underset{\mid}{C}HCl\ (Me) \xrightarrow[\text{pentane}]{LiCH_2CH{=}CH_2} C_6H_{13}\underset{\mid}{C}HCH_2CH{=}CH_2\ (Me)$$

59%

JOC (1970) 35 22
(1954) 19 934

$(MeO)_2CH(CH_2)_2\underset{Me}{C}=CHCH_2Cl$ $\xrightarrow[\text{THF HMPA}]{\overset{Me}{\underset{}{CH_2=CCH_2MgCl}}}$ $(MeO)_2CH(CH_2)_2\underset{Me}{C}=CHCH_2CH_2\underset{Me}{C}=CH_2$ 95%

Tetr Lett (1969) 1393

PhCH$_2$Cl $\xrightarrow[\text{2 BuOTs}]{\text{1 Mg Et}_2\text{O}}$ PhCH$_2$Bu   50-59%

Org Synth (1943) Coll Vol 2 47
(1932) Coll Vol 1 471

$EtCH(CH_2)_{10}I$
$\underset{Me}{}$ $\xrightarrow{C_{18}H_{37}I \quad Na \quad Et_2O}$ $EtCH(CH_2)_{10}C_{18}H_{37}$
$\underset{Me}{}$

JACS (1965) 87 5452

BuBr $\xrightarrow[\substack{\text{2 (ICuPBu}_3)_4 \\ \text{3 O}_2}]{\text{1 Li THF}}$ BuBu   84%

JACS (1967) 89 5302

t-BuCl $\xrightarrow{\text{FeCl}_3 \quad C_6H_6}$ t-BuPh   80%

Org React (1946) 3 1

PrBr $\xrightarrow[\text{2 t-BuCl HgCl}_2]{\text{1 Mg Et}_2\text{O}}$ PrBu-t   21%

JACS (1929) 51 1483
(1938) 60 2598

$$Me_2C=CH(CH_2)_2\overset{Me}{\underset{|}{C}}=CH(CH_2)_2\overset{Me}{\underset{|}{C}}=CHCH_2Cl$$

1 $(NH_2)_2CS$

2 Base O_2

3 Ph_3P

4 Benzyne

5 Li NH_3 Chem Comm (1969) 99

$$Me_2C=CH(CH_2)_2\overset{Me}{\underset{|}{C}}=CH(CH_2)_2\overset{Me}{\underset{|}{C}}=CHCH_2$$
$$Me_2C=CH(CH_2)_2\underset{\overset{|}{Me}}{C}=CH(CH_2)_2\underset{\overset{|}{Me}}{C}=CHCH_2$$

$Me_2C=CHCH_2Br$ $Ni(CO)_4$

DMF

$CH_2CH_2CH=CMe_2$

88%

JACS (1967) 89 2755

$MeCH=CHCH_2Cl$

$\xrightarrow{\quad Ni(CO)_4 \quad MeOH \quad}$

$$MeCH=CHCH_2\underset{\overset{|}{Me}}{C}HCH=CH_2$$

JACS (1951) 73 2654

$$\underset{\overset{|}{O}\;\overset{|}{O}}{CH_2CH(CH_2)_2I}$$

$\xrightarrow{\quad LiC\equiv CMe \quad}$

$$\underset{\overset{|}{O}\;\overset{|}{O}}{CH_2CH(CH_2)_2C\equiv CMe}$$

$\xrightarrow{\quad H_2 \quad Pd \quad}$

$$\underset{\overset{|}{O}\;\overset{|}{O}}{CH_2CH(CH_2)_2Pr}$$

E. J. Corey, in press

$I(CH_2)_{10}COOH$

$\xrightarrow{\quad Bu_2CuLi \; Et_2O \quad}$

$Bu(CH_2)_{10}COOH$

76%

JACS (1968) 90 5615

RI $\xrightarrow[\quad 2 \; O_2 \quad]{\quad 1 \; Me_2CuLi \quad Et_2O \quad}$ RMe   (R=C_5H_{11} or Ph)   98-99%

JACS (1969) 91 4871

$$PhCH=CHBr \xrightarrow{\quad Me_2CuLi \quad Et_2O \quad} PhCH=CHMe \qquad\qquad 81\%$$

JACS (1967) <u>89</u> 3911 4245

$$\underset{\underset{Pr}{|}}{Me_2CCl} \xrightarrow{\quad Me_3Al \quad} \underset{\underset{Pr}{|}}{Me_2CMe} \qquad\qquad 100\%$$

JOC (1970) <u>35</u> 532

$$C_{10}H_{21}I \xrightarrow{\quad Me_3MnLi \quad Et_2O \quad} C_{10}H_{21}Me \qquad\qquad 55\%$$

Tetr Lett (1970) 315

$$PhCl \xrightarrow{\quad NaH \quad Me_2SO \quad} \underset{41\%}{PhCH_2SOMe} \dashrightarrow^{\quad Al-Hg \quad} PhMe$$

JACS (1965) <u>87</u> 1345

$$\underset{(X=Br\ or\ OTs)}{C_{11}H_{23}X} \xrightarrow{\quad Me_2SO \quad NaH \quad} C_{11}H_{23}CH_2SOMe \xrightarrow[EtOH]{\quad Ni \quad} C_{11}H_{23}Me \qquad 62\%$$

Aust J Chem (1966) <u>19</u> 521

$$\xrightarrow{\quad Me_2CuLi \quad Et_2O \quad} \qquad\qquad 65\%$$

JACS (1967) <u>89</u> 3911 4245

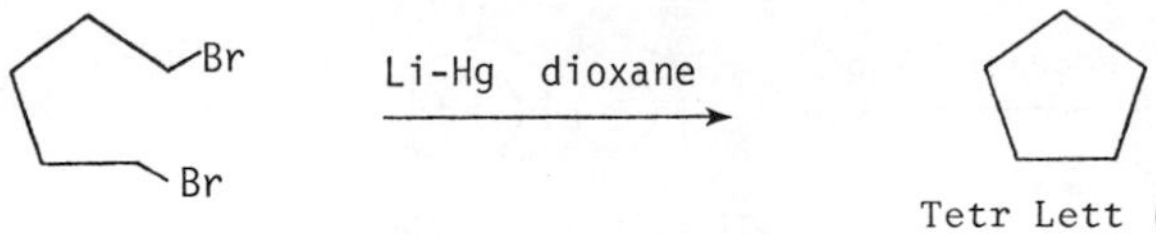
Br
Br
Li-Hg dioxane

76%

Tetr Lett (1967) 4925

Section 71 Alkyls and Aryls from Hydrides (RH)

This section lists examples of the reaction RH → RR' (R,R'=alkyl or aryl)

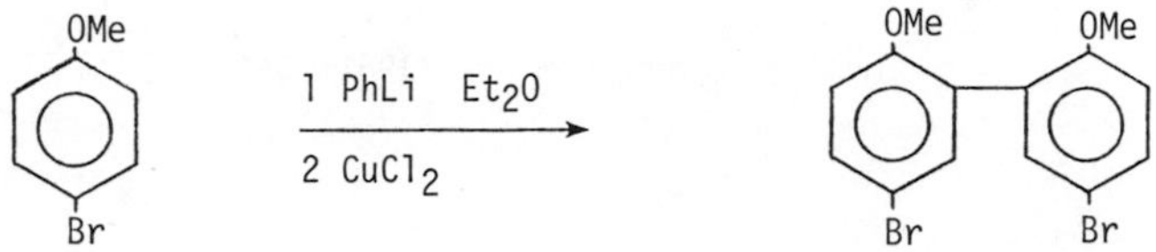
OMe
Br
1 PhLi Et_2O
2 $CuCl_2$
OMe
OMe
Br
Br

19%

JACS (1940) 62 1963

PhH —(1 $Tl(OCOCF_3)_3$ CF_3COOH; 2 hν C_6H_6)→ PhPh 90%

JACS (1970) 92 6088

PhH —(n-PrCl $AlCl_3$)→ PhPr-n + PhPr-i

JOC (1940) 5 253
Org React (1946) 3 1

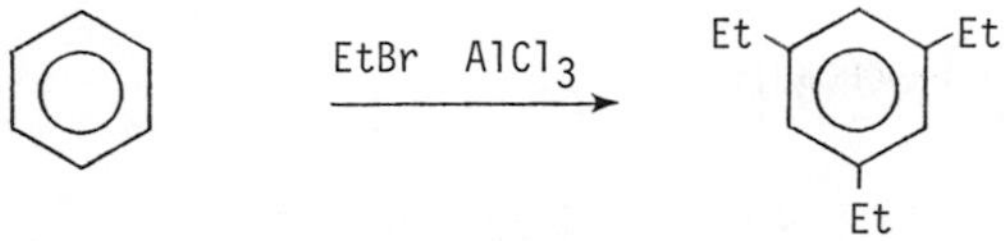
EtBr $AlCl_3$
Et
Et
Et

86%

Org React (1946) 3 1

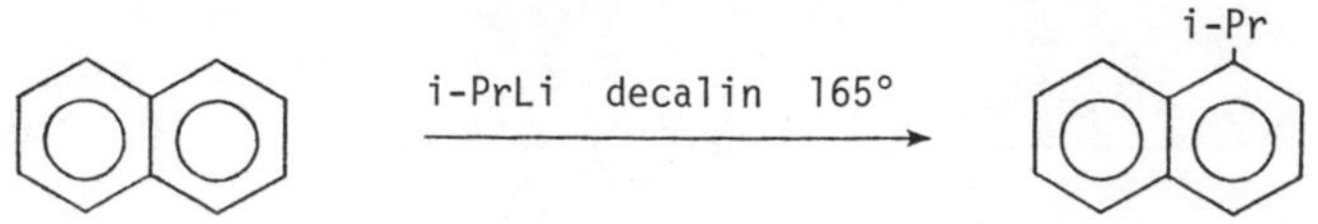
i-PrLi decalin 165°
i-Pr

20%

JACS (1963) 85 1356

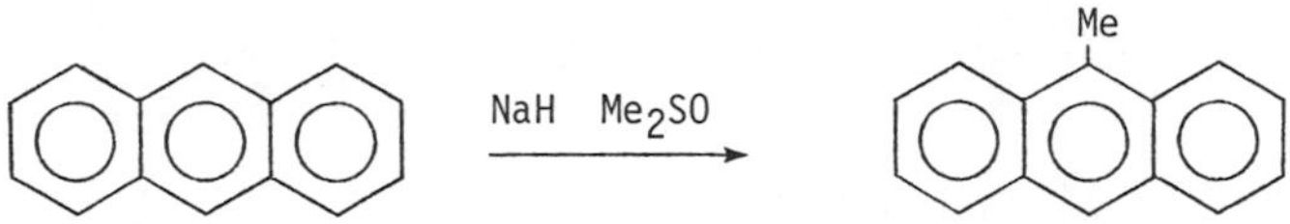
NaH Me2SO
Me

33%

JOC (1966) 31 248

PhH $\xrightarrow{\text{MeHgI } h\nu}$ PhMe

Proc Chem Soc (1961) 240

RH $\xrightarrow{\text{MeF-SbF}_3}$ RMe (R=alkyl or aryl)

JACS (1969) 91 2112

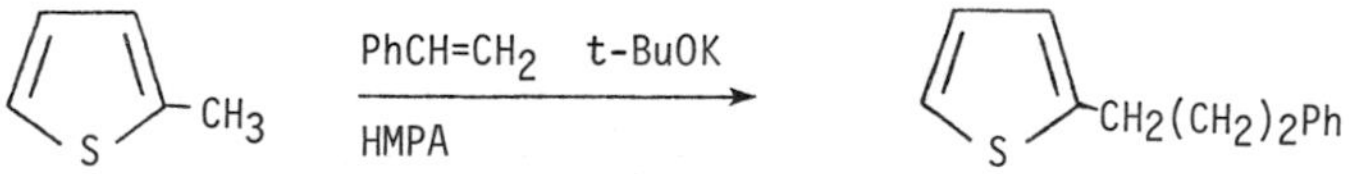
CH3
S
PhCH=CH2 t-BuOK
HMPA
S
CH2(CH2)2Ph

Tetr Lett (1968) 3723

Ph_2CH_2 $\xrightarrow[\text{2 } C_8H_{17}Br]{\text{1 } NaNH_2 \text{ } NH_3}$ $Ph_2CHC_8H_{17}$ 99%

JACS (1957) 79 3142

$PhCH_3$ $\xrightarrow{\text{PrCl Na}}$ $PhCH_2Pr$ 43%

JACS (1941) 63 327

Me_2CHMe $\xrightarrow{\text{CH}_2\text{=CH}_2 \text{ AlCl}_3 \text{ HCl}}$ $Me_2CHCHMe_2$

JACS (1945) 67 1778

Section 72   Alkyls, Methylenes and Aryls from Ketones

The conversions $R_2CO \longrightarrow RR$, R_2CH_2, R_2CHR, etc. (R=alkyl or aryl) are
included in this section. For the conversion $R_2CO \longrightarrow RH$ see section 162
(Hydrides from Ketones)

JACS (1968) 90 817   44%

$EtCO(CH_2)_{14}COEt$ $\xrightarrow[\text{2 HI P}_4 \text{ 270°}]{\text{1 C}_{16}\text{H}_{33}\text{C}\equiv\text{CMgBr}}$ $\underset{\overset{|}{C_{18}H_{37}} \quad \overset{|}{C_{18}H_{37}}}{EtCH(CH_2)_{14}CHEt}$ 26%

Helv (1937) 20 1179

Me_2CO $\xrightarrow[\text{2 I}_2]{\text{1 BuMgBr Et}_2\text{O}}$ $Me_2C=CHPr$ $\xrightarrow[\text{catalyst}]{\text{H}_2}$ Me_2CHBu

JACS (1929) 51 1483
(1949) 71 819
JOC (1948) 13 239

Further examples of the conversion of ketones into olefins and of the
hydrogenation of olefins are included in section 207 (Olefins from Ketones)
and section 74 (Alkyls and Methylenes from Olefins)

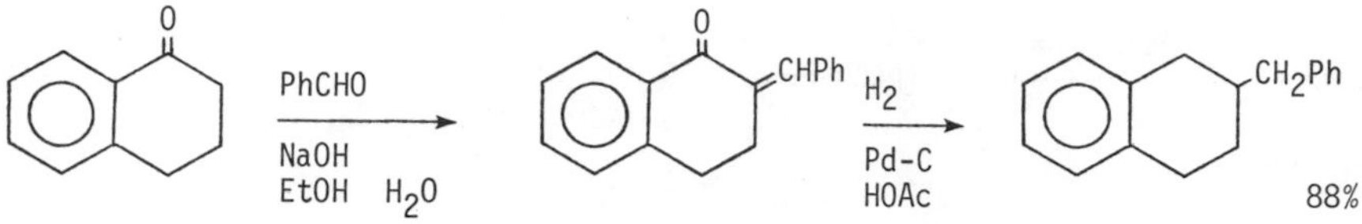

Org Prep and Procedures (1970) 2 37

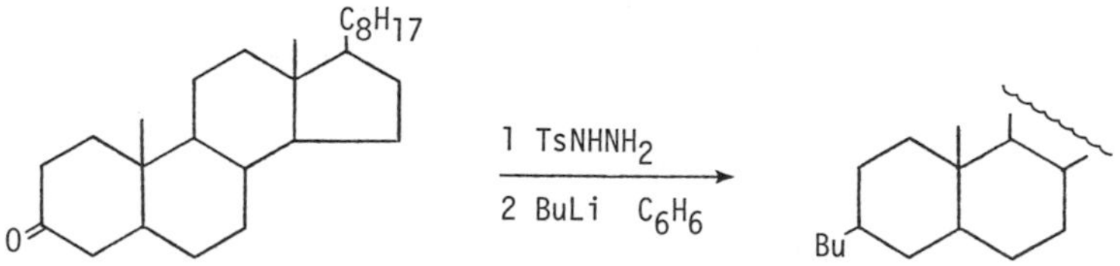

Chem Comm (1969) 1395

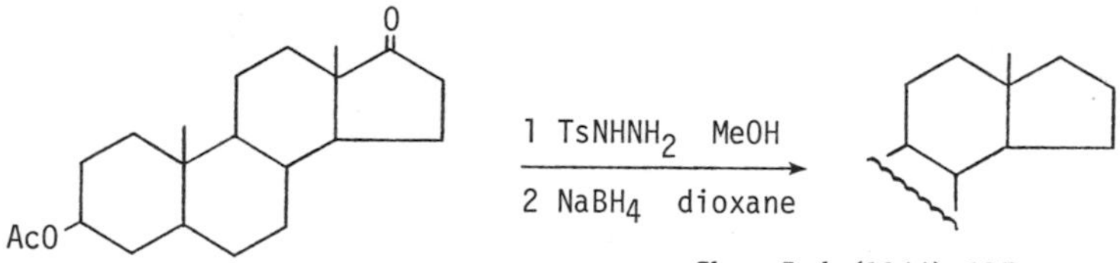

60-70%

Chem Ind (1964) 153
Tetrahedron (1963) 19 1127

PhCOEt $\xrightarrow[\text{diethylene glycol}]{\text{N}_2\text{H}_4 \quad \text{KOH}}$ PhCH$_2$Et   82%

JACS (1946) 68 2487
Org React (1948) 4 378

For reduction of hindered ketones with N$_2$H$_4$·HCl etc. see

Chem Ind (1964) 1194

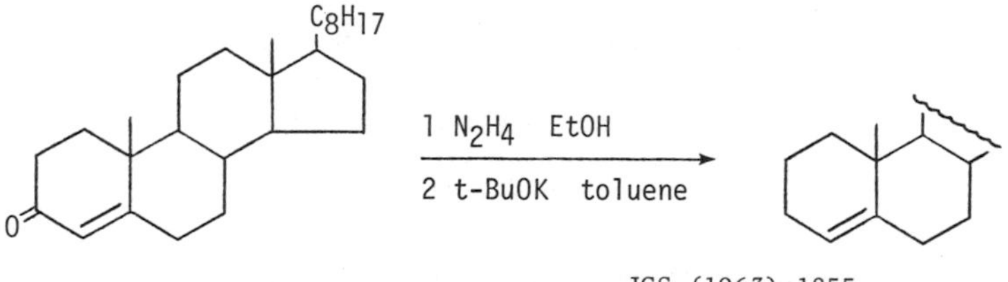

~65%

JCS (1963) 1855
Tetrahedron (1970) 26 649

$$\begin{array}{c} \text{1 } N_2H_4 \\ \hline \text{2 t-BuOK   Me}_2\text{SO   25}° \end{array}$$

80%

JACS (1962) 84 1734

For reduction of hydrazones with BuLi see JACS (1968) 90 7287

$$\begin{array}{c} N_2H_4 \cdot H_2O \quad KOH \\ \hline \text{ethylene glycol} \end{array}$$

JACS (1959) 81 5834
      (1962) 84 2938
JCS (1960) 4413

$$C_6H_{13}COMe \xrightarrow[Na_2SO_4]{EtSH \quad ZnCl_2} C_6H_{13}\underset{Me}{C}(SEt)_2 \xrightarrow[EtOH]{Ni \quad H_2O} C_6H_{13}CH_2Me$$

50%

JACS (1944) 66 909
Org React (1962) 12 356

$$(C_{10}H_{21})_2CO \xrightarrow{\text{Zn-Hg   HCl   H}_2O} (C_{10}H_{21})_2CH_2$$

32%

Rec Trav Chim (1936) 55 903
Org React (1942) 1 155
JACS (1947) 69 2350

$$\xrightarrow{\text{Zn   HCl   Et}_2O}$$

75%

Chem Comm (1969) 919

$$\begin{array}{c} \text{1 } B_2H_6 \quad \text{diglyme} \\ \hline \text{2 Ac}_2O \end{array}$$

85%

Tetrahedron (1964) 20 957

$$\xrightarrow{\begin{array}{l}\text{1 Li   NH}_3\text{   Et}_2\text{O}\\\text{2 (EtO)}_2\text{POCl   Et}_2\text{O}\\\text{3 Li   EtNH}_2\text{   t-BuOH}\end{array}}$$

50%

Tetr Lett (1969) 2145

$$\xrightarrow{\text{LiAlH}_4\text{   AlCl}_3\text{   Et}_2\text{O}}$$

Tetrahedron (1965) 21 2641

For application to phenyl ketones see JCS (1957) 3755

$$\text{MeO} - \text{C}_6\text{H}_4 - \text{CO(CH}_2)_2\text{COOH} \xrightarrow[\text{HOAc}]{\text{H}_2\text{   Pd-C}} \text{MeO} - \text{C}_6\text{H}_4 - \text{CH}_2\text{(CH}_2)_2\text{COOH}$$

75%

JACS (1949) 71 1036
Org React (1953) 7 263

$$\xrightarrow[\text{CuSO}_4]{\text{Zn   NH}_3}$$

90%

JOC (1970) 35 711

$$\text{HO} - \text{C}_6\text{H}_4 - \text{COEt} \xrightarrow{\text{Ni-Al   NaOH   H}_2\text{O}} \text{HO} - \text{C}_6\text{H}_4 - \text{CH}_2\text{Et}$$

78%

Org React (1953) 7 263

$$\xrightarrow{\text{HI   I}_2\text{   P}_4\text{   HOAc}}$$

94%

Ber (1959) 92 1705

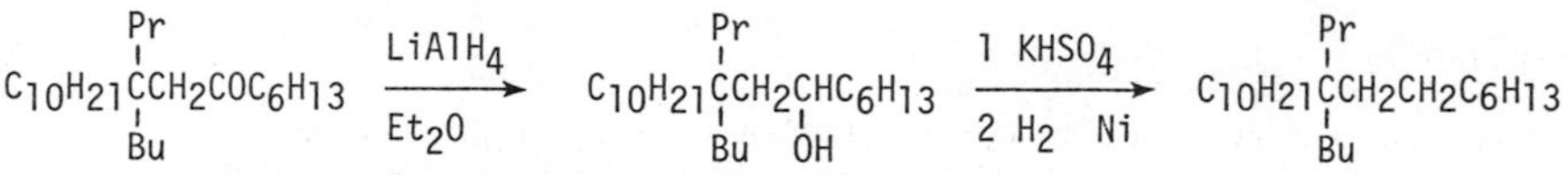

$$C_{10}H_{21}\underset{\underset{Bu}{|}}{\overset{\overset{Pr}{|}}{C}}CH_2COC_6H_{13} \xrightarrow[Et_2O]{LiAlH_4} C_{10}H_{21}\underset{\underset{Bu}{|}}{\overset{\overset{Pr}{|}}{C}}CH_2\underset{\underset{OH}{|}}{C}HC_6H_{13} \xrightarrow[2\ H_2\ \ Ni]{1\ KHSO_4} C_{10}H_{21}\underset{\underset{Bu}{|}}{\overset{\overset{Pr}{|}}{C}}CH_2CH_2C_6H_{13}$$

JOC (1959) 24 1964

Electrolysis H$_2$SO$_4$

dioxane H$_2$O

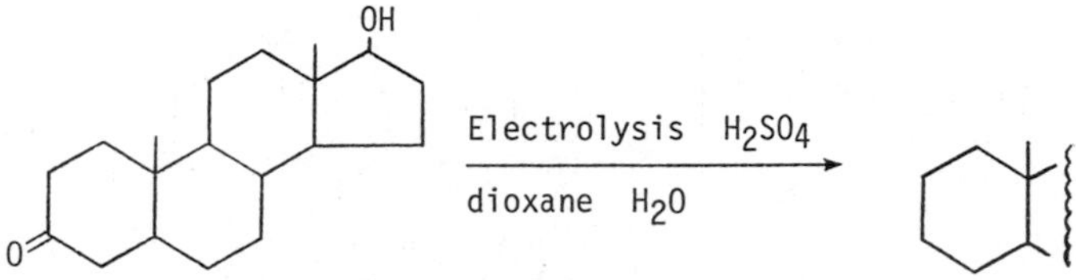

97%

JACS (1967) 89 4789

PhCOMe $\xrightarrow{Et_3SiH\ \ CF_3COOH}$ PhCH$_2$Me   70%

Tetrahedron (1967) 23 2235

(PhCH$_2$)$_2$CO $\xrightarrow{h\nu}$ PhCH$_2$CH$_2$Ph

JACS (1970) 92 6076 6077

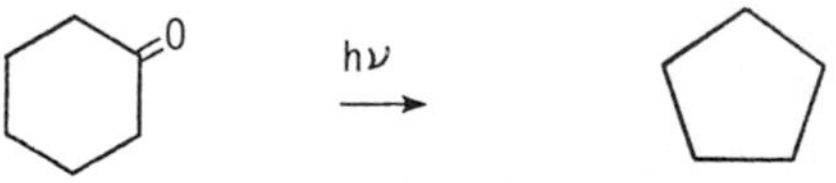

$\xrightarrow{h\nu}$

JACS (1942) 64 80
     (1961) 83 4923

Section 73   Alkyls from Nitriles

The conversion RCN $\longrightarrow$ RMe is included in this section. For the replacement of CN by hydrogen, RCN $\longrightarrow$ RH, see section 163 (Hydrides from Nitriles)

Tetr Lett (1966) 4255
Tetrahedron (1961) 13 287

Section 74   Alkyls, Methylenes and Aryls from Olefins

The hydrogenation, alkylation, arylation, dimerization etc. of olefins, forming alkanes or aryl-substituted alkanes, are included in this section. For the conversion $R_2C=CR_2 \longrightarrow$ RH or R_2CH_2 see section 164 (Hydrides from Olefins)

BuCH=CH$_2$ →(1 NaBH$_4$ BF$_3$·Et$_2$O diglyme / 2 KOH AgNO$_3$ H$_2$O)   Bu(CH$_2$)$_4$Bu   66%

JACS (1961) 83 1002
Chem Comm (1968) 938

EtCH=CHEt   →(C$_6$H$_6$ H$_2$SO$_4$)   EtCHCH$_2$Et
                                              |
                                              Ph

JACS (1939) 61 1002
Org React (1946) 3 1

$C_{10}H_{21}CH=CH_2$ $\xrightarrow{\text{1 Pr}_3\text{Al}}{\text{2 H}_2\text{O}}$ $C_{10}H_{21}\underset{\underset{Pr}{|}}{C}HMe$

Angew (1952) <u>64</u> 323

$\xrightarrow{\text{t-BuLi} \quad \text{Et}_3\text{N}}$

—Bu-t

JOC (1965) <u>30</u> 917

$C_5H_{11}CH=CH_2$ $\xrightarrow[\text{benzoyl peroxide}]{\text{CHCl}_2\text{SO}_2\text{Cl}}$ $C_5H_{11}\underset{\underset{Cl}{|}}{C}HCH_2CHCl_2$ $\xrightarrow[\text{dioxane}]{\text{LiAlH}_4}$ $C_5H_{11}(CH_2)_2Me$

$+ \ C_5H_{11}\underset{\diagdown\diagup}{C}\underset{CH_2}{H}CH_2$

Tetrahedron (1964) <u>20</u> 1613

$PhCH=CH_2$ $\overset{Br_2}{\dashrightarrow}$ $Ph\underset{\underset{Br}{|}}{C}HCH_2Br$ $\xrightarrow[\text{Me}_2\text{SO}]{\text{NaBH}_4}$ $PhCH_2Me$ 64%

Tetr Lett (1969) 3495

$BuCH=CH_2$ $\xrightarrow[\text{2 EtCOOH}]{\text{1 NaBH}_4 \quad \text{BF}_3\cdot\text{Et}_2\text{O} \quad \text{diglyme}}$ $BuCH_2Me$ 91%

JACS (1959) <u>81</u> 4108

$C_7H_{15}CH=CH_2$ $\xrightarrow[\substack{\text{2 Hg(OAc)}_2 \\ \text{3 NaBH}_4}]{\text{1 B}_2\text{H}_6 \quad \text{THF}}$ $C_7H_{15}CH_2Me$

JACS (1970) <u>92</u> 3221

$C_6H_{13}CH=CH_2$ $\xrightarrow[\text{225°}]{\text{H}_2 \text{ (2000 psi)} \quad \text{(i-Bu)}_3\text{B}}$ $C_6H_{13}CH_2Me$

JOC (1962) 27 4368
JACS (1961) 83 4672

$Ph_2C=CHMe$ $\xrightarrow[\text{2 H}_2\text{O}]{\text{1 Et}_2\text{AlCl}}$ Ph_2CHCH_2Me

Tetr Lett (1970) 3471

$C_6H_{13}CH=CH_2$ $\xrightarrow{\text{H}_2 \quad \text{Pt}}$ $C_6H_{13}CH_2Me$

JACS (1962) 84 1495
Org Synth (1943) Coll Vol 2 191

For hydrogenation with Pd   catalyst see JACS (1960) 82 6087
　　　"　　　　　"　　　　"　　Ru　　"　　" JOC (1959) 24 708
　　　"　　　　　"　　　　"　　Re$_2$S$_7$　"　　" JACS (1954) 76 1519
　　　　　　　　　　　　　　　　　　　　　　　　　(1959) 81 3587
　　　"　　　　　"　　　　"　　Ni　　"　　" JOC (1970) 35 1900

$PhSCH_2CH=CH_2$ $\xrightarrow{\text{H}_2 \quad \text{(Ph}_3\text{P)}_3\text{RhCl} \quad \text{C}_6\text{H}_6}$ $PhS(CH_2)_2Me$ 93%

Tetr Lett (1967) 1935
JOC (1967) 32 2013 3074

For hydrogenation with Rh(COOMe)$_4$ catalyst see Chem Comm (1969) 825
　　　"　　　　　"　　　　"　　RuClH(PPh$_3$)$_3$　"　　"　Chem Comm (1967) 305

$C_8H_{17}CH=CH(CH_2)_7COOH$ $\xrightarrow{\text{N}_2\text{H}_4 \quad \text{air} \quad \text{EtOH}}$ $C_8H_{17}(CH_2)_9COOH$ 90%

Chem Ind (1961) 433
JOC (1965) 30 3980
Chem Rev (1965) 65 51
For copper catalyzed diimide reduction see Tetr Lett (1961) 347
and JACS (1970) 92 6635

$HOCH_2CH=CH_2$ $\xrightarrow{\text{TsNHNH}_2 \quad \text{diglyme}}$ $HOCH_2CH_2Me$ 99%

JACS (1961) <u>83</u> 3729

 $\xrightarrow[\text{NCOONa} \quad \text{Pyr} \quad \text{HOAc} \quad \text{H}_2\text{O}]{\overset{\displaystyle \text{NCOONa}}{\|}}$

JACS (1963) <u>85</u> 3297
(1961) <u>83</u> 3725

$BuCH=CH_2$ $\xrightarrow{\text{Na} \quad \text{NH}_3 \quad \text{MeOH}}$ $BuCH_2Me$ ~41%

JACS (1954) <u>76</u> 1258
JOC (1970) <u>35</u> 3565

$Me_2C=CH(CH_2)_2\underset{\underset{\text{Me}}{|}}{C}=CHCH_2CH=\underset{\underset{\text{Me}}{|}}{C}-CH=CH_2$ $\xrightarrow[\text{EtOH}]{\text{Na}}$ $Me_2C=CH(CH_2)_2\underset{\underset{\text{Me}}{|}}{C}=CHCH_2CH_2\underset{\underset{\text{Me}}{|}}{C}=CHMe$

Tetr Lett (1967) 2201
Ber (1956) <u>89</u> 1549

$BuCH=CHMe$ $\xrightarrow{\text{Li} \quad \text{NH}_2\text{CH}_2\text{CH}_2\text{NH}_2}$ $Bu(CH_2)_2Me$ 82%

JOC (1957) <u>22</u> 891
JACS (1959) <u>81</u> 1745

$Me_2C=CHMe$ $\xrightarrow{\text{Et}_3\text{SiH} \quad \text{CF}_3\text{COOH}}$ Me_2CHCH_2Me

($BuCH=CH_2$ is not reduced)

Tetrahedron (1967) <u>23</u> 2235

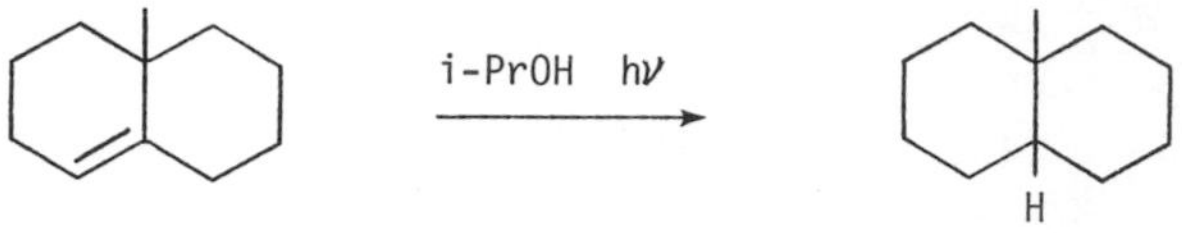

i-PrOH hν

50%

Chem Comm (1968) 296

$Me_2CHCH=CHCHMe_2$ $\xrightarrow{\begin{array}{c} 1\ O_3 \\ 2\ h\nu \end{array}}$ $Me_2CHCHMe_2$ 33%

Tetr Lett (1968) 3291

Section 75   Alkyls and Methylenes from Miscellaneous Compounds
○○

$PrCOCH(CH_2)_{12}COOMe$ $\xrightarrow{\begin{array}{c} 1\ TsNHNH_2 \\ 2\ NaBH_4 \end{array}}$ $PrCH_2CH_2(CH_2)_{12}COOMe$ 70%
$\overset{|}{EtCOO}$

Chem Ind (1967) 2150

H_2 Pt-glass

200°

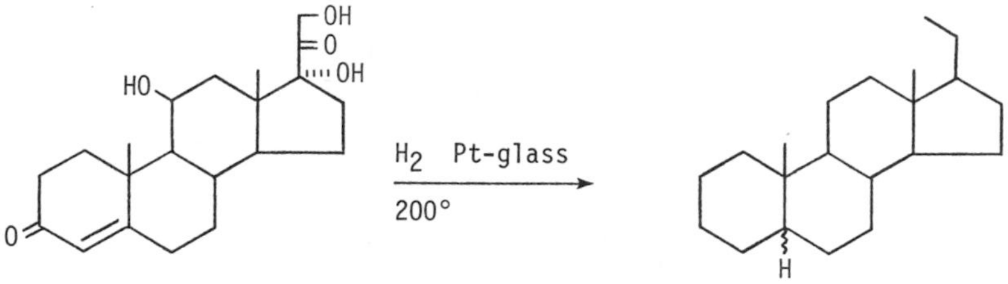

Biochem J (1967) 105 40P

Chapter 6   PREPARATION
                OF
                AMIDES

Section 76   Amides from Acetylenes
              °°°°°°°°°°°°°°°°°°°°°°°°°°

$C_5H_{11}C\equiv CH$ $\xrightarrow[250°]{N_2O \quad cyclohexylamine}$ $C_5H_{11}CH_2CONH$ $\sim 94\%$

JCS (1951) 3016

$C_5H_{11}C\equiv CH$ $\xrightarrow{S \quad NH_3 \quad H_2O \quad Pyr}$ $C_5H_{11}CH_2CONH_2$ 35%

JACS (1946) $\underline{68}$ 2033

$PhC\equiv CMe$ $\xrightarrow{S \quad NH_3 \quad H_2O}$ $PhCH_2CH_2CONH_2$ 72%

JACS (1946) $\underline{68}$ 2029

$C_5H_{11}C\equiv CH$ $\xrightarrow[2 \; Et_2NH \quad t\text{-}BuBr]{1 \; BuLi \quad S}$ $C_5H_{11}CH_2CSNEt_2$ $\dashrightarrow$ $C_5H_{11}CH_2CONEt_2$

Rec Trav Chim (1968) $\underline{87}$ 38

Section 77 Amides from Carboxylic Acids and Acid Halides

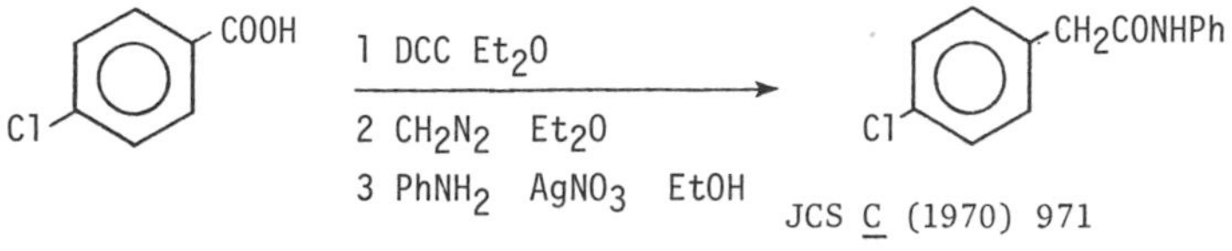
COOH
Cl
1 DCC Et2O
2 CH2N2 Et2O
3 PhNH2 AgNO3 EtOH
CH2CONHPh
Cl

33%

JCS C (1970) 971

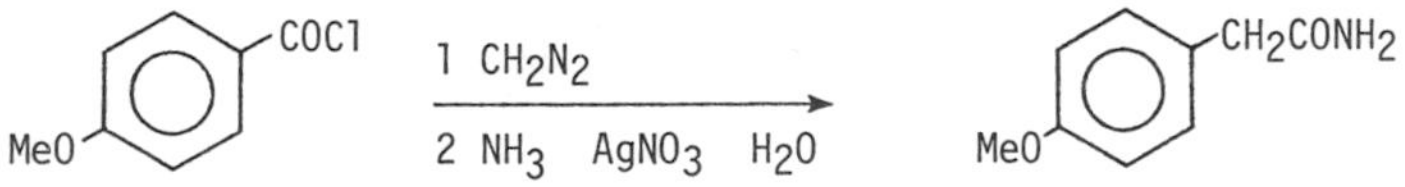
COCl
MeO
1 CH2N2
2 NH3 AgNO3 H2O
CH2CONH2
MeO

57%

Org React (1942) 1 38

$PrCOCl \xrightarrow{CH_2N_2} PrCOCHN_2 \xrightarrow{PhNHMe \quad h\nu \quad C_6H_6} PrCH_2CON(Me)Ph$ 45%

Ber (1959) 92 528
JCS C (1970) 1208

$C_6H_{13}COOH \xrightarrow{NH_3 \quad 190°} C_6H_{13}CONH_2$ 75%

JACS (1931) 53 1879

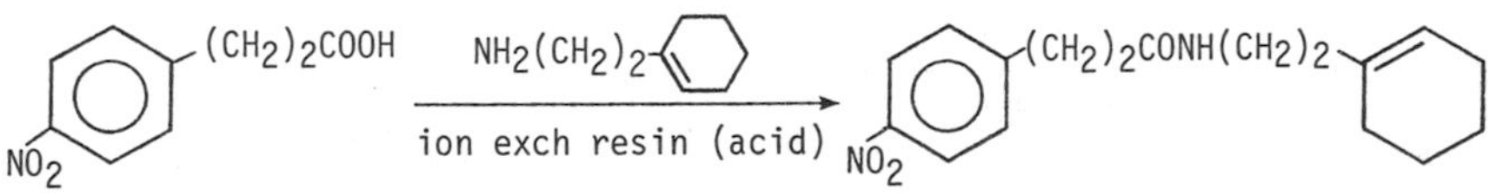
(CH2)2COOH
NO2
NH2(CH2)2
ion exch resin (acid)
(CH2)2CONH(CH2)2
NO2

Helv (1961) 44 1546
JCS C (1969) 874

$C_6H_{13}COOH$ $\xrightarrow{\text{NH}_2\text{CONH}_2 \quad 170\text{-}180°}$ $C_6H_{13}CONH_2$ 68-74%

Org Synth (1963) Coll Vol 4 513

$PhCOCl$ $\xrightarrow{\text{HCONMe}_2 \quad 150°}$ $PhCONMe_2$ 97%

JACS (1954) 76 1372
(1949) 71 2215

$i\text{-}PrCOCl$ $\xrightarrow{\text{NH}_3 \quad \text{H}_2\text{O}}$ $i\text{-}PrCONH_2$ 70%

Org Synth (1955) Coll Vol 3 490
JOC (1954) 19 623

$MeCH=CHCOCl$ $\xrightarrow{\text{NH}_4\text{OAc} \quad \text{Me}_2\text{CO}}$ $MeCH=CHCONH_2$ 63%

JCS (1962) 2824

$C_8H_{17}CH=CH(CH_2)_7COCl$ $\xrightarrow[\text{pet ether}]{\text{BuNH}_2}$ $C_8H_{17}CH=CH(CH_2)_7CONHBu$ 83%

JACS (1949) 71 2215
Helv (1959) 42 2073

$\underset{\overset{|}{CH_2Ph}}{PhCH=CHCH_2CHCOOH}$ $\xrightarrow[\text{2 NH}_3]{\text{1 ClCOOEt} \quad \text{Et}_3\text{N} \quad \text{CHCl}_3}$ $\underset{\overset{|}{CH_2Ph}}{PhCH=CHCH_2CHCONH_2}$ 94%

J Med Chem (1968) 11 534

(furan)-COOH $\xrightarrow[\text{2 p-Toluidine}]{\text{1 TsCl   Pyr}}$ (furan)-CONH-(phenyl)-Me

JACS (1955) $\underline{77}$ 6214   90%

$BrCH_2CH_2COOH$ $\xrightarrow[\text{CH}_2\text{Cl}_2]{\text{MeO-(aniline NH}_2\text{)-OMe   DCC}}$ $BrCH_2CH_2CONH$-(phenyl, MeO, OMe)

JACS (1968) $\underline{90}$ 4706
Compt Rend ($\overline{1965}$) $\underline{260}$ 2249
JACS (1968) $\underline{90}$ 2448

Amides by reaction of acids with amines and $HC\equiv COMe$

Rec Trav Chim (1955) $\underline{74}$ 769

"   "   "   "   "   "   "   " $HC\equiv CCH_2SMe_2Br$

JCS $\underline{C}$ (1969) 1904

"   "   "   "   "   "   "   " $Me_2C(OMe)_2$

Chim Ther (1967) $\underline{2}$ 195
(Chem Abs $\underline{67}$ 108840)

"   "   "   "   "   "   "   " $SiCl_4$

JOC (1969) $\underline{34}$ 2766

"   "   "   "   "   "   "   " $TiCl_4$

Can J Chem (1970) $\underline{48}$ 983

"   "   "   "   "   "   "   " $(PNCl_2)_3$

JOC (1968) $\underline{33}$ 2979

"   "   "   "   "   "   "   " 2-iodo-1-methylpyridinium iodide

JCS (1964) 4650

"   "   "   "   "   "   "   " $SO_3 \cdot DMF$

JOC (1959) $\underline{24}$ 368

Further examples of the reaction $RCOOH + R_2NH \longrightarrow RCONR_2$ are included in section 82 (Amides from Amines)

$C_{11}H_{23}COOH$ $\xrightarrow{(Me_2N)_3PO\ 180-200°}$ $C_{11}H_{23}CONMe_2$ 95%

Chem Ind (1966) 1529

$C_{11}H_{23}COONa$ $\xrightarrow{Bu_2NCOCl}$ $C_{11}H_{23}CONBu_2$ 78%

JOC (1963) 28 232

$PhCH_2COOH$ $\xrightarrow{(BuNH)_3B}$ $PhCH_2CONHBu$ 90%

Tetrahedron (1970) 26 1539

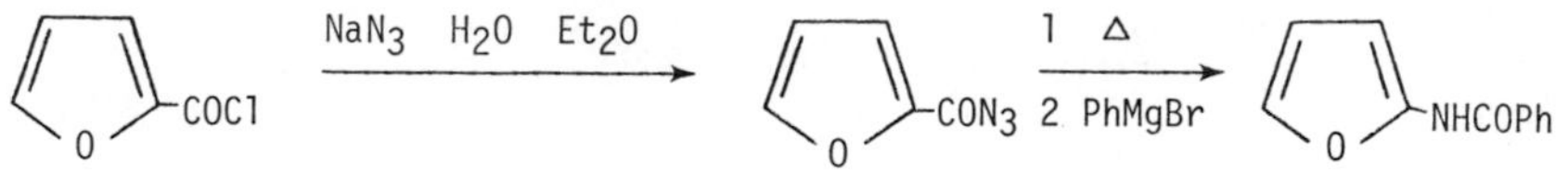

JACS (1938) 60 540   ∼50%

Amides may also be prepared from carboxylic acids via ester intermediates.
See section 83 (Amides from Esters)

Section 78   Amides from Alcohols and Phenols

$$\xrightarrow[\text{xylene}]{\text{MeC(OMe)}_2\text{NMe}_2}$$

CH$_2$CONMe$_2$

30%

Tetr Lett (1968) 1899
Angew (1968) 80 626
(Internat Ed 7 629)

PhCH=CHCH$_2$OH $\xrightarrow[\text{NH}_3 \quad \text{Et}_2\text{O}]{\text{Nickel peroxide}}$ PhCH=CHCONH$_2$ 85%

Chem Comm (1966) 17

PhCHCHMe$_2$ $\xrightarrow{\text{(NH}_4)_2\text{S}_x \quad 200°}$ PhCH$_2$CHCONH$_2$
$|$ $|$
OH   Me

JACS (1946) 68 632 2033

$\xrightarrow{\text{MeCN} \quad \text{H}_2\text{SO}_4 \quad \text{HOAc}}$

JOC (1960) 25 331
Org React (1969) 17 213

Me   Me
$|$ $|$
PhCH$_2$COH $\xrightarrow{\text{NaCN} \quad \text{H}_2\text{SO}_4 \quad \text{HOAc}}$ PhCH$_2$CNHCHO   92%
$|$ $|$
Me   Me

JOC (1961) 26 3002

PhOH $\xrightarrow{\begin{array}{c} 1 \ PhN=CPh \quad MeONa \\ 2 \ 312\text{-}315° \end{array}}$ Ph$_2$NCOPh

$\quad\quad\quad\quad\quad\quad\quad\quad\quad\quad\quad\quad$ Org React (1965) $\underline{14}$ 1

Section 79   Amides from Aldehydes

$\quad\quad\quad\quad\quad\quad\quad\quad$ Arch Pharm (1957) $\underline{290}$ 218

PhCHO $\xrightarrow{\text{PhCH}_2\text{CH}_2\text{N}_3 \quad \text{H}_2\text{SO}_4}$ PhCONHCH$_2$CH$_2$Ph $\quad\quad\quad\quad\quad$ 10%

$\quad\quad\quad\quad\quad\quad\quad\quad$ JACS (1955) $\underline{77}$ 951

PhCHO $\xrightarrow[\text{CCl}_4]{1 \ SO_2Cl_2 \quad \text{benzoyl peroxide}}$ PhCOCl $\xrightarrow{\text{PhNH}_2}$ PhCONHPh $\quad\quad$ 67%

$\quad\quad\quad\quad\quad\quad\quad\quad$ Nippon Kagaku Z (1960) $\underline{81}$ 1450
$\quad\quad\quad\quad\quad\quad\quad\quad$ (Chem Abs $\underline{56}$ 2370)

C$_6$H$_{13}$CHO $- - - \rightarrow$ C$_6$H$_{13}$CH=NOH $\xrightarrow{\text{Ni} \quad 100°}$ C$_6$H$_{13}$CONH$_2$ $\quad\quad$ 100%

$\quad\quad\quad\quad\quad\quad\quad\quad$ Org React (1960) $\underline{11}$ 1
$\quad\quad\quad\quad\quad\quad\quad\quad$ JCS (1946) 599
$\quad\quad\quad\quad\quad\quad\quad\quad$ JACS (1961) $\underline{83}$ 1983

PhCH=CHCHO $\xrightarrow[\text{Et}_2\text{O}]{\text{Nickel peroxide} \quad \text{NH}_3}$ PhCH=CHCONH$_2$ 85%

Chem Comm (1966) 17

Ph(CH$_2$)$_2$CHO $\xrightarrow{\text{S} \quad \text{NH}_3 \quad \text{Pyr} \quad \text{H}_2\text{O}}$ Ph(CH$_2$)$_2$CONH$_2$ 48%

JACS (1946) 68 2029

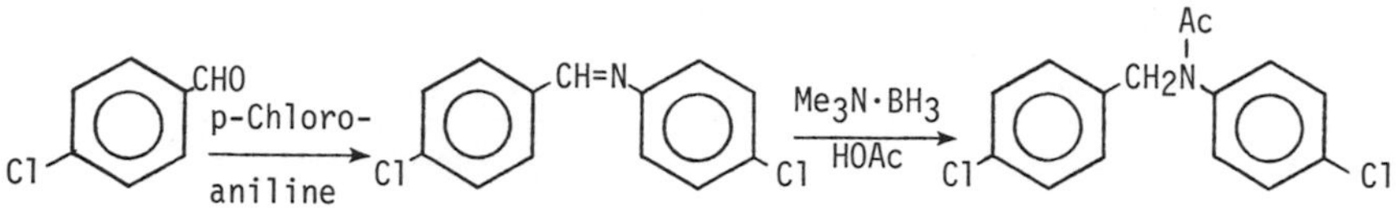

JOC (1962) 27 2640

Section 80   Amides from Alkyls

This section lists the conversion of alkyl groups into amide. For the
conversion RH $\longrightarrow$ RCONR$_2$ see section 86 (Amides from Hydrides)

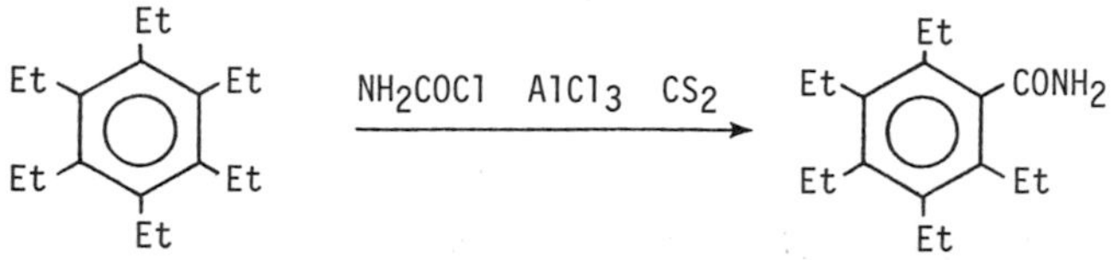

98%

Helv (1960) 43 1473

Section 81   <u>Amides from Amides</u>

$$MeCH_2CONMe_2 \xrightarrow[\text{2 EtBr}]{\text{1 NaNH}_2 \quad \text{NH}_3} MeCHCONMe_2 \quad \overset{|}{Et}$$ 62%

JOC (1966) <u>31</u> 982 989
Ber (1968) <u>101</u> 3113
JACS (1967) <u>89</u> 1647

$$\xrightarrow[\text{Me}_2\text{CO} \quad h\nu]{\text{C}_5\text{H}_{11}\text{CH=CH}_2}$$

31%   14%
Chem Ind (1965) 768

$$MeCONH_2 \xrightarrow[\text{2 H}_2]{\text{1 MeCH=CH}_2 \quad \text{PdCl}_2 \quad \text{Na}_2\text{HPO}_4} MeCONHCHMe_2$$

Proc Chem Soc (1961) 370

$$MeCONH_2 \xrightarrow[\text{H}_2\text{SO}_4 \quad \text{HOAc}]{\text{MeCH(OEt)}_2 \quad \text{H}_2 \quad \text{Pd-C}} MeCONHEt$$ 45%

JOC (1962) <u>27</u> 2205

$$MeCONH_2 \xrightarrow{\text{MeNH}_2 \cdot \text{HCl} \quad \Delta} MeCONHMe$$ 75%

JACS (1943) <u>65</u> 1566

$PhCH_2CONH_2$ $\xrightarrow{PhNH_2\ \ BF_3\ \ \Delta}$ $PhCH_2CONHPh$ 99%

JACS (1937) 59 1202

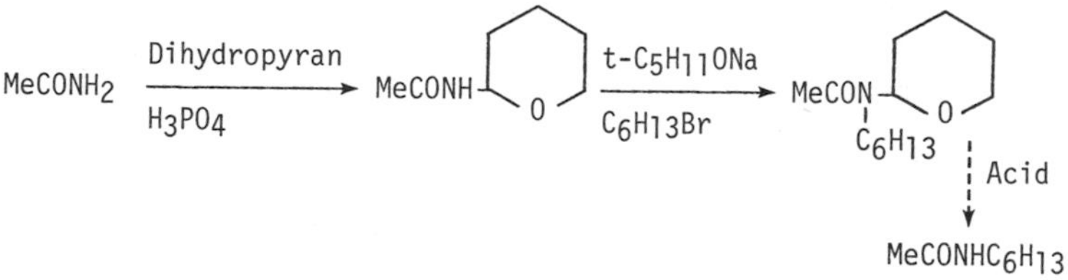
MeCONH2
Dihydropyran
H3PO4
MeCONH
O
t-C5H11ONa
C6H13Br
MeCON
C6H13
O
Acid
MeCONHC6H13

Bull Soc Chim Fr (1964) 292

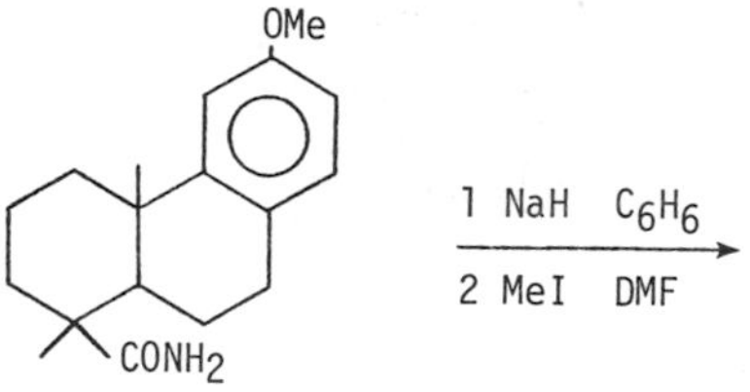
OMe
CONH2
1 NaH C6H6
2 MeI DMF
OMe
CONHMe

> 66%

JACS (1961) 83 1492
JOC (1949) 14 1099
(1967) 32 3679

Examples of the N-alkylation of acetyl- and trifluoroacetyl-amines are included in section 97 (Amines from Amines)

PhCONHPh $\xrightarrow{PCl_5}$ PhC(Cl)=NPh $\xrightarrow{PhONa}$ PhC(OPh)=NPh $\xrightarrow{315°}$ $PhCONPh_2$

Org React (1965) 14 1

$PhCONBu_2$ $\xrightarrow[190\text{-}200°]{\text{Pyridine hydrochloride}}$ PhCONHBu 30%

Ber (1954) 87 1294

PrCONMe$_2$ $\xrightarrow{\text{K}_2\text{S}_2\text{O}_8 \quad \text{K}_2\text{HPO}_4 \quad \text{H}_2\text{O}}$ PrCONHMe   44%

JOC (1964) <u>29</u> 3632

CH$_2$=CH(CH$_2$)$_7$CONH$_2$ $\xrightarrow{\text{Pb(OAc)}_4 \quad \text{C}_6\text{H}_6 \quad \text{HOAc}}$ CH$_2$=CH(CH$_2$)$_7$NHAc   53%

Chem Comm (1965) 161
Aust J Chem (1968) <u>21</u> 185

$\xrightarrow{\text{Ni} \quad \text{EtOH}}$

JACS (1954) <u>76</u> 5774
JOC (1957) <u>22</u> 148

Section 82   Amides from Amines
○○○○○○○○○○○○○○○○○○○

Amides by reaction of amines with carboxylic acids, amides,
    esters and other derivatives of carboxylic acids . . . . page 213-217
Dealkylation of amines by carboxylic acids and anhydrides . . . 217
Oxidation of amines to amides 217-218

$\xrightarrow{\text{MeCOCl} \quad \text{Pyr}}$ 80%

JOC (1970) <u>35</u> 1219

$\xrightarrow[\text{NaOH} \quad \text{H}_2\text{O} \quad \text{Et}_2\text{O}]{\text{Ph}_2\text{C}=\text{CHCOCl}}$

JOC (1970) <u>35</u> 825
Ber (1954) <u>87</u> 1760
For mild procedure see JCS (1962) <u>1445</u>

Tetr Lett (1964) 1597

i-PrCHNH$_2$ $\xrightarrow{\text{HCOOH   Ac}_2\text{O}}$ i-PrCHNHCHO   85-90%
  | |
 COOH   COOH

JACS (1958) <u>80</u> 1154
     (1968) <u>90</u> 3245

PhNH$_2$ $\xrightarrow{\text{PhCOOH   PhSO}_2\text{Cl   Pyr}}$ PhCONHPh   94%

JACS (1955) <u>77</u> 6214

PhCH$_2$NH$_2$ $\xrightarrow[\text{Me}_2\text{SO}]{\overset{+}{\text{PhCOOH   Me}_2\text{SCH}_2\text{C}\equiv\text{CH}}\ \overset{-}{\text{Br}}}$ PhCH$_2$NHCOPh   91%

JCS <u>C</u> (1969) 1904

93%

Compt Rend (1965) <u>260</u> 2249
JACS (1968) <u>90</u> 4706

Further examples of the reaction R$_2$NH + RCOOH $\longrightarrow$ RCONR$_2$ are included
in section 77 (Amides from Carboxylic Acids and Acid Halides) and section
105A (Protection of Amines)

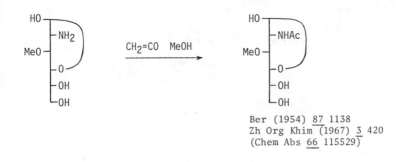

Ber (1954) 87 1138
Zh Org Khim (1967) 3 420
(Chem Abs 66 115529)

PhNH₂ $\xrightarrow[\text{2 PhCOOEt}]{\text{1 LiAlH}_4 \quad \text{THF}}$ PhNHCOPh

JOC (1962) 27 1042
JCS (1954) 1188

Further examples of the reaction R₂NH + RCOOR ⟶ RCONR₂ are included
in section 83 (Amides from Esters)

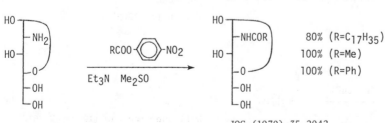

JOC (1961) 26 2563   97%

PhCH₂NH₂·HCl $\xrightarrow{\text{MeCONH}_2 \quad \Delta}$ PhCH₂NHAc

JACS (1943) 65 1566

80% (R=C₁₇H₃₅)
100% (R=Me)
100% (R=Ph)

JOC (1970) 35 2042
Bull Chem Soc Jap (1963) 36 754

$HOCH_2CH_2NH_2 \xrightarrow{CH_2=C(CN)OAc \quad CHCl_3} HOCH_2CH_2NHAc$

Chem Zvesti (1964) 18 218
(Chem Abs 61 14773)

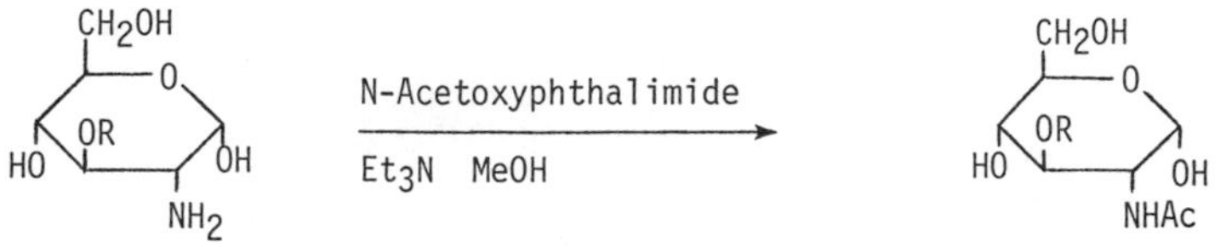
CH2OH
O
OR
HO
OH
NH2
N-Acetoxyphthalimide
Et3N MeOH
CH2OH
O
OR
HO
OH
NHAc

70%

JOC (1965) 30 448

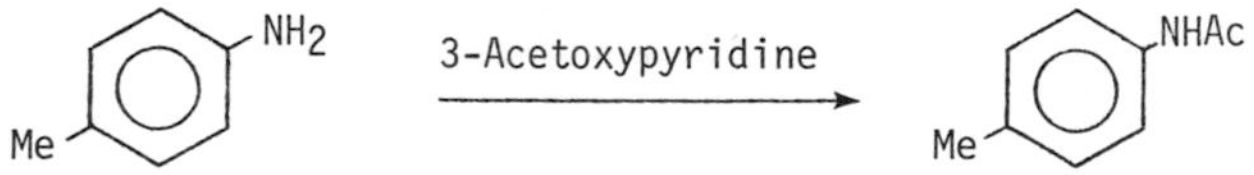
NH2
Me
3-Acetoxypyridine
NHAc
Me

88%

Bull Chem Soc Jap (1964) 37 864

$BuNH_2 \xrightarrow{CCl_3CHO \quad CHCl_3} BuNHCHO$ 83%

JACS (1952) 74 3933

$BuNH_2 \xrightarrow{CO \quad RhCl_2(CO)_4 \quad Me_3P \quad C_6H_6} BuNHCHO$ 96%

Tetr Lett (1969) 2329

$Me_2NH \xrightarrow{CO \quad CuCl} Me_2NCHO$ 73%

Bull Chem Soc Jap (1969) 42 2610

MeO—(ring)—$(CH_2)_2NH_2$　N_2CHCO—(ring)—OMe　$\xrightarrow{h\nu \quad THF}$　MeO—(ring)—$(CH_2)_2NHCOCH_2$—(ring)—OMe

Chem Ind (1969) 493　　　　　72%
JCS C (1970) 1208
Org React (1942) 1 38

$PhNMe_2$　$\xrightarrow[210-220°]{Ph(CH_2)_2COOH}$　$Ph(CH_2)_2CONMe$
　　　　　　　　　　　　　　　　　　$\underset{Ph}{|}$　　　　　15%

Ber (1930) 63B 489

$PhNEt_2$　$\xrightarrow{Pb(OAc)_4 \quad Ac_2O \quad CHCl_3}$　$PhNAc$
　　　　　　　　　　　　　　　　　　　　$\underset{Et}{|}$　　　　90%

Ber (1959) 92 288

$\xrightarrow{Ac_2O}$

28%

Bull Soc Chim Fr (1964) 234

Ph_2NCH_2Me　$\xrightarrow{KMnO_4 \quad NaHCO_3 \quad Me_2CO}$　Ph_2NAc　　　　70%

JCS (1946) 454
JACS (1968) 90 1648

$PhNMe$　$\xrightarrow{MnO_2 \quad CHCl_3}$　$PhNCHO$　　　　　　　　83% (R=H)
$\underset{R}{|}$　　　　　　　　　　　$\underset{R}{|}$　　　　　　　　　　80% (R=Me)

JCS (1957) 3032
(1966)　995

CrO$_3$ Pyr

Tetrahedron (1967) $\underline{23}$ 4691

PhCH$_2$NMe$_2$ $\xrightarrow{\text{O}_2 \quad \text{Pt} \quad \text{C}_6\text{H}_6}$ PhCH$_2$NCHO   85%

Me

Tetr Lett (1968) 4085
(1970) 5049
Chem Comm (1969) 639

PrCH$_2$NBu$_2$ $\xrightarrow{\text{O}_3 \quad \text{pentane}}$ PrCONBu$_2$ 44%

JCS (1964) 711

PhCH$_2$CH$_2$NH$_2$ $\xrightarrow{\text{S} \quad \text{(NH}_4)_2\text{S}_x \quad \text{dioxane}}$ PhCH$_2$CONH$_2$ 32%

JACS (1953) $\underline{75}$ 740 5392

Section 83   Amides from Esters

C$_8$H$_{17}$CH=CH(CH$_2$)$_7$COOMe $\xrightarrow[230°]{\text{C}_{12}\text{H}_{25}\text{NH}_2}$ C$_8$H$_{17}$CH=CH(CH$_2$)$_7$CONHC$_{12}$H$_{25}$ 69%

JACS (1949) $\underline{71}$ 2215

For catalysis by 2-hydroxypyridine see JCS $\underline{C}$ (1969) 89
     "   "   " NaNH$_2$ " Chem Ind (1956) 277
     "   "   " MeONa   " JOC (1963) $\underline{28}$ 2915

$\xrightarrow{\text{MeNH}_2-\text{LiAlH}_4}$

61%

(Procedure for hindered esters)
Tetr Lett (1969) 1573
JCS (1954) 1188
Tetr Lett (1970) 1791

i-PrCOOMe $\xrightarrow{\text{HCONHR MeONa}}$ i-PrCONHR

53% (R=H)
97% (R=Me)

JOC (1965) 30 2376

BuC≡CCOOMe $\xrightarrow{\text{NH}_3 \quad \text{MeOH}}$ BuC≡CCONH$_2$

JACS (1941) 63 1151

PhCOOEt $\xrightarrow{\text{NH}_3 \quad \text{NH}_4\text{Cl}}$ PhCONH$_2$

JACS (1938) 60 579

For catalysis by NaNH$_2$ see JACS (1955) 77 469
          "   "   "   MeONa   "   JACS (1937) 59 1568
          "   "   "   BuLi   "   Tetr Lett (1970) 1791

$\xrightarrow[\text{2 NaNO}_2 \quad \text{HOAc}]{\text{1 N}_2\text{H}_4 \cdot \text{H}_2\text{O}}$

$\xrightarrow{\text{Ac}_2\text{O}}$

Helv (1944) 27 883
Org React (1946) 3 337

HCOOCPh$_3$ $\xrightarrow{\text{MeCN} \quad \text{H}_2\text{SO}_4}$ AcNHCPh$_3$

JCS (1964) 5609
Org React (1969) 17 213

Section 84   Amides from Ethers and Epoxides

t-BuOMe $\xrightarrow{\overset{\text{CN}}{\underset{}{(\text{CH}_2)_4\text{CN}}} \ \text{H}_2\text{SO}_4 \ \ \text{HOAc}}$ $\overset{\text{CONHBu-t}}{\underset{}{(\text{CH}_2)_4\text{CONHBu-t}}}$ 75%

Org React (1969) 17 213

Electrolysis   Et$_4$NCN

MeCN   H$_2$O   9%

JACS (1969) 91 4181

$\xrightarrow{\text{S} \ \ (\text{NH}_4)_2\text{S}_x \ \ \text{dioxane}}$ PhCH$_2$CONH$_2$ 87%

JACS (1953) 75 740

Section 85   Amides from Halides

PhCH$_2$Cl $\xrightarrow{\begin{array}{c}1 \ \overset{\text{COOEt}}{\underset{}{\text{MeCHCONMe}_2}} \ \ \text{NaNH}_2 \ \ \text{HMPA} \\ 2 \ \text{LiI} \ \ \text{lutidine}\end{array}}$ $\overset{\text{Me}}{\underset{}{\text{PhCH}_2\text{CHCONMe}_2}}$ 46%

Ber (1968) 101 4230

C$_5$H$_{11}$Br $\xrightarrow{\text{LiCH}_2\text{CONMe}_2 \ \ \text{THF}}$ C$_5$H$_{11}$CH$_2$CONMe$_2$ 64%

Ber (1968) 101 3113
JOC (1966) 31 982 989

$$\text{PhCH}_2\text{Cl} \xrightarrow[\text{LiCH}_2\text{CONPh} \quad \text{Et}_2\text{O}]{\overset{\text{Li}}{|}} \text{PhCH}_2\text{CH}_2\text{CONHPh} \qquad 69\%$$

JACS (1967) <u>89</u> 1647

$$\text{PhCH=CHBr} \xrightarrow[\text{MeOH}]{\text{Pyrrolidine} \quad \text{Ni(CO)}_4} \text{PhCH=CHCON}\bigcirc \qquad 82\%$$

JACS (1969) <u>91</u> 1233

$$\text{C}_7\text{H}_{15}\text{Br} \xrightarrow[\text{2 PhNCO}]{\text{1 Mg}} \text{C}_7\text{H}_{15}\text{CONHPh} \qquad 88\%$$

JACS (1947) <u>69</u> 2007
(1938) <u>60</u> 540

1 Mg

2 Ethyl cyclohexylidene-
carbamate

Bull Res Council Israel (1952) <u>2</u> 72
(Chem Abs <u>48</u> 8727)

$$\text{PhCH}_2\text{CH}_2\text{Br} \xrightarrow[]{\text{S} \quad \text{NH}_3 \quad \text{H}_2\text{O} \quad 170°} \text{PhCH}_2\text{CONH}_2 \qquad 80\%$$

JACS (1953) <u>75</u> 740 5395

$$\text{C}_8\text{H}_{17}\text{Br} \xrightarrow[]{\text{HCONH}_2 \quad \text{NH}_3} \text{C}_8\text{H}_{17}\text{NHCHO} \qquad 91\%$$

JOC (1969) <u>34</u> 3204

For examples of the reaction RCONHR' + R"Hal $\longrightarrow$ RCONR'R" (R'=H or
alkyl) see section 81 (Amides from Amides)

Section 86   Amides from Hydrides (RH)

The conversions RH $\longrightarrow$ RCH$_2$CONH$_2$, RCONR'R" or RNHAc are included in this section. For the reaction RR' $\longrightarrow$ RCONH$_2$ (R=Ar, R'=alkyl) see section 80 (Amides from Alkyls)

Tetr Lett (1969) 2387   21%

Ber (1955) 88 301
JCS (1931) 2323

JOC (1970) 35 2104   27%

Tetr Lett (1963) 77   20%

PhCH$_3$ $\xrightarrow{\text{HCONH}_2 \quad h\nu \quad \text{Me}_2\text{CO}}$ PhCH$_2$CONH$_2$ 23%

Tetr Lett (1963) 77

PhCH$_3$ $\xrightarrow[\text{LiClO}_4 \quad \text{H}_2\text{O}]{\text{Electrolysis   MeCN}}$ PhCH$_2$NHAc

Tetr Lett (1968) 2411

$\xrightarrow[\text{H}_2\text{SO}_4 \quad \text{hexane}]{\text{MeCN   t-BuOH}}$

—NHAc   36%

Ber (1964) 97 3234
Org React (1969) 17 213

Section 87   Amides from Ketones
∘∘∘∘∘∘∘∘∘∘∘∘∘∘∘∘∘∘∘∘∘∘

$\xrightarrow[\text{2 HClO}_4]{\text{1 Pyrrolidine}}$

ClO$_4^-$
Helv (1967) 50 1759

$\xrightarrow[\text{h}\nu]{\text{HCONH}_2}$

CONH$_2$ 16%

MeCO   - - - ->   MeC=NOH $\xrightarrow{\text{TsCl   Pyr}}$ NHAc   ~92%

JACS (1962) 84 1064

PhCOMe   - - ->   PhC=NOH $\xrightarrow{\text{CF}_3\text{COOH}}$ PhNHAc   ~91%
                         |
                         Me

Org React (1960) 11 1

PhCOCMe$_2$ - - -➤ PhCCMe$_2$ $\xrightarrow{\text{Polyphosphoric acid}}$ PhCONH$_2$ ~63%
  | ‖ NOH
  Ph   Ph

JOC (1963) 28 278

$\xrightarrow{h\nu \quad iPrOH}$ C$_5$H$_{11}$CONH$_2$ 44%

$\xrightarrow{h\nu \quad MeOH}$

Can J Chem (1968) 46 3381

1 Br$_2$
2 Ph$_3$P
3 Δ
4 H$_2$O   ~81%

Tetr Lett (1965) 4541

PhNH$_2$

1 MeCOO$_2$H   Et$_2$O
2 Xylene   reflux

Ber (1958) 91 1057
Tetr Lett (1969) 2281

$\xrightarrow{\text{S \quad NH}_3 \quad \text{Pyr} \quad \text{H}_2\text{O}}$ 40%

Org React (1946) 3 83

R$_2$CO $\xrightarrow[\text{EtOH}]{\text{N}_2\text{H}_4 \cdot \text{H}_2\text{O}}$ R$_2$C=NNH$_2$ $\xrightarrow[\text{H}_2\text{SO}_4 \quad \text{H}_2\text{O}]{\text{NaNO}_2}$ RCONHR   40% (R=i-Pr)
                                                                                                72% (R=Ph)

JACS (1953) 75 5905

PhCH$_2$COMe $\xrightarrow{\text{NaN}_3 \quad \text{polyphosphoric acid}}$ PhCH$_2$NHAc   50%

JOC (1958) <u>23</u> 1330
JCS (1942) <u>61</u>
Org React (1946) <u>3</u> 307

$\xrightarrow{\text{NaNH}_2 \quad \text{toluene}}$

88%

JACS (1953) <u>75</u> 369
Org React (1957) <u>9</u> 1

Amides may also be prepared by conversion of ketones into amines
followed by acylation. See section 102 (Amines from Ketones)

Section 88   Amides from Nitriles
              ○○○○○○○○○○○○○○○○○○○○○

PhCH$_2$CN $\xrightarrow{\text{Me}_2\text{C=CHEt} \quad \text{H}_2\text{SO}_4 \quad \text{H}_2\text{O}}$ PhCH$_2$CONHCMe$_2$ 69%
                                                        |
                                                        Pr

JACS (1948) <u>70</u> 4045
Org React (1969) <u>17</u> 213

MeCN $\xrightarrow[\text{EtCHOH BF}_3]{\text{Me}}$ MeCONHCHEt   65%
                                          (Me above CHEt)

Acta Chem Scand (1968) <u>22</u> 1787

PhCN $\xrightarrow{\text{(i-PrO)}_2\text{CH}^+ \text{ BF}_4^- \quad \text{CH}_2\text{Cl}_2}$ PhCONHPr-i

JOC (1969) <u>34</u> 627

PhCH$_2$CN $\xrightarrow{\text{HCl} \quad \text{H}_2\text{O}}$ PhCH$_2$CONH$_2$

Org Synth (1963) Coll Vol 4 760
JACS (1948) 70 3091
Annalen (1968) 713 212

PhCH$_2$CN $\xrightarrow{\text{BF}_3\cdot\text{AcOH}}$ PhCH$_2$CONH$_2$ 95%

JOC (1955) 20 1448

BuCN $\xrightarrow{\overset{+ \quad -}{\text{Et}_3\text{O BF}_4} \quad \text{CH}_2\text{Cl}_2}$ BuCONH$_2$ 93%

JOC (1969) 34 627

RCN $\xrightarrow{\text{MnO}_2 \quad \text{CH}_2\text{Cl}_2}$ RCONH$_2$ 40% (R=Me)
72% (R=Ph)

Chem Comm (1966) 121

89%

Bull Chem Soc Jap (1966) 39 8
(1964) 37 1325

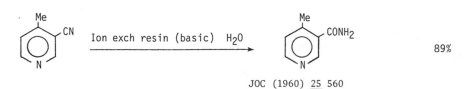

90-92%

Org Synth (1943) Coll Vol 2 586
JOC (1950) 15 800

JOC (1960) 25 560

89%

$$PhCN \xrightarrow{\text{NaOH} \quad Me_2SO} PhCONH_2 \qquad 96\%$$

JCS (1965) 1290

$$PhCH_2CN \xrightarrow{\text{H}_2 \quad Ni \quad NaOAc \quad Ac_2O} PhCH_2CH_2NHAc \qquad 97\%$$

JOC (1960) 25 1658

Amides may also be prepared by reduction of nitriles to amines followed by acylation. See section 103 (Amines from Nitriles)

Section 89   Amides from Olefins
ooooooooooooooooooooo

$$BuCH=CH_2 \xrightarrow[\substack{+- \\ 2 \; Me_2SCHCONEt_2}]{1 \; B_2H_6 \quad THF} Bu(CH_2)_3CONEt_2 \qquad \sim50\%$$

JACS (1967) 89 6804

$$C_5H_{11}CH=CH_2 \xrightarrow[\text{di-t-butyl peroxide}]{CH_3COONHEt} C_5H_{11}(CH_2)_3CONHEt$$

Dokl (1964) 158 1127
(Chem Abs 62 2703)
JCS (1965) 1918

$$C_6H_{13}CH=CH_2 \xrightarrow[\text{di-t-butyl peroxide}]{HCONMe_2} C_6H_{13}(CH_2)_2CONMe_2 \qquad 56\%$$

Tetr Lett (1961) 238

BuCH=CH$_2$ $\xrightarrow[\text{Me}_2\text{CO} \quad \text{t-BuOH}]{\text{HCONH}_2 \quad h\nu}$ Bu(CH$_2$)$_2$CONH$_2$ 50%

JOC (1964) <u>29</u> 1855
(1965) <u>30</u> 3361
Angew (1961) <u>73</u> 621

$\xrightarrow{\text{PhNH}_2 \quad \text{CO} \quad \text{Co}}$ 70%

JACS (1952) <u>74</u> 4496

PhCH$_2$CH=CH$_2$ $\xrightarrow{\text{(NH}_4\text{)}_2\text{S}_x \quad \text{H}_2\text{O}}$ Ph(CH$_2$)$_2$CONH$_2$

JACS (1946) <u>68</u> 632 2025 2029

$\xrightarrow[\text{2 PhSH} \quad \text{Pyr} \quad \text{Me}_2\text{CO}]{\text{1 ClSO}_2\text{NCO}}$ 66%

JOC (1968) <u>33</u> 370
(1970) <u>35</u> 2043
Annalen (1968) <u>718</u> 94

BuCH=CH$_2$ $\xrightarrow[\text{2 NaBH}_4 \quad \text{NaOH} \quad \text{H}_2\text{O}]{\text{1 Hg(NO}_3\text{)}_2 \quad \text{MeCN}}$ BuCHMe   70%
                                                                  |
                                                                NHAc

JACS (1969) <u>91</u> 5647

C$_{12}$H$_{25}$CH=CH$_2$ $\xrightarrow{\text{NaCN} \quad \text{H}_2\text{SO}_4 \quad \text{MeCN}}$ C$_{12}$H$_{25}$CHMe   68%
                                                                              |
                                                                            NHAc

J Am Oil Chem Soc (1964) <u>41</u> 78
(Chem Abs <u>60</u> 6733)
JACS (1948) <u>70</u> 4045
Org React (1969) <u>17</u> 213

$MeCH=CH_2$ $\xrightarrow{\begin{array}{c} 1 \ MeCONH_2 \quad PdCl_2 \\ \hline Na_2HPO_4 \\ 2 \ H_2 \end{array}}$ $Me_2CHNHAc$

Proc Chem Soc (1961) 370

Amides may also be prepared by conversion of olefins into amines followed by acylation. See section 104 (Amines from Olefins)

Section 90   <u>Amides from Miscellaneous Compounds</u>
ᴏᴏ

H_2 Pd-BaSO$_4$ EtOH

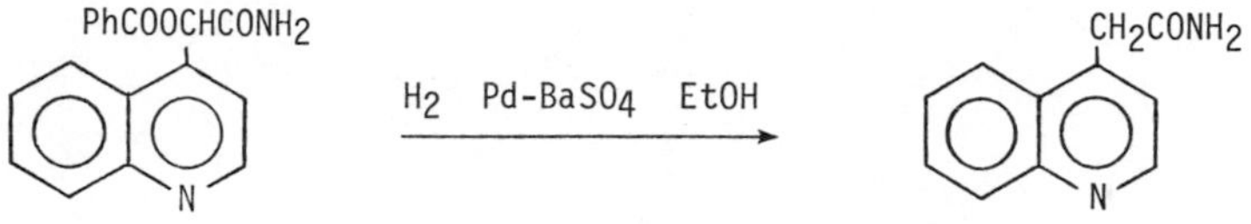

90%

Arch Pharm (1957) <u>290</u> 218

$C_6H_{13}CH=CHCOOH$ $\xrightarrow{\text{S} \quad NH_3 \quad Pyr \quad H_2O}$ $C_6H_{13}CH_2CONH_2$ 41%

JOC (1947) <u>12</u> 76

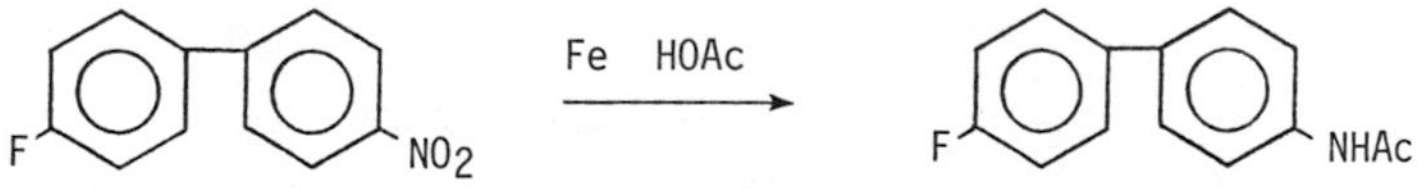

Fe   HOAc

71%

JACS (1966) <u>88</u> 3318

Chapter 7   PREPARATION
OF
AMINES

Section 91   <u>Amines from Acetylenes</u>

$$MeC\equiv CH \xrightarrow[\text{Zn(OAc)}_2]{\text{Me}_2\text{NH}\quad\text{Cd(OAc)}_2} \underset{\underset{\text{NMe}_2}{|}}{\text{Me}_2\text{CC}\equiv\text{CMe}} \xrightarrow[\text{HOAc}]{\text{H}_2\quad\text{Pt}} \underset{\underset{\text{NMe}_2}{|}}{\text{Me}_2\text{CCH}_2\text{CH}_2\text{Me}}$$

JACS (1961) <u>83</u> 213 216

Section 92   <u>Amines from Carboxylic Acids and Acid Halides</u>

$$C_{11}H_{23}COOH\cdot NH_3 \xrightarrow[\text{copper chromite}]{\text{H}_2\ (300\ \text{atmos})} (C_{11}H_{23}CH_2)_2NH \qquad 79\%$$

JACS (1934) <u>56</u> 2419

$$C_9H_{19}COOH \xrightarrow{\text{Li}\quad\text{MeNH}_2} C_9H_{19}CH=NMe \xrightarrow[\text{MeNH}_2]{\text{H}_2\quad\text{Pd-C}} C_9H_{19}CH_2NHMe \qquad 68\%$$

JACS (1970) <u>92</u> 5774

$C_{17}H_{35}COOH$ $\xrightarrow{\text{HN}_3 \quad \text{H}_2\text{SO}_4 \quad \text{C}_6\text{H}_6}$ $C_{17}H_{35}NH_2$ 96%

Org React (1946) <u>3</u> 307

$PhCH_2COCl$ $\xrightarrow[\text{2 HCl} \quad \text{H}_2\text{O}]{\text{1 NaN}_3 \quad \text{C}_6\text{H}_6}$ $PhCH_2NH_2$ 94%

Org React (1946) <u>3</u> 337
Nature (1963) <u>197</u> 787

Ph—△—COOH
1 ClCOOEt Et₃N Me₂CO
2 NaN₃
3 Toluene 100°
4 HCl H₂O
→ Ph—△—NH₂   77%

JOC (1961) <u>26</u> 3511

1 $N_2H_4 \cdot H_2O$
2 $NaNO_2$ HOAc H_2O
3 HCl H_2O

Rec Trav Chim (1921) <u>40</u> 285

$\xrightarrow[\text{polyphosphoric acid}]{\text{NH}_2\text{OH} \cdot \text{HCl}}$ 76%

JACS (1953) <u>75</u> 2014
Org React (1946) <u>3</u> 337
Chem Rev (1943) <u>33</u> 209

$C_5H_{11}COOH$ $\xrightarrow[\text{170-180°}]{\text{NH}_2\text{OSO}_3\text{H} \quad \text{mineral oil}}$ $C_5H_{11}NH_2$ 25%

JOC (1964) <u>29</u> 2576

$$PhCOOH \xrightarrow{\quad MeNO_2 \quad polyphosphoric\ acid \quad} PhNH_2 \qquad\qquad 68\%$$

JOC (1964) <u>29</u> 2576

+

43%   23%

Ber (1963) <u>96</u> 3359
Org React (1969) <u>17</u> 213

Section 93   Amines from Alcohols and Phenols

$$C_{12}H_{25}OH \xrightarrow[\text{Cu-Ba-Cr oxide}]{\text{Et}_3\text{N} \quad \text{H}_2 \text{ (380 atmos)}} C_{12}H_{25}NEt_2 \qquad 70\%$$

JACS (1952) <u>74</u> 4287

JCS (1946) 393
JACS (1933) <u>55</u> 345

Further examples of the preparation of amines from tosylates are
included in section 100 (Amines from Halides and Sulfonates)

$$\underset{\text{Me}}{PhCHOH} \xrightarrow{\text{KCN} \quad \text{H}_2\text{SO}_4 \quad \text{Bu}_2\text{O}} \underset{\text{Me}}{PhCHNCHO} \xrightarrow{\text{NaOH}} \underset{\text{Me}}{PhCHNH_2} \qquad 60\%$$

Org React (1969) <u>17</u> 213
JOC (1961) <u>26</u> 3002

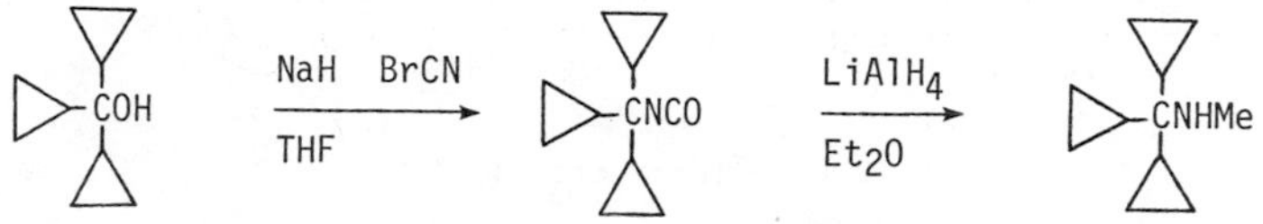

JOC (1968) 33 4054
JACS (1964) 86 4732

PhCHOH
|
Me

1 NaH MeOCH$_2$CH$_2$OMe

2 Me$_2$NSO$_2$Cl

3 60°

PhCHNMe$_2$
|
Me

JACS (1965) 87 5261

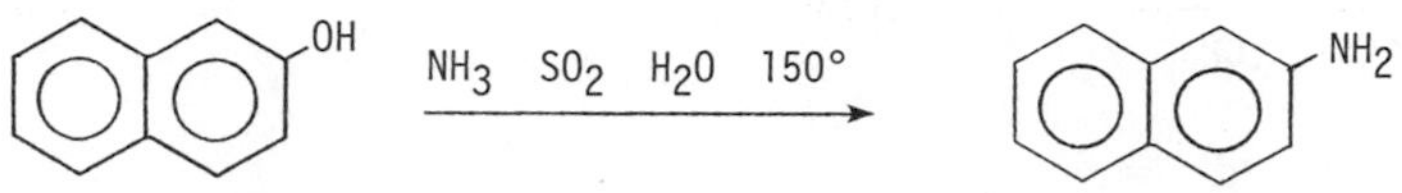

NH$_3$ SO$_2$ H$_2$O 150°

94-96%

Org React (1942) 1 105

Section 94   Amines from Aldehydes
○○○○○○○○○○○○○○○○○○○○○○

PhCHO

$\xrightarrow[C_6H_6]{MeNH_2}$

PhCH=NMe

$\xrightarrow[Et_2O]{PhCH_2MgBr}$

PhCHNHMe
|
CH$_2$Ph

84%

Org Synth (1963) Coll Vol 4 605

PhCHO

$\xrightarrow[KOH]{HCN}$

PhCHCN
|
OH

$\xrightarrow[HCl\ \ EtOH]{H_2\ \ Pd-C}$

PhCH$_2$CH$_2$NH$_2$

52%

JACS (1928) 50 3370

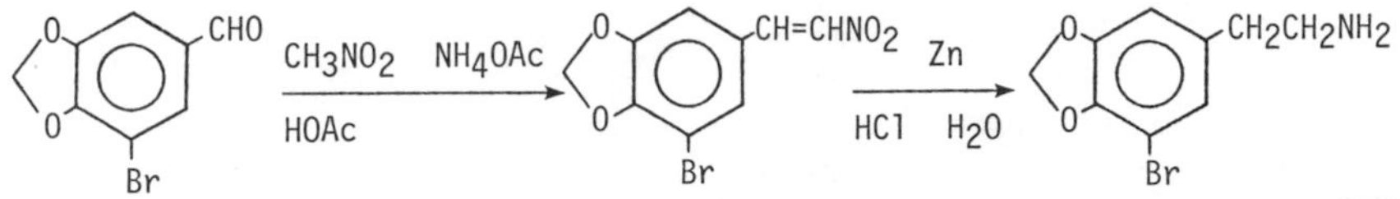

Tetr Lett (1967) 1201   50%

$$BuCHO \xrightarrow[\text{}]{HCOONH_4 \quad HCOOH} (BuCH_2)_3N$$

Org React (1949) 5 301

$$C_6H_{13}CHO \xrightarrow[NaOAc \quad EtOH]{PhNH_2 \quad H_2 \quad Ni} C_6H_{13}CH_2NHPh$$ 65%

Org React (1948) 4 174

$$PhCHO \xrightarrow[EtOH]{NH_3 \quad H_2 \quad Ni} PhCH_2NH_2$$ 89%

Org React (1948) 4 174

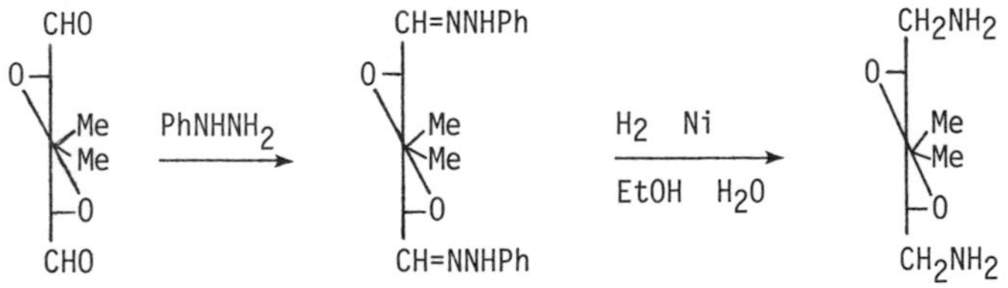

JOC (1958) 23 571

$$PhCHO \xrightarrow[\text{2 } H_2 \quad PdO]{\text{1 } NH_2COOCH_2Ph} PhCH_2NH_2$$ 63%

JOC (1941) 6 878

PhCHO $\xrightarrow[\text{MeOH \quad pH 5-6}]{\text{EtNH}_2 \quad \text{LiBH}_3\text{CN}}$ PhCH$_2$NHEt   72%

JACS (1969) <u>91</u> 3996

PhCHO $\xrightarrow[\text{HOAc \quad EtOH}]{\text{PhNH}_2 \quad \text{NaBH}_4 \quad \text{NaOAc}}$ PhCH$_2$NHPh   83%

JOC (1963) <u>28</u> 3259

PhCHO $\xrightarrow{\text{o-Chloroaniline}}$ PhCH=N⟨⟩Cl $\xrightarrow[\text{HOAc}]{\text{Me}_2\text{NH-BH}_3}$ PhCH$_2$NH⟨⟩Cl   84%

JOC (1961) <u>26</u> 1437

$\xrightarrow[\text{EtOH}]{\text{NH}_2\text{OH}}$ $\xrightarrow[\text{Et}_2\text{O}]{\text{LiAlH}_4}$ 87%

J Biol Chem (1954) <u>211</u> 725

For reduction of oximes with H$_2$ and Pd-C see JACS (1928) <u>50</u> 3370

    "   "   "   "   "   Ni   " JCS <u>C</u> (1966) 531

    "   "   "   "   "   Na and EtOH " Org Synth (1943) Coll Vol 2 318

    "   "   "   "   "   Na-Hg   " JACS (1949) <u>71</u> 2257

PhCHO $\xrightarrow{\text{NH}_2\text{OMe}}$ PhCH=NOMe $\xrightarrow[\text{2 KOH \quad H}_2\text{O}]{\text{1 B}_2\text{H}_6 \quad \text{THF}}$ PhCH$_2$NH$_2$ 92%

JOC (1969) <u>34</u> 1817

PhCHO $\xrightarrow{\text{BuN}_3 \quad \text{H}_2\text{SO}_4 \quad \text{C}_6\text{H}_6}$ PhNHBu   22%

JOC (1959) <u>24</u> 561

Section 95   Amines from Alkyls, Methylenes and Aryls

No examples of the reaction RR' $\longrightarrow$ RNR$_2$ (R,R'=alkyl, aryl etc.) occur
in the literature. For the conversion RH $\longrightarrow$ RNR$_2$ see section 101
(Amines from Hydrides)

Section 96   Amines from Amides

$$(i\text{-}Pr)_2NCHO \xrightarrow{\text{BuMgBr Et}_2O} (i\text{-}Pr)_2NCHBu_2 \qquad 67\%$$

Monatsh (1951) 82 330

$$C_6H_{13}CON \langle \quad \xrightarrow[\text{dioxane 250°}]{\text{H}_2 \text{ (300 atmos) copper chromite}} \quad C_6H_{13}CH_2N \langle \quad 92\%$$

JACS (1934) 56 2419

$$\xrightarrow{\text{LiAlH}_4 \text{ Et}_2O} \qquad 84\%$$

Helv (1948) 31 1397

$$PhOCH_2CONH_2 \xrightarrow{\text{LiAlH}_4 \text{ Et}_2O} PhOCH_2CH_2NH_2 \qquad 80\%$$

Helv (1948) 31 1397
Org React (1951) 6 469
Org Synth (1963) Coll Vol 4 564

$i\text{-Pr}(CH_2)_2CONEt_2$ $\xrightarrow[\text{Et}_2O]{\text{LiAlH}_4 \quad \text{AlCl}_3}$ $i\text{-Pr}(CH_2)_2CH_2NEt_2$

Z. Chem (1966) 6 224

$PhNHAc$ $\xrightarrow[C_6H_6]{\text{NaAlH}_2(OCH_2CH_2OMe)_2}$ $PhNHEt$ 84%

Tetr Lett (1968) 3303

$PrCONH_2$ $\xrightarrow{\text{NaBH}_4 \quad \text{CoCl}_2 \quad \text{MeOH}}$ $PrCH_2NH_2$ 70%

Tetr Lett (1969) 4555

$C_7H_{15}CONMe_2$ $\xrightarrow{\text{NaBH}_4 \quad \text{Pyr}}$ $C_7H_{15}CH_2NMe_2$ 41%

Chem Pharm Bull (1969) 17 98

$PhCONHEt$ $\xrightarrow{\begin{array}{c} 1 \; Et_3O^+ \; BF_4^- \; CH_2Cl_2 \\ 2 \; NaBH_4 \quad EtOH \end{array}}$ $PhCH_2NHEt$ 92%

Tetr Lett (1968) 61
Tetrahedron (1970) 26 803

$C_6H_{13}CONMe_2$ $\xrightarrow{i\text{-Bu}_2\text{AlH} \quad \text{Et}_2O}$ $C_6H_{13}CH_2NMe_2$ 75-95%

Izv (1959) 2146
(Chem Abs 54 10932)

$PrCONEt_2$ $\xrightarrow{\text{Et}_3\text{SiH} \quad \text{ZnCl}_2}$ $PrCH_2NEt_2$ 70%

Compt Rend (1962) 254 2357

$C_5H_{11}CONHMe$ $\xrightarrow{\quad B_2H_6 \quad THF \quad}$ $C_5H_{11}CH_2NHMe$ 98%

JACS (1964) 86 3566
JOC (1968) 33 3637

$\xrightarrow{\begin{array}{c} 1 \ P_2S_5 \\ \hline 2 \ Na_2S \end{array}}$

$\xrightarrow{\begin{array}{c} Ni \quad EtOH \\ \hline H_2O \end{array}}$ 38%

JOC (1951) 16 131

$\xrightarrow{\quad Ca \quad NH_3 \quad}$

JACS (1962) 84 2018

Et_2NCHO $\xrightarrow{\quad NaH \quad diglyme \quad}$ Et_2NH 40%

Tetr Lett (1965) 1713

MeCHCONHCH$_2$COOH $\xrightarrow{\quad H_2O_2 \quad H_2O \quad}$ MeCHCONHCH$_2$COOH
| |
HCONH   NH$_2$

Annalen (1960) 636 140

95-97%

Org Synth (1955) Coll Vol 3 661

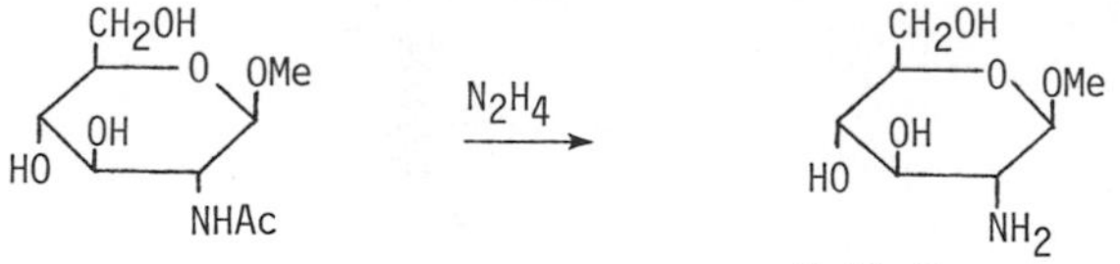

55%

Bull Chem Soc Jap (1966) *39* 185

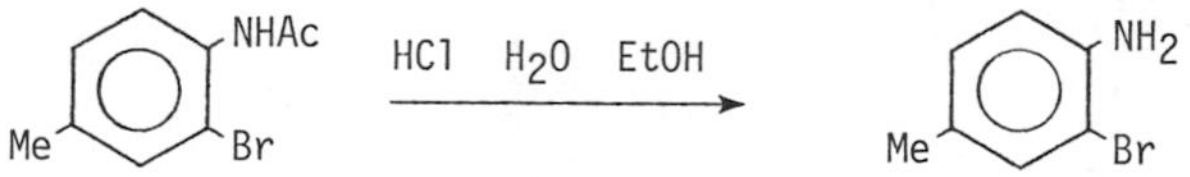

Org Synth (1932) Coll Vol 1 111

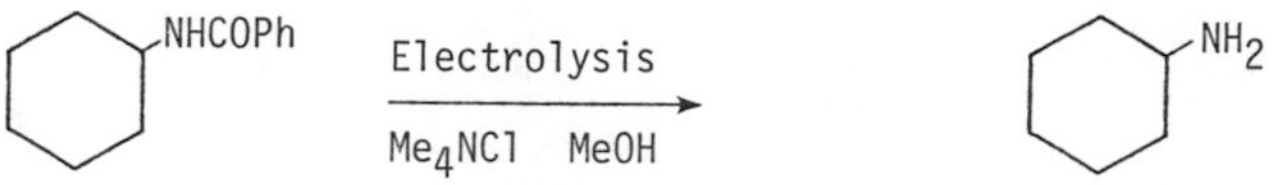

71%

Ber (1965) *98* 3462

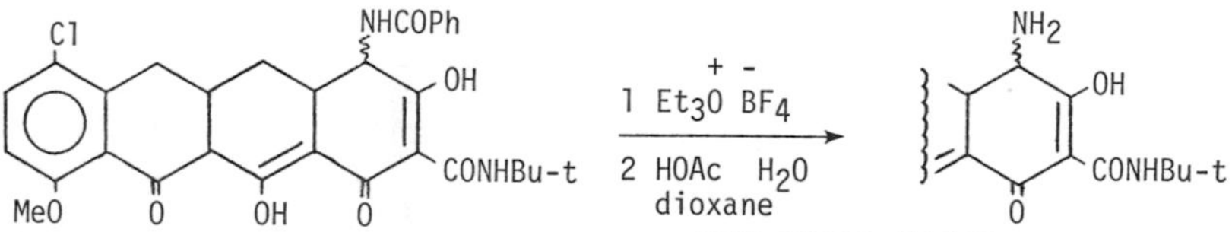

JACS (1965) *87* 933
Helv (1968) *51* 1108

$C_{15}H_{31}CONH_2$ $\xrightarrow[\text{2 CaO } H_2O]{\text{1 Br}_2 \text{ MeONa MeOH}}$ $C_{15}H_{31}NH_2$ 100%

Org React (1946) *3* 267

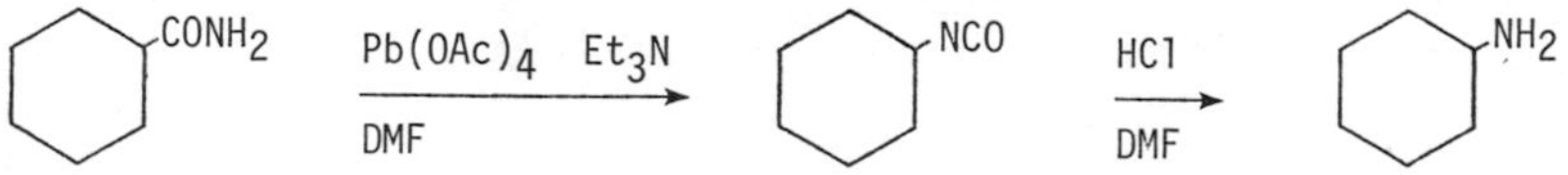

49%

JACS (1965) *87* 1141
Tetr Lett (1965) 4039

$$PhCONHPh \xrightarrow[\text{polyphosphoric acid}]{NH_2OH \cdot HCl} PhNH_2$$

                                                76%

JACS (1953) <u>75</u> 2014

Section 97   Amines from Amines

$$Et_2NCH_2Me \xrightarrow[\text{hexane}]{BuLi \quad BuI} Et_2NCHMe$$
                                                $\underset{Bu}{|}$

                                                25%

JOC (1966) <u>31</u> 2061

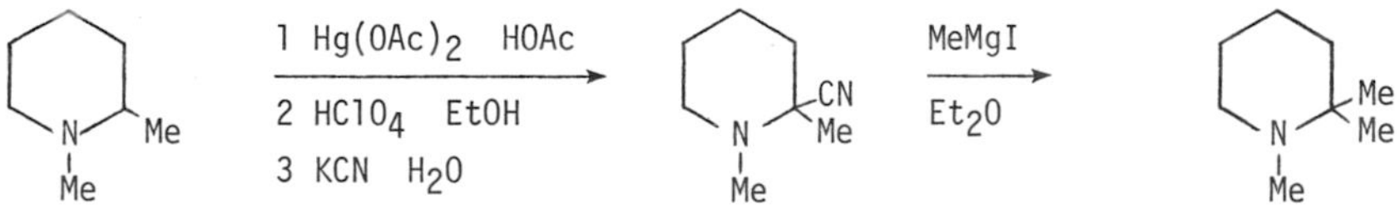

JACS (1957) <u>79</u> 5279

$$PrNH_2 \dashrightarrow[PrCHO]{} PrN=CHPr \xrightarrow{PhLi \quad Et_2O} PrNHCHPr$$
                                                $\underset{Ph}{|}$

                                                60%

Bull Soc Chim Fr (1964) 952

$(i\text{-}Pr)_2NH \cdot HCl \xrightarrow[\text{KCN} \quad H_2O]{\text{MeCHO}} (i\text{-}Pr)_2NCHCN \xrightarrow[\text{Et}_2O]{\text{MeMgCl}} (i\text{-}Pr)_2NCHMe_2$
$\qquad\qquad\qquad\qquad\qquad\qquad\qquad\quad |$
$\qquad\qquad\qquad\qquad\qquad\qquad\qquad\;\; Me$

Monatsh (1962) <u>93</u> 476

$(i\text{-}Pr)_2NH \xrightarrow{\text{HCOOH}} (i\text{-}Pr)_2NCHO \xrightarrow[\text{Et}_2O]{\text{EtMgCl}} (i\text{-}Pr)_2NCHEt_2$ 23%

Monatsh (1962) <u>93</u> 476

EtOH Ni   60-67%

Org Synth (1963) Coll Vol 4 283
JACS (1954) <u>76</u> 6174

$PhCH_2NH_2 \xrightarrow{\text{HCOOH} \quad \text{HCHO}} PhCH_2NMe_2$ 80%

Org React (1949) <u>5</u> 301

$MeNHCO(CH_2)_4NH_2 \xrightarrow[\text{H}_2O]{\text{HCHO} \quad H_2 \quad \text{Pd-C}} MeNHCO(CH_2)_4NMe_2$ 60%

JCS <u>C</u> (1969) 1358

Chem Pharm Bull (1967) <u>15</u> 1339

$$PrNH_2 \xrightarrow{\text{PrCHO} \quad \text{KOH}} PrN=CHPr \xrightarrow{\text{H}_2 \quad \text{PtO}_2 \quad \text{EtOH}} PrNHCH_2Pr \qquad 65\%$$

Org React (1948) <u>4</u> 174
Tetr Lett (1968) 2639

$$C_5H_{11}NH_2 \xrightarrow[\substack{\text{2 H}_2 \quad \text{PtO}_2 \quad \text{HOAc} \\ \text{3 BuBr} \quad \text{MeOH}}]{\text{1 PhCHO} \quad C_6H_6} C_5H_{11}\underset{\underset{CH_2Ph}{|}}{N}Bu \xrightarrow[\text{HOAc}]{\text{H}_2 \quad \text{PtO}_2} C_5H_{11}NHBu$$

JACS (1941) <u>63</u> 1964

$$\xrightarrow[\text{HCOOH} \quad \text{HOAc}]{\text{Cyclohexanone}} \qquad 86\%$$

Rev Chim (Bucharest) (1968) <u>19</u> 360
(Chem Abs <u>69</u> 105967)
Org React (1949) <u>5</u> 301

$$BuNH_2 \xrightarrow[\text{NaOAc} \quad \text{HOAc} \quad \text{H}_2\text{O}]{\text{Me}_2\text{CO} \quad \text{NaBH}_4} BuNHCHMe_2 \qquad 63\%$$

JOC (1963) <u>28</u> 3259

For further examples of the reductive alkylation of amines with aldehydes and ketones see section 94 (Amines from Aldehydes) and section 102 (Amines from Ketones)

$$\xrightarrow[\text{2 NaNH}_2 \quad \text{NH}_3]{\text{1 Ph}_3\text{PBr}_2 \quad \text{Et}_3\text{N}} \quad N=PPh_3 \xrightarrow[\text{2 KOH EtOH}]{\text{1 EtI}} NHEt \qquad \sim 67\%$$

JOC (1970) <u>35</u> 2826

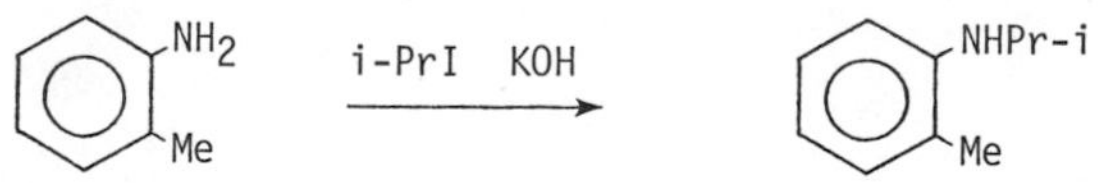

JACS (1960) **82** 6163
Ber (1952) **85** 1056

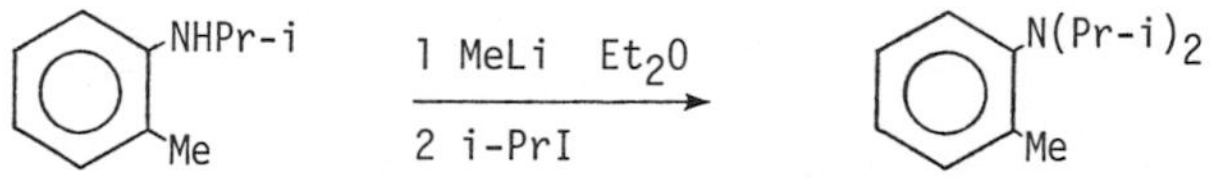

92%

JACS (1960) **82** 6163
(1964) **86** 2813

$$t\text{-BuNH}_2 \xrightarrow[\substack{2\ Me_2CC\equiv CH \\ Cl}]{1\ PhLi\ \ Et_2O} t\text{-BuN}=CHCH=CMe_2 \xrightarrow[\substack{EtOH}]{H_2\ \ Ni} t\text{-BuNHCH}_2CH_2CHMe_2$$

JOC (1961) **26** 3772

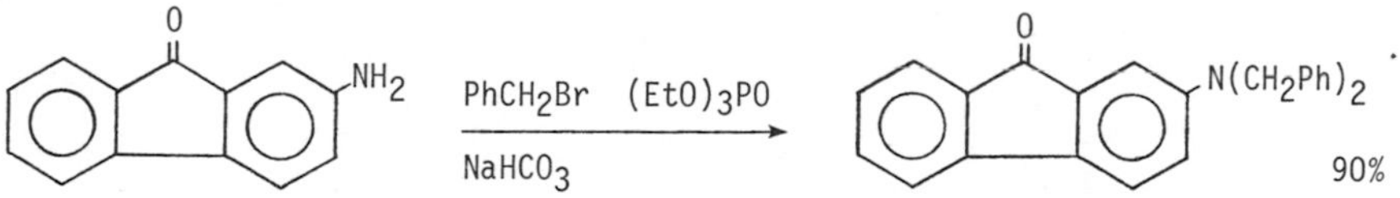

JOC (1962) **27** 3639

90%

$$PhCH_2NH_2 \xrightarrow[\substack{MeOH}]{MeI\ \ NaHCO_3} PhCH_2\overset{+}{N}Me_3\overset{-}{I} \xrightarrow[\substack{THF}]{LiAlH_4} PhCH_2NMe_2$$

72%

JACS (1960) **82** 4651
Ber (1957) **90** 395

$$Ph_2NH \xrightarrow[\substack{PhNO_2}]{PhI\ \ Cu\ \ K_2CO_3} Ph_2NPh$$

82%

Org Synth (1941) Coll Vol 1 544
JCS (1946) 5

PhCH=CH—⟨C6H4⟩—NH₂ $\xrightarrow[\substack{2\ Na\ toluene \\ 3\ Me_2SO_4}]{1\ TsCl\ Pyr}$ Ts-NMe $\xrightarrow[\substack{H_2O}]{HCl\ HOAc}$ PhCH=CH—⟨C6H4⟩—NHMe   56%

Annalen (1956) <u>598</u> 174
JOC (1968) <u>33</u> 1142
Ber (1953) <u>86</u> 1246

BuNH₂ $\xrightarrow[\substack{2\ MeI\ K_2CO_3\ Me_2CO}]{1\ PhCOCH_2SO_2Cl\ CH_2Cl_2\ Pyr}$ BuNSO₂CH₂COPh (Me) $\xrightarrow[\substack{HOAc}]{Zn\ HCl}$ BuNHMe

Tetr Lett (1970) 345

PhNH₂ $\xrightarrow[\substack{2\ EtI\ KOH\ Me_2CO}]{1\ (CF_3CO)_2O}$ PhNCOCF₃ (Et) $\xrightarrow[\substack{}]{KOH\ H_2O}$ PhNHEt   83%

JCS <u>C</u> (1969) 2223

MeO—⟨C6H4⟩—NH₂ $\xrightarrow[\substack{2\ Me_2SO_4\ NaNH_2 \\ toluene}]{1\ Ac_2O\ NaOH\ H_2O}$ Me-NAc $\xrightarrow[\substack{EtOH}]{KOH\ H_2O}$ MeO—⟨C6H4⟩—NHMe

Ber (1954) <u>87</u> 1760
JOC (1949) <u>14</u> 1099

⟨NH⟩ $\xrightarrow[\substack{2\ CH_2=CH_2}]{1\ Na\ Pyr}$ ⟨NEt⟩   77-83%

Org Synth (1963) <u>43</u> 45
Bull Chem Soc Jap (1967) <u>40</u> 2991

BuNH₂ $\xrightarrow[\substack{Na_2HPO_4\ THF \\ 2\ H_2}]{1\ MeCH=CH_2\ PdCl_2}$ BuNHCHMe₂

Proc Chem Soc (1961) 370

PhNHMe $\xrightarrow{\text{1 HgCl}_2 \quad \text{CH}_2=\text{CH}_2}{\text{2 LiAlH}_4}$ PhNMe
$\qquad\qquad\qquad\qquad\qquad\qquad$ |
$\qquad\qquad\qquad\qquad\qquad\qquad$ Et

Compt Rend (1966) C 262 1591
Tetr Lett (1967) 5165
(1969) 2289

cyclohexyl-NH$_2$ $\xrightarrow{\text{HC}\equiv\text{CH}}$ cyclohexyl-N=CHMe $\xrightarrow{\text{H}_2 \quad \text{Pt}}$ cyclohexyl-NHEt

JACS (1961) 83 213

Cl-phenyl-NH$_2$ $\xrightarrow{\text{HC(OEt)}_3 \quad \text{H}_2\text{SO}_4}$ Cl-phenyl-N(Et)CHO $\xrightarrow{\text{HCl} \quad \text{H}_2\text{O}}$ Cl-phenyl-NHEt   ~74%

Org Synth (1963) Coll Vol 4 420
JACS (1956) 78 4778

2-Cl-phenyl-NH$_2$ $\xrightarrow{\text{(EtO)}_3\text{PO}}$ 2-Cl-phenyl-NEt$_2$ JACS (1946) 68 895   91%

3,4,5-(MeO)-phenyl-CH(OH)CH$_2$NH$_2$ $\xrightarrow{\text{ClCOOEt} \quad \text{NaOH}}{\text{H}_2\text{O}}$ CH(OH)CH$_2$NHCOOMe $\xrightarrow{\text{LiAlH}_4 \quad \text{Et}_2\text{O}}$ (MeO)$_2$-phenyl-CH(OH)CH$_2$NHMe (OMe)

J Med Chem (1963) 6 227
JOC (1965) 30 2483

indanyl-NH$_2$ $\xrightarrow{\text{HCOOH}}{\text{Ac}_2\text{O}}$ indanyl-NHCHO $\xrightarrow{\text{LiAlH}_4 \quad \text{THF}}$ indanyl-NHMe

J Med Chem (1966) 9 830

Further examples of the preparation of N-alkyl amines by reduction of amides are included in section 96 (Amines from Amides)

$$\text{(piperidine NH)} \xrightarrow[\text{2 MeCOOEt}]{\text{1 LiAlH}_4 \quad \text{THF}} \text{(piperidine NEt)} \qquad 80\%$$

JOC (1962) <u>27</u> 1042

$$\text{(cyclohexyl-NH}_2) \xrightarrow{\text{Ni toluene}} \text{(dicyclohexyl-NH)} \qquad 82\%$$

Annalen (1961) <u>644</u> 23

$$\text{PhNMe}_2 \xrightarrow[\text{150-160}°]{\text{C}_5\text{H}_{11}\text{Br}} \underset{\underset{\text{C}_5\text{H}_{11}}{|}}{\text{PhNMe}}$$

Ber (1881) <u>14</u> 622
Org React (1953) <u>7</u> 198

$$\text{BuNH(CH}_2)_3\text{CH}_3 \xrightarrow[\text{2 H}_2\text{SO}_4 \quad \text{H}_2\text{O}]{\text{1 Cl}_2 \quad \text{NaOH} \quad \text{H}_2\text{O}} \text{BuN (pyrrolidine)} \qquad 70\text{-}80\%$$

Org Synth (1955) Coll Vol 3 159
Chem Rev (1963) <u>63</u> 55

$$\text{(PhNHC}_{16}\text{H}_{33}) \xrightarrow{\text{CoCl}_2 \quad 212°} \text{(C}_{16}\text{H}_{33}\text{-phenyl-NH}_2)$$

JCS (1937) 1119
Tetrahedron (1961) <u>14</u> 208

$$\text{(C}_6\text{H}_{13})_2\text{NCH}_2\text{Ph} \xrightarrow{\text{H}_2 \quad \text{PtO}_2 \quad \text{HOAc}} \text{(C}_6\text{H}_{13})_2\text{NH} \qquad 100\%$$

Org React (1953) <u>7</u> 263
JACS (1950) <u>72</u> 3410
        (1941) <u>63</u> 1964

$$\xrightarrow[\text{CHCl}_3]{\text{BrCN} \quad \text{K}_2\text{CO}_3}$$

$$\xrightarrow[]{\text{LiAlH}_4 \quad \text{THF}}$$

49%

JACS (1967) 89 1942
      (1955) 77 4079
Org Synth (1955) Coll Vol 3 608
Org React (1953) 7 198

$$\text{Bu}_3\text{N} \xrightarrow[]{\text{ClCOOPh} \quad \text{CH}_2\text{Cl}_2} \text{Bu}_2\text{NCOOPh} \xrightarrow[]{\text{Acid}} \text{Bu}_2\text{NH}$$

85%

JCS C (1967) 2015

$$(\text{i-Pr})_2\text{NEt} \xrightarrow[\text{CuCl}]{\text{i-PrONO} \quad \text{O}_2} (\text{i-Pr})_2\text{NNO} \dashrightarrow (\text{i-Pr})_2\text{NH}$$

Angew (1970) 82 876
(Internat Ed 9 892)
JACS (1967) 89 1147

$$\text{PhNMe}_2 \xrightarrow[]{\text{HBr} \quad 150°} \text{PhNHMe}$$

JOC (1963) 28 3144

$$\text{Pr}_3\text{N} \xrightarrow[]{\text{NBS} \quad \text{dioxane} \quad \text{H}_2\text{O}} \text{Pr}_2\text{NH}$$

87%

JCS (1957) 4905
JACS (1968) 90 3502
Chem Rev (1963) 63 21

$$\xrightarrow[\text{MeOH}]{\text{O}_2 \quad h\nu \quad \text{methylene blue}}$$

80%

Tetr Lett (1970) 3649

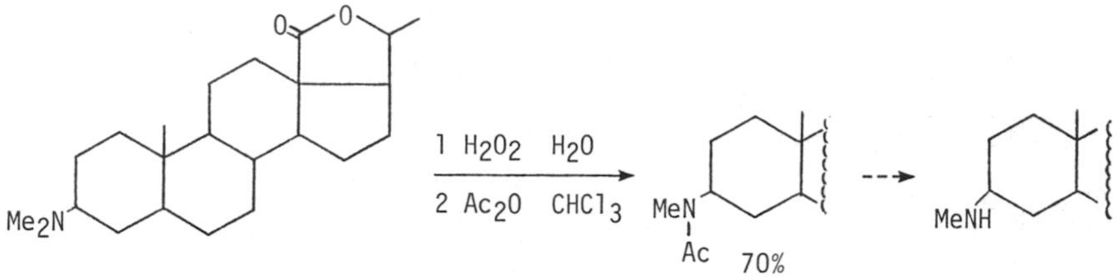

Tetrahedron (1967) 23 4681

$$(i\text{-Pr})_2NH \xrightarrow{\text{t-BuOOH}} i\text{-PrNH}_2$$

JOC (1960) 25 2114   88%

Further reactions which may be used for the dealkylation of amines are included in section 82 (Amides from Amines) and section 81 (Amides from Amides)

$$Et_3NCH_2Ph \overset{+}{} \overset{-}{Cl} \xrightarrow{\text{PhSH NaOH}} Et_3N$$

J Med Chem (1969) 12 694   22%
Tetr Lett (1966) 1375

$$BuNMe_3 \overset{+}{} \overset{-}{I} \xrightarrow{\text{HOCH}_2\text{CH}_2\text{NH}_2} BuNMe_2$$

Ber (1957) 90 395

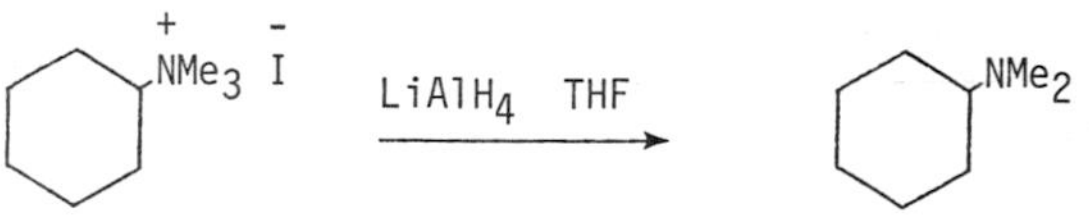

JACS (1960) 82 4651

Section 98   Amines from Esters
              ○○○○○○○○○○○○○○○○○○○○

$PhCH_2COOR$ $\xrightarrow[\text{2 PhCOCl \ MeONa}]{\text{1 NH}_2\text{OH \ base}}$ $PhCH_2CONHOCOPh$ $\xrightarrow[\text{THF}]{\text{LiAlH}_4 \ \ \text{AlCl}_3}$ $PhCH_2CH_2NH_2$

Bull Soc Chim Fr (1960) 509

Amines may also be prepared by conversion of esters into amides followed
by reduction. See section 83 (Amides from Esters) and section 96 (Amines
from Amides)

$PhCH_2COOEt$ $\xrightarrow[\text{2 NaNO}_2 \ \ \text{HCl} \ \ \text{H}_2\text{O}]{\text{1 N}_2\text{H}_4 \cdot \text{H}_2\text{O} \ \ \text{EtOH}}$ $PhCH_2CON_3$ $\xrightarrow[\text{H}_2\text{O}]{\text{HCl} \ \ \text{HOAc}}$ $PhCH_2NH_2$ ~50%

Org React (1946) 3 337

$t\text{-BuOAc}$ $\xrightarrow{\text{HCN} \ \ \text{H}_2\text{SO}_4}$ $t\text{-BuNH}_2$

Org React (1969) 17 213

Section 99   Amines from Ethers
              ○○○○○○○○○○○○○○○○○○○○

54%

Rec Trav Chim (1966) 85 56

Section 100   Amines from Halides and Sulfonates

$$PhBr \xrightarrow[\text{2 ClCH}_2\text{CH}_2\text{NMe}_2]{\text{1 Mg Et}_2\text{O}} PhCH_2CH_2NMe_2$$

J Med Chem (1966) 9 790   13%

$$PhBr \xrightarrow[\text{2 EtCN}]{\text{1 Mg Et}_2\text{O}} \underset{\underset{Et}{|}}{PhC}=NH \xrightarrow[]{\text{LiAlH}_4 \quad \text{THF}} \underset{\underset{Et}{|}}{PhCH}NH_2$$

JACS (1953) 75 5898   80%

$$BuBr \xrightarrow[\text{2 (i-Pr)}_2\text{NCHO}]{\text{1 Mg Et}_2\text{O}} (i\text{-Pr})_2NCHBu_2$$

Monatsh (1951) 82 330   67%

$$MeCl \xrightarrow[\underset{\underset{CN}{|}}{\text{2 (i-Pr)}_2\text{NCHMe}}]{\text{1 Mg Et}_2\text{O}} (i\text{-Pr})_2NCHMe_2$$

Monatsh (1962) 93 476   54%

$$CH_2=CHCH_2Cl \xrightarrow[\text{2 BuOCH}_2\text{NEt}_2]{\text{1 Mg Et}_2\text{O}} CH_2=CHCH_2CH_2NEt_2$$

JCS (1923) 123 532

$$C_5H_{11}Br \xrightarrow[\text{2 NH}_2\text{OMe}]{\text{1 Mg Et}_2\text{O}} C_5H_{11}NH_2$$

JCS (1946) 781   65%

BuBr $\xrightarrow[\text{2 ClNH}_2]{\text{1 Mg Et}_2\text{O dioxane}}$ BuNH$_2$ JACS (1941) $\underline{63}$ 1692   97%
                                                                                   (1936) $\underline{58}$ 27

$\xrightarrow[\substack{\text{2 TsN}_3 \\ \text{3 Ni-Al NaOH H}_2\text{O}}]{\text{1 Mg THF}}$

82%

JOC (1969) $\underline{34}$ 3430

C$_7$H$_{15}$Br $\xrightarrow{\text{NH}_3\ \text{MeOH}}$ C$_7$H$_{15}$NH$_2$ 47%

JACS (1932) $\underline{54}$ 1499 3441

i-PrBr $\xrightarrow{\text{Et}_2\text{NH HOCH}_2\text{CH}_2\text{OH}}$ Et$_2$NPr-i   60%

JACS (1932) $\underline{54}$ 4457

PrBr $\xrightarrow{\text{PhNH}_2}$ PrNHPh   70%

JCS (1930) 992

$\xrightarrow[\text{H}_2\text{O 195°}]{\text{NH}_3\ \text{CuCl Cu}}$

79%

Org Synth (1955) Coll Vol 3 307

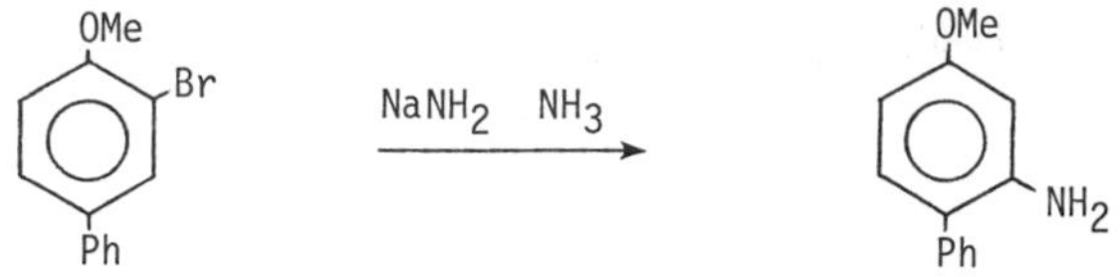

60%

JOC (1957) 22 500
Ber (1964) 97 1994
Chem Rev (1962) 62 81

PhI $\xrightarrow{\text{PhLi}\quad\text{Me}_3\text{N}}$ PhNMe$_2$ 60%

Ber (1943) 76B 109

Further examples of the reaction RHal + R$_2$NH $\longrightarrow$ RNR$_2$ (R=alkyl, aryl etc.)
are included in section 97 (Amines from Amines)

C$_{16}$H$_{33}$OTs $\xrightarrow{\text{BuNH}_2\quad\text{toluene}}$ C$_{16}$H$_{33}$NHBu   +   (C$_{16}$H$_{33}$)$_2$NBu

                                                51%   33%

JACS (1933) 55 345
JCS (1955) 694

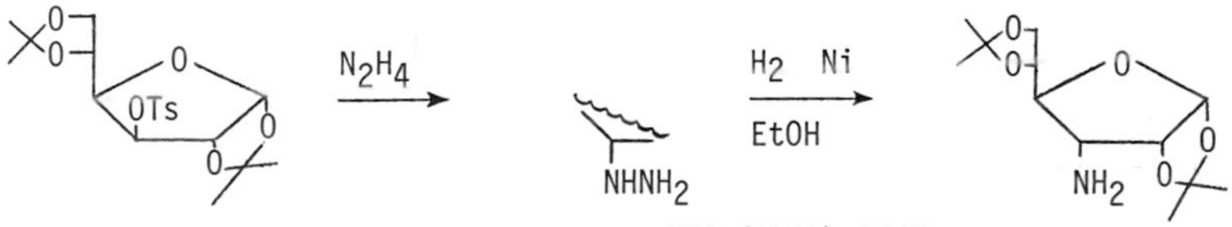

JCS (1961) 1643
Ber (1970) 103 475

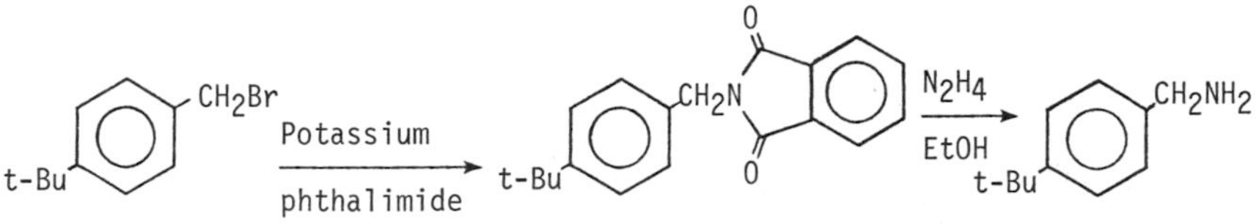

JCS (1935) 1847
JACS (1950) 72 2787
Angew (1968) 80 986
(Internat Ed 7 919)

Chem Comm (1969) 578

PhCH$_2$Br $\xrightarrow{\text{Saccharin} \quad K_2CO_3}$ [structure] $\xrightarrow[\substack{\text{2 BuOTs} \\ \text{3 HCl} \quad H_2O}]{\text{1 KOH}}$ PhCH$_2$NHBu   74%

J Pharm Soc Jap (1953) 73 1319
                 (1955) 75 153 159

24%

JCS (1946) 5

Coll Czech (1967) 32 2826

BuX $\xrightarrow[\text{2 HCl} \quad H_2O]{\text{1 (PhS)}_2\text{NLi} \quad \text{THF}}$ BuNH$_2$ 60% (X=Br)
                                                                                                      78% (X=OTs)
                                                                                                      Tetr Lett (1970) 3411

PhCH$_2$Cl $\xrightarrow[\text{NaI} \quad \text{EtOH}]{\text{Hexamethylenetetramine}}$ PhCH$_2$NH$_2$ 82%

JACS (1939) 61 3585
Org React (1954) 8 197

$$C_5H_{11}Br \xrightarrow[\text{2 NaOH}]{\text{1 } (NH_2)_2C=NH \quad EtOH \quad H_2O} C_5H_{11}NH_2 \qquad 71\%$$

Tetr Lett (1969) 13

$$Me_2C=CH(CH_2)_3Br \xrightarrow[\substack{\text{2 KCN MeOH} \\ \text{3 HOAc } H_2O}]{\text{1 } NH_2CN \quad MeSOCH_2^- \quad Na^+ \quad Me_2SO} [Me_2C=CH(CH_2)_3]_2NH \qquad 55\%$$

Tetr Lett (1969) 3327
Tetrahedron (1970) 26 1275

$$\xrightarrow{NaN_3 \quad H_2O \quad CCl_4} \xrightarrow{LiAlH_4 \quad Et_2O}$$

JACS (1970) 92 6302
(1951) 73 5865
Helv (1958) 41 181

$$\xrightarrow[DMF]{NaN_3} \xrightarrow[Et_2O]{LiAlH_4}$$ 34%

JOC (1962) 27 2925
JACS (1969) 91 2961

$$\xrightarrow[\text{2 NaOH}]{\text{1 MeCN } H_2SO_4}$$ 65%

J Med Chem (1963) 6 760
Org React (1969) 17 213

$$t\text{-}BuCl \xrightarrow{NCl_3 \quad AlCl_3} t\text{-}BuNH_2 \qquad 90\%$$

JOC (1969) 34 911

Chem Eng News (1970) March 9 39
JACS (1964) <u>86</u> 4732

$Bu(CH_2)_4Cl \xrightarrow{NaN_3} Bu(CH_2)_4N_3 \xrightarrow{h\nu \quad THF}$

JCS (1962) 622

Section 101   <u>Amines from Hydrides (RH)</u>

$PhH \xrightarrow{CH_2=CHCH_2NHBu \quad AlCl_3} PhCHCH_2NHBu$
                                            |
                                            Me

66%

JACS (1943) <u>65</u> 674 762

$PhH \xrightarrow{MeCH-CH_2 \quad AlCl_3}$

(with NH bridge on MeCH-CH_2)

$PhCHCH_2NH_2$ + $PhCH_2CHNH_2$
      | |
      Me 12%   Me 4%

J Heterocyclic Chem (1968) <u>5</u> 339

$EtCHMe_2 \xrightarrow{HCN \quad t\text{-}BuOH \quad H_2SO_4} EtCMe_2$
                                              |
                                              NH_2

45%

Ber (1964) <u>97</u> 3234
Org React (1969) <u>17</u> 213

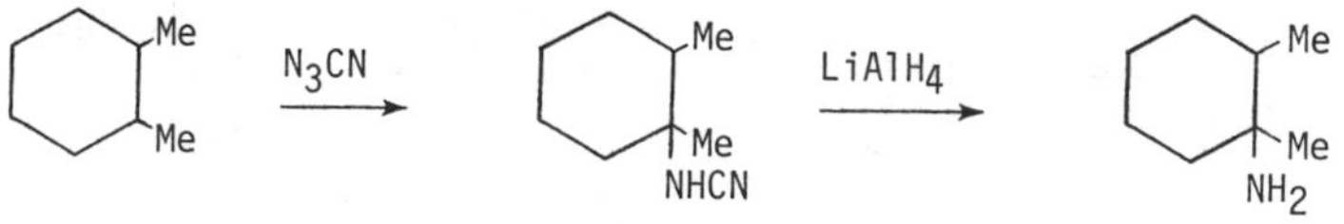

JACS (1967) 89 3177

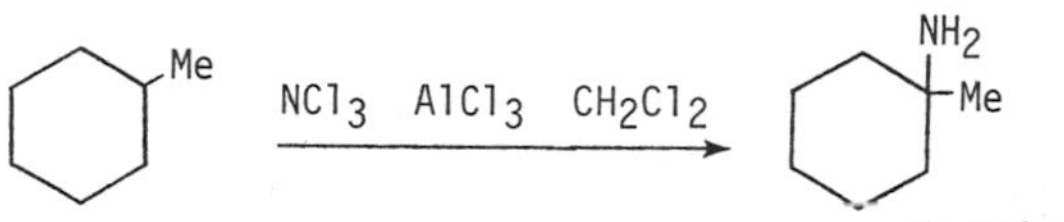

Tetrahedron (1967) 23 3563

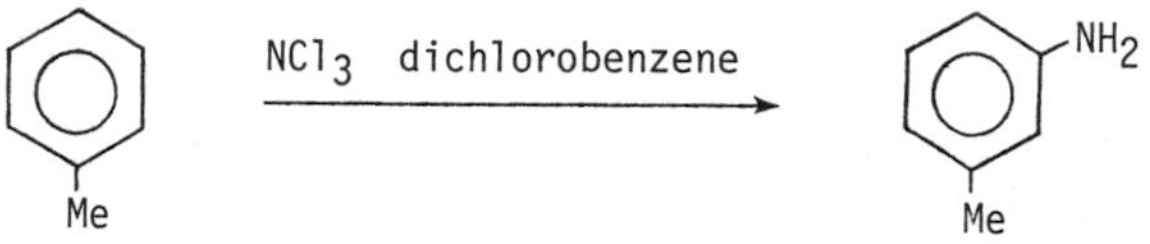

68%

JACS (1966) 88 100
     (1964) 86 1650

36%

PhH $\xrightarrow{\text{Me}_2\text{NCl} \quad \text{H}_2\text{SO}_4 \quad \text{Na}_2\text{SO}_4}$ PhNMe$_2$

Ber (1966) 99 1347 1361

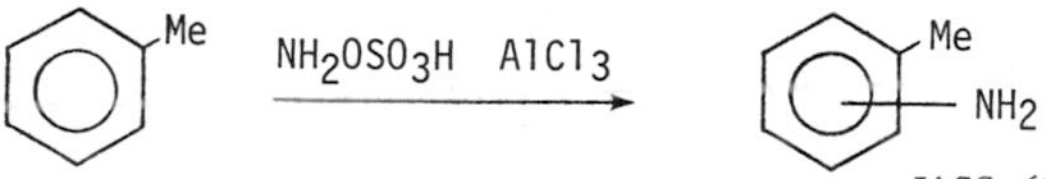

73%

JACS (1961) 83 221 743

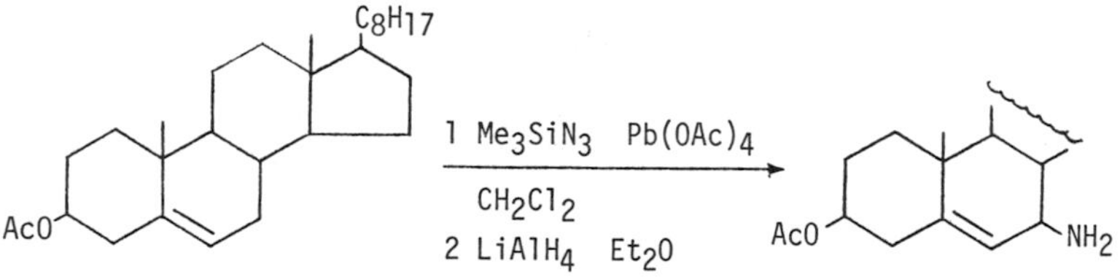

50%

Tetrahedron (1970) 26 1417

49%

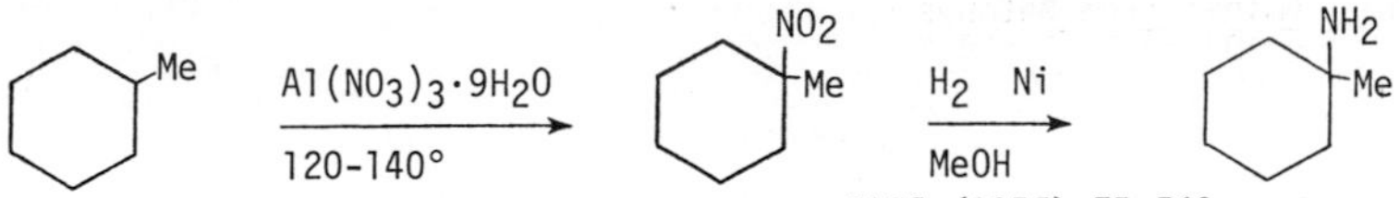
Me
$Al(NO_3)_3 \cdot 9H_2O$
120-140°
NO_2
Me
H_2 Ni
MeOH
NH_2
Me

18%

JACS (1953) 75 369

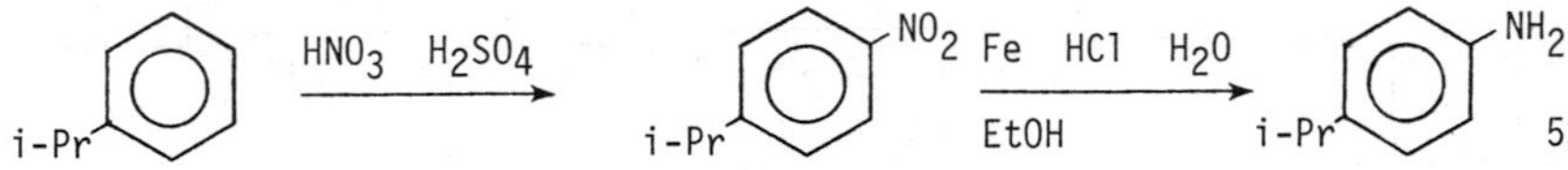
i-Pr
HNO_3 H_2SO_4
NO_2
i-Pr
Fe HCl H_2O
EtOH
NH_2
i-Pr

52%

JCS (1939) 1299

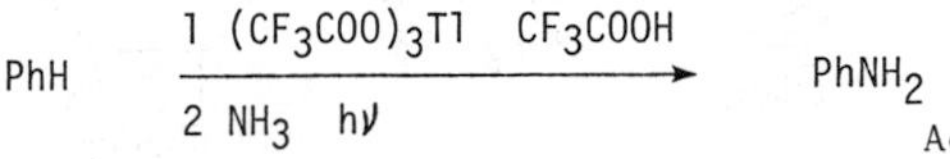
PhH
1 $(CF_3COO)_3Tl$ CF_3COOH
2 NH_3 hν
$PhNH_2$

Acc Chem Res (1970) 3 338

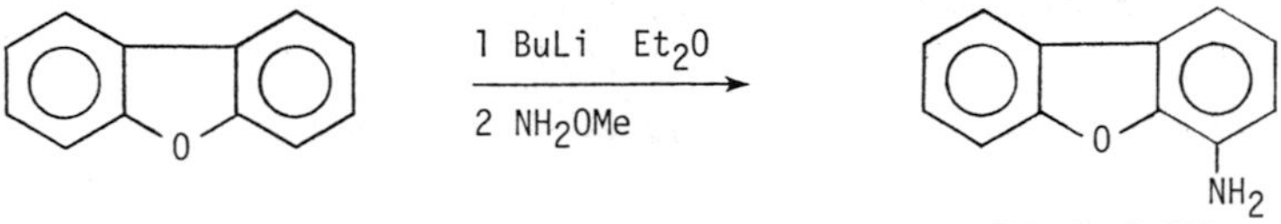
O
1 BuLi Et_2O
2 NH_2OMe
O
NH_2

54%

Org React (1954) 8 258

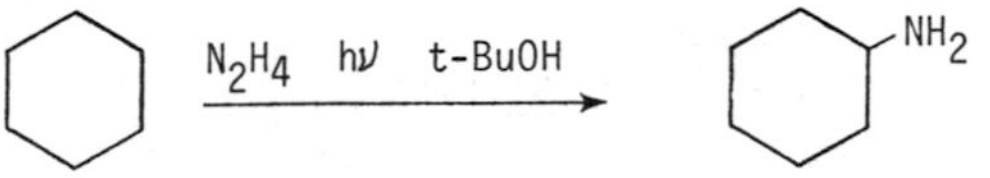
N_2H_4 hν t-BuOH
NH_2

45%

Tetrahedron (1966) 22 483

Amines may also be prepared by conversion of hydrides (RH) into amides (RNHCOR, $RCONR_2$ etc.) followed by hydrolysis or reduction. See section 86 (Amides from Hydrides) and section 96 (Amines from Amides)

Section 102   <u>Amines from Ketones</u>

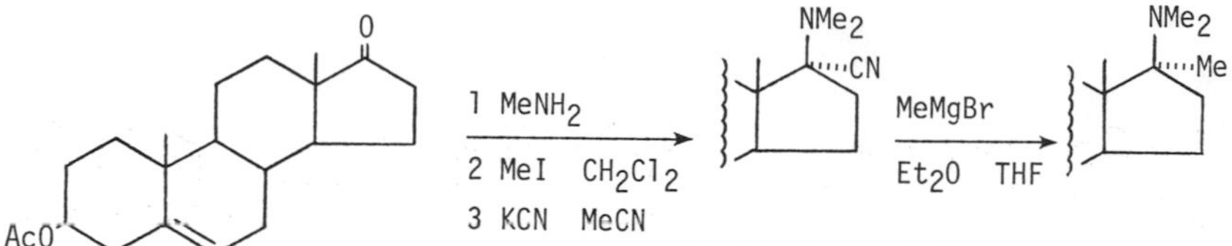

JOC (1962) <u>27</u> 2541
(1965) <u>30</u> 3203

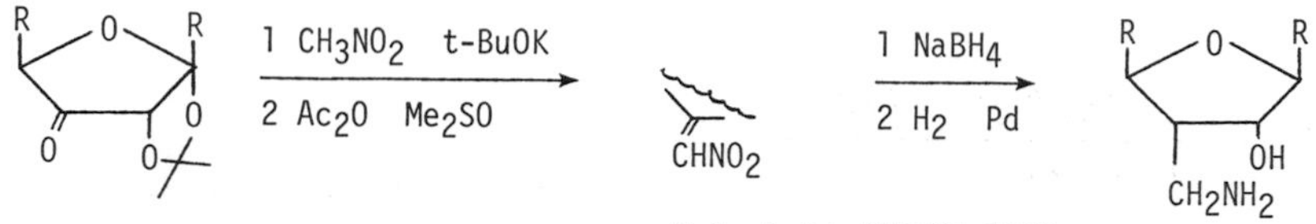

Tetr Lett (1970) 1063

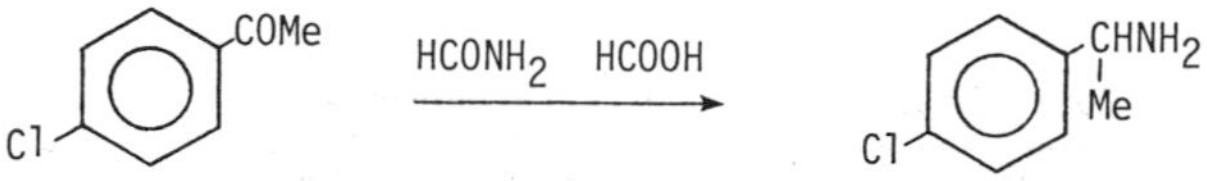

Org React (1949) <u>5</u> 301

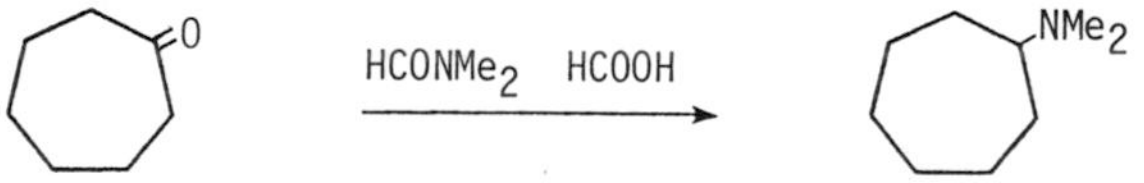

82%

JOC (1968) <u>33</u> 1647

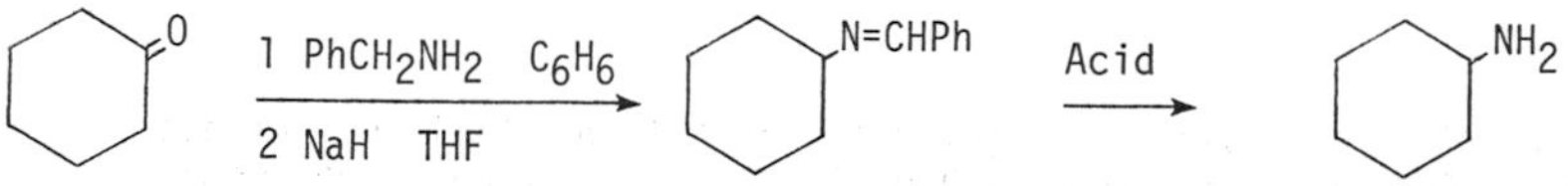

~71%

Can J Chem (1970) <u>48</u> 570

57%

Me_2CO $\xrightarrow[\text{HOAc EtOH}]{\text{PhNH}_2 \quad \text{NaBH}_4 \quad \text{NaOAc}}$ $Me_2CHNHPh$ 91%

JOC (1963) <u>28</u> 3259

$(CH_2)_{11}$ CO $\xrightarrow[\text{MeOH pH 5-6}]{\text{Me}_2\text{NH} \quad \text{LiBH}_3\text{CN}}$ $(CH_2)_{11}$ CHNMe$_2$ 96%

JACS (1969) <u>91</u> 3996

$\xrightarrow[\text{MeOH pH 5-6}]{\text{NH}_3 \quad \text{LiBH}_3\text{CN}}$ 48%

NH$_2$

JACS (1969) <u>91</u> 3996

OH

$\xrightarrow{\text{Pyrrolidine}}$ $\xrightarrow[\text{2 MeOH}]{\text{1 B}_2\text{H}_6 \quad \text{THF}}$

Tetr Lett (1970) 2849
JOC (1965) <u>30</u> 3203

OH

$\xrightarrow[\text{2 NaBH}_4]{\text{1 PhCH}_2\text{NH}_2}$ PhCH$_2$NH— $\xrightarrow[\text{EtOH}]{\text{H}_2 \quad \text{Pd-C}}$ NH$_2$

25%

Bull Soc Chim Fr (1963) 798

$$Ph_2CO \xrightarrow{\quad PhNH_2 \quad ZnCl_2 \quad} Ph_2C=NPh \xrightarrow{\quad LiAlH_4 \quad Et_2O \quad} Ph_2CHNHPh \qquad 62\%$$

JOC (1958) 23 535

For reduction of imines with H_2 and Pt see Bull Soc Chim Fr (1964) 753

   "   "   "   "   " H_2 " Pd " JCS C (1970) 1303

   "   "   "   "   " H_2 " Ni " JOC (1962) 27 2209

   "   "   "   "   by electrolysis " JOC (1970) 35 261

For review of reduction of Schiff's bases see Org React (1948) 4 174

$$Ph_2CO \dashrightarrow Ph_2CN_2 \xrightarrow{\quad Et_2NH \quad h\nu \quad} Ph_2CHNEt_2 \qquad 23\%$$

Annalen (1958) 614 19

J Heterocyclic Chem (1964) 1 53

$$PhCOEt \xrightarrow{\quad NH_2OH \quad} \underset{Et}{PhC=NOH} \xrightarrow[\quad EtOH \quad H_2O \quad]{\quad Ni \quad NaH_2PO_2 \quad} \underset{Et}{PhCHNH_2} \qquad 78\%$$

JCS C (1966) 531

For reduction of oximes with Ni-Al and NaOH see JCS C (1966) 655

   "   "   "   "   "   Zn   " NH_3 " Monatsh (1963) 94 677

   "   "   "   "   "   Na and NH_3 see Zh Obshch Khim (1965) 35 125
(Chem Abs 62 13068)

   "   "   "   "   " $LiAlH_4$ see JOC (1952) 17 294

   "   "   "   "   " B_2H_6 " JOC (1969) 34 1817

   "   "   "   "   " H_2 and Rh " JOC (1962) 27 2209

   "   "   "   "   by electrolysis   " JACS (1967) 89 6374

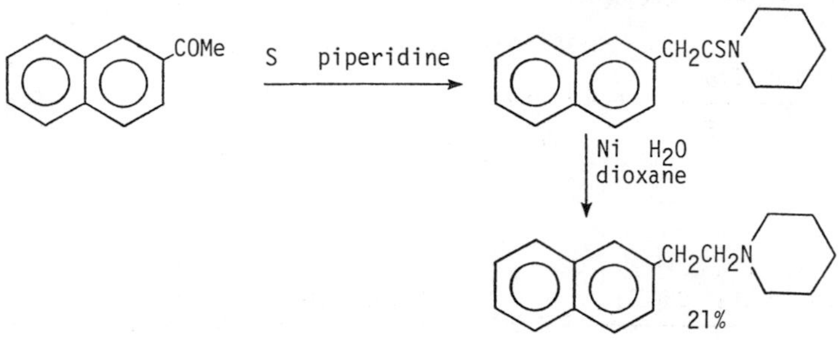

JOC (1951) 16 131

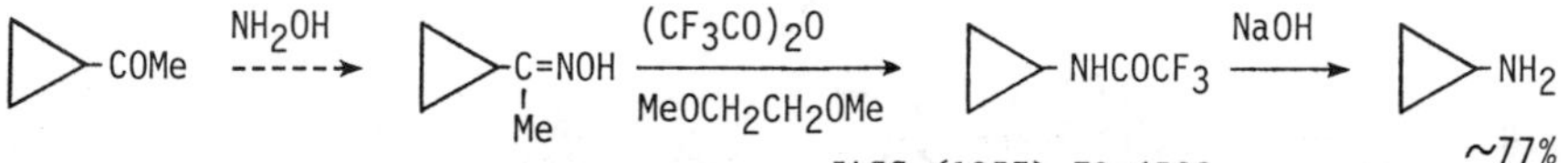

JACS (1957) 79 6522   ~77%

Ph_2CO $\xrightarrow[\text{polyphosphoric acid}]{NH_2OH \cdot HCl}$ $PhNH_2$ 66%

JACS (1953) 75 2014

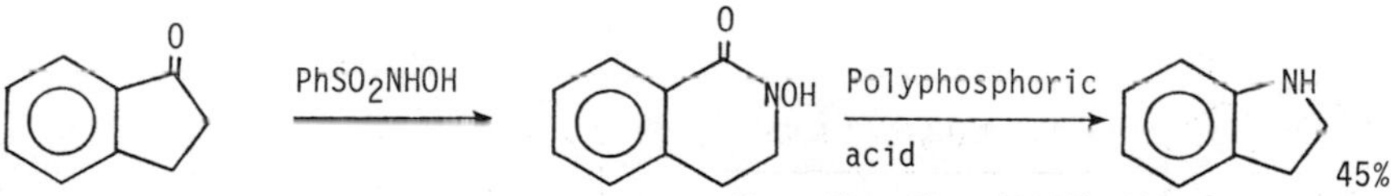

Proc Chem Soc (1963) 224   45%

Amines may also be prepared by conversion of ketones into amides followed by hydrolysis or reduction. See section 87 (Amides from Ketones) and section 96 (Amines from Amides)

Some of the reactions listed in section 94 (Amines from Aldehydes) may also be applied to the preparation of amines from ketones

Section 103   Amines from Nitriles

$$C_5H_{11}CN \xrightarrow[\text{2 LiAlH}_4 \quad \text{THF}]{\text{1 PhMgBr} \quad \text{Et}_2\text{O}} C_5H_{11}\underset{\overset{|}{Ph}}{C}HNH_2 \qquad 54\%$$

JACS (1953) 75 5898

$$PhCN \xrightarrow[\substack{\text{2 EtOH} \\ \text{3 NaOH} \quad H_2\text{O}}]{\substack{\text{1 (i-PrO)}_2\text{CH}\ \overset{+}{B}\overset{-}{F}_4 \quad CH_2Cl_2}} \underset{\overset{|}{OEt}}{PhC=NPr-i} \xrightarrow{\text{NaBH}_4 \quad \text{EtOH}} PhCH_2NHPr-i \qquad 94\%$$

JOC (1969) 34 627

$$PhCN \xrightarrow[\text{HCl} \quad \text{EtOH}]{\text{H}_2 \quad \text{Pd-C}} PhCH_2NH_2$$

JACS (1928) 50 3370
JCS (1942) 426

$$C_6H_{13}O(CH_2)_2CN \xrightarrow[\text{NH}_3 \quad \text{EtOH}]{\text{H}_2 \quad \text{Rh-Al}_2\text{O}_3} C_6H_{13}O(CH_2)_2CH_2NH_2 \qquad 90\%$$

JACS (1960) 82 2386

$$\xrightarrow[\text{EtOH} \quad H_2\text{O}]{\text{Ni} \quad \text{NaH}_2\text{PO}_2 \quad \text{NaOH}}$$

82%

JCS C (1966) 531
Bull Chem Soc Jap (1967) 40 1548

$$BuCN \xrightarrow{\text{Na} \quad \text{EtOH} \quad \text{toluene}} BuCH_2NH_2 \qquad 76\%$$

JACS (1934) 56 1614
Ber (1942) 75B 991

$$\text{(o-Me-C}_6\text{H}_4\text{-CN)} \xrightarrow{\text{LiAlH}_4 \quad \text{Et}_2\text{O}} \text{(o-Me-C}_6\text{H}_4\text{-CH}_2\text{NH}_2)$$

JACS (1948) 70 3738
Org React (1951) 6 469

88%

$$\text{PhCH}_2\text{CN} \xrightarrow{\text{LiAlH}_4 \quad \text{AlCl}_3 \quad \text{Et}_2\text{O}} \text{PhCH}_2\text{CH}_2\text{NH}_2$$

JACS (1955) 77 2544

83%

$$\text{PhCN} \xrightarrow{\text{(i-Bu)}_2\text{AlH} \quad \text{C}_6\text{H}_6} \text{PhCH=NH} \xrightarrow{\text{H}_2 \quad \text{Pd}} \text{PhCH}_2\text{NH}_2$$

JOC (1959) 24 627

57%

$$\text{(p-Me-C}_6\text{H}_4\text{-CN)} \xrightarrow[\text{diglyme}]{\text{NaBH}_4 \quad \text{AlCl}_3} \text{(p-Me-C}_6\text{H}_4\text{-CH}_2\text{NH}_2)$$

JACS (1956) 78 2582

85%

$$\text{C}_7\text{H}_{15}\text{CN} \xrightarrow{\text{NaBH}_4 \quad \text{CoCl}_2 \quad \text{MeOH}} \text{C}_7\text{H}_{15}\text{CH}_2\text{NH}_2$$

Tetr Lett (1969) 4555

80%

$$\xrightarrow{\text{B}_2\text{H}_6 \quad \text{THF}}$$

JACS (1960) 82 681

85%

$$\xrightarrow[\text{H}_2\text{O}]{\text{H}_2\text{S} \quad \text{NaOH}} \qquad \xrightarrow{\text{H}_2 \quad \text{Pd-C}}$$

Chem Pharm Bull (1961) 9 119
JOC (1951) 16 131

PhCN $\xrightarrow{\text{MeOH HCl}}$ PhC=NH·HCl $\xrightarrow[\text{H}_2\text{SO}_4 \quad \text{H}_2\text{O}]{\text{Electrolysis}}$ PhCH$_2$NH$_2$ 76%

OMe

JACS (1935) <u>57</u> 772

PhCN $\xrightarrow[\text{polyphosphoric acid}]{\text{NH}_2\text{OH·HCl}}$ PhNH$_2$ JACS (1953) <u>75</u> 2014   20%

Amines may also be prepared by conversion of nitriles into amides followed by hydrolysis or reduction. See section 88 (Amides from Nitriles) and section 96 (Amines from Amides)

Section 104   Amines from Olefins

MeCH$_2$NEt$_2$ hν

Chem Comm (1969) 753   11%

Me$_2$NH paraformaldehyde

H$_2$SO$_4$ HOAc

CH$_2$NMe$_2$

Tetr Lett (1966) 6483   19%

1 NaBH$_4$ BF$_3$·Et$_2$O diglyme

2 NH$_2$OSO$_3$H

Me

NH$_2$ (trans)   45%

JACS (1966) <u>88</u> 2870
JOC (1967) <u>32</u> 3199

MeCH=CH$_2$ $\xrightarrow{\text{1 Hg(OAc)}_2 \quad \text{PhNH}_2}{\text{2 NaBH}_4}$ Me$_2$CHNHPh   70%

Tetr Lett (1967) 5165
(1969) 2289
Compt Rend (1966) C 262 1591

$\xrightarrow{\text{1 Hg(NO}_3)_2 \quad \text{MeCN}}{\text{2 NaBH}_4 \quad \text{NaOH} \quad \text{H}_2\text{O}}$

JACS (1969) 91 5647

$\xrightarrow{\text{1 NaCN} \quad \text{H}_2\text{SO}_4 \quad \text{HOAc}}{\text{2 H}_2\text{O}}$ 58%

Tetrahedron (1967) 23 3563
Org React (1969) 17 213

$\xrightarrow{\text{1 AgCNO} \quad \text{I}_2 \quad \text{Et}_2\text{O}}{\text{2 MeOH}}$ $\xrightarrow{\text{Zn} \quad \text{HOAc}}{\text{Et}_2\text{O}}$

JACS (1970) 92 1326

MeCH=CH$_2$ $\xrightarrow{\text{1 BuNH}_2 \quad \text{PdCl}_2 \quad \text{Na}_2\text{HPO}_4 \quad \text{THF}}{\text{2 H}_2}$ Me$_2$CHNHBu

Proc Chem Soc (1961) 370

C$_8$H$_{17}$CH=CH$_2$ $\xrightarrow{\text{1 O}_3 \quad \text{MeOH}}{\text{2 H}_2 \quad \text{Ni} \quad \text{NH}_3}$ C$_8$H$_{17}$CH$_2$NH$_2$ 60%

JOC (1962) 27 2392

Amines may also be prepared by conversion of olefins into amides followed
by hydrolysis or reduction. See section 89 (Amides from Olefins) and
section 96 (Amines from Amides)

Section 105   Amines from Miscellaneous Compounds

$$EtCHNO_2 \xrightarrow{\ LiAlH_4 \quad Et_2O\ } EtCHNH_2$$

(Me below $EtCHNO_2$; Me below $EtCHNH_2$)

85%

JACS (1948) <u>70</u> 3738
(1951) <u>73</u> 1293

$$PrNO_2 \xrightarrow[Et_2O]{\ Pd \quad cyclohexene\ } PrNH_2$$

55%

JCS (1954) 3586

For reduction of nitro compounds with H_2 and Pd-C see
Ber (1953) <u>86</u> 939
"   "   "   "   "   " H_2 and Ni see
Org Synth (1955) Coll Vol 3 59
"   "   "   "   "   " Fe and HCl see
Org Synth (1943) Coll Vol 2 160
"   "   "   "   "   " Al-Hg see
JACS (1968) <u>90</u> 3245

Section 105A   Protection of Amines

Amine protecting groups which are generally applicable are included in this section. For a review of the protection of amino acids and peptides see
Advances in Org Chem (1963) <u>3</u> 159

$$\underset{K_2CO_3 \quad MeOH \quad H_2O}{\overset{(CF_3CO)_2O}{\rightleftarrows}}$$

JOC (1965) <u>30</u> 1287
JACS (1953) <u>75</u> 3473

$$NH_2CH_2CONHR \underset{HCl \quad MeOH}{\overset{HCOOH \quad Ac_2O}{\rightleftarrows}} HCONHCH_2CONHR$$

Chem Ind (1953) 107
Helv (1960) <u>43</u> 1751

PhCH$_2$CHNH$_2$ → Phthalic anhydride → PhCH$_2$CHN (phthalimide) (Stable to acid)

$$\text{ClCH}_2\text{COCl}$$

Reaction: 4-nitroaniline + ClCH$_2$COCl → 4-nitro-NHCOCH$_2$Cl

reagents: 1 (NH$_2$)$_2$CS EtOH / 2 H$_2$O

JACS (1968) 90 4508

HCl·NH$_2$CHCONHCH$_2$COOH (i-Bu)
→ Diketene Et$_3$N EtOH → MeCOCH$_2$CONHCHCONHCH$_2$COOH (i-Bu)
← PhNHNH$_2$ HOAc ←

Tetr Lett (1965) 605

Further examples of the preparation and cleavage of acyl derivatives of
amines are included in section 82 (Amides from Amines) and section 96
(Amines from Amides)

PhCH$_2$CHNH$_2$ (COOH)
→ Phthalic anhydride →
← N$_2$H$_4$·H$_2$O EtOH ←
PhCH$_2$CHN (phthalimide) (COOH)   (Stable to acid)

JACS (1952) 74 3822
Rec Trav Chim (1960) 79 688

(2-R-4-Cl-aniline) NH$_2$
→ Ac$_2$O →
← NaOH MeOH ←
(2-R-4-Cl) NAc$_2$ (Stable to NBS)

Can J Chem (1955) 33 1819

HN‿NH (piperazine)
→ ClCOOEt NaOH H$_2$O →
← HCl H$_2$O ←
HN‿NCOOEt

JCS (1929) 39
JACS (1954) 76 1164

$$PhNH_2 \xrightleftharpoons[HBr]{ClCOOCH_2Ph} PhNHCOOCH_2Ph$$

JOC (1952) 17 1564

For the hydrogenolysis of benzyloxycarbonyl groups see
JACS (1957) 79 1636
and Org React (1953) 7 263
For the photolytic cleavage of benzyloxycarbonyl groups see
Tetr Lett (1962) 697

$$HCl \cdot NH_2CH_2COOMe \xrightleftharpoons[HBr\ HOAc]{N_3COOBu\text{-}t\ \ Et_3N\ \ EtOAc} t\text{-}BuOCONHCH_2COOMe$$

Helv (1959) 42 2622
JACS (1957) 79 4686

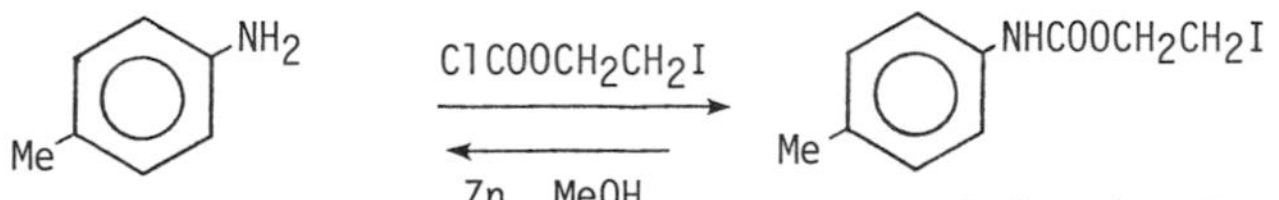

JCS (1965) 7136

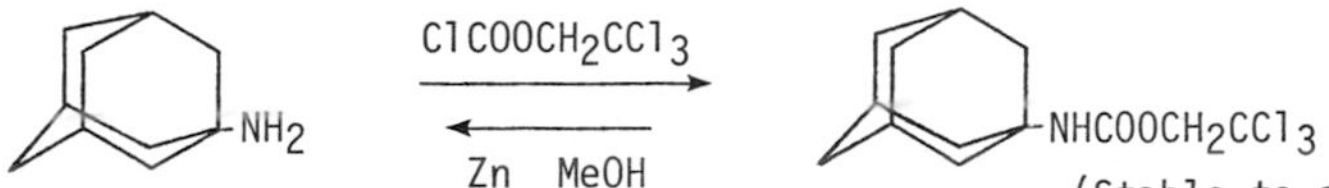

(Stable to acid, base and
CrO₃)

Tetr Lett (1967) 2555

$$PhNH_2 \xrightleftharpoons[KOH\ EtOH]{ClCOOCH_2CH_2Ts} PhNHCOOCH_2CH_2Ts$$

Proc Chem Soc (1962) 363

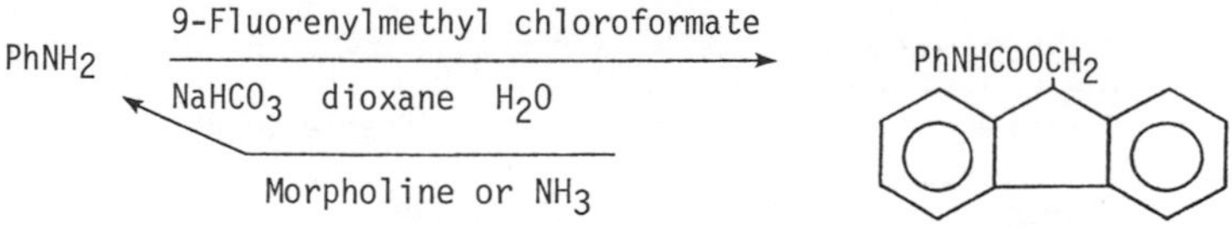

PhNH$_2$

9-Fluorenylmethyl chloroformate →

NaHCO$_3$ dioxane H$_2$O ←

Morpholine or NH$_3$

PhNHCOOCH$_2$

JACS (1970) 92 5748

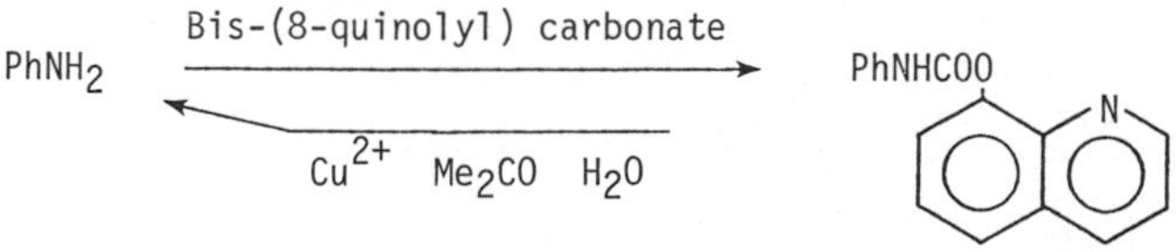

PhNH$_2$

Bis-(8-quinolyl) carbonate →

Cu^{2+} Me$_2$CO H$_2$O ←

PhNHCOO

JACS (1962) 84 4899

(C$_6$H$_{13}$)$_2$NH - - - - - - → (C$_6$H$_{13}$)$_2$NCH$_2$Ph

H$_2$ Pt HOAc ←

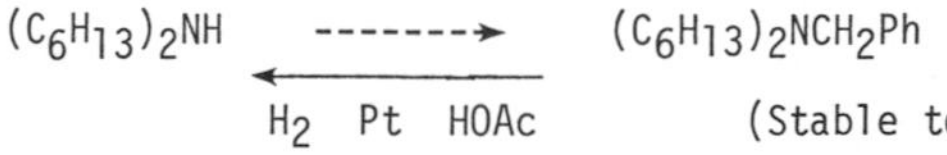

(Stable to acid, base and RMgX)

JCS (1940) 1307
Org React (1953) 7 263
JACS (1950) 72 3410

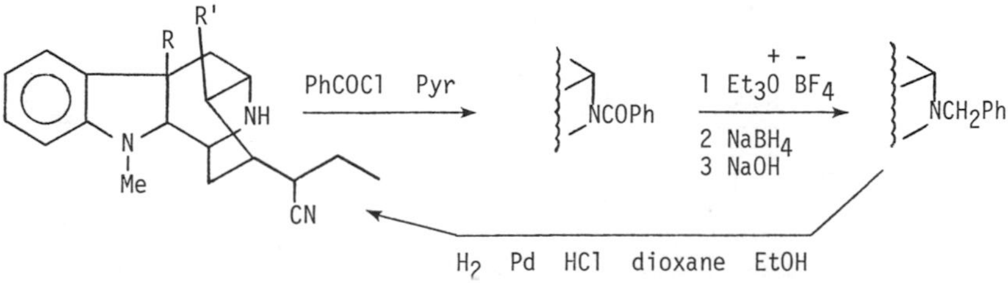

PhCOCl Pyr →

NCOPh

1 Et$_3$O$^+$ BF$_4^-$ →

2 NaBH$_4$

3 NaOH

NCH$_2$Ph

H$_2$ Pd HCl dioxane EtOH ←

Tetrahedron (1970) 26 803

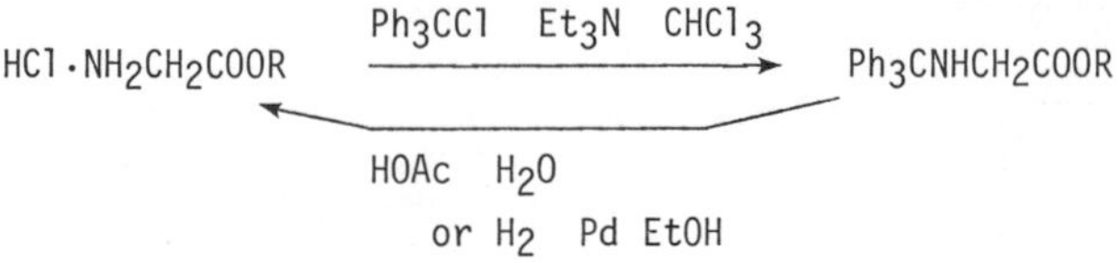

HCl·NH$_2$CH$_2$COOR

Ph$_3$CCl Et$_3$N CHCl$_3$ →

HOAc H$_2$O
or H$_2$ Pd EtOH ←

Ph$_3$CNHCH$_2$COOR

(Stable to base)

JACS (1956) 78 1359

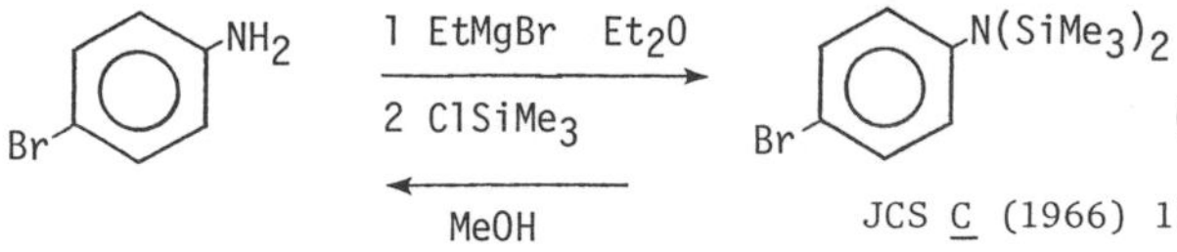

JCS C (1966) 1706

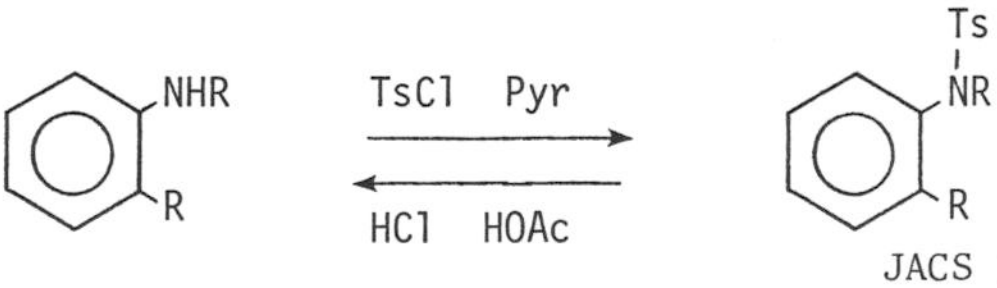

JACS (1949) 71 1901
(1952) 74 2006

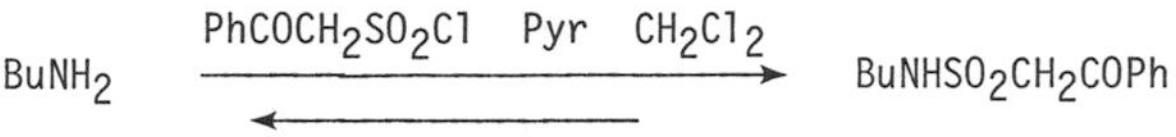

BuNH₂   →[PhCOCH₂SO₂Cl Pyr CH₂Cl₂ / Zn HCl HOAc]   BuNHSO₂CH₂COPh

(Stable to acid, base and CrO₃)
Tetr Lett (1970) 345

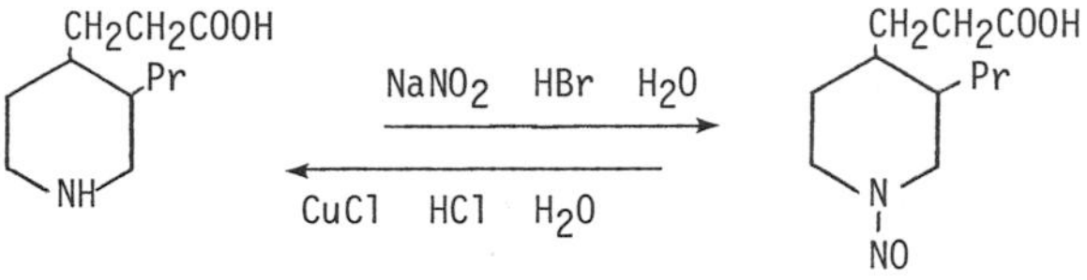

JACS (1946) 68 146
(1957) 79 2215

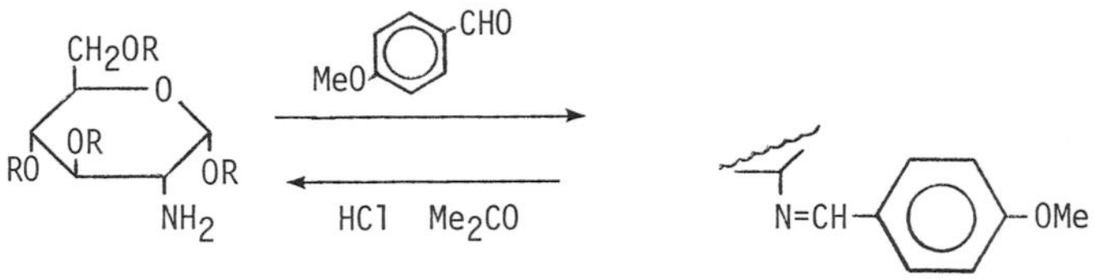

JACS (1956) 78 1393

Chapter 8   PREPARATION
OF
ESTERS

   Esters from Acetylenes
ooooooooooooooooooooooo

$BuC \equiv CH$ $\xrightarrow{\begin{array}{c} 1 \text{ LiNH}_2 \text{ dioxane} \\ \hline 2 \text{ ClCOOEt} \end{array}}$ $BuC \equiv CCOOEt$ $\xrightarrow[]{\text{H}_2 \text{ Pd}}$ $BuCH_2CH_2COOEt$

$\qquad\qquad\qquad\qquad\qquad\qquad\qquad\qquad\quad$ 25%   Annalen (1958) <u>614</u> 37

$C_5H_{11}C \equiv CH$ $\xrightarrow{\begin{array}{c} 1 \text{ Na} \\ \hline 2 \text{ TsCl} \end{array}}$ $C_5H_{11}C \equiv CCl$ $\xrightarrow{\text{EtONa}}$ $C_5H_{11}CH_2COOEt$

$\qquad\qquad\qquad\qquad\qquad\qquad\qquad\qquad\quad$ Annalen (1931) <u>16</u> 309

$C_5H_{11}C \equiv CH$ $\xrightarrow[250°]{\text{N}_2\text{O} \quad \text{MeOH}}$ $C_5H_{11}CH_2COOMe$ $\qquad\qquad\qquad\qquad\qquad$ 57%

$\qquad\qquad\qquad\qquad\qquad\qquad\qquad\qquad\quad$ JCS (1951) 3016

$PhC \equiv CH$ $\xrightarrow[\text{BF}_3 \cdot \text{Et}_2\text{O}]{\text{HOAc} \quad \text{HgO}}$ $PhCH = CHOAc$ $\xrightarrow{\begin{array}{c} \text{NCOOK} \\ \| \\ \text{NCOOK} \end{array}}$ $PhCH_2CH_2OAc$

$\qquad\qquad\qquad\qquad\qquad\qquad\qquad\qquad\quad$ JOC (1965) <u>30</u> 3985
$\qquad\qquad\qquad\qquad\qquad\qquad\qquad\qquad\quad$ JACS (1934) <u>56</u> 1802

271

Section 107   <u>Esters from Carboxylic Acids, Acid Halides and Anhydrides</u>

$C_5H_{11}COOH$ $\xrightarrow[\text{electrolysis}]{\begin{array}{c}\text{COOH}\\|\\(CH_2)_4COOMe \quad MeOH \quad H_2O\end{array}}$ $C_5H_{11}(CH_2)_4COOMe$ 48%

JCS (1950) 3326
Advances in Org Chem (1960) <u>1</u> 1

$\xrightarrow[\text{2 Ag}_2\text{O EtOH}]{\text{1 CH}_2\text{N}_2 \text{ Et}_2\text{O}}$ 73-82%

Org React (1942) <u>1</u> 38

$\xrightarrow[\text{2 hν MeOH}]{\text{1 Me}_2\overset{+}{\text{S}}\text{OCH}_2^{-} \text{ THF}}$ 80%

JACS (1964) <u>86</u> 1640

CH_3COOH $\xrightarrow{\begin{array}{c}NH_2\\|\\Me_2CCH_2OH\end{array}}$ $CH_3C\overset{O}{\underset{N}{\diagdown}}$ $\xrightarrow[\begin{array}{c}\text{2 BuBr}\\\text{3 H}_2\text{SO}_4 \text{ EtOH}\end{array}]{\text{1 BuLi}}$ $BuCH_2COOEt$

JACS (1970) <u>92</u> 6644

Further methods for the alkylation and homologation of carboxylic acids
and esters are included in section 17 (Carboxylic Acids, Acid Halides
and Anhydrides from Carboxylic Acids and Acid Halides) and section 113
(Esters from Esters)

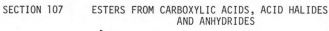

JACS (1951) <u>73</u> 5487

55%

t-BuCH$_2$COOH $\xrightarrow{\text{EtOH   H}_2\text{SO}_4}$ t-BuCH$_2$COOEt   77%

JACS (1938) <u>60</u> 2790

C$_8$H$_{17}$COOH $\xrightarrow[\text{molecular sieves}]{\text{MeOH   H}_2\text{SO}_4}$ C$_8$H$_{17}$COOMe

Chem Ind (1968) 1568
(1967)   825

i-Pr, COOH $\xrightarrow[\text{2 MeOH}]{\text{1 H}_2\text{SO}_4}$ i-Pr, COOMe   81%

i-Pr   Pr-i   i-Pr   Pr-i

JACS (1941) <u>63</u> 2431

C$_{12}$H$_{25}$COOH $\xrightarrow{\text{EtOH   HCl}}$ C$_{12}$H$_{25}$COOEt

Org Synth (1943) Coll Vol 2 292
(1932) Coll Vol 1 246

HO, COOH $\xrightarrow[\text{xylene}]{\text{C}_{12}\text{H}_{25}\text{OH   TsOH}}$ HO, COOC$_{12}$H$_{25}$

HO   HO

OH   OH

Rec Trav Chim (1951) <u>70</u> 277
JACS (1945) <u>67</u> 902
Acta Chem Scand (1959) <u>13</u> 1407

ROH $BF_3 \cdot Et_2O$ (R=Me, Et, i-Pr)   75-81%

Tetr Lett (1970) 4011
JCS (1965) 5770

$$\underset{CH_2COOH}{CH_2COOH} \quad \xrightarrow{PhOH \quad POCl_3} \quad \underset{CH_2COOPh}{CH_2COOPh}$$

62-67%

Org Synth (1963) Coll Vol 4 390 178

$$PhCOOH \quad \xrightarrow[CHCl_3]{PhOH \quad P_2O_5\text{-}Et_2O} \quad PhCOOPh$$

90%

Chem Ind (1964) 2102

$$\underset{NH_2}{PhCHCOOH} \quad \xrightarrow{PhCH_2OH \quad SOCl_2} \quad \underset{NH_2}{PhCHCOOCH_2Ph}$$

90%

JOC (1965) 30 3575
Annalen (1961) 640 136

MeCHCH$_2$CH$_2$COOH

MeCHCH$_2$CH$_2$COOMe

$MeOH$ $MeCOCl$

100%

Org Synth (1955) Coll Vol 3 237

$$C_{15}H_{31}COOH \quad \xrightarrow[\underset{\underset{\times}{O \quad O}}{HOCH_2CHCH_2}]{(CF_3CO)_2O} \quad C_{15}H_{31}COOCH_2CHCH_2$$

Applicable to hindered acids

JCS (1965) 4594
JOC (1965) 30 927

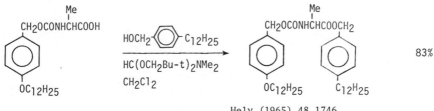

NO₂⟨benzene⟩COOH   →(1 TsCl Pyr / 2 EtOH)→   NO₂⟨benzene⟩COOEt   90%

JACS (1955) 77 6214

OMe⟨benzene⟩COOH, OMe   →(1 ClCOOEt Et₃N Et₂O / 2 MeOH)→   OMe⟨benzene⟩COOMe, OMe

Rev Roumaine Chim (1966) 11 1237
(Chem Abs 66 65249)
JACS (1958) 80 5714
     (1969) 91 3931

HO⟨benzene⟩COOH   →(MeOH DCC Et₂O)→   HO⟨benzene⟩COOMe   45%

Tetrahedron (1965) 21 3531
Compt Rend (1963) 256 1804
            (1962) 255 945

PhCH=CHCOOH   →(MeOH NN'-carbonyldiimidazole / MeONa)→   PhCH=CHCOOMe   83%

Ber (1962) 95 1284
JCS C (1969) 1610

Me
CH₂OCONHCHCOOH⟨benzene⟩OC₁₂H₂₅   →(HOCH₂⟨benzene⟩C₁₂H₂₅ / HC(OCH₂Bu-t)₂NMe₂ / CH₂Cl₂)→   Me
CH₂OCONHCHCOOCH₂⟨benzene⟩OC₁₂H₂₅ ⟨benzene⟩C₁₂H₂₅   83%

Helv (1965) 48 1746
JACS (1969) 91 5674

$(PhO)_2CHCOOH$ $\xrightarrow{\quad EtOH \quad Me_2C(OMe)_2 \quad}$ $(PhO)_2CHCOOEt$

Bull Soc Chim Fr (1964) 381
JOC (1959) 24 261

Further examples of the reaction RCOOH + ROH ⟶ RCOOR are included in
section 108 (Esters from Alcohols and Phenols) and section 30A (Protection
of Carboxylic Acids)

MeCHCOOH ... $\xrightarrow{\quad MeI \quad NaHCO_3 \quad DMA \quad}$ ... MeCHCOOMe

JOC (1968) 33 2143   91%

COONa / COONa $\xrightarrow{\quad PhCH_2Cl \quad EtOH \quad}$ $COOCH_2Ph$ / $COOCH_2Ph$

J Med Chem (1967) 10 706

COONa, Me $\xrightarrow{\quad Ph_3CBr \quad C_6H_6 \quad}$ $COOCPh_3$, Me   84%

JOC (1962) 27 3595
JCS C (1966) 1191

CH_2COOH, NO_2 $\xrightarrow{\quad Et_2SO_4 \quad Na_2CO_3 \quad DMF \quad}$ CH_2COOEt, NO_2 88%

(Applicable to hindered acids)

Monatsh (1968) 99 103
Helv (1943) 26 2283

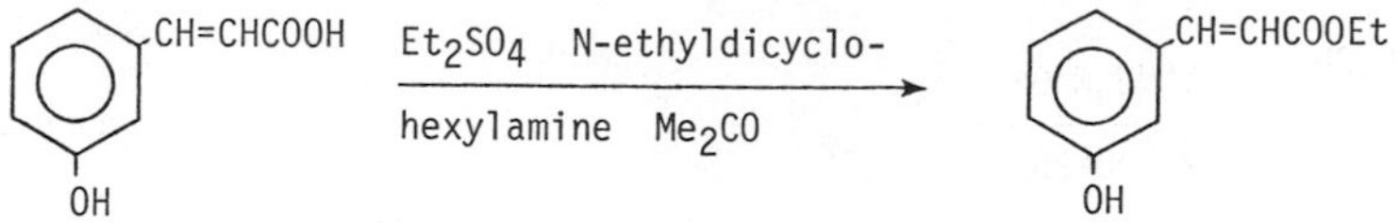

(Applicable to hindered acids)

JOC (1964) 29 2490
JACS (1969) 91 3544

92%

$PhCOOH \xrightarrow[\text{xylene}]{C_6H_{13}Cl \quad Et_3N} PhCOOC_6H_{13}$

JOC (1961) 26 5180

47%

$CH_2=CCH_2COOH \xrightarrow[\text{2 AgNO}_3]{1 \text{ NaOH } H_2O} CH_2=CCH_2COOAg \xrightarrow{MeI \quad Et_2O} CH_2=CCH_2COOMe$
(with Me substituent)

JACS (1949) 71 3214
(1951) 73 5487
(1952) 74 3944

47%

$Bu_3CCOONa \xrightarrow{BuOSOCl \quad C_6H_6} Bu_3CCOOBu$

JACS (1947) 69 1046

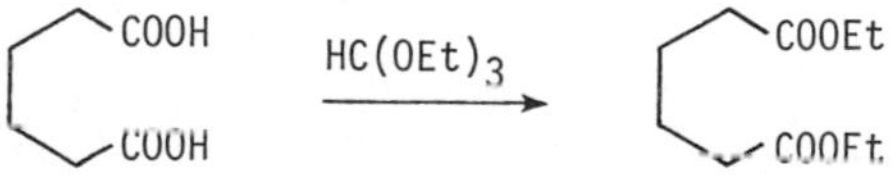

82%

(Applicable to hindered acids)
Chem Ind (1965) 349

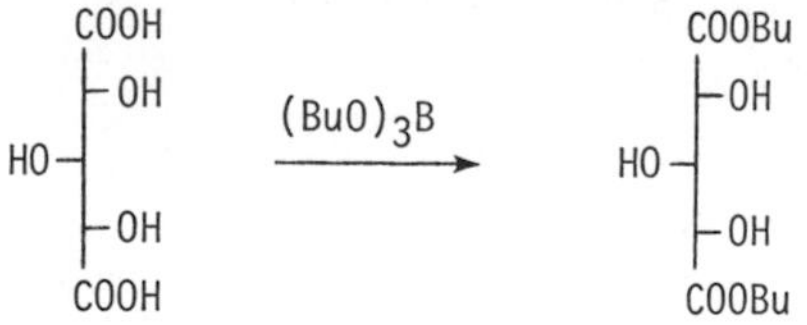

89%

Zh Obshch Khim (1967) 37 1481
(Chem Abs 68 22157)

98%

$PhCOOH \xrightarrow{Me_2NCH(OR)_2 \quad C_6H_6} PhCOOR$ (R=Me, Et, CH_2Ph)

Helv (1965) 48 1746

80-95%

2,4,6-triethylbenzoic acid →

$\xrightarrow{\quad 1\ Me_3N^+\ OH^-\quad}$
$\xrightarrow{\quad 2\ 200\text{-}250°\quad}$

methyl 2,4,6-triethylbenzoate   63-90%

JACS (1939) $\underline{61}$ 1290
Z Physiol Chem (1936) $\underline{244}$ 56

4-nitrobenzoic acid (COOH)

$\xrightarrow{\quad Me_2S^+OCH_2^-\quad Me_2SO\quad}$

methyl 4-nitrobenzoate (COOMe)   69%

Tetr Lett (1964) 867

$CH_2(COOH)_2$ $\xrightarrow{\quad Me_2C=CH_2\quad H_2SO_4\quad}$ $CH_2(COOBu\text{-}t)_2$ 58-60%

Org Synth (1963) Coll Vol 4 261
JOC (1963) $\underline{28}$ 1251

$PhCONHCH_2COOH$ $\xrightarrow{\quad Et_3O^+\ BF_4^-\quad NaHCO_3\quad H_2O\quad}$ $PhCONHCH_2COOEt$ 82%

Tetr Lett (1969) 1819
Ber (1956) $\underline{89}$ 209 2060

$PhCOOH$ $\xrightarrow[\quad (EtO)_3P\quad NCOOEt\quad Et_2O\quad]{\overset{\displaystyle NCOOEt}{\|}}$ $PhCOOEt$ 95%

Bull Chem Soc Jap (1967) $\underline{40}$ 2380

$PhCOOH$ $\xrightarrow{\quad CH_2N_2\quad Et_2O\quad}$ $PhCOOMe$

JACS (1954) $\underline{76}$ 4481

PhCOOH $\xrightarrow{\text{Ph}_2\text{CN}_2 \quad \text{Et}_2\text{O}}$ PhCOOCHPh$_2$ 100%

                                        Org Synth (1955) Coll Vol 3 351

   For preparation of CH$_2$N$_2$ see Org Synth (1943) Coll Vol 2 165
                                        and Org React (1942) 1 50
     "   "   "   Me$_2$CN$_2$ and Pr$_2$CN$_2$ see JCS C (1966) 467
     "   "   "   PhCHN$_2$ "   JACS (1964) 86 658
     "   "   "   anthryldiazomethane   "   Bull Chem Soc Jap (1967) 40 691

C$_{15}$H$_{31}$COOH $\xrightarrow[\text{2 I}_2]{\text{1 Pb(OAc)}_4 \quad \text{tetrachloroethane}}$ C$_{15}$H$_{31}$COOC$_{15}$H$_{31}$ 54%

                              JOC (1963) 28 65

(PrCH$_2$CO)$_2$O $\xrightarrow{\text{HgO} \quad \text{I}_2 \quad \text{BrCH}_2\text{CH}_2\text{Br}}$ PrCH$_2$COOCH$_2$Pr   50%

                              JOC (1970) 35 3167

BuCHCOCl $\xrightarrow[\text{Pyr \quad hexane}]{\text{m-Chloroperbenzoic acid}}$ ~73%
   |
   Et

                              JOC (1965) 30 3760
                              JACS (1950) 72 67

Ph$_2$CHCOOH $\xrightarrow{\text{Electrolysis \quad HOAc}}$ Ph$_2$CHOAc   37%

                              JCS (1952) 3624

Section 108   Esters from Alcohols and Phenols
○○○○○○○○○○○○○○○○○○○○○○○○○○○○○○○○○○○○○○○

$$CH_2=C(CH_2)_2CH=C(CH_2)_2\!\!-\!\!OH \xrightarrow[\text{EtCOOH}]{\text{MeC(OEt)}_3} CH_2=C(CH_2)_2CH=C(CH_2)_2\!\!-\!\!(CH_2)_2COOEt \quad >91\%$$

with Me substituents below both C positions

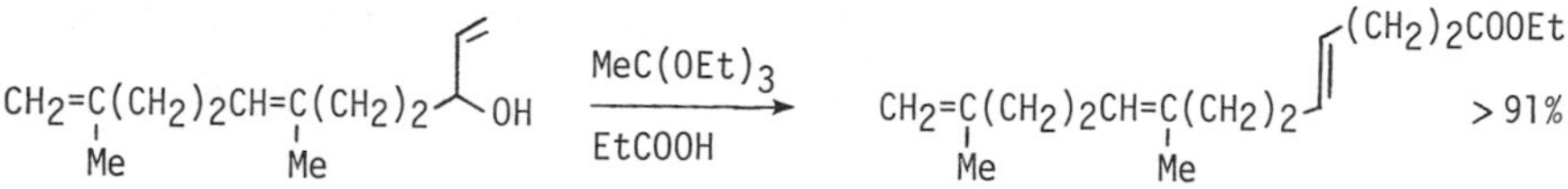

JACS (1970) 92 741
Chem Comm (1970) 1512 1513

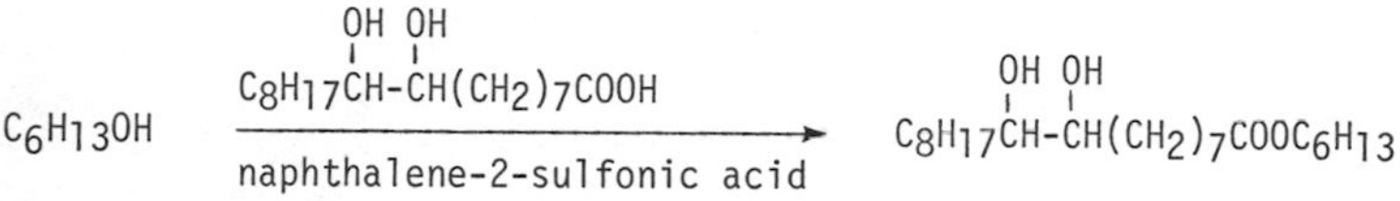

$$C_6H_{13}OH \xrightarrow[\text{naphthalene-2-sulfonic acid}]{C_8H_{17}\overset{OH}{CH}-\overset{OH}{CH}(CH_2)_7COOH} C_8H_{17}\overset{OH}{CH}-\overset{OH}{CH}(CH_2)_7COOC_6H_{13}$$

JACS (1945) 67 902
Rec Trav Chim (1951) 70 277

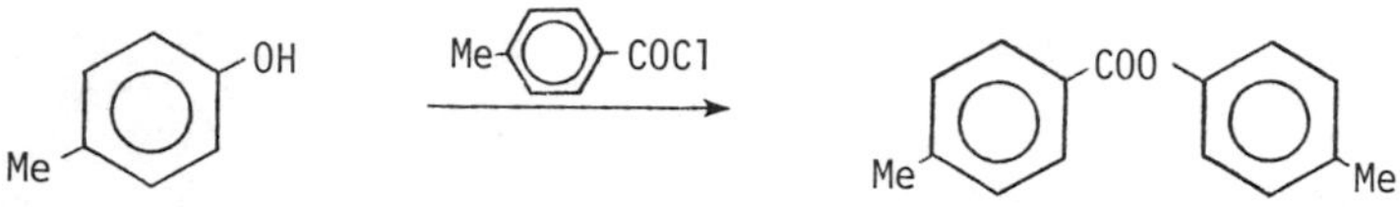

Annalen (1969) 726 216

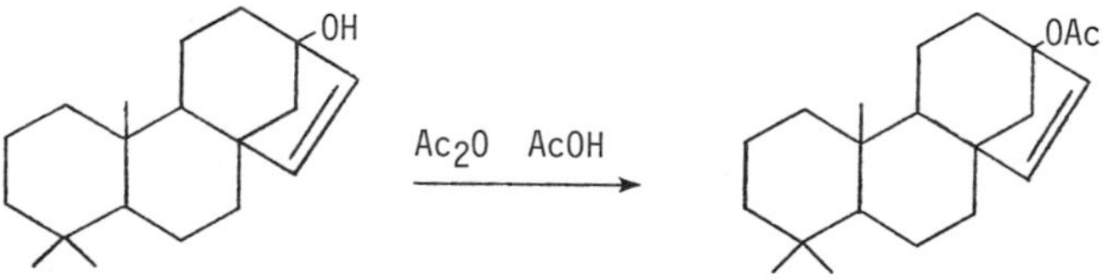

Ac₂O   AcOH

JCS C (1969) 2115

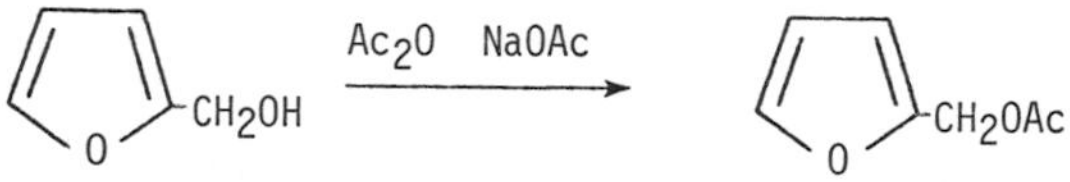

Ac₂O   NaOAc → furan-CH₂OAc   87-93%

Org Synth (1941) Coll Vol 1 285
JOC (1965) 30 3480

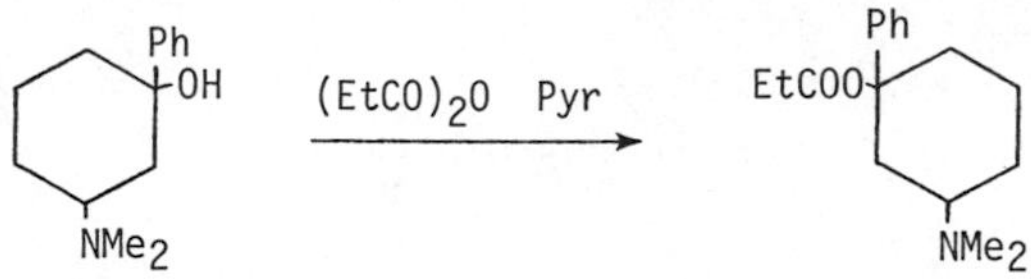
Ph
OH
$(EtCO)_2O$ Pyr
NMe2
Ph
EtCOO
NMe2

60%

J Med Chem (1970) 13 125

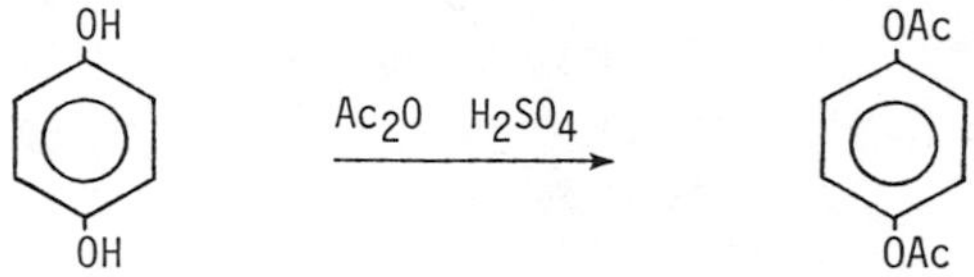
OH
Ac_2O H_2SO_4
OH
OAc
OAc

96-98%

Org Synth (1955) Coll Vol 3 452

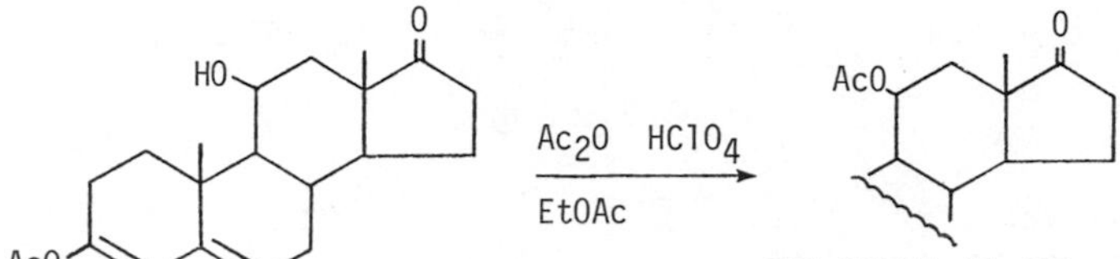
O
HO
AcO
Ac_2O $HClO_4$
EtOAc
O
AcO

JOC (1966) 31 324

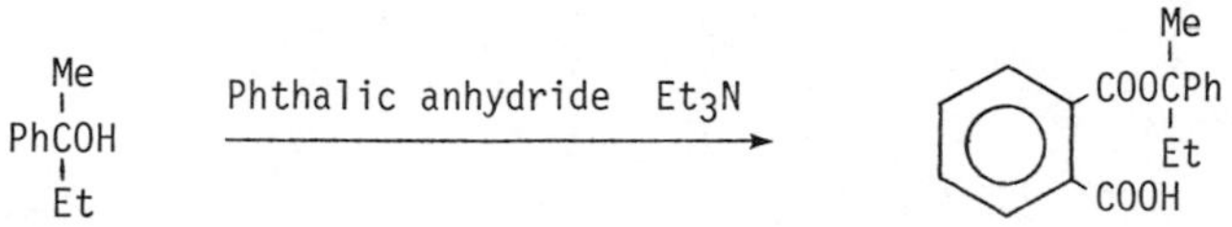
Me
PhCOH
Et
Phthalic anhydride Et_3N
Me
COOCPh
Et
COOH

JCS (1957) 3148

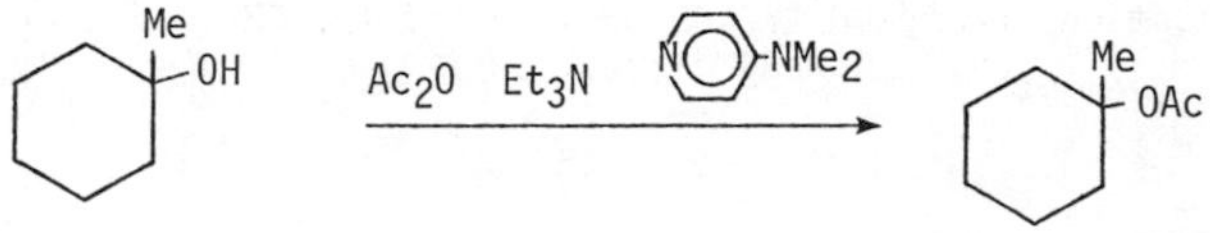
Me
OH
Ac_2O Et_3N
N
NMe2
Me
OAc

86%

Angew (1969) 81 1001
(Internat Ed 8 981)

t-BuOH $\xrightarrow{EtCOCl \quad PhNMe_2 \quad Et_2O}$ EtCOOBu-t

JACS (1943) 65 986
Helv (1944) 27 513
JCS (1952) 4883

t-BuOH $\xrightarrow[\text{Et}_2\text{O}]{\text{MeCOCl \quad tetramethylurea}}$ t-BuOAc   50-60%

Tetr Lett (1967) 3267

Et_3COH $\xrightarrow[\text{2 PhCOCl}]{\text{1 BuLi \quad THF}}$ PhCOOCEt_3 94%

JOC (1970) <u>35</u> 1198

Esters from alcohols, Ph$_3$CNa and (RCO)$_2$O JOC (1963) <u>28</u> 582

  "   "   "   EtMgBr "   "   JACS (1936) <u>58</u> 1384
JCS (1963) 3578

  "   "   "   K   "   "   JACS (1961) <u>83</u> 423

  "   "   "   CaC$_2$ "   "   Monatsh (1966) <u>97</u> 62

  "   "   "   Mg   and RCOCl Org Synth (1955) Coll Vol 3 144

  "   "   "   NaH   "   "   Chem Comm (1967) 259

$\text{C}_{12}\text{H}_{25}\text{OH}$ $\xrightarrow[\text{TsCl \quad Pyr}]{\text{3,5-Dinitrobenzoic acid}}$

82%

(Applicable to hindered alcohols)
JACS (1955) <u>77</u> 6214
(1963) <u>85</u> 2446

PhOH $\xrightarrow{\text{PhCOOH \quad (CF}_3\text{CO)}_2\text{O}}$ PhCOOPh   70%

(Applicable to hindered alcohols and phenols)
Tetrahedron (1965) <u>21</u> 3531
Tetr Lett (1964) 1285
JCS (1965) 4594

For selective esterification of 2ry-OH in the presence of 1ry-OH
see JOC (1961) <u>26</u> 177

JACS (1969) <u>91</u> 3931
(1958) <u>80</u> 5714

PhOH $\xrightarrow{\text{PhCOOH DCC } Et_2O}$ PhCOOPh

Tetrahedron (1965) 21 3531
Compt Rend (1963) 256 1804
(1962) 255 945

$PhCH_2OH$ $\xrightarrow{\text{PhCOOH N,N'-carbonyl diimidazole}}$ $PhCOOCH_2Ph$ 89%

Ber (1962) 95 1284
JCS C (1969) 1610

Further examples of the reaction ROH + RCOOH ⟶ RCOOR are included in section 107 (Esters from Carboxylic Acids, Acid Halides and Anhydrides) and section 45A (Protection of Alcohols and Phenols)

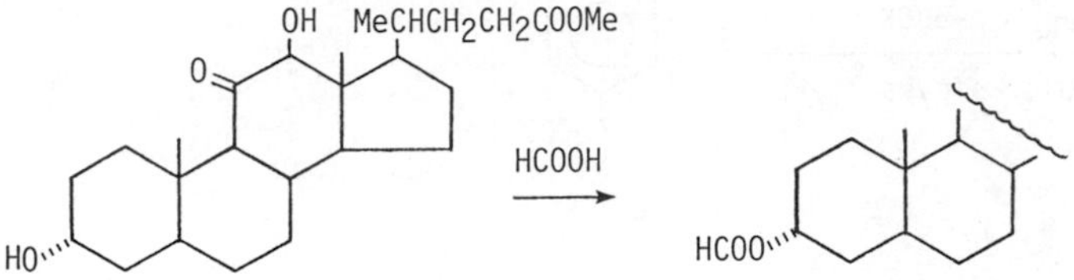

91%

JACS (1954) 76 4915
(1945) 67 740

$Me_2C{-}CH(CH_2)_2C(Me){=}CH(CH_2)_2C(OH)(Me)CH{-}CH_2$ $\xrightarrow[Ac_2O]{HCOOH}$ $Me_2C{=}CH(CH_2)_2C(Me){=}CH(CH_2)_2C(OOCH)(Me)CH{=}CH_2$

Bull Soc Chim Fr (1965) 484
Rec Trav Chim (1964) 83 1287 1294

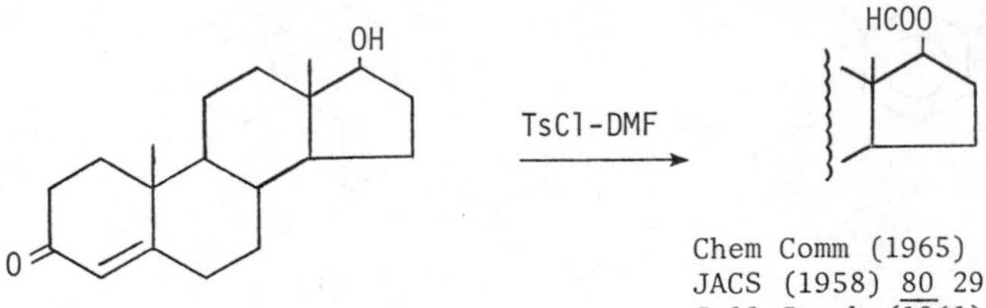

79%

Chem Comm (1965) 413
JACS (1958) 80 2906
Coll Czech (1961) 26 1723

$C_8H_{17}OH$ $\xrightarrow{\text{HCOF} \quad \text{Et}_3\text{N} \quad \text{Et}_2\text{O}}$ $HCOOC_8H_{17}$ 73%

JACS (1960) <u>82</u> 2380

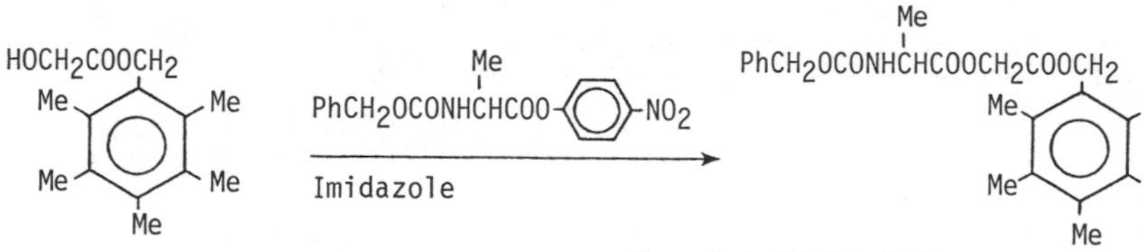

Chem Ind (1967) 1960

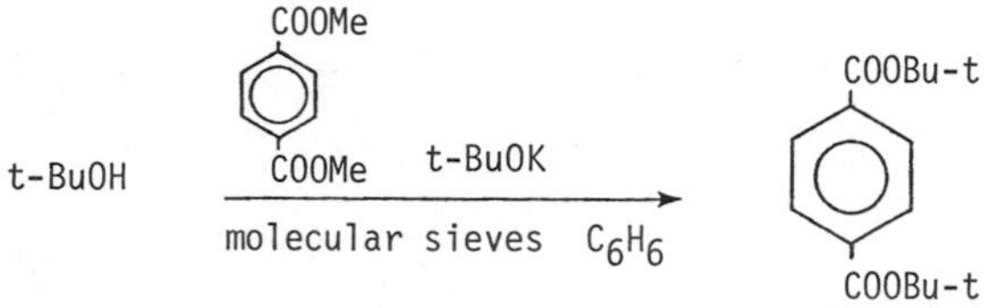

Chem Ind (1966) 1622

$Me(CH_2)_7CH=CH(CH_2)_7CH_2OH$ $\xrightarrow{\text{EtOAc} \quad \text{MeONa}}$ $Me(CH_2)_7CH=CH(CH_2)_7CH_2OAc$

Lipids (1967) <u>2</u> 437
JACS (1956) <u>78</u> 6347

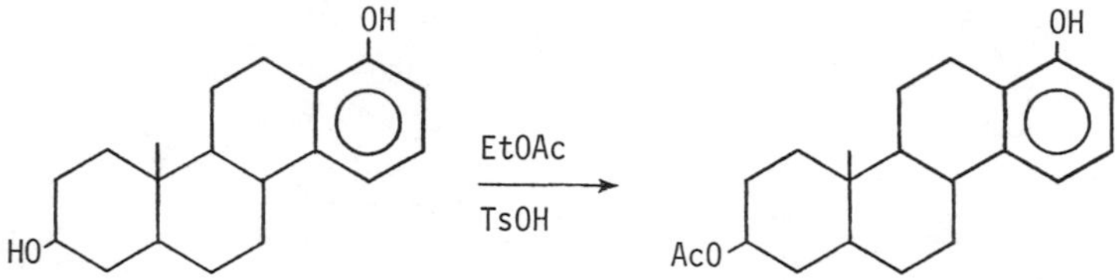

78%

JACS (1956) <u>78</u> 6322

$$\text{CH}_2\text{-COAc} \quad \text{TsOH}$$

with Me substituent

59%

J Med Chem (1969) **12** 563

$$\text{EtOH} \xrightarrow{\overset{+}{\text{MeCO}}\ \overset{-}{\text{SbF}_6}\ \ \text{MeCN}} \text{EtOAc}$$

(Applicable to hindered alcohols)
JACS (1962) **84** 2733
Gazz (1967) **97** 442
(Chem Abs **67** 117049)

$$\text{CH}_2\text{=CO} \quad \text{C}_6\text{H}_6$$

95%

(Applicable to hindered alcohols)
J Med Chem (1969) **12** 432
JOC (1968) **33** 3695
Org React (1946) **3** 108

For preparation of $CH_2=CO$ see

$$\text{t-BuOH} \xrightarrow{\text{Me}_2\text{C}\overset{\text{COO}}{\underset{\text{COO}}{}}\text{C=CMe}_2 \quad \text{K}_2\text{CO}_3} \text{Me}_2\text{CHCOOBu-t}$$

95%

Angew (1963) **75** 841
(Internat Ed **2** 608)

with reagent

$$\text{MeCN}$$

~91%

JACS (1969) **91** 2162

BuOH $\xrightarrow{\quad\quad}$ BuOAc   93%

Bull Chem Soc Jap (1964) <u>37</u> 864
Chem Ind (1965) 1498

$\xrightarrow[\text{toluene}]{\text{(EtCO)}_2\text{NH \quad HBr}}$ 85%

Roczniki Chem (1952) <u>26</u> 692
(Chem Abs <u>49</u> 2464)

$\dashrightarrow$ TsO $\xrightarrow[\text{Me}_2\text{CO}]{\text{Bu}_4\text{NOAc}}$ AcO

JCS (1969) 1605

i-PrCH$_2$CH$_2$OH $\xrightarrow{\text{t-Butylhypochlorite \quad CCl}_4}$ i-PrCH$_2$COOCH$_2$CH$_2$Pr-i   89%

Helv (1953) <u>36</u> 1763

PrCH$_2$OH $\xrightarrow{\text{Na}_2\text{Cr}_2\text{O}_7 \quad \text{H}_2\text{SO}_4 \quad \text{H}_2\text{O}}$ PrCOOCH$_2$Pr   41-47%

Org Synth (1941) Coll Vol 1 138

Section 109   Esters from Aldehydes

JOC (1949) 14 1013

$Me_2C=CH(CH_2)_2\overset{Me}{\underset{|}{C}}=CHCHO$ $\xrightarrow[\text{HOAc   MeOH}]{\text{NaCN   MnO}_2}$ $Me_2C=CH(CH_2)_2\overset{Me}{\underset{|}{C}}=CHCOOMe$

(Applicable to unsaturated and aromatic aldehydes only)
JACS (1968) 90 5616
Ber (1970) 103 3774

$PrCHO$ $-\rightarrow$ $PrCH(OEt)_2$ $\xrightarrow{\text{MeCOO}_2\text{H   H}_2\text{SO}_4}$ $PrCOOEt$

JOC (1960) 25 1699

$PhCHO$ $\xrightarrow{\text{MeOH   (NH}_4)_2\text{S}_2\text{O}_8\text{   H}_2\text{SO}_4}$ $PhCOOMe$ 100%

JOC (1968) 33 2525

Tetr Lett (1963) 2189

$C_5H_{11}CH_2CHO$ $\xrightarrow[\text{2 NaI   Me}_2\text{CO}]{\text{1 SO}_2\text{Cl}_2\text{   CH}_2\text{Cl}_2}$ $C_5H_{11}\overset{}{\underset{\underset{I}{|}}{C}HCHO}$ $\xrightarrow[\text{Ag}^+\text{   Et}_2\text{O}]{\text{EtOH}}$ $C_5H_{11}CH_2COOEt$

Tetr Lett (1968) 4415
JACS (1954) 76 2695

PhCHCHO $\xrightarrow{\text{NaN}_3 \quad \text{HCl} \quad \text{EtOH}}$ PhCHCOOEt   51%
| |
Me   Me   JACS (1954) 76 4564

PrCHO $\xrightarrow{\text{Al(OEt)}_3}$ PrCOOCH$_2$Pr   81%

                          JACS (1947) 69 2605

PhCHO $\xrightarrow{\text{NaH} \quad \text{C}_6\text{H}_6}$ PhCOOCH$_2$Ph   92%

                          JACS (1946) 68 2647

For examples of the reaction RCHO ⟶ HCOOR see page 84 (Section 34,
Alcohols and Phenols from Aldehydes)

Esters may also be prepared by oxidation of aldehydes to carboxylic acids
followed by esterification. See section 19 (Carboxylic Acids and Acid
Halides from Aldehydes)

Section 110   Esters from Alkyls, Methylenes and Aryls

No examples of the reaction RR ⟶ RCOOR' or R'COOR (R=alkyl, aryl etc.)
occur in the literature. For the reaction RH ⟶ RCOOR' or R'COOR see
section 116 (Esters from Hydrides)

Section 111   Esters from Amides

PhCONH$_2$ $\xrightarrow[\text{105°}]{\text{MeOH BF}_3}$ PhCOOMe

Chem Comm (1969) 414   100%

JOC (1960) <u>25</u> 560
Arch Pharm (1957) <u>290</u> 218
JOC (1970) <u>35</u> 125   85%

PhCONH$_2$ $\xrightarrow[\text{170°}]{\text{C}_8\text{H}_{17}\text{Br H}_2\text{O}}$ PhCOOC$_8$H$_{17}$ JOC (1969) <u>34</u> 3204   79%

MeCONHBu $\xrightarrow[\text{2 CCl}_4\text{ 77°}]{\text{1 HNO}_3\text{ Ac}_2\text{O}}$ MeCOOBu

JACS (1955) <u>77</u> 6011
JOC (1969) <u>34</u> 3834   75%

JACS (1955) <u>77</u> 6008 6011
(1954) <u>76</u> 4497
Tetr Lett (1965) 2627   83%

JCS (1965) 181

Section 112   Esters from Amines

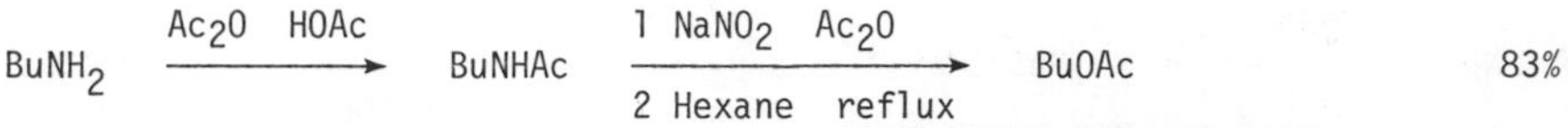

$$BuNH_2 \xrightarrow[\text{}]{Ac_2O \quad HOAc} BuNHAc \xrightarrow[\text{2 Hexane \quad reflux}]{1 \quad NaNO_2 \quad Ac_2O} BuOAc \qquad 83\%$$

JACS (1955) <u>77</u> 6008 6011
(1954) <u>76</u> 4497
JOC (1969) <u>34</u> 3834
Chem Pharm Bull (1960) <u>8</u> 266

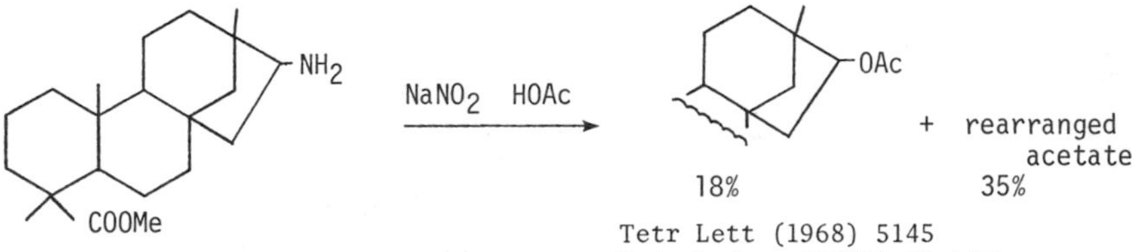

18%   35%

Tetr Lett (1968) 5145
Tetrahedron (1970) <u>26</u> 147

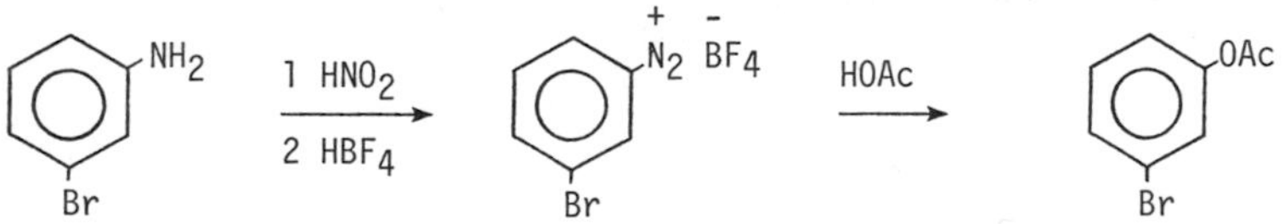

JACS (1939) <u>61</u> 143
(1933) <u>55</u> 4954

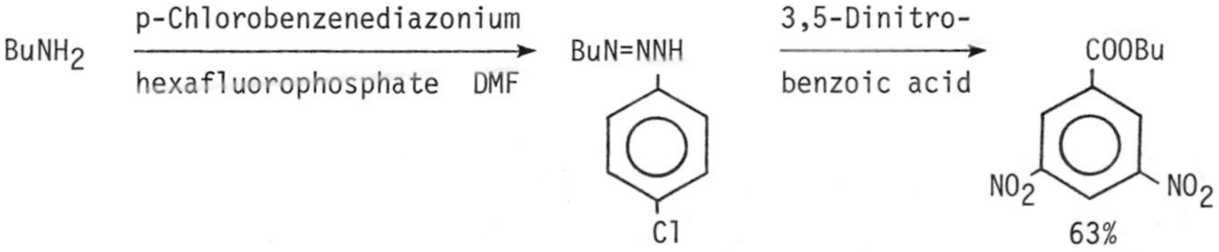

Tetr Lett (1961) 758

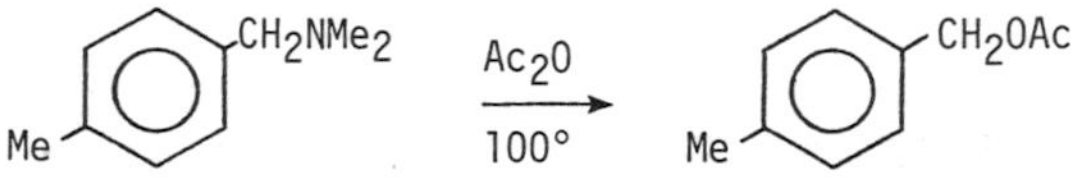

Bull Soc Chim Fr (1914) <u>15</u> 162

Section 113   Esters from Esters

PhCH$_2$COOBu-t $\xrightarrow[\text{monoglyme}]{\text{BuBr NaH}}$ PhCHCOOBu-t   66%
$\qquad\qquad\qquad\qquad\qquad\qquad\qquad$ |
$\qquad\qquad\qquad\qquad\qquad\qquad\qquad$ Bu   JOC (1965) <u>30</u> 2937
$\qquad\qquad\qquad\qquad\qquad\qquad\qquad\qquad\qquad\qquad$ (1964) <u>29</u> 2990

Me$_2$CHCOOEt $\xrightarrow[\text{2 EtI}]{\text{1 Ph}_3\text{CNa Et}_2\text{O}}$ Me$_2$CCOOEt   58%
$\qquad\qquad\qquad\qquad\qquad\qquad\qquad\qquad$ |
$\qquad\qquad\qquad\qquad\qquad\qquad\qquad\qquad$ Et   JACS (1940) <u>62</u> 2457

MeCH$_2$COOBu-t $\xrightarrow[\text{2 PhBr}]{\text{1 NaNH}_2\text{ NH}_3}$ MeCHCOOBu-t   38%
$\qquad\qquad\qquad\qquad\qquad\qquad\qquad\qquad\qquad$ |
$\qquad\qquad\qquad\qquad\qquad\qquad\qquad\qquad\qquad$ Ph   JACS (1959) <u>81</u> 1627

$\qquad\qquad\qquad\qquad\qquad\qquad$ Me
$\qquad\qquad\qquad\qquad\qquad\qquad$ |
$\qquad\qquad\qquad$ i-Pr(CH$_2$)$_3$CHCH=CH$_2$
C$_{16}$H$_{33}$CH$_2$COOMe $\xrightarrow[\text{di-t-butyl peroxide}]{}$ C$_{16}$H$_{33}$CHCOOMe   42%
$\qquad\qquad\qquad\qquad\qquad\qquad\qquad\qquad\qquad$ i-Pr(CH$_2$)$_3$CH(CH$_2$)$_2$
$\qquad\qquad\qquad\qquad\qquad\qquad\qquad\qquad\qquad\qquad\qquad\qquad$ Me
$\qquad\qquad\qquad\qquad\qquad\qquad\qquad$ J Amer Oil Chem Soc (1968) <u>45</u> 453
$\qquad\qquad\qquad\qquad\qquad\qquad\qquad$ (Chem Abs <u>70</u> 3210)

$\bigcirc$—COOPh $\xrightarrow[\text{2 h}\nu\text{ MeOH}]{\text{1 Me}_2\overset{+}{\text{S}}\text{OCH}_2\overset{-}{\text{ }}\text{THF}}$ $\bigcirc$—CH$_2$COOMe

$\qquad\qquad\qquad\qquad\qquad\qquad\qquad$ JACS (1964) <u>86</u> 1640

Further examples of the alkylation and homologation of carboxylic acids
and esters are included in section 17 (Carboxylic Acids, Acid Halides
and Anhydrides from Carboxylic Acids and Acid Halides) and section 107
(Esters from Carboxylic Acids, Acid Halides and Anhydrides)

$$\begin{array}{c}COOMe \\ \hline \\ COOMe\end{array} \xrightarrow[\text{t-BuOH } C_6H_6]{\text{t-BuOK molecular sieves}} \begin{array}{c}COOBu\text{-}t \\ \hline \\ COOBu\text{-}t\end{array}$$

> 90%

Chem Ind (1966) 1622

$$C_{17}H_{35}COOEt \xrightarrow[\text{MeOH}]{\text{Ion exch resin (basic)}} C_{17}H_{35}COOMe$$

80-90%

JOC (1969) 34 2032

$$CH_2{=}CHCOOMe \xrightarrow{C_{10}H_{21}OH \quad H_2SO_4} CH_2{=}CHCOOC_{10}H_{21}$$

83%

JACS (1944) 66 1203
Org Synth (1955) Coll Vol 3 146 605

$$RCOOBu \xrightarrow{HBr \quad Et_2O} RCOOEt$$

J Pharm Sci (1969) 58 949

$$PhCOOBu\text{-}t \xrightarrow[\text{reflux}]{\text{MeOH}} PhCOOMe$$

62%

JACS (1941) 63 3382

$$\begin{array}{c}COOMe \\ \hline \\ Cl\end{array} \xrightarrow{PhCH_2NMe_2 \quad 170°} \begin{array}{c}COOCH_2Ph \\ \hline \\ Cl\end{array}$$

65%

JACS (1952) 74 547

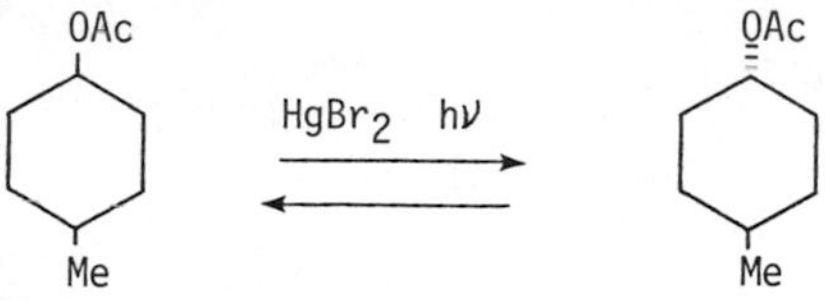

HgBr$_2$ hν

JACS (1970) <u>92</u> 1094

PhCH(COOEt)$_2$ $\xrightarrow{470°}$ PhCH$_2$COOEt

JOC (1964) <u>29</u> 1249

36%

EtCH(COOEt)$_2$ $\xrightarrow{\text{NaCN   Me}_2\text{SO}}$ EtCH$_2$COOEt

Tetr Lett (1967) 215

80%

Section 114   Esters from Ethers
ooooooooooooooooooo

PhCH$_2$OBu-t $\xrightarrow{\text{O}_3 \quad \text{Freon 11}}$ PhCOOBu-t

JACS (1968) <u>90</u> 6777
Annalen (1929) <u>476</u> 233

62%

PrCH$_2$OBu $\xrightarrow{\text{Trichloroisocyanuric acid   H}_2\text{O}}$ PrCOOBu

Tetr Lett (1968) 5819

100%

PrCH$_2$OBu $\xrightarrow{\text{RuO}_4 \quad \text{CCl}_4}$ PrCOOBu

JACS (1958) <u>80</u> 6682

100%

$RCH_2OC_{15}H_{31}$ $\xrightarrow{\quad CrO_3 \quad HOAc \quad CH_2Cl_2 \quad}$ $RCOOC_{15}H_{31}$ 48% (R=H)
                                                                                  9% (R=Me)

Chem Comm (1966) 752

50%

Carbohydrate Res (1970) 12 147

$CH_2=CH(CH_2)_3CH=CHCH_2OPh$ $\xrightarrow[\text{HOAc}]{PdCl_2(Ph_3P)_2}$ $CH_2=CH(CH_2)_3CH=CHCH_2OAc$

Chem Comm (1970) 1392

$\begin{array}{l} C_{17}H_{35}COOCH_2 \\ C_{17}H_{35}COOCH \\ \quad Ph_3COCH_2 \end{array}$ $\xrightarrow{\quad C_{15}H_{31}COCl \quad AgClO_4 \quad MeCN \quad}$ $\begin{array}{l} C_{17}H_{35}COOCH_2 \\ C_{17}H_{35}COOCH \\ C_{15}H_{31}COOCH_2 \end{array}$ 70%

Ber (1961) 94 812
JCS (1950) 1992

Bu_2O $\xrightarrow{\quad TsOAc \quad MeCN \quad}$ $BuOAc$ 50%

JACS (1968) 90 3878

93%

JOC (1965) 30 1734
Chem Rev (1954) 54 615

BuOBu-t $\xrightarrow{\text{HOAc TsOH}}$ BuOAc   Tetr Lett (1970) 4269   94%

Section 115   Esters from Halides and Sulfonates
○○○○○○○○○○○○○○○○○○○○○○○○○○○○○○○○○○○

PrBr $\xrightarrow{\begin{array}{l}\text{1 Mg Et}_2\text{O}\\ \text{2 MeCOCH}_2\text{CH}_2\text{COOEt}\\ \quad \text{C}_6\text{H}_6\\ \text{3 SOCl}_2\\ \text{4 185-195°}\end{array}}$ PrC=CHCH$_2$COOEt
|
Me $\xrightarrow{\begin{array}{l}\text{H}_2 \quad \text{PtO}_2\\ \text{EtOH}\end{array}}$ PrCHCH$_2$CH$_2$COOEt
|
Me

JACS (1944) <u>66</u> 1764

BuBr $\xrightarrow{\begin{array}{l}\text{1 Mg Et}_2\text{O}\\ \text{2 MeCH=CHCOOBu-s}\end{array}}$ BuCHCH$_2$COOBu-s
|
Me   ~60%

Org Synth (1961) <u>41</u> 60

PhBr $\xrightarrow{\begin{array}{l}\text{1 Mg THF}\\ \text{2 CdCl}_2\\ \text{3 BrCH}_2\text{COOEt}\end{array}}$ PhCH$_2$COOEt   53%

JOC (1968) <u>33</u> 1675

PhBr $\xrightarrow{\begin{array}{l}\text{1 Li Et}_2\text{O}\\ \text{2 CuBr}\\ \text{3 N}_2\text{CHCOOEt}\end{array}}$ PhCH$_2$COOEt   51%

Chem Comm (1969) 515

73%

JACS (1969) <u>91</u> 4304

$$\text{BuBr} \quad \xrightarrow[\text{diglyme}]{\overset{\displaystyle \overset{\text{Me}}{|}}{\text{EtCHCOOEt}} \quad \text{NaH}} \quad \overset{\text{Me}}{\underset{\text{Et}}{\overset{|}{\underset{|}{\text{BuCCOOEt}}}}} \qquad\qquad 38\%$$

JOC (1964) <u>29</u> 2990

Further examples of the alkylation of esters with halides are included in section 113 (Esters from Esters)

$$\text{PhBr} \quad \xrightarrow{\text{CH}_3\text{COOEt} \quad \text{NaNH}_2 \quad \text{NH}_3} \quad \text{PhCH}_2\text{COOEt} \qquad\qquad 42\%$$

JACS (1959) <u>81</u> 1627

$$\text{EtI} \quad \xdashrightarrow{\text{CH}_2(\text{COOEt})_2} \quad \text{EtCH(COOEt)}_2 \quad \xrightarrow{\text{NaCN} \quad \text{Me}_2\text{SO}} \quad \text{EtCH}_2\text{COOEt}$$

Tetr Lett (1967) 215

$$\text{C}_{10}\text{H}_{21}\text{Br} \quad \xrightarrow[\text{EtONa} \quad \text{EtOH}]{\text{MeCOCH}_2\text{COOEt}} \quad \underset{\underset{\text{MeCO}}{|}}{\text{C}_{10}\text{H}_{21}\text{CHCOOEt}} \quad \xrightarrow[\text{EtOH}]{\text{EtONa}} \quad \text{C}_{10}\text{H}_{21}\text{CH}_2\text{COOEt} \qquad 84\%$$

JOC (1962) <u>27</u> 622

$$\xrightarrow[\text{2 (EtO)}_2\text{CO}]{\text{1 Mg} \quad \text{Et}_2\text{O}}$$

68-73%

Org Synth (1943) Coll Vol 2 282

$$\xrightarrow[\text{2 C(OEt)}_4]{\text{1 Mg} \quad \text{Et}_2\text{O}} \qquad \xdashrightarrow{\text{Acid}}$$

Ber (1905) <u>38</u> 561

PhBr $\xrightarrow[\text{2 ClCOOEt}]{\text{1 Mg Et}_2\text{O}}$ PhCOOEt   75%

Ber (1903) <u>36</u> 3087

$C_7H_{15}I$ $\xrightarrow[\text{t-BuOH}]{\text{Ni(CO)}_4 \quad \text{t-BuOK}}$ $C_7H_{15}COOBu\text{-}t$ 66%

JACS (1969) <u>91</u> 1233
          (1963) <u>85</u> 2779

RBr $\xrightarrow[\text{MeOH}]{\text{Ni(CO)}_4 \quad \text{MeONa}}$ RCOOMe   88% (R=Ph)
                                                                 95% (R= PhCH=CH)

JACS (1969) <u>91</u> 1233

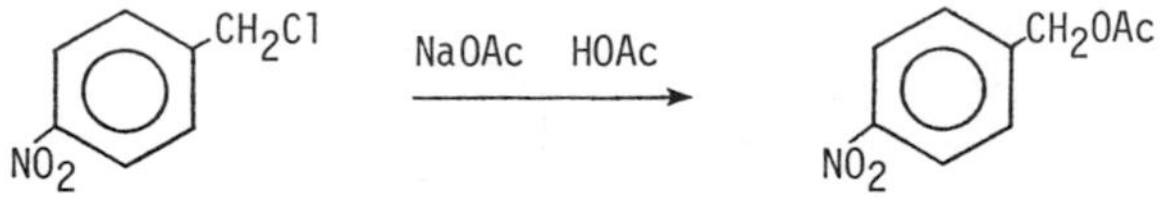

NaOAc HOAc

Org Synth (1955) Coll Vol 3 650
Ber (1956) <u>89</u> 1732

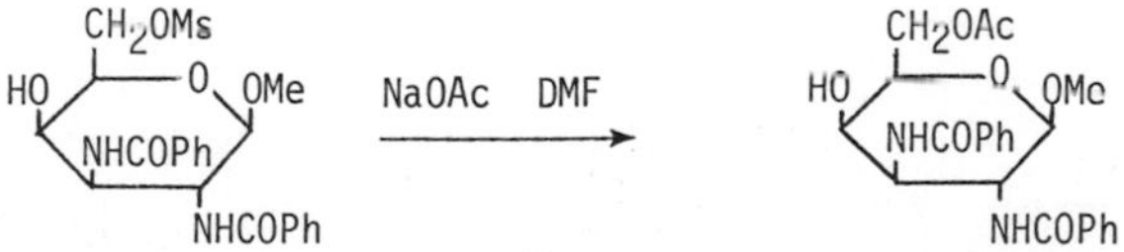

NaOAc DMF

Ber (1970) <u>103</u> 37

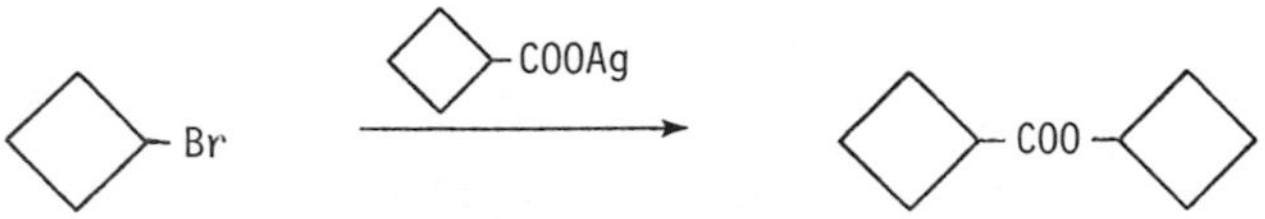

JACS (1951) <u>73</u> 5487
          (1949) <u>71</u> 3214

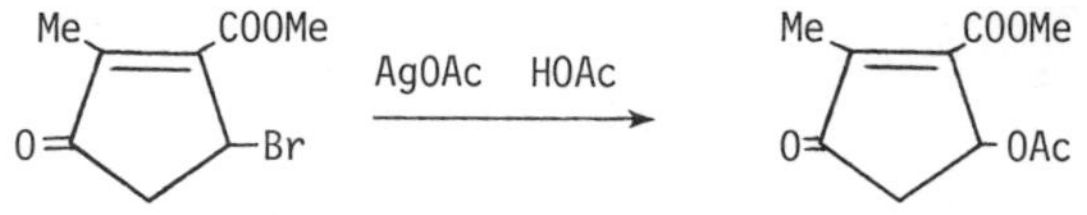

Tetrahedron (1968) 24 5421
Ber (1964) 97 443

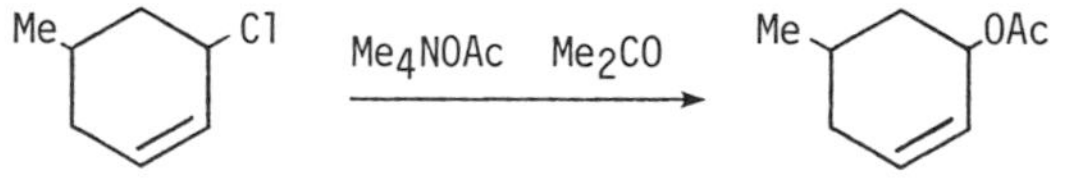

JACS (1955) 77 4042
(1967) 89 2758

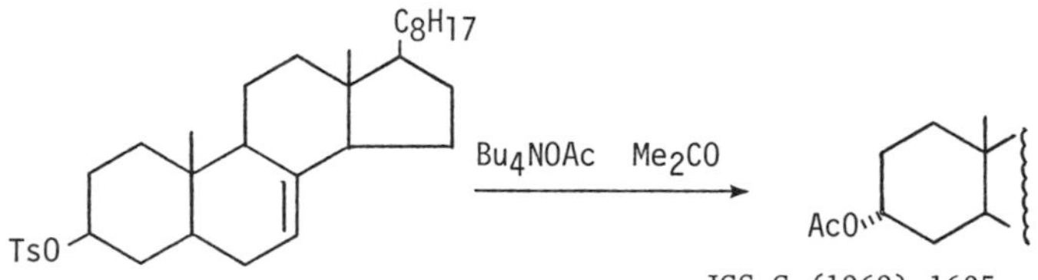

76%

JCS C (1969) 1605

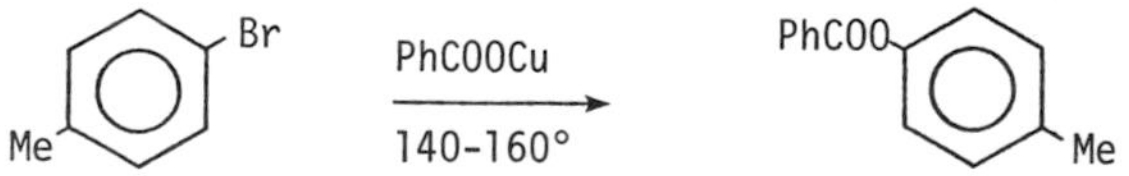

JACS (1966) 88 4521

79%

BuI →(MeCOO2H HOAc xylene)→ BuOAc   78%

JOC (1969) 34 3974

Section 116   Esters from Hydrides (RH)

This section contains examples of the reaction RH $\longrightarrow$ RCOOR' or R'COOR
(R=alkyl, allyl, vinyl, aryl etc.) and ArH $\longrightarrow$ Ar-X-COOR (X=alkyl chain)

$CH_2=CH(CH_2)_2COOEt$ $AlCl_3$

81%

JOC (1953) 18 1499
Tetrahedron (1967) 23 2481

PhH $\xrightarrow{\text{ClCH}_2\text{CH}_2\text{COOEt}\ \ \text{AlCl}_3}$ $PhCH_2CH_2COOEt$ 68%

Dokl (1945) 50 257
(Chem Abs 43 4638)

PhH $\xrightarrow{\text{ClCOOPh}\ \ \text{AlCl}_3}$ PhCOOPh   64%

JOC (1957) 22 325

$PhCH=CH_2$ $\xrightarrow[\text{MeONa}\quad\text{MeOH}]{\text{Electrolysis}\quad \text{CO (70 kg/cm}^2)}$ PhCH=CHCOOMe

Bull Chem Soc Jap (1965) 38 2122
JACS (1965) 87 3525

$PhCH=CH_2$ $\xrightarrow[\text{2 CO}\quad\text{Et}_3\text{N}\quad\text{MeOH}]{\text{1 PdCl}_2}$ PhCH=CHCOOMe   41%

JOC (1969) 34 738

$\xrightarrow{\text{PhLi}\quad\text{Et}_2\text{O}}$ $\xrightarrow{\text{CO(OEt)}_2}$ 44%

JACS (1953) 75 3843

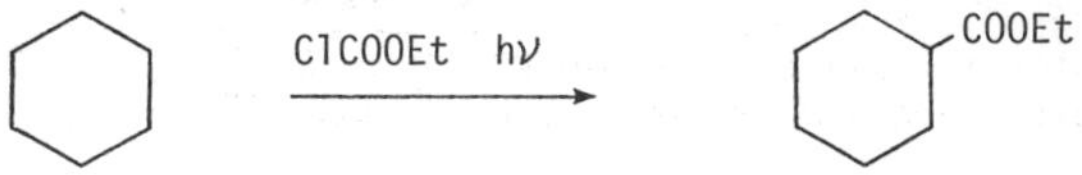

ClCOOEt hν

COOEt

15%

Bull Chem Soc Jap (1966) _39_ 2463

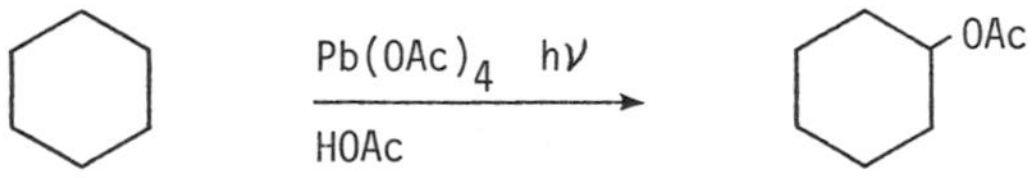

$Pb(OAc)_4$ hν

HOAc

OAc

24%

Nature (1966) _209_ 395

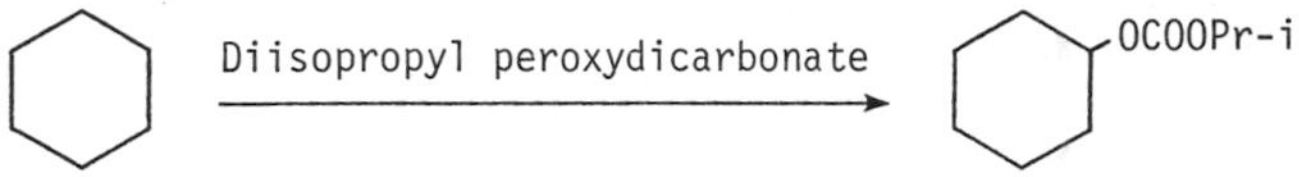

Diisopropyl peroxydicarbonate

OCOOPr-i

JCS (1965) 1932

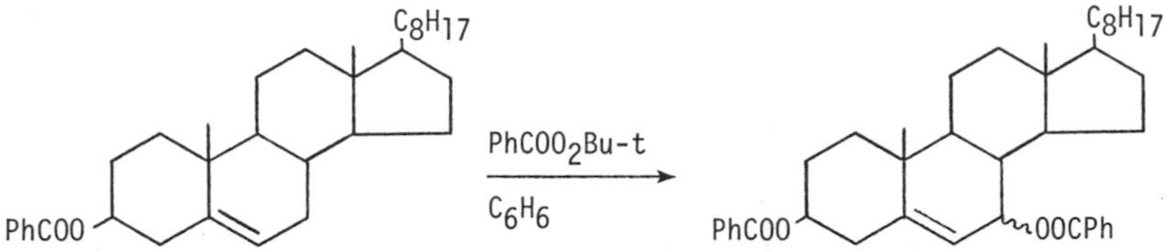

C_8H_{17}

$PhCOO_2Bu-t$

C_6H_6

PhCOO

C_8H_{17}

PhCOO   OOCPh

Proc Chem Soc (1962) 63
JACS (1958) _80_ 756

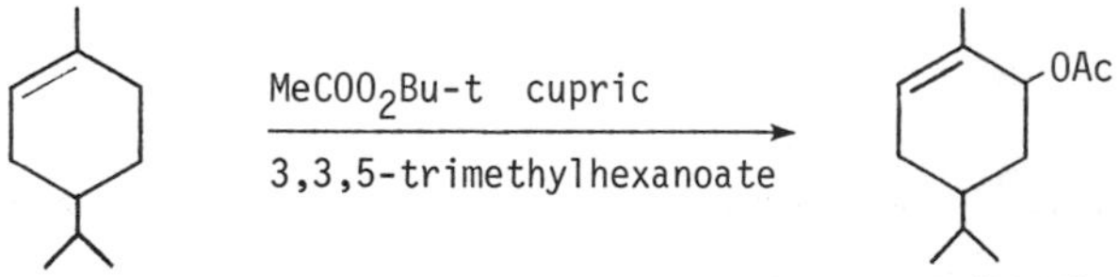

$MeCOO_2Bu-t$ cupric

3,3,5-trimethylhexanoate

OAc

65%

JACS (1962) _84_ 4969 774

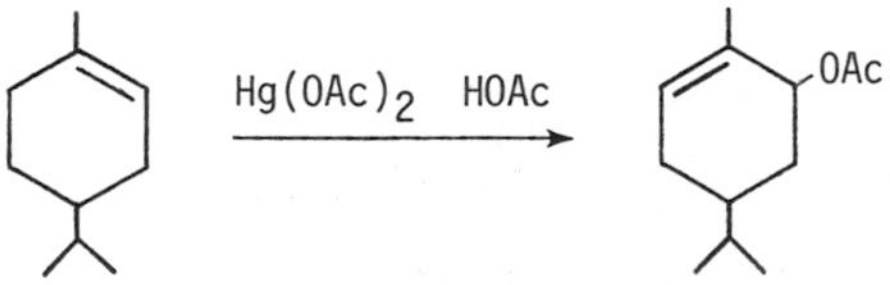

$Hg(OAc)_2$ HOAc

OAc

Annalen (1953) _581_ 59

Tetrahedron (1964) <u>20</u> 1017
Org React (1949) <u>5</u> 331

EtCH=CHCH$_3$ $\xrightarrow{\text{Pd(OAc)}_2}$ EtCHCH=CH$_2$
$\qquad\qquad\qquad\qquad\qquad\quad$ |
$\qquad\qquad\qquad\qquad\qquad\;$ OAc

JACS (1966) <u>88</u> 2054

C$_5$H$_{11}$CH$_2$CH=CH$_2$ $\xrightarrow[\text{HOAc}]{\text{Electrolysis Et}_4\text{NTs}}$ C$_5$H$_{11}$CHCH=CH$_2$
$\qquad\qquad\qquad\qquad\qquad\qquad\qquad\qquad\qquad\qquad\qquad$ |
$\qquad\qquad\qquad\qquad\qquad\qquad\qquad\qquad\qquad\qquad\;$ OAc

Tetr Lett (1968) 6207

PhCH$_3$ $\xrightarrow[\text{(PhCOO)}_2\text{Cu   RCOOH}]{\text{Di-t-butyl peroxide}}$ RCOOCH$_2$Ph   (R=Ph, Me)

Proc Chem Soc (1962) 63

JOC (1966) <u>31</u> 2033   90%

JACS (1968) <u>90</u> 1082

JOC (1969) <u>34</u> 3302   86%

Bull Chem Soc Jap (1969) 42 2733
Rec Trav Chim (1927) 46 54

Section 117   Esters from Ketones

Ph_2CO ---→ Ph_2CN_2 $\xrightarrow[\text{EtOH}]{\text{Ni(CO)}_4 \quad \text{CO}}$ $Ph_2CHCOOEt$ 74%

Ber (1960) 93 1840

Ph_2CO $\xrightarrow{N_2H_4 \quad EtOH}$ $Ph_2C=NNH_2$ $\xrightarrow{Pb(OAc)_4 \quad CH_2Cl_2}$ Ph_2CHOAc 43%

JCS C (1970) 1033

PhCOMe $\xrightarrow{Ph_3SnH \quad MeCOCl \quad C_6H_6}$ PhCHOAc   100%
                                                    |
                                                    Me

JACS (1966) 88 1833 4970

EtCHCOPr $\xrightarrow{Br_2}$ EtCCOPr $\xrightarrow[Et_2O]{MeONa}$ EtCCOOMe   <34%
| | |
Me   Me   Me

with Br above EtCCOPr carbon and Pr above EtCCOOMe carbon

JACS (1942) 64 300
Org React (1960) 11 261

$Ph_2CHCOMe$ ---→ Ph_2CCOMe $\xrightarrow[EtOH]{EtONa}$ Ph_2CHCH_2COOEt
            Cl_2 |
                         Cl

Org React (1960) 11 261

KOCl MeOH

80%

JACS (1944) <u>66</u> 208

1 i-PrCH$_2$ONO HCl

2 NaOH Me$_2$SO$_4$ H$_2$O

Compt Rend (1942) <u>214</u> 113

Et$_2$CO $\xrightarrow[\text{Na}_2\text{HPO}_4 \quad \text{CH}_2\text{Cl}_2]{\text{H}_2\text{O}_2 \quad (\text{CF}_3\text{CO})_2\text{O}}$ EtCOOEt   78%

JACS (1955) <u>77</u> 2287

i-PrCH$_2$COMe $\xrightarrow[\text{CH}_2\text{Cl}_2]{\text{HOOCCH=CHCO}_2\text{OH}}$ i-PrCH$_2$OAc   72%

Tetrahedron (1962) <u>17</u> 31

PhCOO$_2$H CHCl$_3$

67%

JACS (1949) <u>71</u> 14
Org React (1957) <u>9</u> 73
For use of m-chloroperbenzoic acid see JACS (1967) <u>89</u> 4530
and JOC (1964) <u>29</u> 2914

C$_6$H$_{13}$COMe $\xrightarrow{\text{H}_2\text{O}_2 \quad \text{BF}_3 \cdot \text{Et}_2\text{O}}$ C$_6$H$_{13}$OAc   62%

JOC (1962) <u>27</u> 24

$$C_5H_{11}COMe \xrightarrow[PhNO_2]{\overset{\overset{\displaystyle OOH}{|}}{(CF_3)_2COH} \quad CH_2Cl_2} C_5H_{11}OAc \qquad\qquad 81\%$$

Tetr Lett (1970) 2741

Esters may also be prepared by conversion of ketones into carboxylic acids followed by esterification. See section 27 (Carboxylic Acids from Ketones)

Section 118   Esters from Nitriles

$$PhCH_2CN \xrightarrow{C_7H_{15}OH \quad TsOH} PhCH_2COOC_7H_{15} \qquad\qquad 70\%$$

JOC (1958) 23 1225
(1960) 25 560

$$PhCH_2CN \xrightarrow{EtOH \quad H_2O \quad H_2SO_4} PhCH_2COOEt \qquad\qquad 83\text{-}87\%$$

Org Synth (1941) Coll Vol 1 270

$$PhCH_2CN \xrightarrow{HOCH_2CH_2OH} PhCH_2COOCH_2CH_2OH$$

JCS (1963) 2417

$$CH_2=CHCN \xrightarrow{Ph_3P=CHCOOEt \quad C_6H_6} CH_2=CHCOOEt$$

Tetr Lett (1967) 2407

Esters may also be prepared by hydrolysis of nitriles to carboxylic acids followed by esterification. See section 28 (Carboxylic Acids from Nitriles)

Section 119   Esters from Olefins
                oooooooooooooooooooo

1 B$_2$H$_6$ THF
―――――――――――――――→   [cyclopentane ring]—CH$_2$COOEt   75%
2 BrCH$_2$COOEt THF
  potassium di-t-butylphenoxide

            JACS (1969) 91 6855 2146
                 (1968) 90 818 1911

                1 B$_2$H$_6$ THF
BuCH=CH$_2$ ―――――――――――→   Bu(CH$_2$)$_3$COOEt   83%
                2 N$_2$CHCOOEt
                        JACS (1968) 90 6891

                    1 B$_2$H$_6$ THF
C$_5$H$_{11}$CH=CH$_2$ ―――――――――――→   C$_5$H$_{11}$(CH$_2$)$_3$COOEt   45%
                       +-
                    2 Me$_2$SCHCOOEt   JACS (1967) 89 6804

                        CH$_3$COO—[hexagon]
C$_6$H$_{13}$CH=CH$_2$ ―――――――――――――――→   C$_6$H$_{13}$(CH$_2$)$_3$COO—[hexagon]   59%
                    di-t-butyl peroxide
                            JCS (1965) 1918

Ni(CO)$_4$ EtOH
―――――――――――→   [norbornane]   72%
HOAc H$_2$O   COOEt
                JCS (1963) 410

CO Co i-PrOH
―――――――――――→   [cyclohexane]—COOPr-i   12%

            JACS (1952) 74 4496

$$C_{12}H_{25}CH=CH_2 \xrightarrow[\text{MeOH}]{\text{CO} \quad H_2PtCl_6\text{-}SnCl_2} C_{12}H_{25}(CH_2)_2COOMe \qquad 65\%$$

JOC (1970) 35 2846

$$\xrightarrow[\text{MeOH}]{\text{CO} \quad (Ph_3P)_2PdCl_2}$$

Angew (1968) 80 352
(Internat Ed 7 329)

$$\xrightarrow{\text{HCOOH} \quad HClO_4}$$

69%

JACS (1953) 75 6212
Org Synth (1965) 45 74

$$\xrightarrow{\text{HOAc} \quad h\nu}$$

JACS (1967) 89 5199

$$\begin{array}{c} \text{COOEt} \\ | \\ (CH_2)_8CH=CH_2 \end{array} \xrightarrow[\text{2 EtOH} \quad 150\text{-}155°]{\text{1 O}_3 \quad \text{EtOH}} \begin{array}{c} \text{COOEt} \\ | \\ (CH_2)_8COOEt \end{array} \qquad 50\%$$

Can J Chem (1965) 43 319

Esters may also be prepared by conversion of olefins into carboxylic acids or alcohols followed by esterification. See section 29 (Carboxylic Acids from Olefins) and section 44 (Alcohols from Olefins)

Section 120   Esters from Miscellaneous Compounds

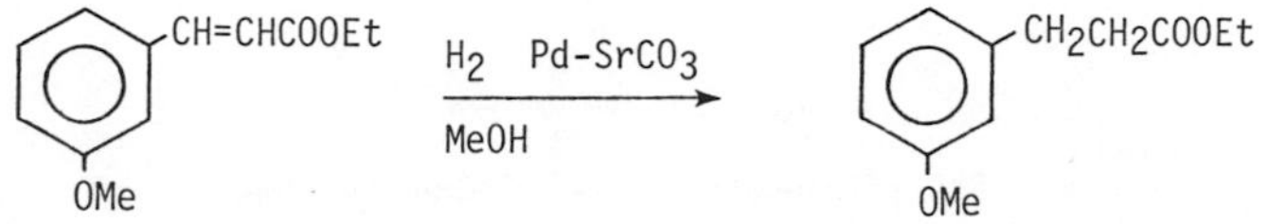

JCS (1936) 192
JACS (1962) 84 4951

$$PhCH=CHCOOCH_2Ph \xrightarrow{\text{H}_2 \quad (Ph_3P)_3RhCl} PhCH_2CH_2COOCH_2Ph \qquad 65\%$$

Compt Rend (1966) 263 251

$$Me_2C=CHCOOEt \xrightarrow{Ph_3SnH \quad h\nu \quad MeOH} Me_2CHCH_2COOEt \qquad 40\%$$

Tetr Lett (1967) 4805

$$PhCH=CHCOOMe \xrightarrow[\text{NCOOK}]{\overset{\text{NCOOK}}{\underset{\text{Pyr} \quad \text{HOAc}}{\|}}} PhCH_2CH_2COOMe \qquad 96\%$$

JOC (1965) 30 3985

$$PhCH=CHOAc \xrightarrow[\text{NCOOK}]{\overset{\text{NCOOK}}{\underset{\text{dioxane} \quad \text{HOAc}}{\|}}} PhCH_2CH_2OAc \qquad 50\%$$

JOC (1965) 30 3985

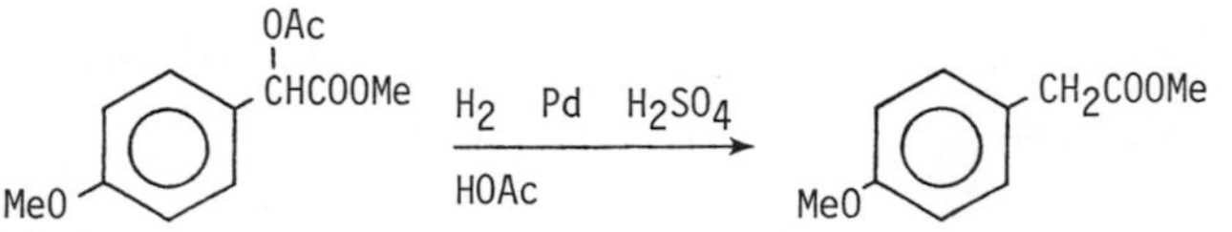

Ber (1941) 74B 315
Org React (1953) 7 263

ClCH$_2$COOPr-i $\xrightarrow{\text{Mg \quad i-PrOH}}$ MeCOOPr-i   63%

Proc Chem Soc (1963) 219

PhCH=CHCHO $\xrightarrow[\text{2 NaOH \quad MeOH}]{\text{1 PhNH}_2 \quad \text{NaCN}}$ PhCH$_2$CH$_2$$\overset{\text{OMe}}{\underset{|}{\text{C}}}$=NPh $\dashrightarrow$ PhCH$_2$CH$_2$COOMe

Chem Comm (1967) 1290
Chem Ind (1967) 583

Chapter 9   PREPARATION
                OF
                ETHERS
                AND EPOXIDES

Section 121   Ethers and Epoxides from Acetylenes
∘∘∘∘∘∘∘∘∘∘∘∘∘∘∘∘∘∘∘∘∘∘∘∘∘∘∘∘∘∘∘∘∘∘∘∘∘∘

No examples

Section 122   Ethers from Carboxylic Acids
∘∘∘∘∘∘∘∘∘∘∘∘∘∘∘∘∘∘∘∘∘∘∘∘∘∘∘∘∘∘∘∘∘∘

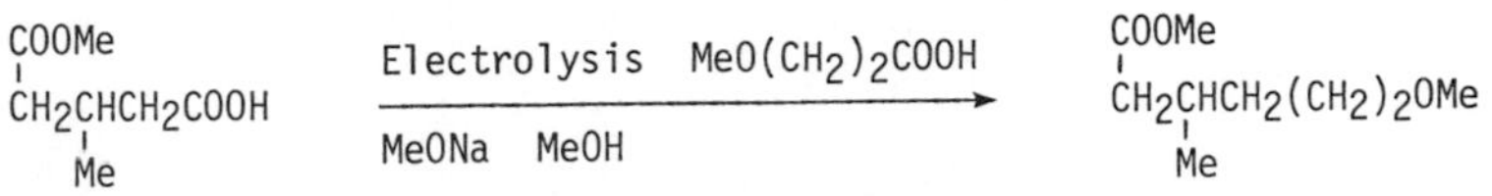

$$\begin{array}{c}COOMe\\|\\CH_2CHCH_2COOH\\|\\Me\end{array} \quad \xrightarrow[\text{MeONa} \quad \text{MeOH}]{\text{Electrolysis} \quad MeO(CH_2)_2COOH} \quad \begin{array}{c}COOMe\\|\\CH_2CHCH_2(CH_2)_2OMe\\|\\Me\end{array}$$

JCS (1956) 1620
Advances in Org Chem (1960) 1 1

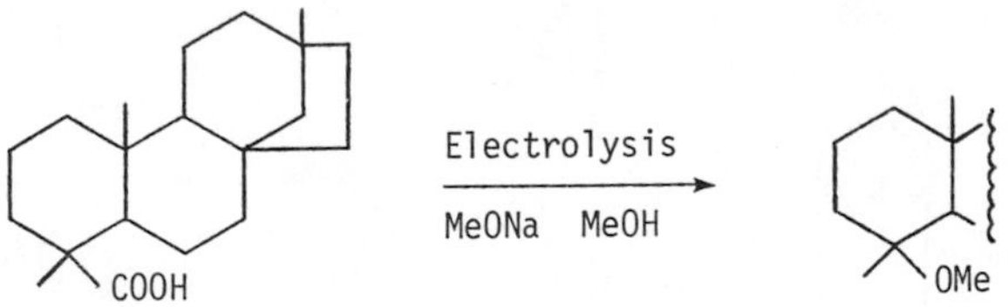

Electrolysis

MeONa   MeOH

JOC (1962) 27 4689
    (1960) 25 136
JACS (1960) 82 2645

Section 123   <u>Ethers from Alcohols and Phenols</u>

For the preparation of allyl, t-butyl, methoxymethyl, tetrahydropyranyl and triphenylmethyl ethers see section 45A (Protection of Alcohols and Phenols)

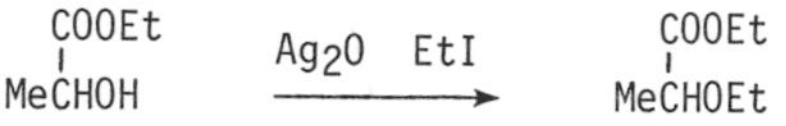

$$\underset{\text{MeCHOH}}{\overset{\text{COOEt}}{|}} \quad \xrightarrow{\text{Ag}_2\text{O} \quad \text{EtI}} \quad \underset{\text{MeCHOEt}}{\overset{\text{COOEt}}{|}}$$

JACS (1932) <u>54</u> 3732
JCS (1959) 2<u>5</u>94

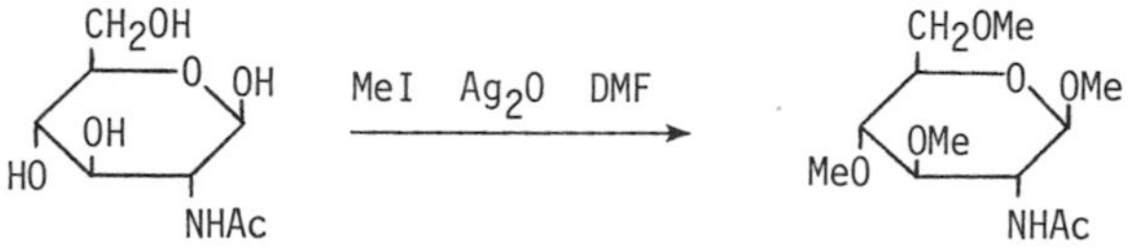

Angew (1955) <u>67</u> 32
JCS <u>C</u> (1969) <u>23</u>72
Can <u>J</u> Chem (1963) <u>41</u> 1801

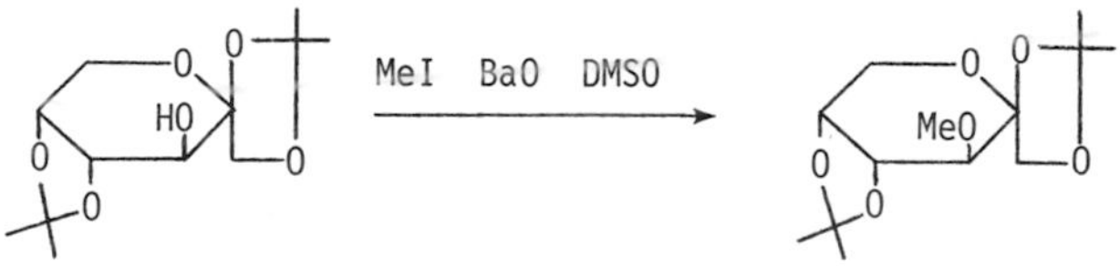

JCS (1967) 2681

59%

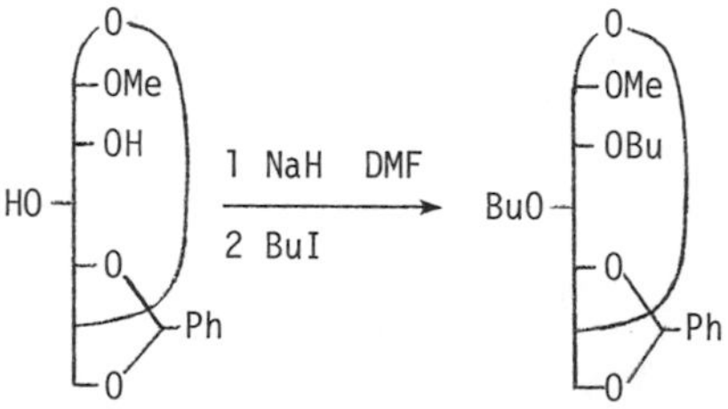

97%

Carbohydrate Res (1966) <u>2</u> 167
Can J Chem (1966) <u>44</u> 159<u>1</u>

$$\text{(pyranyl-CH}_2\text{OH)} \xrightarrow[\text{MeOCH}_2\text{CH}_2\text{OMe}]{\text{MeI \quad NaH}} \text{(pyranyl-CH}_2\text{OMe)}$$

91%

Can J Chem (1966) 44 1591

$$\underset{\overset{|}{\text{Et}}}{\text{Me}_2\text{COH}} \xrightarrow[\text{2 EtBr}]{\text{1 RMgBr \quad HMPA}} \underset{\overset{|}{\text{Et}}}{\text{Me}_2\text{COEt}}$$

40%

Compt Rend (1968) C 266 1178

$$\xrightarrow[\text{2 EtI}]{\text{1 K \quad toluene}}$$

~41%

JOC (1961) 26 4553

$$\xrightarrow[\text{NaNH}_2 \quad \text{Et}_2\text{O}]{\text{Ph} - \text{C}_6\text{H}_4 - \text{CH}_2\text{Cl}}$$

41%

JOC (1958) 23 1700

$$\underset{\overset{\displaystyle |}{\text{OCH}_2}}{\overset{\overset{\displaystyle \text{OCH}_2}{|}}{\text{PhCH} \quad \text{CHOH}}} \xrightarrow[\text{2 C}_{18}\text{H}_{37}\text{OTs}]{\text{1 K \quad C}_6\text{H}_6} \underset{\overset{\displaystyle |}{\text{OCH}_2}}{\overset{\overset{\displaystyle \text{OCH}_2}{|}}{\text{PhCH} \quad \text{CHOC}_{18}\text{H}_{37}}}$$

99%

JOC (1959) 24 409
(1966) 31 498

$$\xrightarrow{\text{MeOH \quad H}_2\text{SO}_4}$$

Tetrahedron (1961) 16 85
J Med Chem (1965) 8 57

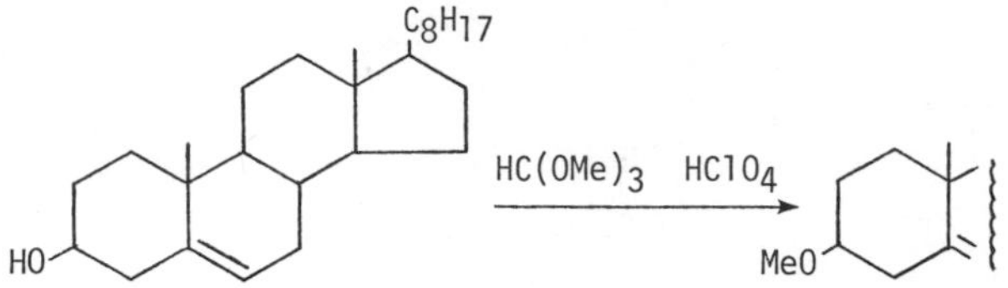

52%

Steroids (1966) 8 495

$C_8H_{17}OH$ $\xrightarrow{\text{CH}_2\text{N}_2 \quad \text{HBF}_4 \quad \text{Et}_2\text{O}}$ $C_8H_{17}OMe$ 84%

Tetrahedron (1959) 6 36
Annalen (1963) 662 38
(1964) 677 55

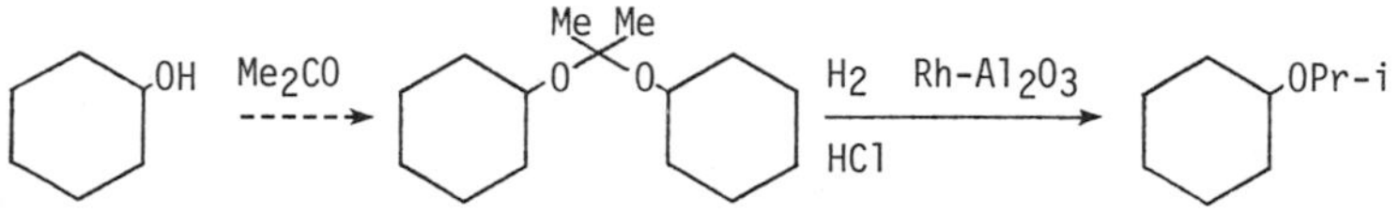

JOC (1961) 26 1026

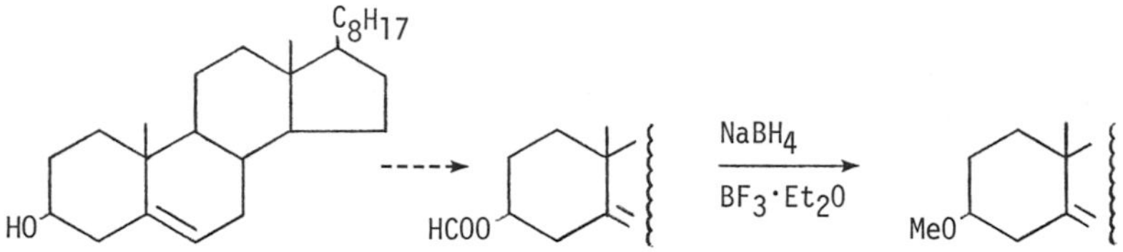

JOC (1961) 26 1685
(1962) 27 2127

PhOH $\xrightarrow[\text{2 } C_5H_{11}Br]{\text{1 NaNH}_2 \quad \text{NH}_3}$ $PhOC_5H_{11}$ 45%

JACS (1935) 57 510

PhOH $\xrightarrow[\text{150-190}°]{\text{PhCH}_2\text{Cl} \quad \text{Et}_3\text{N}}$ $PhOCH_2Ph$ 46%

JOC (1961) 26 5180

$$\text{(2-nitrophenol)} \xrightarrow{\text{BuBr} \quad K_2CO_3 \quad Me_2CO} \text{(2-nitro-butoxybenzene)} \qquad 75\text{-}80\%$$

Org Synth (1955) Coll Vol 3 140 418
Tetr Lett (1964) 1431
JACS (1950) 72 3396

$$\text{PhOH} \xrightarrow[]{\overset{\text{Me}}{\underset{|}{\text{BuCHOTs}}} \quad \text{KOH} \quad \text{DMF}} \overset{\text{Me}}{\underset{|}{\text{PhOCHBu}}} \qquad 38\%$$

JACS (1969) 91 1376

$$\text{PhOH} \xrightarrow{\text{Me}_2SO_4 \quad \text{NaOH} \quad H_2O} \text{PhOMe} \qquad 72\text{-}75\%$$

Org Synth (1941) Coll Vol 1 58

$$\text{PhOH} \xrightarrow{\text{PrBr} \quad \text{NaOH} \quad H_2O} \text{PhOPr} \qquad 63\%$$

JACS (1947) 69 2451

$$\xrightarrow{CH_2N_2 \quad Et_2O \quad MeOH}$$

JACS (1950) 72 3396
     (1962) 84 2972

$$\xrightarrow{CH_2N_2 \quad HBF_4 \quad CH_2Cl_2} \qquad 41\%$$

Tetrahedron (1959) 6 36
Annalen (1963) 662 38

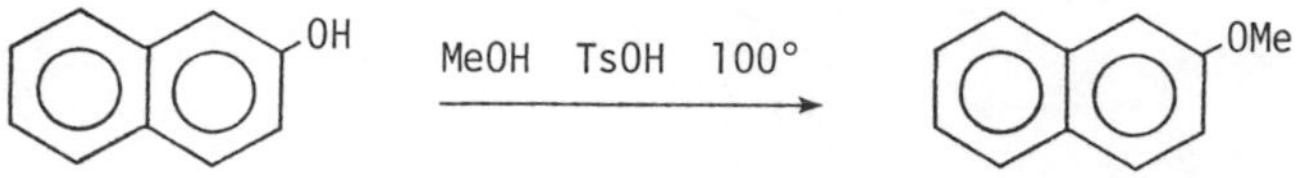

70%

JOC (1960) 25 832

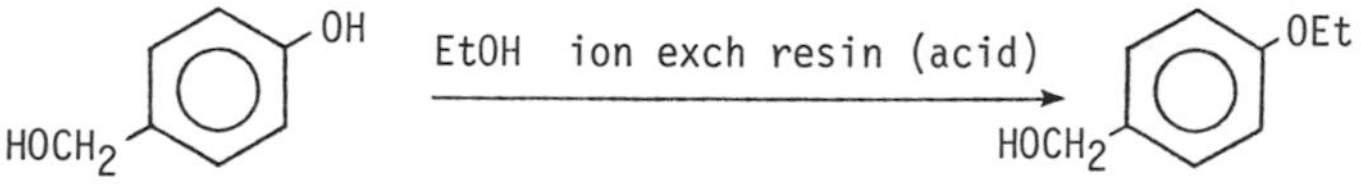

37%

JOC (1962) 27 2662

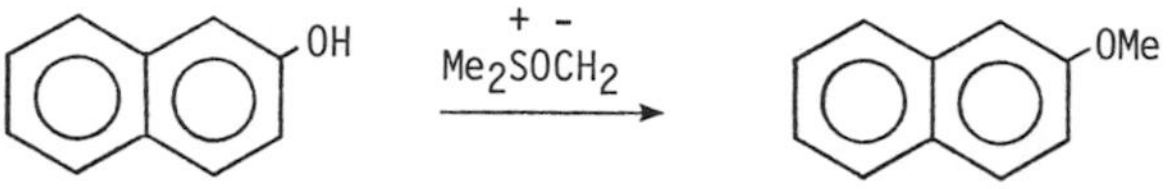

77%

Tetr Lett (1964) 867

$$\text{PhOH} \xrightarrow{\overset{+}{Me_3}O \; \overset{-}{BF_4}} \text{PhOMe}$$

73%

J Prakt Chem (1937) 147 257

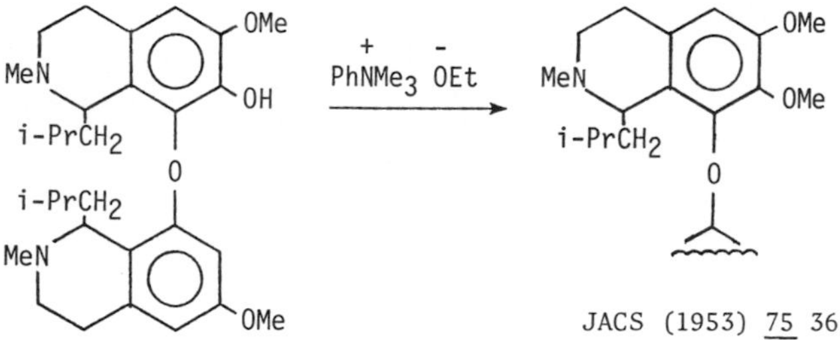

53%

JACS (1953) 75 3632

$$\text{PhOH} \xrightarrow{C(OEt)_4} \text{PhOEt}$$

85%

Acta Chem Scand (1956) 10 1006

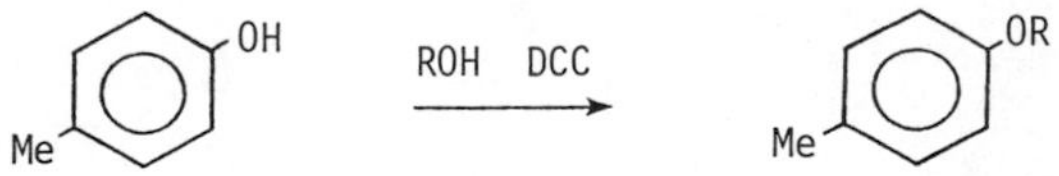

87% (R=Me)
90% (R=CH$_2$Ph)

Ber (1962) 95 2997
Angew (1963) 75 377
(Internat Ed 2 218)

PhOH $\xrightarrow{\text{HC(OCH}_2\text{Ph)}_2\text{NMe}_2}$ PhOCH$_2$Ph   64%

Angew (1963) 75 296
(Internat Ed 2 211)

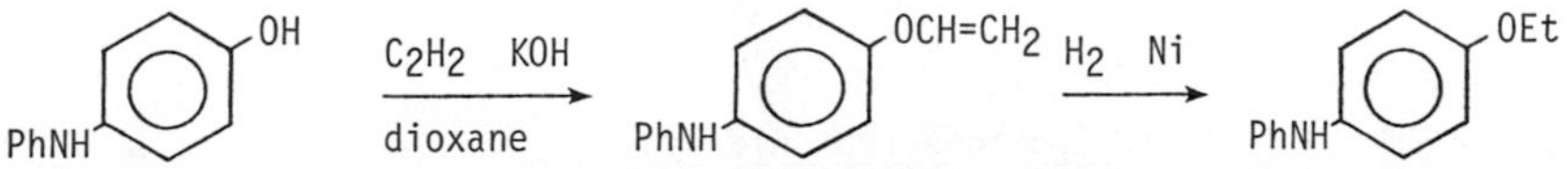

Zh Obshch Khim (1964) 34 3424
(Chem Abs 62 2727)

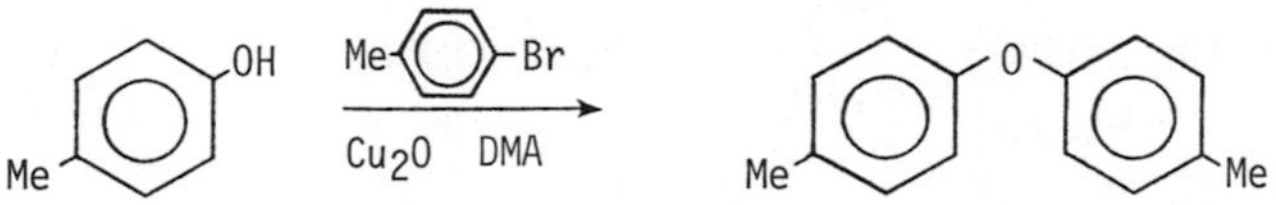

81%

JCS (1965) 4953
Chem Rev (1946) 38 405

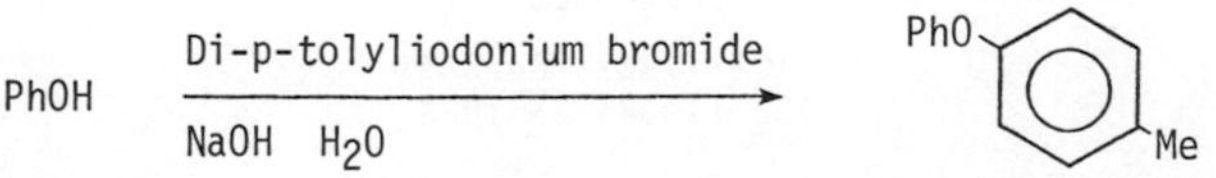

PhOH $\xrightarrow[\text{NaOH   H}_2\text{O}]{\text{Di-p-tolyliodonium bromide}}$

86%

JCS (1963) 4578

C$_7$H$_{15}$OH $\xrightarrow{\text{Pb(OAc)}_4 \quad \text{C}_6\text{H}_6}$ C$_3$H$_7$—

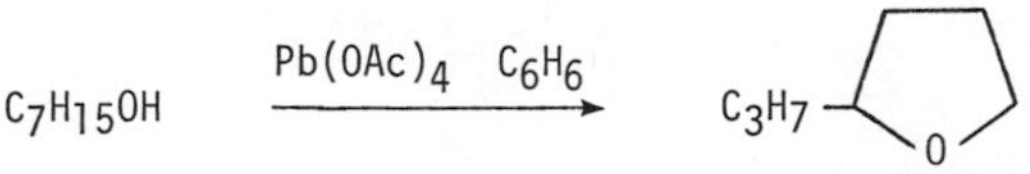

37%

Tetr Lett (1963) 2091
Synthesis (1970) 209

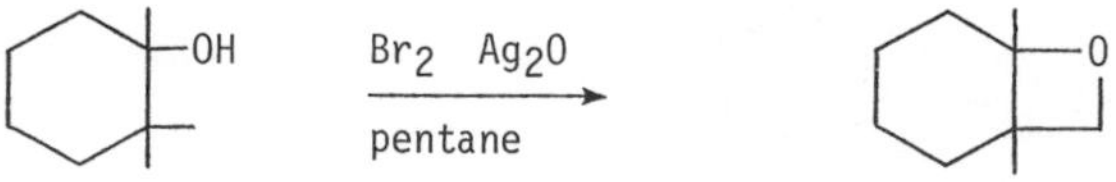

JACS (1964) 86 3905 1528

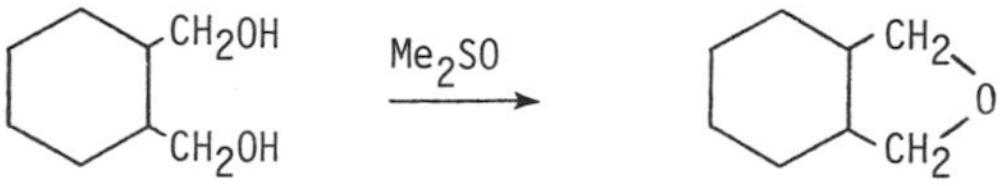

66%

JOC (1963) 28 1388

Section 124   Ethers and Epoxides from Aldehydes

Pr_2NCH_2CHO --> $Pr_2NCH_2CH(OMe)_2$ $\xrightarrow[\text{xylene}]{\text{PhMgBr}}$ $Pr_2NCH_2\underset{\underset{Ph}{|}}{C}HOMe$ 65%

JACS (1951) 73 4893

$PrCHO$ --> $PrCH(OEt)_2$ $\xrightarrow[\text{Et}_2\text{O}]{\text{LiAlH}_4 \quad \text{AlCl}_3}$ $PrCH_2OEt$ 49%

JACS (1962) 84 2371
JOC (1958) 23 1088

$i\text{-}PrCH_2CHO$ --> $i\text{-}PrCH_2CH(OEt)_2$ $\xrightarrow[\text{C}_6\text{H}_6]{(i\text{-Bu})_2\text{AlH}}$ $i\text{-}PrCH_2CH_2OEt$ 83%

Izv (1959) 2255
(Chem Abs 54 10837)

$PhCHO$ --> $PhCH(OEt)_2$ $\xrightarrow{\text{Et}_3\text{SiH} \quad \text{ZnCl}_2}$ $PhCH_2OEt$

Compt Rend (1962) 254 1814

i-PrCHO $\xrightarrow[\text{MeOH}]{\text{H}_2 \quad \text{PtO}_2 \quad \text{HCl}}$ i-PrCH$_2$OMe

JCS (1963) 5598

PhCHO $\xrightarrow[\text{Co}_2\text{CO}_8]{\text{H}_2 \quad \text{(3200 psi)} \quad \text{CO}}$ PhCH$_2$OCH$_2$Ph   43%

JACS (1950) 72 4375

$\xrightarrow{(\text{Me}_2\text{N})_3\text{P}}$ 50-55%

Org Synth (1966) 46 31
JACS (1963) 85 1884

$\xrightarrow[]{\overset{+-}{\text{Ph}_2\text{SCHR}} \quad \text{THF}}$ 72% (R=Ph)
40% (R=Bu)

JACS (1964) 86 918

PhCHO $\xrightarrow{\overset{+ \quad -}{\text{Me}_3\text{SO I}} \quad \text{NaH} \quad \text{Me}_2\text{SO}}$ $\overset{O}{\overset{\diagup\diagdown}{\text{PhCHCH}_2}}$ 56%

JACS (1965) 87 1353

$\xrightarrow{\overset{+ \quad -}{\underset{\text{Me}_2\text{SO}}{\overset{\text{NMe}_2}{\text{PhSOMe BF}_4}}} \quad \text{NaH}}$ 74%

JACS (1970) 92 6594

$$C_{11}H_{23}CHO \xrightarrow{\text{CH}_2\text{Br}_2 \quad \text{Li} \quad \text{THF}} C_{11}H_{23}\overset{\displaystyle O}{\overset{\displaystyle /\backslash}{C}}HCH_2 \qquad 35\%$$

Chem Comm (1969) 1047

Some of the methods listed in section 132 (Ethers and Epoxides from Ketones) may also be applied to the preparation of ethers and epoxides from aldehydes

Section 125   Ethers and Epoxides from Alkyls, Methylenes and Aryls

No examples of the preparation of ethers and epoxides by replacement of alkyl, methylene and aryl groups occur in the literature. For the conversion RH ⟶ ROR' (R=alkyl, aryl etc.) see section 131 (Ethers from Hydrides)

Section 126   Ethers and Epoxides from Amides

No examples

Section 127   Ethers from Amines

$$PhNH_2 \dashrightarrow PhN_2^+ Cl^- \xrightarrow{\text{MeOH}} PhOMe \qquad 93\%$$

JACS (1955) 77 1745
(1958) 80 6072
Org React (1944) 2 262

Section 128   Ethers from Esters

$$\xrightarrow[\text{MeOH}]{\text{PhCH}_2\text{Cl} \quad \text{NaHCO}_3}$$

48%

JCS (1948) 376
Ber (1943) 76 466

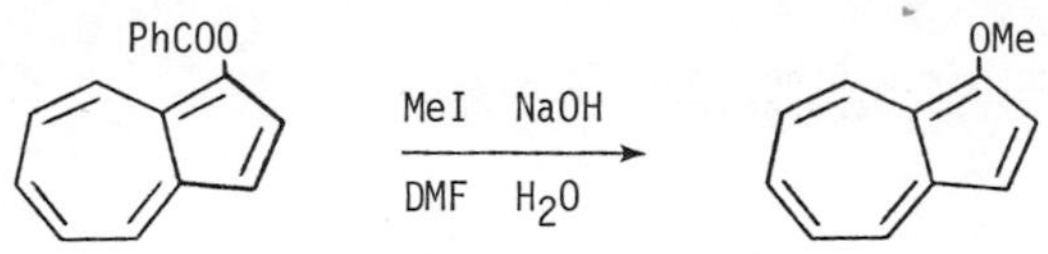

81%

JOC (1964) 29 2805

HCOOBu $\xrightarrow[\text{MeOCH}_2\text{CH}_2\text{OMe}]{\text{PhCH}_2\text{Br} \quad \text{NaH}}$ BuOCH$_2$Ph   43%

Tetr Lett (1965) 1713

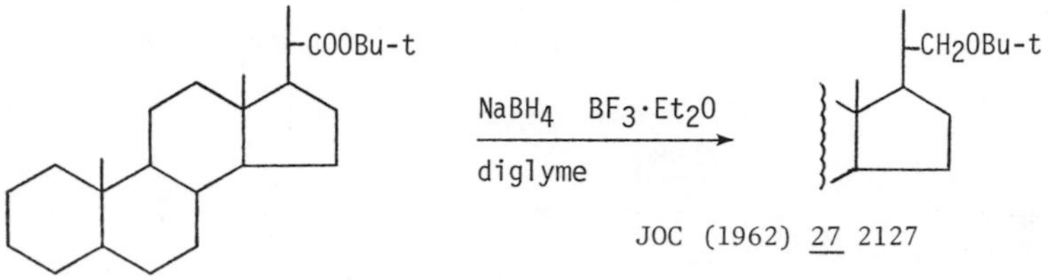

76%

JOC (1962) 27 2127

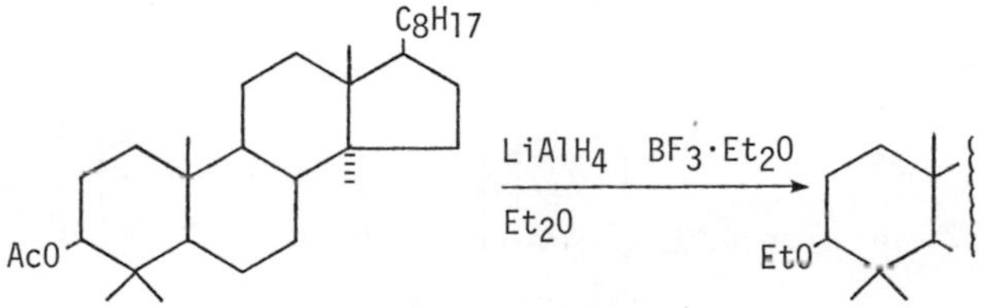

38%

JOC (1961) 26 4553 1685

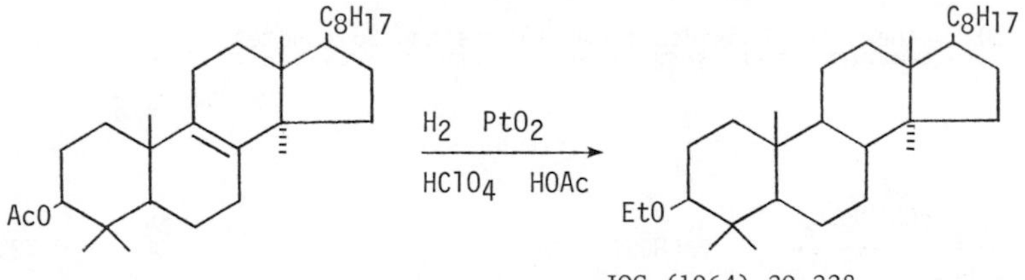

24%

JOC (1964) 29 228
Chem Ind (1964) 975

Section 129 Ethers and Epoxides from Ethers and Epoxides

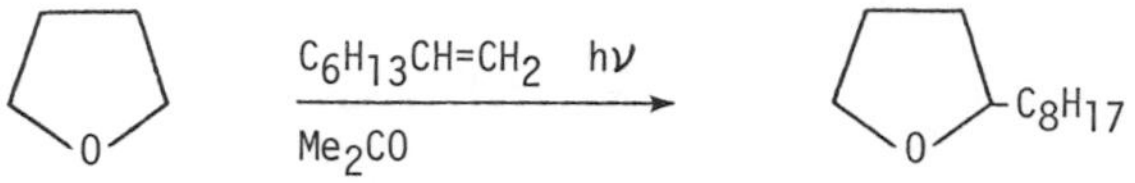
C6H13CH=CH2 hν
Me2CO
C8H17

JOC (1964) 29 2031
Tetrahedron (1967) 23 3193

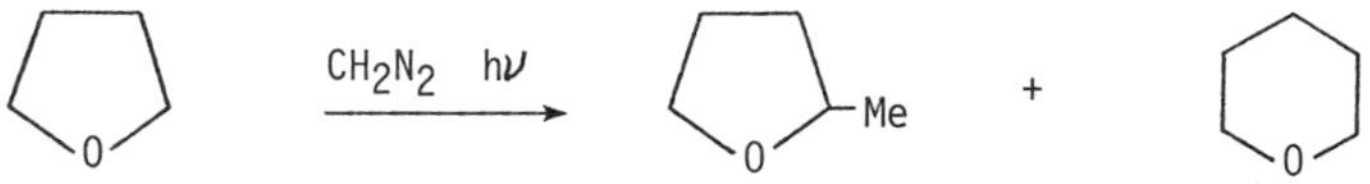
CH2N2 hν
Me
+

Chem Comm (1966) 454

$$Et_2O \underset{}{\overset{MeOSO_2F}{\rightleftarrows}} EtOMe$$

Chem Comm (1968) 1533

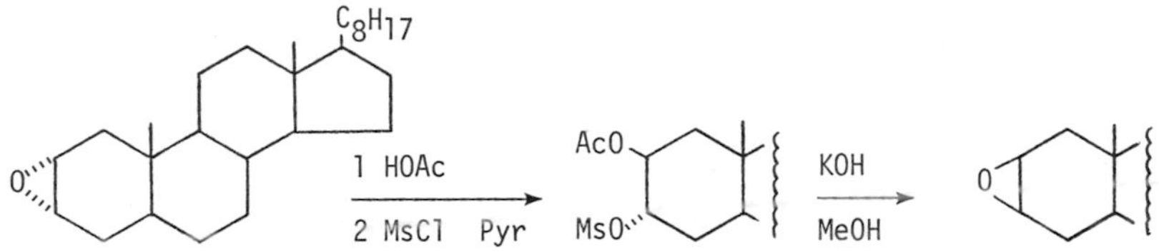
C8H17
1 HOAc
2 MsCl Pyr
AcO
MsO
KOH
MeOH

Helv (1949) 32 275

Section 130 Ethers and Epoxides from Halides and Sulfonates

$$PrBr \xrightarrow[PhPr\text{-}i]{Mg \quad MeCH(OEt)_2} PrCH(Me)OEt$$ 73%

Rec Trav Chim (1962) 81 238
JACS (1951) 73 4893

$$BuBr \quad ---\!\!\rightarrow \quad BuLi \quad \xrightarrow{ClCH_2OBu} \quad BuCH_2OBu \qquad\qquad 70\%$$

Ber (1964) <u>97</u> 636

Cyclohexanol   NaNH$_2$

Et$_2$O

44%

JOC (1958) <u>23</u> 1700

$$C_{18}H_{37}OTs \quad \xrightarrow{\text{t-BuOK} \quad \text{t-BuOH}} \quad C_{18}H_{37}OBu\text{-}t \qquad\qquad 99\%$$

JACS (1964) <u>86</u> 3072

$$PhBr \quad \xrightarrow[\text{Me}_2\text{SO}]{\text{t-BuOK} \quad \text{t-BuOH}} \quad PhOBu\text{-}t \qquad\qquad 42\text{-}46\%$$

Org Synth (1965) <u>45</u> 89
JOC (1968) <u>33</u> 259

$$RBr \quad \xrightarrow[\text{2 PhCOO}_2\text{Bu-}t]{\text{1 Mg} \quad \text{Et}_2\text{O}} \quad ROBu\text{-}t$$

74% (R=C$_8$H$_{17}$)
65% (R=Ph)
JACS (1959) <u>81</u> 4230
Org Synth (1963) <u>43</u> 55

PhOH   Cu$_2$O

collidine

67%

JCS (1965) 4953
Chem Rev (1946) <u>38</u> 405

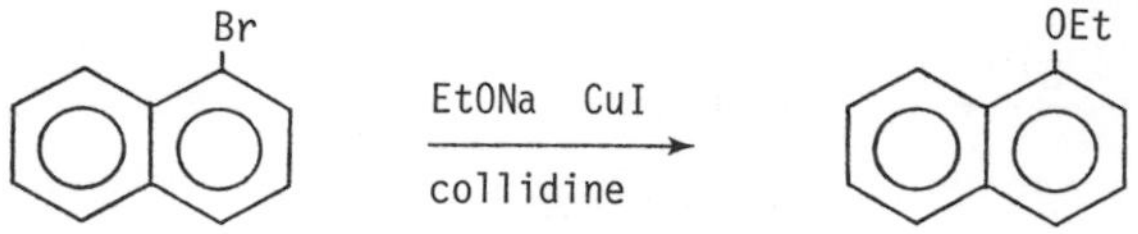

EtONa   CuI

collidine

100%

JCS C (1969) 312

Further examples of the preparation of ethers by the reaction
ROH + RHal ⟶ ROR are included in section 123 (Ethers from Alcohols and
Phenols)

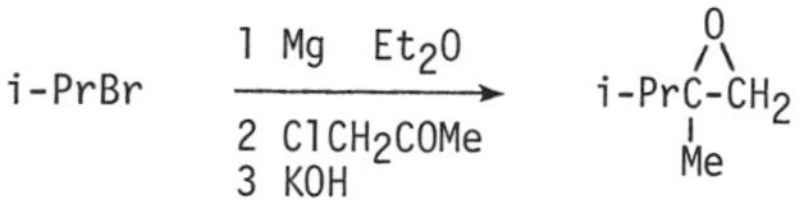

i-PrBr

1 Mg Et$_2$O
⟶
2 ClCH$_2$COMe
3 KOH

i-PrC-CH$_2$
      |
      Me

JACS (1946) 68 2339

Section 131   Ethers from Hydrides (RH)

This section lists examples of the reaction RH ⟶ ROR

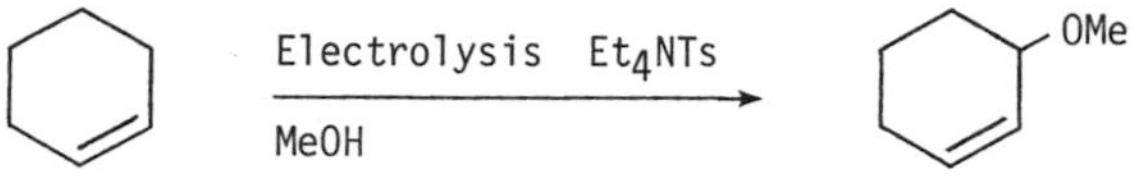

Electrolysis Et$_4$NTs

MeOH

OMe

21%

Tetr Lett (1968) 6207

Electrolysis MeONa

MeOH

OMe

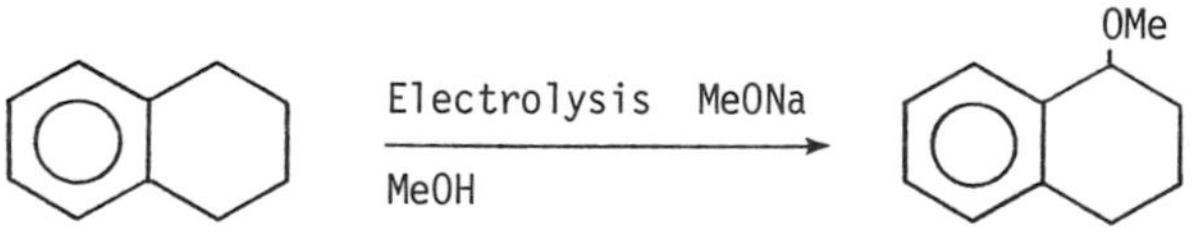

24%

Bull Chem Soc Jap (1964) 37 1597
Tetr Lett (1968) 2415

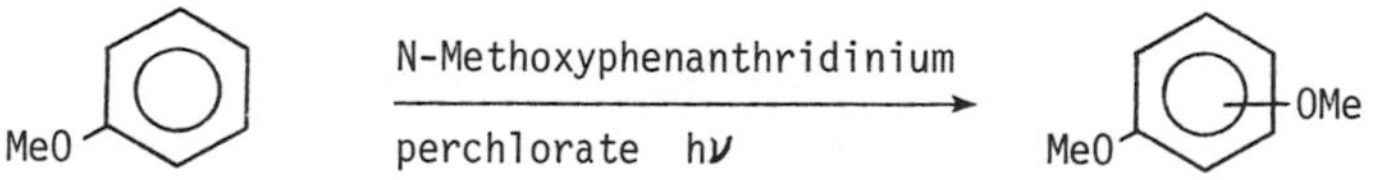

N-Methoxyphenanthridinium

perchlorate hν

MeO⟨ ⟩OMe

JACS (1970) 92 5814

Section 132   <u>Ethers and Epoxides from Ketones</u>

MeCO $\xrightarrow{\text{1 MeMgI}}$ Me$_2$COBu   40%
| 2 BuBr HMPA |
Et   Et

Compt Rend (1968) <u>C</u> <u>266</u> 1178

Bu$_2$CO $\xrightarrow[\text{MeOH}]{\text{H}_2 \quad \text{PtO}_2 \quad \text{HCl}}$ Bu$_2$CHOMe   54%

JCS (1963) 5598
Chem Comm (1967) 422

Me$_2$CO $\dashrightarrow$ Me$_2$C(OBu)$_2$ $\xrightarrow{\text{H}_2 \quad \text{Rh-Al}_2\text{O}_3}$ Me$_2$CHOBu   < 47%

JOC (1961) <u>26</u> 1026

LiAlH$_4$ AlCl$_3$
$\xrightarrow{\hspace{2cm}}$
Et$_2$O

< 92%

JACS (1962) <u>84</u> 2371
JOC (1958) <u>23</u> 1088

Pr$_3$SiH
$\xrightarrow{\hspace{2cm}}$
ZnCl$_2$

< 50%

Compt Rend (1962) <u>C</u> <u>254</u> 1814

$$Ph_2CO \quad \dashrightarrow \quad Ph_2CN_2 \quad \xrightarrow{h\nu \quad MeOH} \quad Ph_2CHOMe$$

Annalen (1958) <u>614</u> 19

Some of the methods listed in section 124 (Ethers and Epoxides from Aldehydes) may also be applied to the preparation of ethers from ketones

$$PhCOBu\text{-}t \quad \xrightarrow{LiCHCl_2} \quad \underset{Bu\text{-}t}{PhC\text{-}CHCl} \quad \xrightarrow{BuLi} \quad \underset{Bu\text{-}t}{PhC\text{-}CHBu}$$

Tetr Lett (1969) 2181

Me₃S⁺ I⁻   NaH
THF

77%

JACS (1965) <u>87</u> 1353

Me₃SO⁺ I⁻   NaH
Me₂SO

72%

JACS (1965) <u>87</u> 1353
(1970) <u>92</u> 6594

Me₂CHSOCMe₂⁻ Na⁺
‖
NTs
Me₂SO

63%

JACS (1970) <u>92</u> 5753

$$C_6H_{13}COMe \quad \xrightarrow{CH_2Br_2 \quad Li \quad THF} \quad \underset{Me}{C_6H_{13}C\text{-}CH_2}$$

52%

Chem Comm (1969) 1047

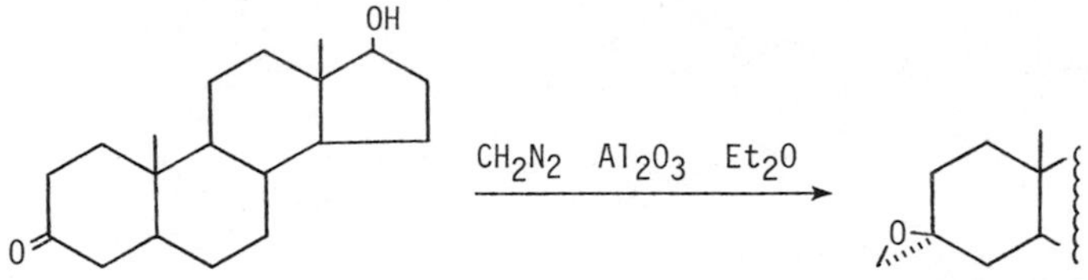
OH
O
CH_2N_2 Al_2O_3 Et_2O
O

50%

Tetr Lett (1969) 305
Org React (1954) 8 364

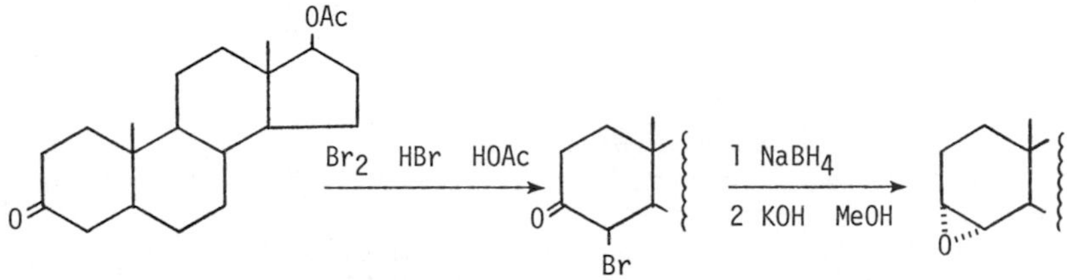
OAc
O
Br_2 HBr HOAc
O
Br
1 $NaBH_4$
2 KOH MeOH
O

20%

JACS (1953) 75 1704

Section 133 Ethers and Epoxides from Nitriles

No examples

Section 134 Ethers and Epoxides from Olefins

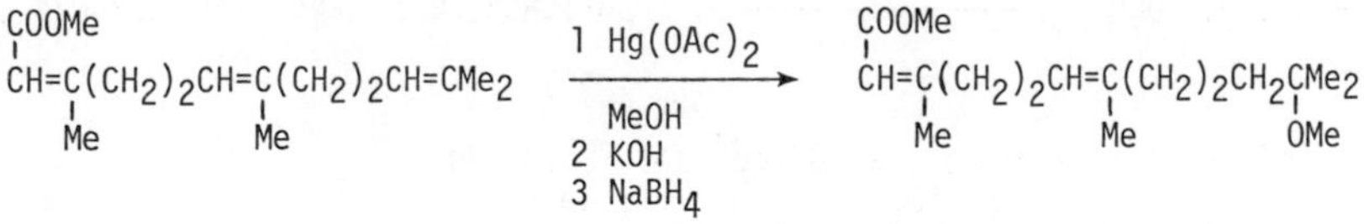
COOMe
CH=C(CH2)2CH=C(CH2)2CH=CMe2
Me
Me
1 Hg(OAc)2
MeOH
2 KOH
3 NaBH4
COOMe
CH=C(CH2)2CH=C(CH2)2CH2CMe2
Me
Me
OMe

74%

J Med Chem (1969) 12 191
JACS (1969) 91 5646

(cyclohexene with Et substituent) $\xrightarrow{h\nu \quad MeOH}$ (cyclohexane with Et and OMe)

60%

JACS (1967) 89 5199
Acc of Chem Res (1969) 2 33

Review: Epoxidation and Hydroxylation of Ethylenic Compounds with
       Organic Peracids

Org React (1953) 7 378

$HOCH_2CH_2CH(CH_2)_2CH=CMe_2$ (Me substituent) $\xrightarrow[CH_2Cl_2]{MeCOO_2H \quad NaOAc}$ $HOCH_2CH_2CH(CH_2)_2\overset{O}{\overset{\triangle}{C}}HCMe_2$ (Me substituent)

87%

Helv (1965) 48 182
JACS (1960) 82 4328
Org Synth (1963) Coll Vol 4 860

$C_{10}H_{21}CH=CH_2$ $\xrightarrow{CF_3COO_2H \quad Na_2CO_3 \quad CH_2Cl_2}$ $C_{10}H_{21}\overset{O}{\overset{\triangle}{C}}HCH_2$

90%

JACS (1955) 77 89

$C_6H_{13}CH=CH_2$ $\xrightarrow{HOOCCH=CHCOO_2H \quad CH_2Cl_2}$ $C_6H_{13}\overset{O}{\overset{\triangle}{C}}HCH_2$

80%

Tetrahedron (1962) 17 31

$C_8H_{17}CH=CH(CH_2)_7COOH$ $\xrightarrow{PhCOO_2H \quad Me_2CO}$ $C_8H_{17}\overset{O}{\overset{\triangle}{C}}HCH(CH_2)_7COOH$

62-67%

Org React (1953) 7 378

For use of m-chloroperbenzoic acid see JOC (1970) 35 251
      and Tetr Lett (1965) 849
For preparation and use of monoperphthalic acid see Org React (1953) 7 378
      and JACS (1960) 82 6373
For preparation of various aliphatic and aromatic peracids see
      JOC (1962) 27 1336

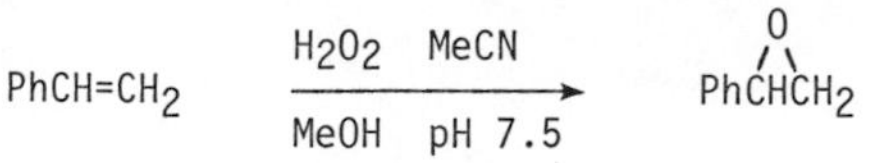

PhCH=CH₂ $\xrightarrow[\text{MeOH \quad pH 7.5}]{\text{H}_2\text{O}_2 \quad \text{MeCN}}$ PhCHCH₂ (epoxide)

JOC (1961) 26 659
JCS C (1970) 731

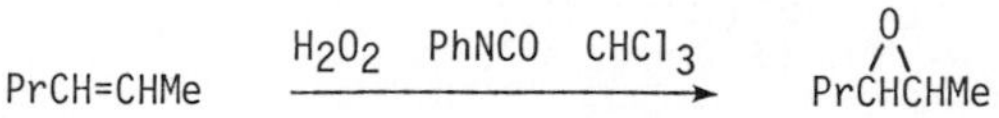

PrCH=CHMe $\xrightarrow{\text{H}_2\text{O}_2 \quad \text{PhNCO} \quad \text{CHCl}_3}$ PrCHCHMe (epoxide)   39%

Tetr Lett (1970) 2029

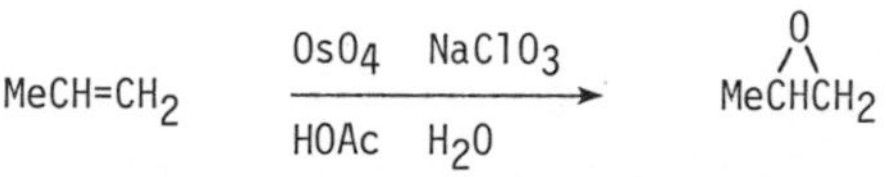

MeCH=CH₂ $\xrightarrow[\text{HOAc \quad H}_2\text{O}]{\text{OsO}_4 \quad \text{NaClO}_3}$ MeCHCH₂ (epoxide)

Chem Comm (1968) 1610

$\xrightarrow[\text{2 MeONa \quad MeOH}]{\text{1 I}_2 \quad \text{HgO \quad H}_2\text{O}}$

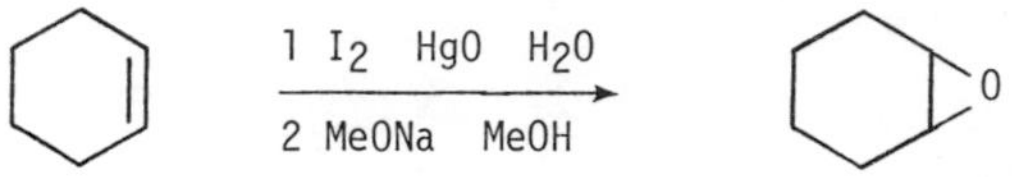

Tetrahedron (1964) 20 1017

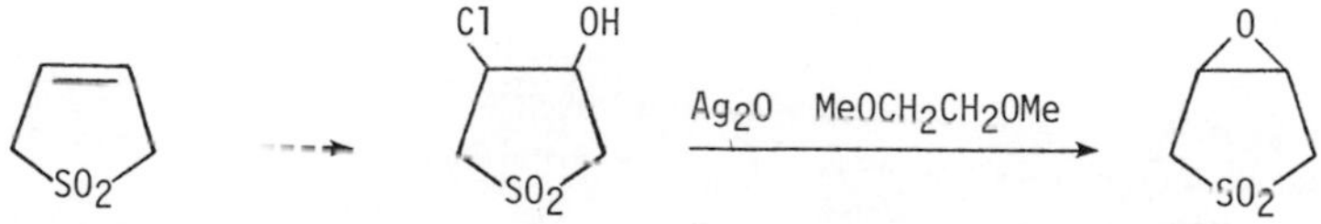

$\xrightarrow{\text{Ag}_2\text{O} \quad \text{MeOCH}_2\text{CH}_2\text{OMe}}$

JOC (1967) 32 3888

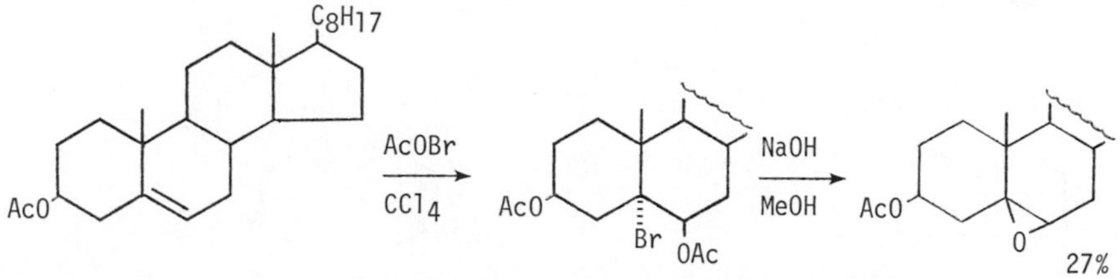

$\xrightarrow[\text{CCl}_4]{\text{AcOBr}}$ $\xrightarrow[\text{MeOH}]{\text{NaOH}}$ 27%

JACS (1959) 81 2826

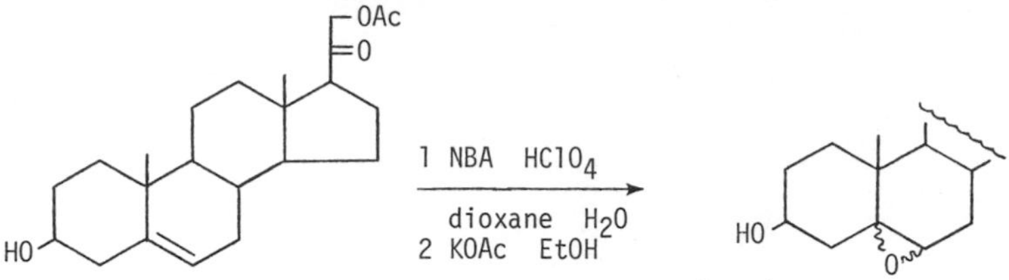

JCS (1956) 4417
JACS (1959) 81 2195

Section 135   Ethers and Epoxides from Miscellaneous Compounds
°°°

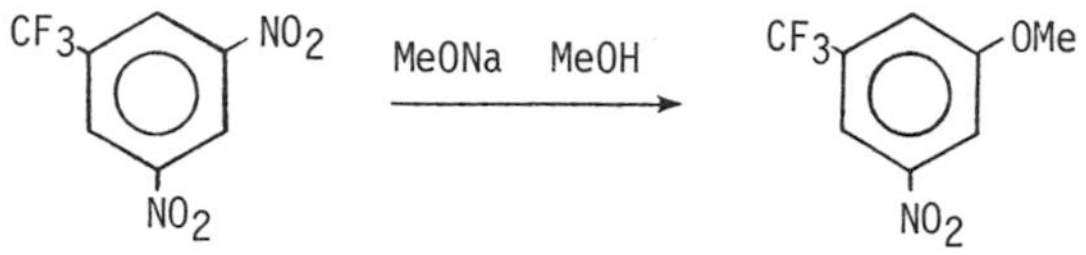

91%

JCS (1951) 2013

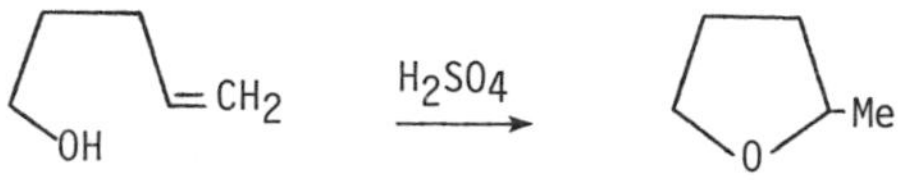

Bull Soc Chim Fr (1944) 11 365

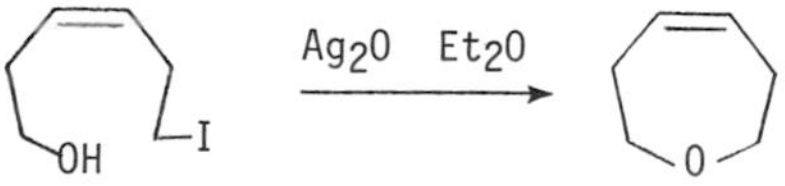

JACS (1958) 80 3132

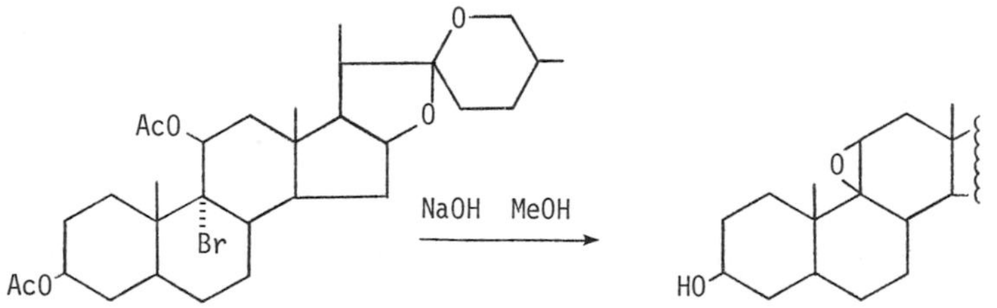

62%

JACS (1959) 81 2826

Chapter 10   PREPARATION
OF
HALIDES
AND SULFONATES

Section 136   Halides from Acetylenes

$$RC\equiv CH \xrightarrow{HBr} RCH=CHBr \xrightarrow[\text{dioxane}]{\overset{\displaystyle NCOOK}{\underset{\displaystyle NCOOK}{\|}} \; HOAc} RCH_2CH_2Br$$

JOC (1965) <u>30</u> 3985
JACS (1936) <u>58</u> 1806

Section 137   Halides from Carboxylic Acids

$$\underset{(CH_2)_8COOH}{\overset{COOMe}{|}} \xrightarrow[\text{MeONa   MeOH}]{\text{Electrolysis   } F(CH_2)_9COOH} \underset{(CH_2)_{17}F}{\overset{COOMe}{|}}$$

JACS (1956) <u>78</u> 2255
Advances in Org Chem (1960) <u>1</u> 1

$$\underset{Me}{\overset{|}{i\text{-}Pr(CH_2)_3CHCH_2COOH}} \xrightarrow[\text{2 Br}_2 \quad CCl_4]{\text{1 NH}_3 \quad AgNO_3 \quad H_2O} \underset{Me}{\overset{|}{i\text{-}Pr(CH_2)_3CHCH_2Br}} \qquad 55\%$$

JCS (1953) 132
Chem Rev (1956) <u>56</u> 219

$$RCOOH \xrightarrow{Br_2 \quad HgO \quad CCl_4} RBr$$

46% (R=C$_{11}$H$_{23}$)
<83% (R=Ph)
JOC (1965) <u>30</u> 415

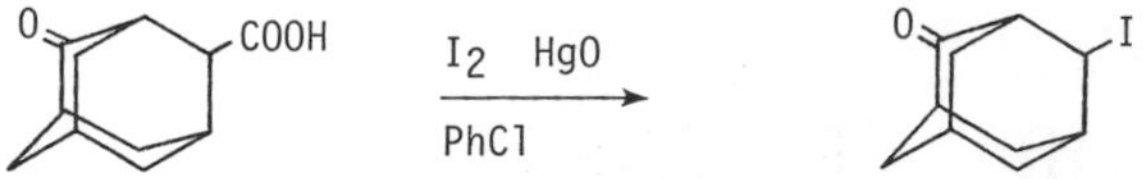

38%

Ber (1968) 101 2010

$C_8H_{17}COONa$ $\xrightarrow{\text{F}_2 \quad \text{H}_2\text{O}}$ $C_8H_{17}F$ ~50%

JOC (1969) 34 2446

$C_{17}H_{35}COOH$ $\xrightarrow{\text{I}_2 \quad \text{benzoyl peroxide}}$ $C_{17}H_{35}I$

J Amer Oil Chem Soc (1969) 46 615

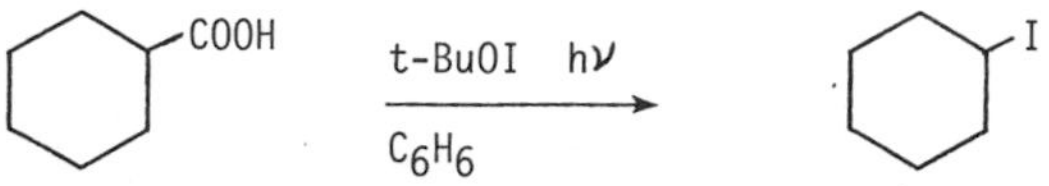

70%

JCS (1965) 2438

RCOOH $\xrightarrow[\text{CCl}_4]{\text{I}_2 \quad \text{Pb(OAc)}_4 \quad h\nu}$ RI   100% (R=C_5H_{11})
                                                          61% (R=Ph)

JCS (1965) 2438
JOC (1963) 28 65

$C_{15}H_{31}COOH$ $\xrightarrow[\text{HOAc}]{\text{Pb}_3\text{O}_4 \quad \text{Ac}_2\text{O}}$ $(C_{15}H_{31}COO)_4Pb$ $\xrightarrow[\text{C}_2\text{H}_2\text{Cl}_4]{\text{I}_2}$ $C_{15}H_{31}I$ 72%

JOC (1963) 28 65
Annalen (1970) 735 47

BuCOOH $\xrightarrow{\text{Pb(OAc)}_4 \quad \text{LiCl} \quad \text{C}_6\text{H}_6}$ BuCl   96%

JOC (1965) <u>30</u> 3265

$\xrightarrow[\text{2 Pyr·Br}_2]{\text{1 Hg(OAc)}_2 \quad \text{N-methylpyrrolidone}}$

JOC (1969) <u>34</u> 1904

RCOOH $\xrightarrow{\text{SOCl}_2}$ RCOCl $\xrightarrow[\text{CH}_2\text{Cl}_2]{\text{RhCl(Ph}_3\text{P)}_3}$ RCl   86% (R=PhCH$_2$)

61% (R=Ph)

JACS (1968) <u>90</u> 99 94
(1966) <u>88</u> 3452

Section 138   <u>Halides and Sulfonates from Alcohols and Phenols</u>

$C_{12}H_{25}OH$ $\xrightarrow{\text{HBr} \quad \text{H}_2\text{SO}_4}$ $C_{12}H_{25}Br$ 91%

Org Synth (1941) Coll Vol 1 29
JCS (1953) 132

$HO(CH_2)_6OH$ $\xrightarrow{\text{KI} \quad \text{polyphosphoric acid}}$ $I(CH_2)_6I$ 83-85%

Org Synth (1963) Coll Vol 4 323

Et_3COH $\xrightarrow{\text{HCl}}$ Et_3CCl 94%

JOC (1966) <u>31</u> 1090

EtCHOH $\xrightarrow{\text{HCl \quad MeCN}}$ EtCHOC=NH·HCl $\xrightarrow{130°}$ EtCHCl   56%
 | | | |
 Me   Me Me   Me

JACS (1955) <u>77</u> 2341

$\xrightarrow{\text{HCl \quad CCl}_3\text{CN \quad CHCl}_3}$ 85-90%

Tetr Lett (1970) 2517

$\xrightarrow{\text{SOCl}_2 \quad \text{Pyr}}$ 91%

JCS (1942) 684
    (1943)   99
    (1950) 3650
Tetr Lett (1970) 2931
        (1971)   87

PhCH$_2$CH$_2$OH $\xrightarrow[\text{2 SOBr}_2 \quad \text{Pyr·HBr}]{\text{1 SOCl}_2 \quad \text{Pyr}}$ PhCH$_2$CH$_2$Br   91%

Chem Ind (1954) 931
JOC (1961) <u>26</u> 3645

C$_{16}$H$_{33}$OH $\xrightarrow{\text{I}_2 \quad \text{P}_4}$ C$_{16}$H$_{33}$I   85%

Org Synth (1943) Coll Vol 2 322

CH$_2$=CCH$_2$CH$_2$C≡CCH$_2$OH $\xrightarrow[\text{collidine}]{\text{PBr}_3 \quad \text{LiBr} \quad \text{Et}_2\text{O}}$ CH$_2$=CCH$_2$CH$_2$C≡CCH$_2$Br   87%
         | |
         Me   Me

Chem Comm (1969) 611
JCS (1950) 3650

Et$_2$CHOH $\xrightarrow[\text{Et}_2\text{O}]{\text{PBr}_3 \text{ or } \text{PBr}_5}$ Et$_2$CHBr   22-47%

JOC (1961) 26 3645
(1959) 24 143
J Lipid Res (1966) 7 453

$\underset{\text{(CH}_2)_9\text{OH}}{\overset{\text{COOH}}{|}}$ $\xrightarrow[\text{2 H}_2\text{O}]{\text{1 PCl}_5}$ $\underset{\text{(CH}_2)_9\text{Cl}}{\overset{\text{COOH}}{|}}$

JCS (1936) 1605
(1970) 1124

C$_7$H$_{15}$OH $\xrightarrow[\text{2 I}_2 \quad \text{CH}_2\text{Cl}_2]{1 \quad \text{PCl} \quad \text{Pyr} \quad \text{Et}_2\text{O}}$ C$_7$H$_{15}$I   77%

JOC (1967) 32 4160
(1968) 33 300

BuOH $\xrightarrow{\text{(PhO)}_3\text{P} \quad \text{MeI}}$ BuI

JCS (1953) 2224   90%
(1954) 2281

$\xrightarrow{\text{Ph}_3\text{P}\cdot\text{I}_2 \quad \text{DMF}}$

59%

JACS (1964) 86 2093

MeCH=C=CHCH$_2$OH $\xrightarrow{\text{Ph}_3\text{P}\cdot\text{Br}_2 \quad \text{Pyr}}$ MeCH=C=CHCH$_2$Br

JCS (1967) 2260
JACS (1969) 91 7405
JOC (1965) 30 3469

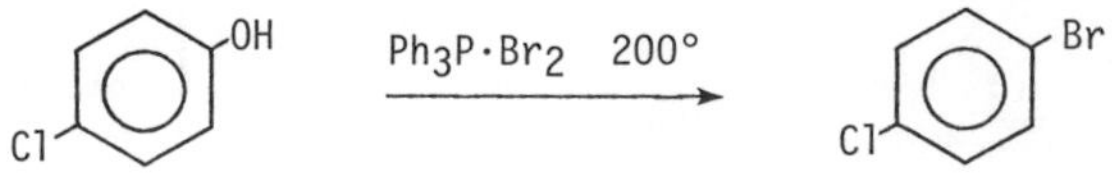

Ph$_3$P·Br$_2$ 200°

90%

JACS (1964) 86 964

PhCH$_2$OH $\xrightarrow{\text{Ph}_3\text{P·BrCN}}$ PhCH$_2$Br

Annalen (1959) 626 26

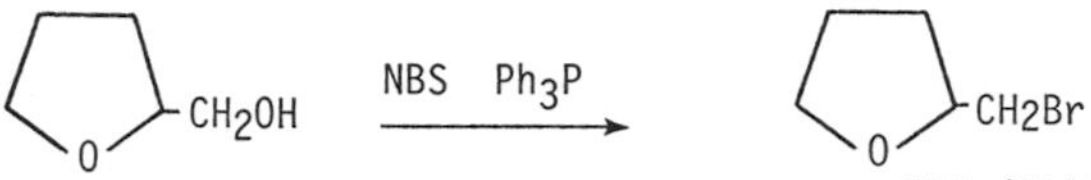

NBS Ph$_3$P

92%

JOC (1969) 34 212

PrCHOH $\xrightarrow{\text{Ph}_3\text{P·F}_2}$ PrCHF
  | |
  Me   Me

52%

Chem Pharm Bull (1968) 16 1009 1784

C$_8$H$_{17}$OH $\xrightarrow{\text{R}_3\text{P}\quad \text{CCl}_4}$ C$_8$H$_{17}$Cl   (R=Ph or C$_8$H$_{17}$)   94%

Can J Chem (1968) 46 86
Chem Comm (1968) 1358
Chem Ind (1966) 900

C$_5$H$_{11}$OH $\xrightarrow{\text{CBr}_4\quad \text{R}_3\text{P}\quad \text{Et}_2\text{O}}$ C$_5$H$_{11}$Br   (R=Ph or C$_8$H$_{17}$)   97%

Can J Chem (1968) 46 86

$$\text{CCl}_4 \quad (\text{Me}_2\text{N})_3\text{P}$$

70%

Chem Comm (1968) 1350

$$\text{ROH} \xrightarrow{\text{COCl}_2} \text{ROCOCl} \xrightarrow{\text{Ph}_3\text{P}} \text{RCl}$$

65% (R=Et)
67% (R=Ph)

Annalen (1966) 698 106

$$\xrightarrow{\text{COCl}_2}$$

ClCOO

$$\xrightarrow[\text{Me}_2\text{CO}]{\text{NaX}}$$

67% (X=Cl)
49% (X=I)

JOC (1967) 32 2633

$$\text{EtCHOH} \xrightarrow[\text{2 TlF}_2]{\text{1 COCl}_2} \text{EtCHOCOF} \xrightarrow{\text{Pyr}} \text{EtCHF}$$
$$\text{Me} \qquad\qquad\qquad \text{Me} \qquad\qquad\quad \text{Me}$$

64%

JACS (1955) 77 3099

$$\xrightarrow[\text{CH}_2\text{Cl}_2]{\text{Et}_2\text{NCF}_2\text{CHClF}}$$

35%

Tetr Lett (1962) 1065

$$\xrightarrow[\text{C}_2\text{H}_2\text{Cl}_4]{\overset{+}{\text{Me}_2\text{N}}=\text{CHCl} \ \overset{-}{\text{Cl}}}$$

~70%

Chem Comm (1967) 1152

$$\underset{\underset{Me}{|}}{EtCHOH} \xrightarrow{\overset{\overset{Cl}{|}}{Cl_2C=CNEt_2}} \underset{\underset{Me}{|}}{EtCHCl} \qquad 69\%$$

JACS (1960) $\underline{82}$ 909

$$C_{10}H_{21}OH \xrightarrow[\substack{2\ HF \\ 3\ KF_2\ \ 250\text{-}300°}]{1\ Ph_2NCN\ \ t\text{-}BuOK} C_{10}H_{21}F \qquad 68\%$$

Can J Chem (1965) $\underline{43}$ 3173

$$\underset{\underset{Me}{|}}{C_6H_{13}CHOH} \xrightarrow[\substack{2\ Ph_2CHCl}]{1\ Naphthyl\ isothiocyanate\ \ t\text{-}BuOK} \underset{\underset{Me}{|}}{C_6H_{13}CHCl} \qquad 47\%$$

JCS (1953) 3572

$$MeOH \xrightarrow{I_2\ \ B_2H_6} MeI \qquad 70\%$$

Chem Ind (1964) 1582
(1965) 223

$$\underset{\underset{Me}{|}}{EtCHOH} \xrightarrow[\substack{Et_2O}]{\overset{\overset{Br}{|}}{C_6H_{13}CHCH_2CH_2COBr}} \underset{\underset{Br}{|}\qquad\underset{Me}{|}}{C_6H_{13}CHCH_2CH_2COOCHEt} \xrightarrow{175\text{-}180°} \underset{\underset{Me}{|}}{EtCHBr}$$

40%

JACS (1956) $\underline{78}$ 4967

$$\underset{\underset{Me}{|}}{(MeO)_2CHCH_2CH_2C=CHCH_2OH} \xrightarrow[\substack{2\ TsCl\ \ LiCl}]{1\ MeLi\ \ HMPA\ \ Et_2O} \underset{\underset{Me}{|}}{(MeO)_2CHCH_2CH_2C=CHCH_2Cl}$$

Tetr Lett (1969) 1393

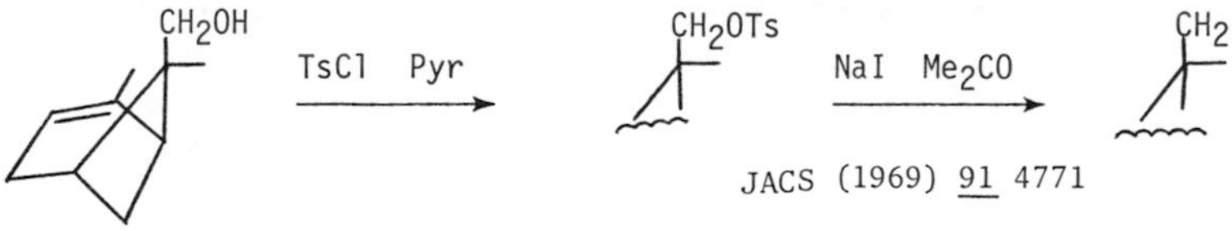

JACS (1969) 91 4771

Further examples of the preparation of halides from sulfonates are included in section 145 (Halides and Sulfonates from Halides and Sulfonates)

$C_{16}H_{33}OH$ $\xrightarrow{\text{TsCl Pyr}}$ $C_{16}H_{33}OTs$

JACS (1933) 55 345
(1970) 92 553

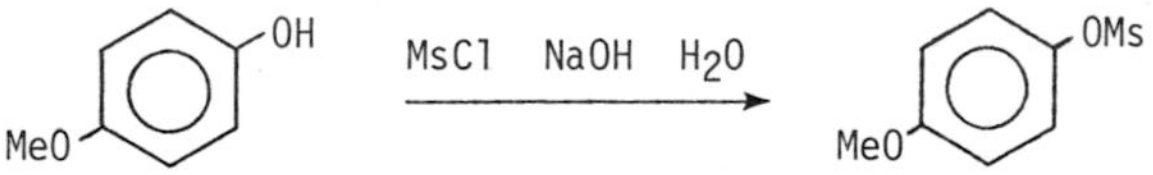

JCS (1955) 522

$PhCH_2OH$ $\xrightarrow[\text{2 TsCl}]{\text{1 NaH Et}_2\text{O}}$ $PhCH_2OTs$ 80%

JACS (1953) 75 3443
(1960) 82 3082

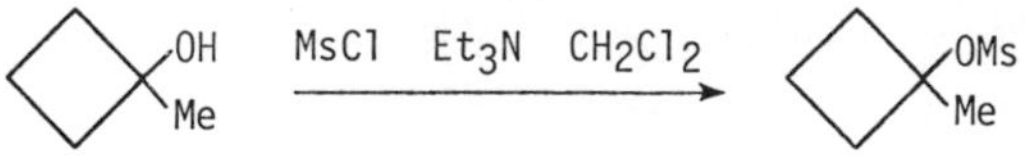

JOC (1970) 35 3195

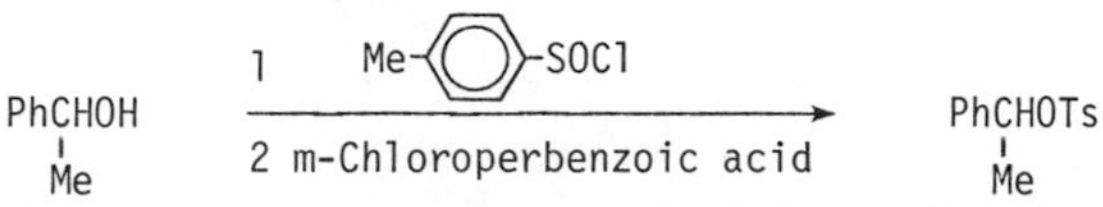

$PhCHOH \atop Me$ $\xrightarrow[\text{2 m-Chloroperbenzoic acid}]{\text{1 Me-} \bigcirc \text{-SOCl}}$ $PhCHOTs \atop Me$ 53%

Tetr Lett (1969) 2705

Section 139   Halides from Aldehydes

$$\text{PhCHO} \xrightarrow[\text{2 HCOOEt}]{\text{1 BuMgBr}} \underset{\underset{\text{Bu}}{|}}{\text{PhCHBr}}$$

Zh Obshch Khim (1955) <u>25</u> 505
(Chem Abs <u>50</u> 3356)   56%

$$\text{i-PrCHO} \xrightarrow[\text{CCl}_4]{\text{Di-t-butyl peroxide}} \text{i-PrCl}$$

JACS (1965) <u>87</u> 2194   45%

Section 140   Halides from Alkyls

The conversion RR' ⟶ RHal (R,R'=alkyl or aryl) is included here. For the reaction RH ⟶ RHal see section 146 (Halides from Hydrides)

$$\xrightarrow{\text{S}_2\text{Cl}_2 \quad \text{SO}_2\text{Cl}_2 \quad \text{Fe}}$$

JACS (1947) <u>69</u> 3146   45%

Section 141   Halides from Amides

$$\underset{\underset{\text{Me}}{|}\quad\underset{\text{Me}}{|}}{\text{BuCH(CH}_2)_3\text{CH(CH}_2)_3\text{NHCOPh}} \xrightarrow{\text{PBr}_3} \underset{\underset{\text{Me}}{|}\quad\underset{\text{Me}}{|}}{\text{BuCH(CH}_2)_3\text{CH(CH}_2)_3\text{Br}}$$

Z Physiol Chem (1942) <u>273</u> 225
Org Synth (1941) Coll Vol 1 428

JACS (1962) 84 769

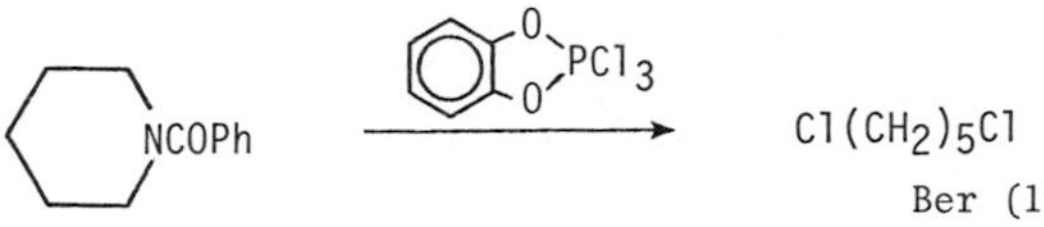

$Cl(CH_2)_5Cl$ 55%

Ber (1963) 96 1387
Org Synth (1941) Coll Vol 1 428

Section 142   Halides from Amines

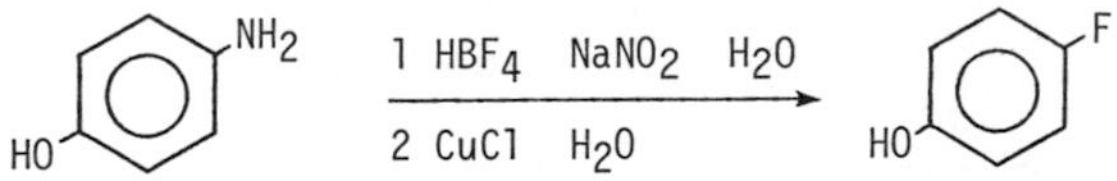

71%

JACS (1956) 78 6037
Chem Ind (1961) 179
Org Synth (1944) Coll Vol 2 295
Org React (1949) 5 193

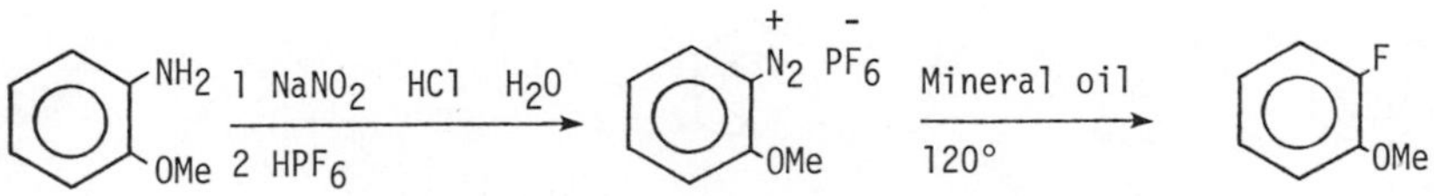

60%

JOC (1961) 26 5149
    (1963) 28 568

$PhNH_2$ $\xrightarrow{\text{CuX}_2 \ \ \text{NO} \ \ \text{MeCN}}$ PhX

73% (X=Cl)
65% (X=Br)

Rec Trav Chim (1966) 85 857
JACS (1947) 69 1221

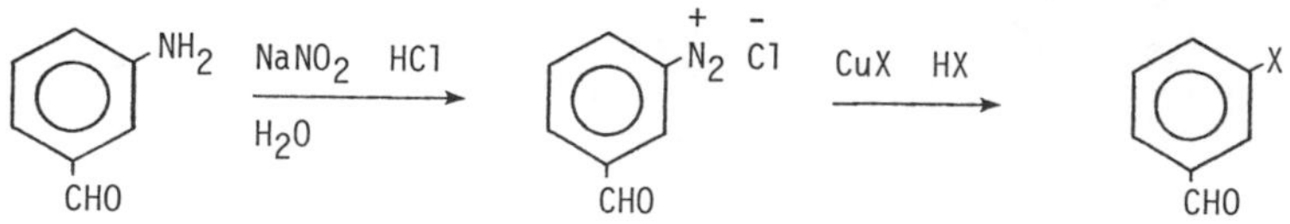

~75% (X=Cl)
~65% (X=Br)
Org Synth (1944) Coll Vol 2 130

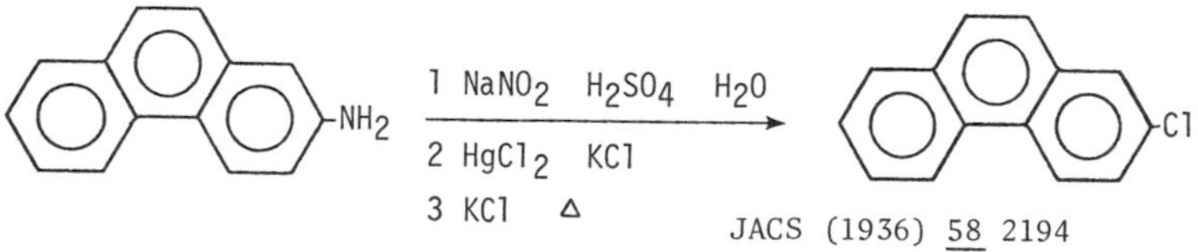

JACS (1936) <u>58</u> 2194

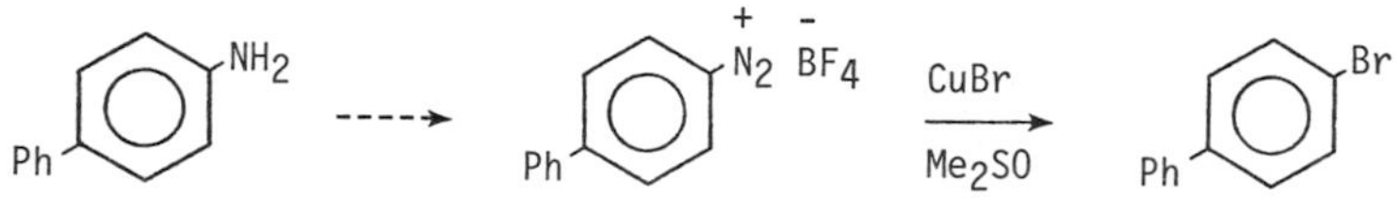

~97%

Org Prep and Procedures (1969) <u>1</u> 221

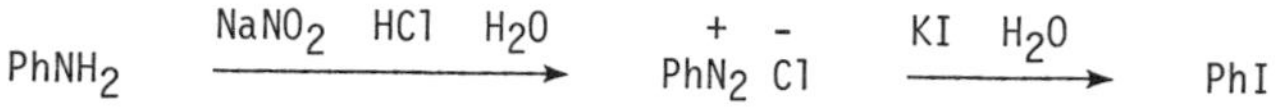

PhNH$_2$ $\xrightarrow{\text{NaNO}_2 \ \text{HCl} \ \text{H}_2\text{O}}$ PhN$_2^+$ Cl$^-$ $\xrightarrow{\text{KI} \ \text{H}_2\text{O}}$ PhI   74-76%

Org Synth (1944) Coll Vol 2 351
Chem Ind (1962) 1760

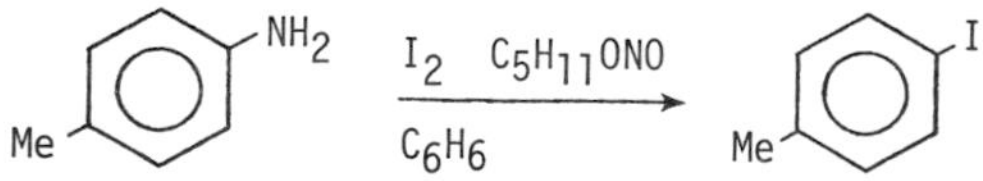

40%

JOC (1968) <u>33</u> 1636

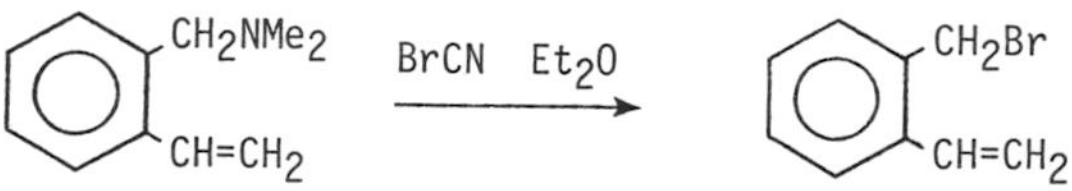

50%

Org React (1953) <u>7</u> 198
Ber (1910) <u>43</u> 3209

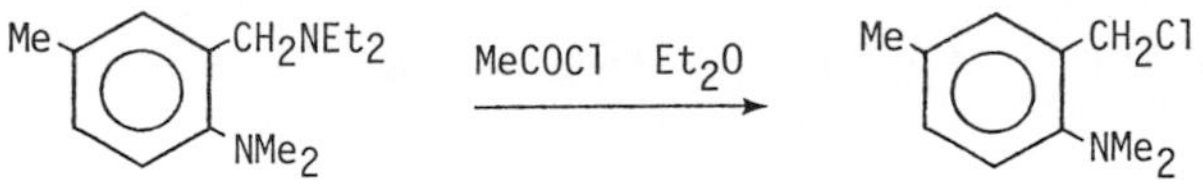

56%

Ber (1960) 93 1310 1305

BuNH$_2$ $\xrightarrow[\text{2 HBr Et}_2\text{O}]{\text{1 p-Chlorobenzenediazonium hexafluorophosphate}}$ BuBr

Tetr Lett (1961) 758

Bu$_3$N $\xrightarrow{\text{ClCOOPh CH}_2\text{Cl}_2}$ BuCl   (Not isolated)

JCS (1967) 2015

Pr$_2$NPh $\xrightarrow{\text{HBr 156°}}$ PrBr

JOC (1963) 28 3144

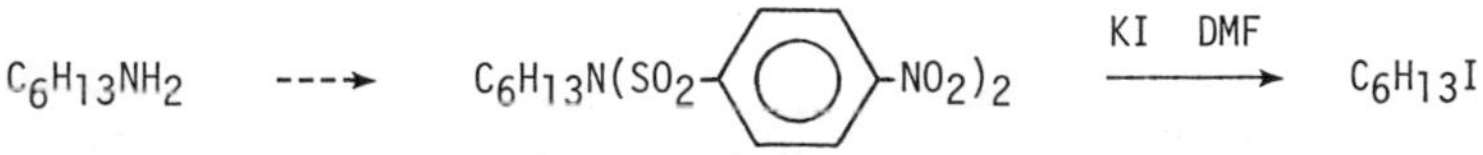

C$_6$H$_{13}$NH$_2$ $\dashrightarrow$ C$_6$H$_{13}$N(SO$_2$-⟨ ⟩-NO$_2$)$_2$ $\xrightarrow{\text{KI DMF}}$ C$_6$H$_{13}$I   79%

JACS (1969) 91 2384

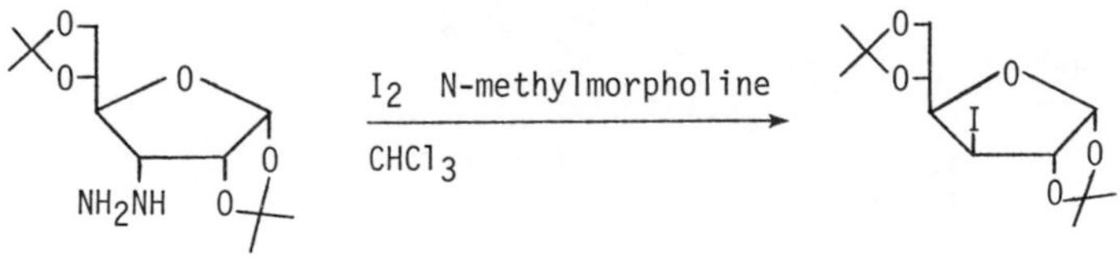

72%

JCS C (1967) 252

Halides may also be prepared from amines via amide intermediates. See section 141 (Halides from Amides)

Section 143   Halides from Esters

For the conversion of sulfonic acid esters to halides see section 145
(Halides and Sulfonates from Halides and Sulfonates)

$$PhCH_2COOBu \xrightarrow{Br_2 \quad P_4} BuBr$$

Arch Pharm (1953) 286 108   80%
JCS B (1967) 1067

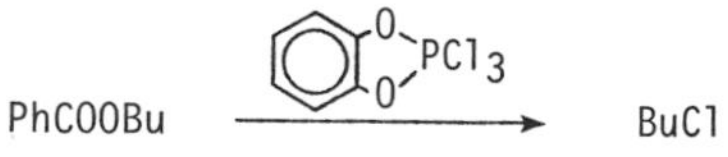

$$PhCOOBu \xrightarrow{} BuCl$$

Ber (1963) 96 1387   73%

$$\xrightarrow{AlCl_3 \quad Et_2O}$$

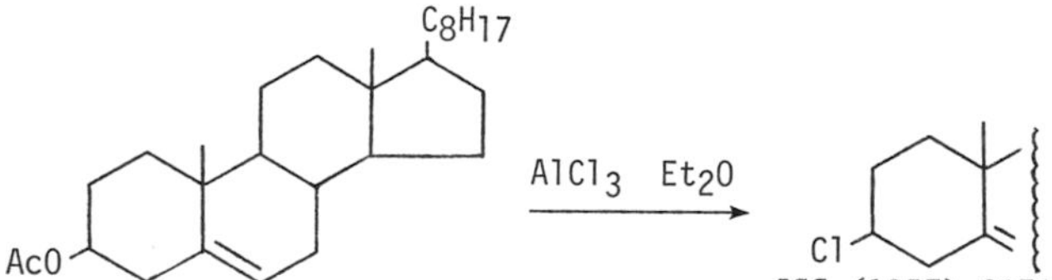

97%

JCS (1957) 2071
Zesz Nauk Pol Slask (1964) 24 193
(Chem Abs 63 11338)
JACS (1940) 62 1432

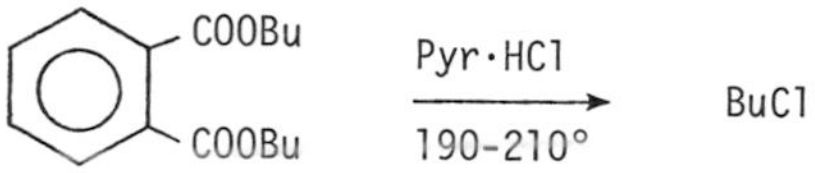

$$\xrightarrow[190-210°]{Pyr \cdot HCl} BuCl$$

66%

Monatsh (1952) 83 1398

$$R'COOR \xrightarrow{LiI \quad DMF} RI \quad (Not\ isolated)$$

JCS (1965) 6655

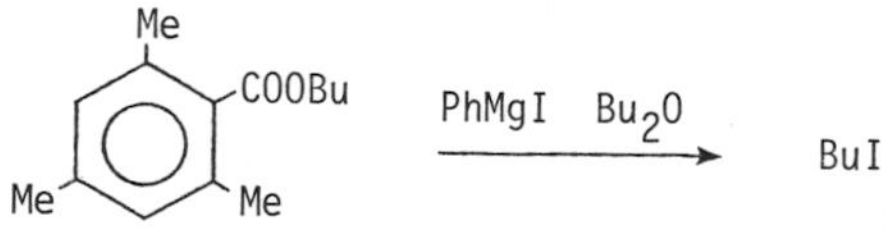

$$\xrightarrow{PhMgI \quad Bu_2O} BuI$$

54%

JACS (1942) 64 1450

Section 144   Halides from Ethers
               °°°°°°°°°°°°°°°°°°°°°°

$$\underset{\underset{Me}{|}}{EtCH}(CH_2)_{10}OMe \xrightarrow{\text{KI   polyphosphoric acid}} \underset{\underset{Me}{|}}{EtCH}(CH_2)_{10}I \qquad \sim 97\%$$

JACS (1965) <u>87</u> 5452

$$MeO(CH_2)_{20}OMe \xrightarrow[180°]{\text{HI   }H_2O} I(CH_2)_{20}I$$

Ber (1942) <u>75B</u> 1715
JCS (1957) <u>463</u>
Chem Rev (1954) <u>54</u> 615

$$\xrightarrow{\text{HBr   }H_2O}$$

81%

Helv (1946) <u>29</u> 1204

$$\xrightarrow{\text{PhCOCl   }ZnCl_2 \text{   }CHCl_3}$$

84%

J Prakt Chem (1938) <u>151</u> 61
JCS (1954) 2819

$$Pr_2O \xrightarrow{\text{MeCOI}} PrI \qquad\qquad 45\%$$

JACS (1933) <u>55</u> 378

$$PhCH_2OMe \xrightarrow{\text{Cl}_2CHOMe \text{   }ZnCl_2} PhCH_2Cl \qquad 88\%$$

Ber (1959) <u>92</u> 83

Et_2O $\xrightarrow{\text{SOCl}_2 \ \ \text{SnCl}_4}$ EtCl

Ber (1936) 69B 1036

ROBu $\xrightarrow{\text{BBr}_3}$ BuBr

77% (R=Bu)
76% (R=Ph)
JACS (1942) 64 1128
Tetrahedron (1968) 24 2289

$(C_5H_{11})_2O$ $\xrightarrow{\text{Ph}_3\text{P·Br}_2 \ \ \text{PhCN}}$ $C_5H_{11}Br$ 78%

JACS (1964) 86 5037

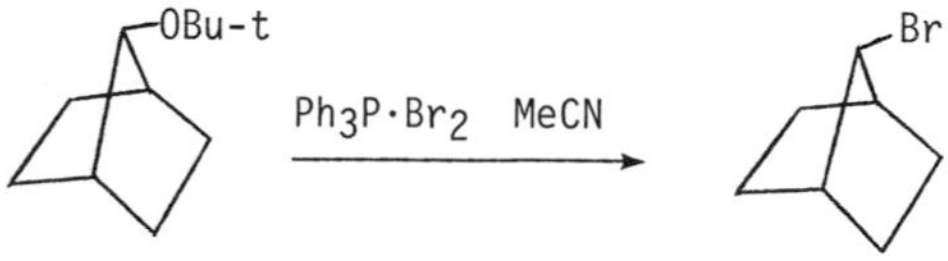

$\xrightarrow{\text{Ph}_3\text{P·Br}_2 \ \ \text{MeCN}}$

83%

Chem Ind (1969) 200

$(n\text{-Bu})_2O$ $\xrightarrow{\text{ICl} \ \ \text{LiBH}_4}$ s-BuI

Chem Ind (1965) 223
(1964) 1582
Tetr Lett (1967) 4131

Et_2O $\xrightarrow{\text{Pyr·BH}_2\text{I} \ \ \text{C}_6\text{H}_6}$ EtI

JACS (1968) 90 6260

Section 145   <u>Halides and Sulfonates from Halides and Sulfonates</u>
○○

$Ph(CH_2)_4Cl$ $\xrightarrow[\text{2 } TsO(CH_2)_3Cl]{\text{1 Mg   Et}_2O}$ $Ph(CH_2)_7Cl$ 24%

JACS (1928) <u>50</u> 1491
(1946) <u>68</u> 1101

$Me_2\overset{|}{\underset{Et}{C}}Cl$ $\xrightarrow{CH_2=CH_2 \ \ AlCl_3}$ $Me_2\overset{|}{\underset{Et}{C}}CH_2CH_2Cl$ 25%

JACS (1945) <u>67</u> 1152

$C_{10}H_{21}I$ $\xrightarrow[\text{2 MeI   NaI   DMF}]{\text{1 PhCH}_2SLi \ \ THF}$ $C_{10}H_{21}CH_2I$

Tetr Lett (1968) 5787

$CH_2=CHCH_2Cl$ $\xrightarrow{CH_2N_2 \ \ Cu}$ $CH_2=CHCH_2CH_2Cl$ + $\underset{\underset{CH_2}{\diagdown\diagup}}{CH_2-CH_2CH_2Cl}$

Ber (1966) <u>99</u> 2855

$C_{11}H_{23}Cl$ $\xrightarrow{KF \ \ HOCH_2CH_2OH}$ $C_{11}H_{23}F$ 48%

JACS (1957) <u>79</u> 2311

$C_6H_{13}Br$ $\xrightarrow{KF \ \ HOCH_2CH_2OH}$ $C_6H_{13}F$ 40-45%

Org Synth (1963) Coll Vol 4 525

$C_{12}H_{25}Br$ —(1 Li THF pet ether; 2 $FClO_3$)→ $C_{12}H_{25}F$ 39%

Ber (1969) 102 1944

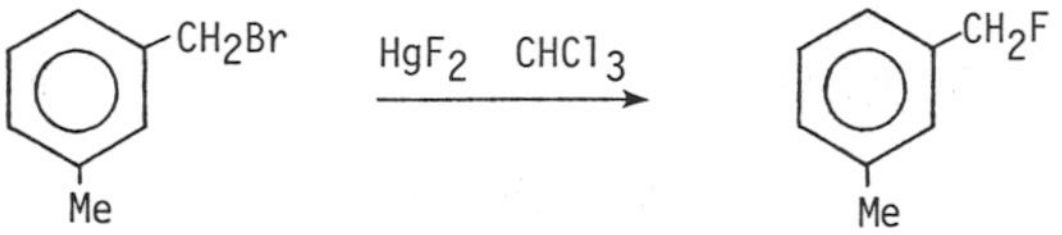

CH2Br
Me
HgF2 CHCl3
CH2F
Me

50%

JACS (1948) 70 2310
Org React (1944) 2 49

$C_6H_{13}OTs$ —(KF diethylene glycol)→ $C_6H_{13}F$ 90%

Chem Ind (1958) 157
Can J Chem (1956) 34 757
JACS (1955) 77 4899

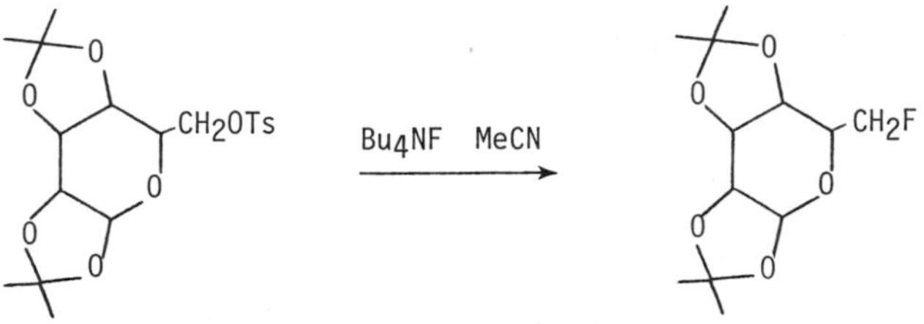

CH2OTs
Bu4NF MeCN
CH2F

69%

Carbohydrate Res (1967) 5 292

$C_8H_{17}Br$ —($ClF_2CCOOAg$)→ $C_8H_{17}Cl$ 89%

Tetr Lett (1970) 3447

$HC{\equiv}CCH_2CH_2OTs$ —($LiCl$ $EtOCH_2CH_2OH$)→ $HC{\equiv}CCH_2CH_2Cl$ 90%

JCS (1950) 3650

BuOTs $\xrightarrow{\text{Pyr·HX   200°}}$ BuX

89% (X=Cl)
66% (X=Br)

Monatsh (1952) 83 1398
JOC (1961) 26 2883

$HOCH_2C\equiv CCH_2Cl$ $\xrightarrow{\text{NaBr   MeOH}}$ $HOCH_2C\equiv CCH_2Br$ 62%

JACS (1955) 77 165

PrI $\xrightarrow[\text{2 Br}_2]{\text{1 Mg   Et}_2O}$ PrBr   JACS (1919) 41 287   30-40%

$C_6H_{13}\underset{\underset{Me}{|}}{C}HOTs$ $\xrightarrow{\text{NaBr   Me}_2SO}$ $C_6H_{13}\underset{\underset{Me}{|}}{C}HBr$ 52%

JOC (1961) 26 3645
    (1962) 27 624
    (1970) 35 2803

$\underset{\underset{\underset{Me}{|}}{(CH_2)_8CHBr}}{\overset{\overset{|}{COOH}}{}}$ $\xrightarrow{\text{NaI   Me}_2CO}$ $\underset{\underset{\underset{Me}{|}}{(CH_2)_8CHI}}{\overset{\overset{|}{COOH}}{}}$

JCS (1936) 1605
JACS (1968) 90 6225

$\xrightarrow{\text{KI   DMF}}$ 82%

JOC (1969) 34 3519

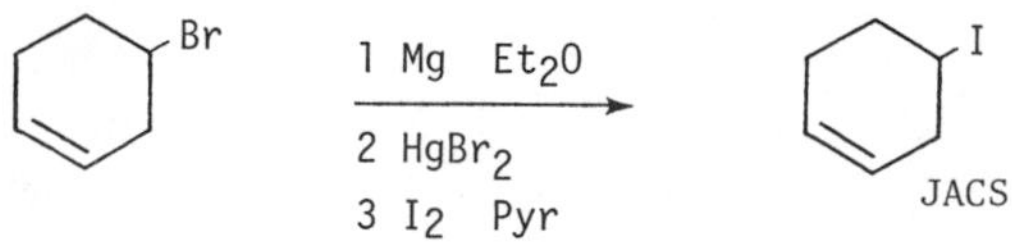

$$\xrightarrow[\begin{array}{l}\text{2 } HgBr_2 \\ \text{3 } I_2 \quad Pyr\end{array}]{\text{1 Mg } Et_2O}$$

JACS (1969) <u>91</u> 5774
(1939) <u>61</u> 1585

13%

$$HC{\equiv}CCH_2CH_2OTs \xrightarrow{\text{NaI } Me_2CO} HC{\equiv}CCH_2CH_2I$$

JCS (1950) 3650

64%

$$C_7H_{15}I \xrightarrow{\text{AgOTs } MeCN} C_7H_{15}OTs$$

JACS (1959) <u>81</u> 4113
JOC (1962) <u>27</u> 2365

$$PhBr \xrightarrow[\text{2 } FClO_3]{\text{1 Li } Et_2O} PhF$$

Ber (1969) <u>102</u> 1944

42%

$$PhI \xrightarrow{h\nu \quad CCl_4} PhCl$$

JOC (1970) <u>35</u> 528

76%

$$PhBr \dashrightarrow PhMgBr \xrightarrow{Ph_3PCl_2 \quad Et_2O} PhCl$$

JOC (1967) <u>32</u> 3710

$$PhI \xrightarrow[\text{2 } Br_2]{\text{1 Mg } Et_2O} PhBr$$

JACS (1919) <u>41</u> 287

30-40%

PhBr $\xrightarrow[\text{2 } I_2]{\text{1 Mg Et}_2O}$ PhI   90%

JACS (1919) <u>41</u> 287
(1945) <u>67</u> 1479
JOC (1938) <u>3</u> 55

PhBr $\xrightarrow[\text{2 ICN}]{\text{1 Mg Et}_2O}$ PhI   Annales de Chimie (1915) <u>4</u> 28   64%

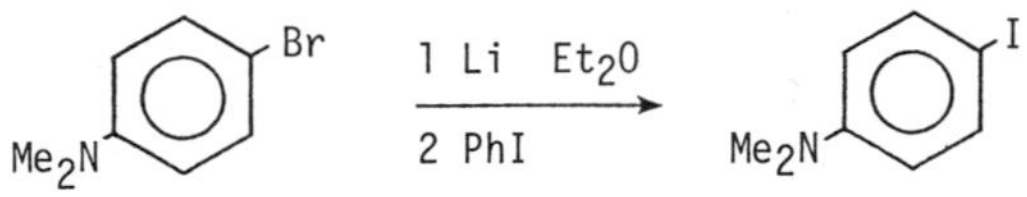

$\xrightarrow[\text{2 PhI}]{\text{1 Li Et}_2O}$ 51%

JACS (1950) <u>72</u> 2767

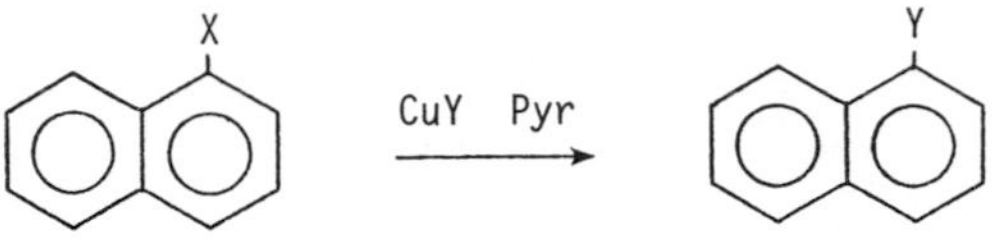

CuY Pyr

X=Br	Y=Cl	84%
I	Cl	100%
I	Br	100%
Br	I	30%

JCS (1964) 1097 1108
Proc Chem Soc (1962) 113

Section 146   <u>Halides from Hydrides (RH)</u>

PhH $\xrightarrow[\text{BF}_3 \text{ or BBr}_3]{\text{FCH}_2\text{CH}_2\text{X}}$ PhCH$_2$CH$_2$X   50-94%

(X=Cl, Br or I)

JOC (1964) <u>29</u> 2317

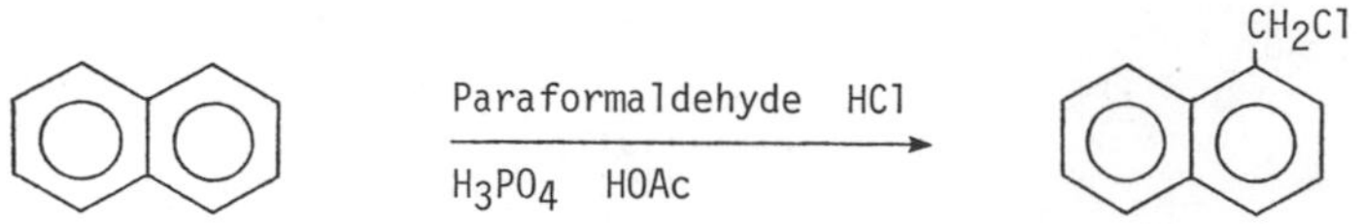

74-77%

Org Synth (1955) Coll Vol 3 195
Org React (1942) 1 63

PhH $\xrightarrow{\text{XeF}_2 \quad \text{HF} \quad \text{CCl}_4}$ PhF   68%

JACS (1969) 91 1563
JOC (1970) 35 723 4020
For aromatic fluorination with CF$_3$OF see JACS (1970) 92 7494

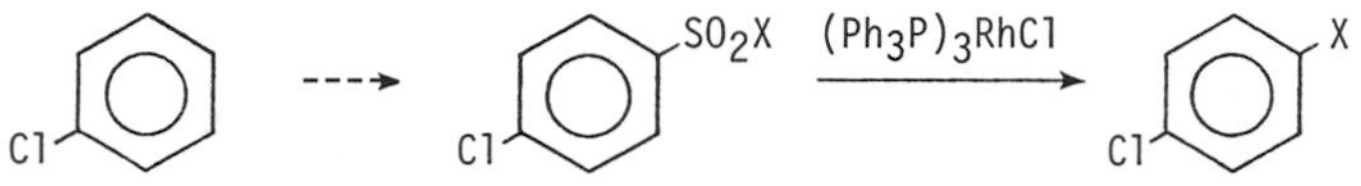

(X=F, Cl or Br)

JOC (1970) 35 1895
Monatsh (1915) 36 719

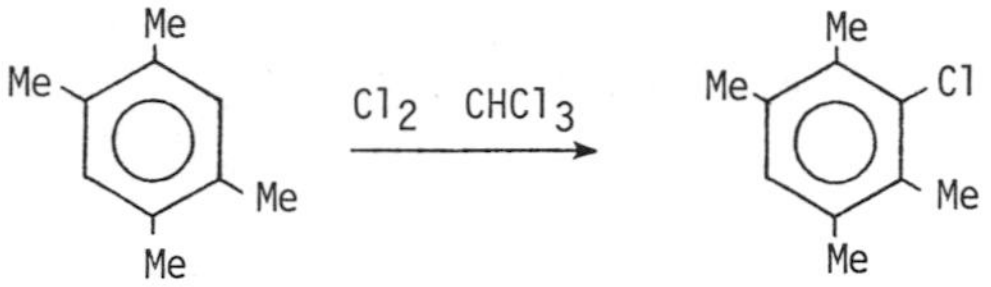

61%

JACS (1936) 58 1

For aromatic chlorination with TiCl$_4$ and CF$_3$COO$_2$H see Tetr Lett (1970) 2611
  "   "   "   " trichlorocyanuric acid see JOC (1970) 35 719

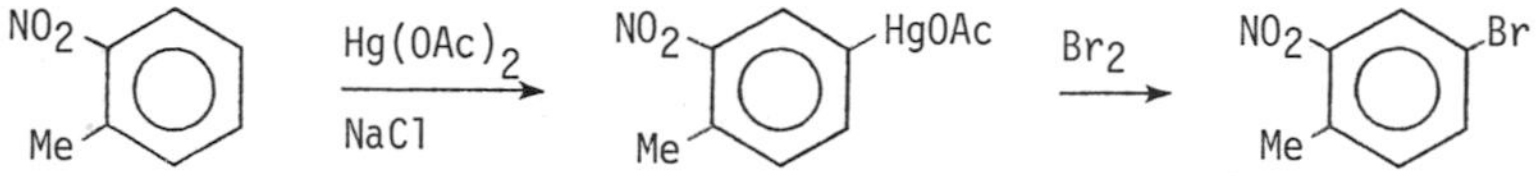

JCS (1926) 637

Zh Org Khim (1969) $\underline{5}$ 387
(Chem Abs $\underline{70}$ 105613)

Org Synth (1955) Coll Vol 3 138

For aromatic bromination with Tl(OAc)$_3$ and Br$_2$ see Tetr Lett (1969) 1623
"   "   "   "   NBS   "   JOC (1965) $\underline{30}$ 304
                                and JACS (1958) $\underline{80}$ 4327
"   "   "   "   CBr$_4$ see JACS (1932) $\underline{54}$ 2025
"   "   "   "   CuBr$_2$ "   JACS (19̲3̲4̲) ̲2̲6̲ 427
"   "   "   "   1,3-dibromo-5,5-dimethylhyɑantoin see
                         An Argentina (1950) $\underline{38}$ 181 188
                         (Chem Abs $\underline{45}$ 2873)

22%

Org React (1954) $\underline{8}$ 258
JACS (1950) $\underline{72}$ 2767

PhH $\xrightarrow{\text{I}_2\ \ \text{HNO}_3}$ PhI   86-87%

Org Synth (1932) Coll Vol 1 323

85-91%

Org Synth (1963) Coll Vol 4 547

For aromatic iodination with I$_2$ and oleum see JCS $\underline{C}$ (1970) 1480
"   "   "   " I$_2$ " KIO$_3$ " JACS (1943) $\underline{65}$ 1273
"   "   "   " I$_2$ " Ag$_2$SO$_4$ " JCS (1952) 150
"   "   "   " I$_2$ " electrolysis see Tetr Lett (1968) 1831
"   "   "   " AlI$_3$ " CuCl$_2$ see JOC (1970) $\underline{35}$ 3436
"   "   "   " (CF$_3$COO)$_3$Tl and KI see Tetr Lett (1969) 2427

PrCH₂CH=CHMe $\xrightarrow[\text{CCl}_4]{\text{NBS \quad dibenzoyl peroxide}}$ PrCHCH=CHMe   58-64%
                                                                         |
                                                                         Br

Org Synth (1963) Coll Vol 4 108

$\xrightarrow[\text{C}_6\text{H}_6]{\text{NBS \quad dibenzoyl peroxide}}$

71-79%

Org Synth (1963) Coll Vol 4 921

Br₂ hν

Org Synth (1963) Coll Vol 4 984

For reviews of allylic and benzylic halogenation see Chem Rev (1948) <u>43</u> 271
and Angew (1959) <u>71</u> 349

The following reagents may also be used for allylic/benzylic halogenation:
N-Chloro-N-cyclohexylbenzenesulfonamide Annalen (1967) <u>703</u> 34

SO₂Cl₂   JACS (1939) <u>61</u> 2142
and JCS (1951) 1851

PCl₅   JOC (1969) <u>34</u> 3655

PhICl₂   JOC (1964) <u>29</u> 3692

1,2-Dibromotetrachloroethane Chem Ind (1963) 1954

BrCCl₃   JACS (1960) <u>82</u> 391

CBr₄   JACS (1932) <u>54</u> 2025

Ph₂C=NBr   JOC (1967) <u>32</u> 223

CCl₃SO₂Br JOC (1965) <u>30</u> 38

1,3-Dibromo-5,5-dimethylhydantoin Helv (1958) <u>41</u> 70

t-BuOI   JACS (1968) <u>90</u> 808

CF₃OF hν

JACS (1970) <u>92</u> 7494

PhICl$_2$ hν

JOC (1964) 29 3692

For chlorination of saturated compounds with SO$_2$Cl$_2$ see JCS (1951) 1851
"   "   "   "   "   " CCl$_3$SO$_2$Cl "   JACS (1960) 82 5246
"   "   "   "   "   "   NCS   "   JOC (1953) 18 649

NBS dibenzoyl peroxide

CCl$_4$

30%

JOC (1953) 18 649
JCS (1952) 2240

For bromination of saturated compounds with Br$_2$ see
                                             Rec Trav Chim (1964) 83 67
"   "   "   "   "   " CCl$_3$SO$_2$Br see
                                                JOC (1965) 30 38
"   "   "   "   "   " Ph$_2$C=NBr see
                                                JOC (1967) 32 223

EtCH$_2$Me $\xrightarrow[\text{CCl}_4]{\text{t-BuOCl HgI}_2 \text{ h}\nu}$ EtCHMe   35%
                                              |
                                              I   JACS (1968) 90 808

Section 147   Halides from Ketones
               ∘∘∘∘∘∘∘∘∘∘∘∘∘∘∘∘∘∘∘∘

NH$_2$OH   PCl$_5$
                  Et$_2$O   64%

Tetr Lett (1965) 525

Section 148   Halides from Nitriles
               ∘∘∘∘∘∘∘∘∘∘∘∘∘∘∘∘∘∘∘∘

No examples

Section 149   Halides from Olefins

The conversion of olefins into saturated halides is considered in this
section. For allylic halogenation see section 146 (Halides from Hydrides)

$C_6H_{13}CH=CH_2$ $\xrightarrow[130°]{CH_3Cl \quad di-t-butyl \ peroxide}$ $C_6H_{13}(CH_2)_3Cl$ 23%

Chem Comm (1966) 258

$\xrightarrow{HF}$ 60%

Org React (1944) 2 49
JOC (1970) 35 4020

$\xrightarrow{HCl \quad HOAc}$ 97%

Helv (1955) 38 1587
JOC (1966) 31 1090
JACS (1957) 79 456

$C_6H_{13}CH=CH_2$ $\xrightarrow[\text{2 N-Chloropiperidine}]{\text{1 } B_2H_6 \quad THF}$ $C_6H_{13}CH_2CH_2Cl \quad + \quad C_6H_{13}\overset{\underset{\textstyle Cl}{|}}{C}HMe$

JOC (1965) 30 4313

$BuCH=CH_2$ $\xrightarrow[\substack{\text{2 } Br_2 \\ \text{3 MeONa}}]{\text{1 } B_2H_6 \quad THF}$ $BuCH_2CH_2Br$ 93%

JACS (1970) 92 6660 7212

$C_7H_{15}CH=CH_2$ $\xrightarrow[\substack{\text{2 } Hg(OAc)_2 \\ \text{3 } Br_2 \quad CCl_4}]{\text{1 } B_2H_6 \quad THF}$ $C_7H_{15}CH_2CH_2Br$ 69%

JACS (1970) 92 3221
Chem Comm (1970) 372

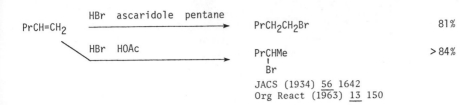

PrCH=CH$_2$ →(HBr ascaridole pentane) PrCH$_2$CH$_2$Br   81%

→(HBr HOAc) PrCHMe >84%
                     |
                     Br

JACS (1934) <u>56</u> 1642
Org React (1963) <u>13</u> 150

AcO(CH$_2$)$_9$CH=CH$_2$ →(HBr dibenzoyl peroxide, C$_6$H$_6$) AcO(CH$_2$)$_9$CH$_2$CH$_2$Br   82%

JOC (1946) <u>11</u> 281
JACS (1946) <u>68</u> 1101

MeCOCH$_2$CH=CH$_2$ →(HBr hν pentane) MeCOCH$_2$CH$_2$CH$_2$Br

Tetrahedron (1969) <u>25</u> 5149

PhC=CH$_2$ →(1 B$_2$H$_6$ THF; 2 I$_2$ NaOH MeOH) PhCHCH$_2$I   63%
  | |
  Me   Me

JACS (1968) <u>90</u> 5038

→(KI H$_3$PO$_4$) cyclohexyl-I   88-90%

Org Synth (1963) Coll Vol 4 543

→(HI C$_6$H$_6$) 83%

(anthracene with CH$_2$CH=CH$_2$ groups → CH$_2$CHMe–I groups)

JCS (1957) 463

BuC=CH$_2$ $\xrightarrow[\text{2 I}_2]{\text{1 Et}_2\text{AlH}}$ BuCHCH$_2$I
|Et |Et   Annalen (1954) $\underline{589}$ 91

Section 150   Halides from Miscellaneous Compounds
             ∘∘∘

                 NCOOK
                 ‖
                 NCOOK   dioxane   HOAc
PhCH=CHBr $\xrightarrow{\hspace{3cm}}$ PhCH$_2$CH$_2$Br   22%

                                      JOC (1965) $\underline{30}$ 3985

         SOCl$_2$ 190-200°
PhNO$_2$ $\xrightarrow{\hspace{2.5cm}}$ PhCl   100%

                                      Monatsh (1915) $\underline{36}$ 723

                 HCl   hν   CHCl$_3$ ~90%

                                      Tetr Lett (1969) 4603

         NH$_2$OH   CuCl$_2$
PhNO $\xrightarrow{\hspace{2.5cm}}$ PhCl   15-20%
         HCl   H$_2$O
                                      JCS (1949) S181

              200-220°
PhCH$_2$CH$_2$SO$_2$Cl $\xrightarrow{\hspace{2cm}}$ PhCH$_2$CH$_2$Cl   59%

                                      Zh Obshch Khim (1953) $\underline{23}$ 204
                                      (Chem Abs $\underline{48}$ 2568)

                 SOCl$_2$ 180°

                                      Monatsh (1915) $\underline{36}$ 719
                                      JOC (1970) $\underline{35}$ 1895

Chapter 11   PREPARATION

OF

HYDRIDES

This chapter lists hydrogenolysis and related reactions by which functional groups are replaced by hydrogen, e.g. $RCH_2X \longrightarrow RCH_2\text{-}H$ or $R\text{-}H$

Section 151   Hydrides from Acetylenes

No examples of the reaction $RC\equiv CR \longrightarrow RH$ occur in the literature. For the hydrogenation of acetylenes see section 61 (Alkyls and Methylenes from Acetylenes)

Section 152   Hydrides from Carboxylic Acids

This section lists examples of the decarboxylation of acids, $RCOOH \longrightarrow RH$ (R=alkyl, aryl, vinyl etc). For the conversion $RCOOH \longrightarrow RCH_3$ see section 62 (Alkyls from Carboxylic Acids)

$$C_{17}H_{35}COOH \quad \xrightarrow{\text{Ni} \quad 350°} \quad C_{17}H_{36}$$

JOC (1944) <u>9</u> 319

MeONa
Δ

64%

JACS (1934) <u>56</u> 715

COOH → $\xrightarrow{\text{1 SOCl}_2 \quad 2 \text{ t-BuOOH} \quad \text{Pyr}}$ → COO$_2$Bu-t → $\xrightarrow{\substack{\text{Diisopropyl-} \\ \text{benzene} \quad 150°}}$ →   29%

JACS (1964) 86 3157
(1961) 83 3998
J Med Chem (1969) 12 192

CHCOOH $\underset{\xleftarrow{\hspace{0.8cm}}}{\overset{200°}{\xrightarrow{\hspace{0.8cm}}}}$ CH$_2$COOH $\xrightarrow{240°}$ CH$_2$ ~60%

JCS (1930) 1603
JACS (1950) 72 4359

CH=CHCOOH / OAc $\xrightarrow{\text{Cu quinoline}}$ CH=CH$_2$ / OAc   37%

JACS (1950) 72 1200
Org Synth (1963) Coll Vol 4 857
Chem Comm (1967) 96

Ph(C≡C)$_3$COOH $\xrightarrow[\text{Me}_2\text{CO}]{\text{Copper tetrammine sulfate}}$ Ph(C≡C)$_3$H   79%

Ber (1964) 97 2586

Me / COOH $\xrightarrow{\text{Cu quinoline}}$ Me   83-89%

Org Synth (1963) Coll Vol 4 628

MeO, MeO, MeO — COOH, OMe $\xrightarrow{\text{CuSO}_4 \text{ quinoline}}$ MeO, MeO, MeO, OMe

JACS (1951) 73 1414

Section 153   Hydrides from Alcohols and Phenols
ooooooooooooooooooooooooooooooooooooooo

This section lists examples of the hydrogenolysis of alcohols and phenols, ROH $\longrightarrow$ RH and RR'CHOH $\longrightarrow$ RH (R=alkyl or aryl, R'=H or aryl). For the conversion ROH $\longrightarrow$ RR' (R'=alkyl or aryl) see section 63 (Alkyls and Aryls from Alcohols)

$$\underset{\overset{|}{Me}}{\overset{\overset{Me}{|}}{i\text{-}PrCCH_2OH}} \quad \xrightarrow[300°]{H_2 \text{ (965 atmos)} \quad Co\text{-}Al_2O_3} \quad \underset{\overset{|}{Me}}{\overset{\overset{Me}{|}}{i\text{-}PrCMe}} \qquad 20\%$$

JACS (1948) 70 3793
       (1951) 73 553
       (1933) 55 1293

$$Me_3COH \quad \xrightarrow{H_2 \quad Pt \quad CF_3COOH} \quad Me_3CH$$

JOC (1964) 29 2325

$$\underset{\overset{|}{OH}}{\overset{C_5H_{11}CHOH}{HOCH_2CH(CHCH_2)_5(CHOH)_3CHMe}} \quad \underset{\overset{|}{Me}}{\overset{|}{\underset{HOCH_2CHCH(CH_2)_9}{OH}}} \quad \xrightarrow[2 \text{ LiAlH}_4]{1 \text{ P}_4 \quad HI \quad H_2O} \quad \underset{\overset{|}{Me}}{\overset{C_5H_{11}CH_2}{\underset{MeCHCH_2(CH_2)_9}{MeCH(CH_2)_{13}CHMe}}} \qquad 13\%$$

JACS (1962) 84 2170
JCS (1959) 1044
Tetrahedron (1970) 26 2199

H₂ Pd-C EtOH

JCS C (1967) 136
Chem Ind (1963) 1354
Org React (1953) 7 263

$$\underset{\overset{|}{OH}}{PhCHMe} \quad \xrightarrow{Na \quad NH_3} \quad PhCH_2Me$$

JCS (1945) 809
    (1949) 2531

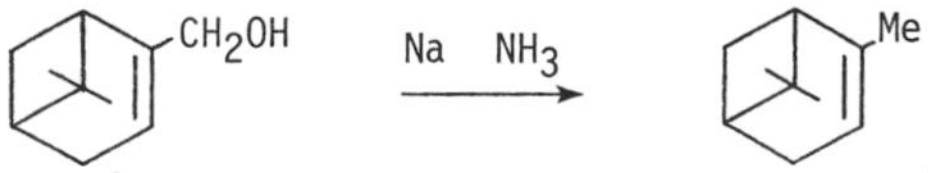
CH2OH
Na NH3
Me

85%

Ber (1956) 89 1549
JCS (1957) 1969
(1945) 809

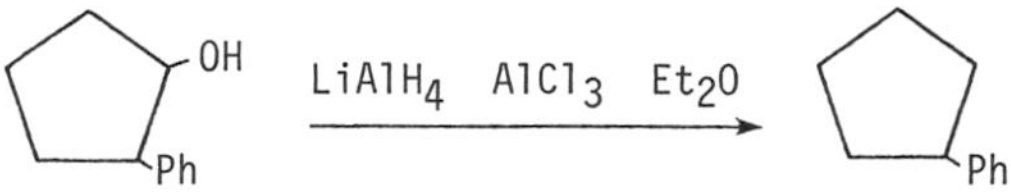
OH
Ph
LiAlH4 AlCl3 Et2O
Ph

78%

JOC (1964) 29 121

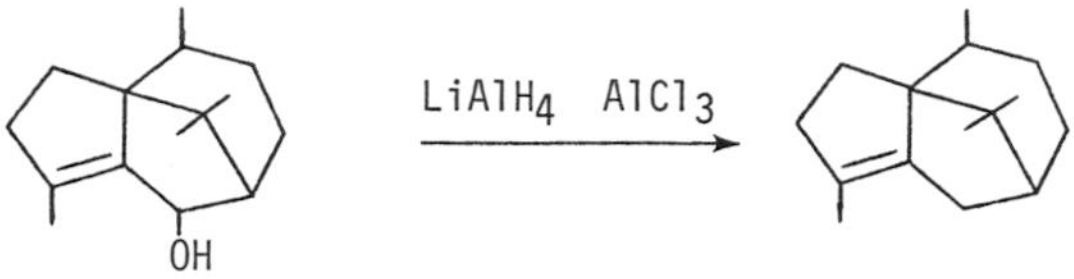
OH
LiAlH4 AlCl3

Tetr Lett (1967) 2447

For reduction of benzylic alcohols see JCS (1957) 3755

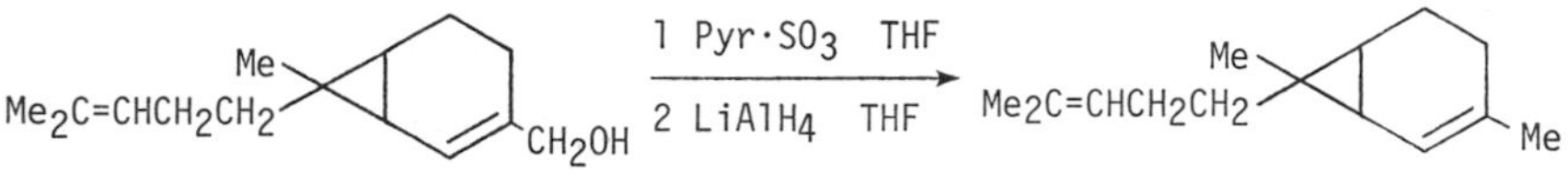
Me
Me2C=CHCH2CH2
CH2OH
1 Pyr·SO3 THF
2 LiAlH4 THF
Me
Me2C=CHCH2CH2
Me

Tetr Lett (1969) 1837
JOC (1969) 34 3667

86%

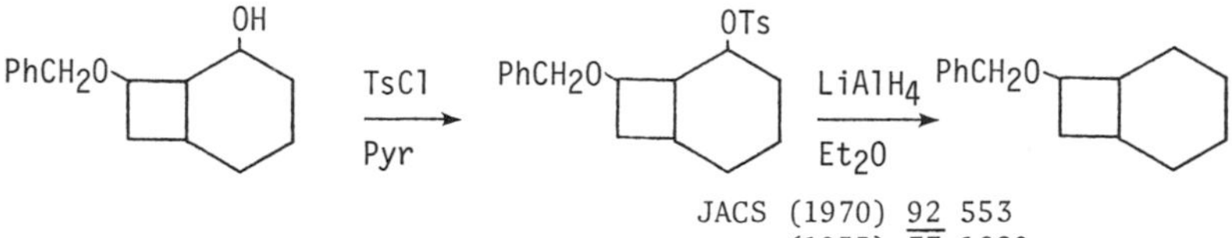
OH
PhCH2O
TsCl
Pyr
OTs
PhCH2O
LiAlH4
Et2O
PhCH2O

79%

JACS (1970) 92 553
(1955) 77 1820

Further examples of the reduction of sulfonates are included in section 160 (Hydrides from Halides and Sulfonates)

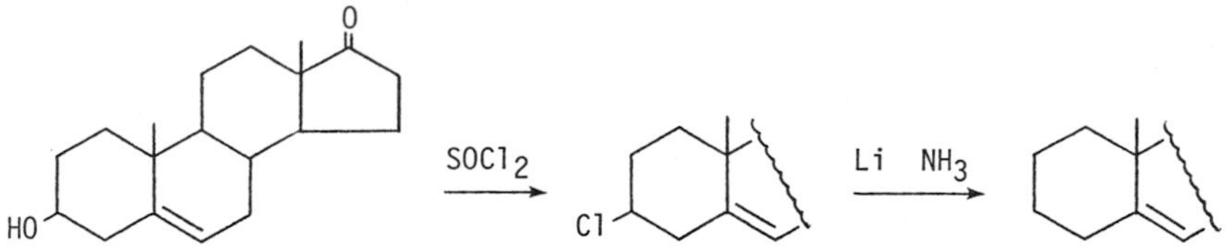

JOC (1968) 33 1196

For further examples of the reaction ROH → RHal and RHal → RH see section 138 (Halides and Sulfonates from Alcohols and Phenols) and section 160 (Hydrides from Halides and Sulfonates)

1 MsCl Pyr

2 PhCH$_2$SNa
  EtOH

Ni   EtOH

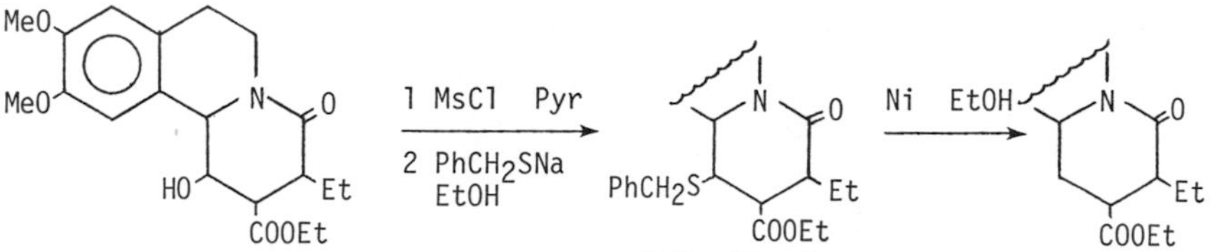

Helv (1959) 42 2431
JACS (1953) 75 384
      (1963) 85 173

77%

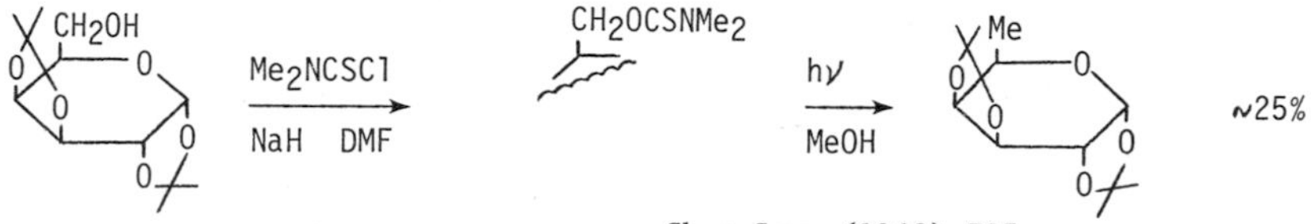

Me$_2$NCSCl

NaH   DMF

hν

MeOH

~25%

Chem Comm (1968) 323
Carbohydrate Res (1967) 4 115

C$_{17}$H$_{35}$CH$_2$OH

$\dfrac{\text{H}_2 \text{ (100-200 atmos)   Ni}}{250°}$

C$_{17}$H$_{36}$

JACS (1933) 55 1293

PhCHCH$_2$OH
 |
 Me

Ni   EtOH

PhCH$_2$
 |
 Me

JACS (1957) 79 1696

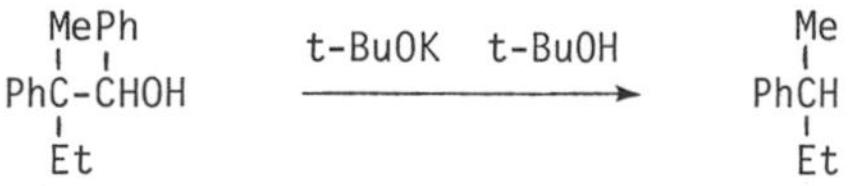
MePh
PhC-CHOH
Et
t-BuOK t-BuOH
Me
PhCH
Et

JACS (1969) 91 1009

Alcohols (ROH) may also be converted into hydrides (RH) via ester or ether intermediates. See section 158 (Hydrides from Esters) and section 159 (Hydrides from Ethers)

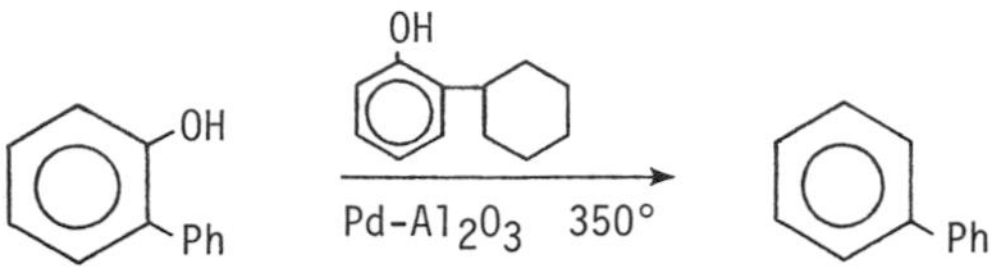
OH
Ph
OH
Pd-Al2O3 350°
Ph

93%

Tetr Lett (1969) 1577

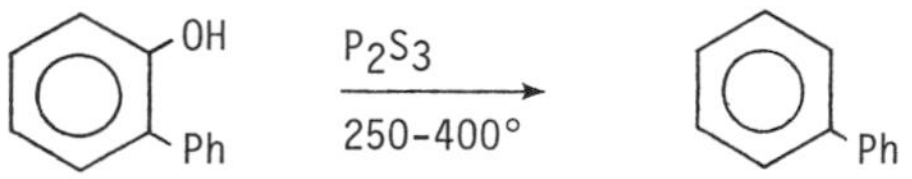
OH
Ph
P2S3
250-400°
Ph

18%

JOC (1961) 26 2528

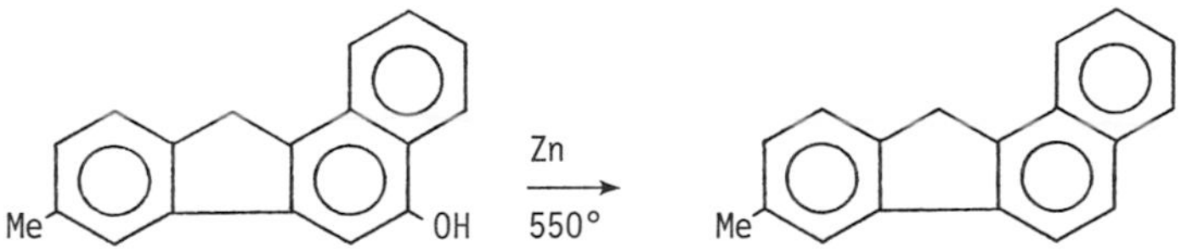
Me
OH
Zn
550°
Me

24%

JACS (1951) 73 3439

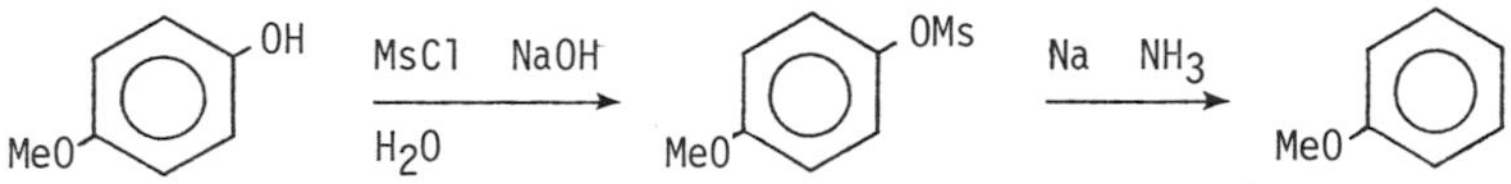
OH
MeO
MsCl NaOH
H2O
OMs
MeO
Na NH3
MeO

JCS (1955) 522

Further examples of the reduction of sulfonates are included in section 160 (Hydrides from Halides and Sulfonates)

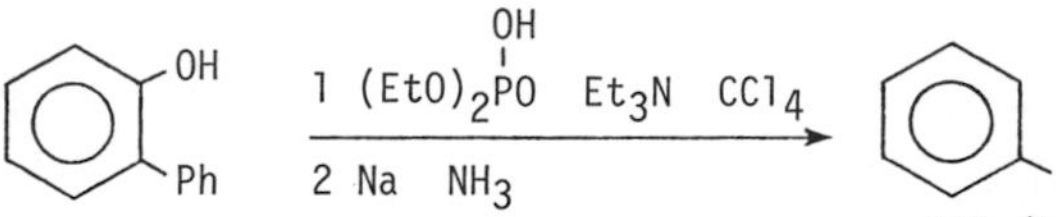
OH
Ph
OH
1 (EtO)2PO Et3N CCl4
2 Na NH3
Ph

48%

JOC (1958) 23 131
J Med Chem (1965) 8 409
JCS (1955) 522

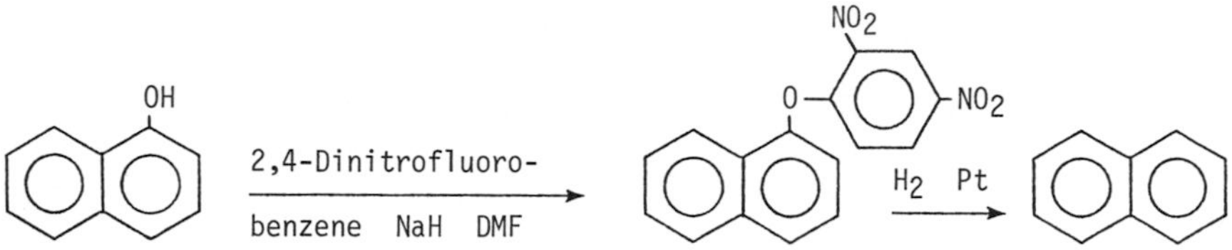
OH
2,4-Dinitrofluoro-
benzene NaH DMF
NO2
O
NO2
H2 Pt

JOC (1964) 29 3124

40%

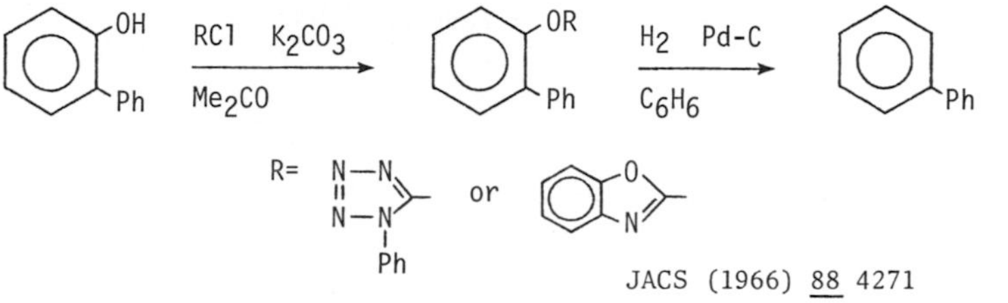
OH
Ph
RCl K2CO3
Me2CO
OR
Ph
H2 Pd-C
C6H6
Ph
R=
N
N
N
N
Ph
or
O
N

70-76%

JACS (1966) 88 4271

Section 154 Hydrides from Aldehydes

This section lists examples of the decarbonylation of aldehydes, RCHO ⟶ RH. For the conversion RCHO ⟶ RMe see section 64 (Alkyls from Aldehydes)

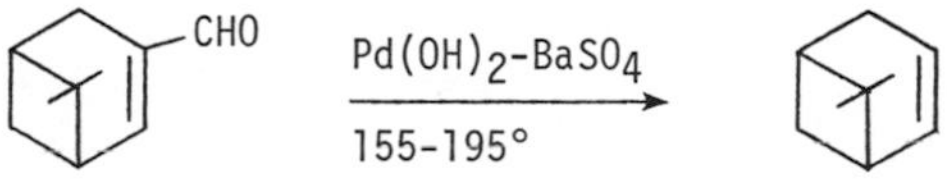
CHO
Pd(OH)2-BaSO4
155-195°

76%

JOC (1959) 24 1369

$$\xrightarrow[210°]{\text{Pd-C}}$$

80%

JOC (1960) 25 2215

$$\underset{\text{Me}}{\text{PhCHCHO}} \xrightarrow{\text{Ni}\quad\text{EtOH}} \underset{\text{Me}}{\text{PhCH}_2}$$

JACS (1957) 79 1696

$$\text{RCHO} \xrightarrow{\text{RhCl(Ph}_3\text{P)}_3} \text{RH}$$

R=C$_6$H$_{13}$, PhCH=C or

$\underset{\text{Et}}{}$

Cl—

JACS (1968) 90 99
Tetr Lett (1970) 823
              (1968) 1899

$$\underset{\text{Me}}{\text{PhCH}_2\text{CHCHO}} \xrightarrow{\text{Di-t-butyl peroxide}} \underset{\text{Me}}{\text{PhCH}_2\text{CH}_2}$$

69%

JOC (1964) 29 1663
JACS (1941) 63 226

$$\xrightarrow{\text{h}\nu\quad\text{EtOH}}$$

90%

Proc Chem Soc (1963) 114

$$\underset{\text{Et}}{\text{BuCHCHO}} \xrightarrow[140\text{-}145°]{\text{h}\nu\quad\text{PhCH}_2\text{SSCH}_2\text{Ph}} \underset{\text{Et}}{\text{BuCH}_2}$$

68%

Tetr Lett (1962) 43

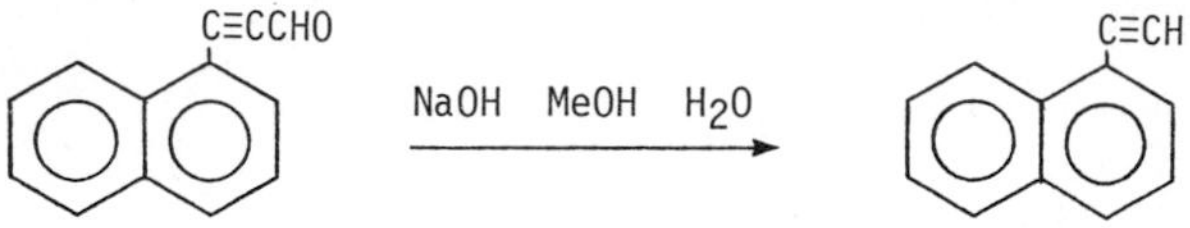

JCS $\underline{C}$ (1969) 2173

Section 155   Hydrides from Alkyls

This section lists examples of the conversion RR' $\longrightarrow$ RH (R,R'=alkyl, aryl etc.)

$Ph_2CHCHPh_2$ $\xrightarrow[\text{methylcyclohexane}]{H_2\quad\text{Cu-Cr oxide}}$ Ph_2CH_2 90%

JACS (1932) $\underline{54}$ 1668

$t\text{-BuCH}_2\text{Pr-i}$ $\xrightarrow{\text{AlCl}_3\quad C_6H_6}$ $t\text{-BuH}$

JACS (1935) $\underline{57}$ 2415

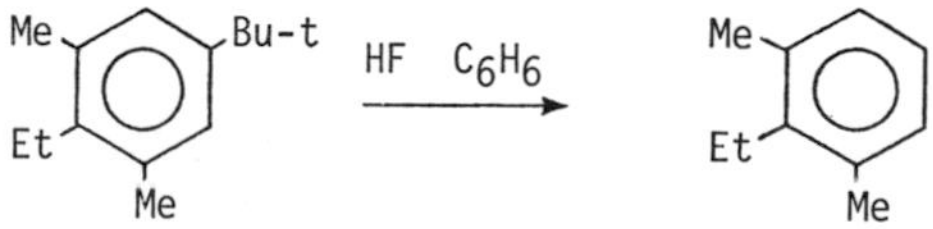

60%

JACS (1954) $\underline{76}$ 4952
(1937) $\underline{59}$ 1417

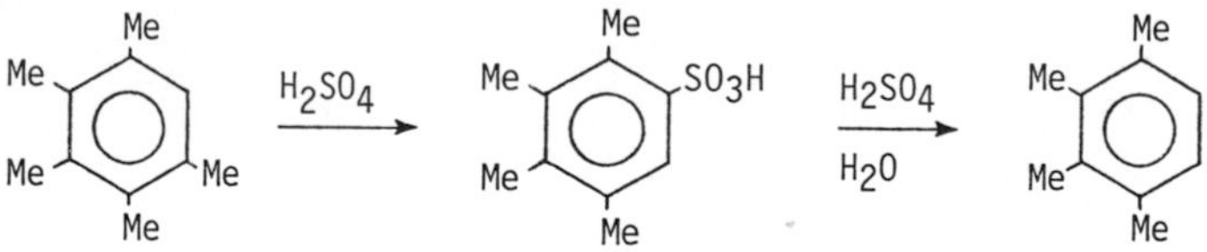

88%

Org React (1942) $\underline{1}$ 370

JCS $\underline{C}$ (1968) 2915

The conversion Ar-Alkyl $\longrightarrow$ ArH may also be achieved via intermediate carboxylic acids ArCOOH. See section 20 (Carboxylic Acids from Alkyls and Aryls) and section 152 (Hydrides from Carboxylic Acids)

Section 156   Hydrides from Amides

This section lists examples of the conversions $RCONH_2 \longrightarrow RH$ and $RNHCOR' \longrightarrow RH$. For the conversions $RCONR'_2 \longrightarrow RR''$ and $RNHCOR' \longrightarrow RR''$ (R"=alkyl or aryl) see section 66 (Alkyls and Aryls from Amides)

JACS (1951) $\underline{73}$ 436   95%

$$Ph_2CHCONH_2 \xrightarrow{\text{BuLi THF hexane}} Ph_2CH_2$$

JACS (1969) $\underline{91}$ 7774   86%

$$PhC\equiv CCONH_2 \xrightarrow{\text{NaNH}_2 \text{ NH}_3 \text{ Et}_2O} PhC\equiv CH$$

JCS (1963) 4402   90%

$$PhNHAc \dashrightarrow \underset{\underset{NO}{|}}{PhNAc} \xrightarrow[25°]{\text{Et}_2O} PhH$$

JACS (1964) $\underline{86}$ 3180

Amides may also be converted into hydrides via intermediate amines or carboxylic acids. See section 157 (Hydrides from Amines) and section 152 (Hydrides from Carboxylic Acids)

Section 157   Hydrides from Amines
             ○○○○○○○○○○○○○○○○○○○○○○○○

This section lists examples of the conversions $RNH_2 \longrightarrow RH$, $RCH_2NH_2 \longrightarrow RH$
and $R'NR_3^+ X^- \longrightarrow R'H$. For the conversion $RNH_2 \longrightarrow RR'$ (R'=alkyl or aryl)
see section 67 (Alkyls and Aryls from Amines)

Review: Replacement of the Aromatic Primary Amino Group by Hydrogen
                              Org React (1944) $\underline{2}$ 262

~75%

JOC (1963) $\underline{28}$ 568

80%

JACS (1952) $\underline{74}$ 3074
Org Synth (1963) Coll Vol 4 947
              (1955) Coll Vol 3 295

For reduction of diazonium salt with $NaBH_4$ see JACS (1961) $\underline{83}$ 1251
   "   "   "   "   "   "   HCHO derivatives see
                                                Angew (1958) $\underline{70}$ 211
   "   "   "   "   "   " H_3PO_2, HCHO, EtOH, Zn or N_2H_4 see
                                                Org React (1944) $\underline{2}$ 262
For selective diazotization of $ArNH_2$ in the presence of $\overline{A}liphNH_2$ see
                                                JACS (1949) $\underline{71}$ 2137

Tetr Lett (1969) 1577

Aromatic amines (RNH_2) may also be converted into hydrides (RH) via
N-acetyl derivatives. See section 156 (Hydrides from Amides)

$C_8H_{17}NH_2 \xrightarrow{\text{TsCl   Pyr}} C_8H_{17}NHTs \xrightarrow[\text{EtOH   H}_2\text{O}]{\text{NH}_2\text{OSO}_3\text{H   NaOH}} C_8H_{18}$ 25-30%

JACS (1964) $\underline{86}$ 1152

$$\text{BuNH}_2 \xrightarrow{\text{HNF}_2} \text{BuH}$$

JACS (1963) <u>85</u> 97
          (1964) <u>86</u> 2233

61%

$$\underset{\overset{|}{\text{Et}}}{\overset{\overset{\text{Me}}{|}}{\text{PhCNHNH}_2}} \xrightarrow{\text{KIO}_4 \quad \text{KOH} \quad \text{H}_2\text{O}} \underset{\overset{|}{\text{Et}}}{\overset{\overset{\text{Me}}{|}}{\text{PhCH}}}$$

JACS (1963) <u>85</u> 1108

68%

$$\text{PhCH}_2\text{CH}_2\text{NH}_2 \xrightarrow{\text{Ni} \quad \text{EtOH}} \text{PhMe} \quad + \quad \text{PhCH}_2\text{Me}$$

JACS (1957) <u>79</u> 1696

Na-Hg
H$_2$O

JACS (1951) <u>73</u> 4122
Org React (1953) <u>7</u> 263

99%

Section 158   Hydrides from Esters

This section lists examples of the hydrogenolysis of esters, R'COOR $\longrightarrow$ RH. For the conversion R'COOR $\longrightarrow$ R'R" or RR" (R"=alkyl) see section 68 (Alkyls and Aryls from Esters). The reduction of esters of sulfonic acids (e.g. ROTs $\longrightarrow$ RH) is included in section 160 (Hydrides from Halides and Sulfonates)

$$\underset{\overset{|}{\text{COOH}}}{\overset{\overset{\text{Me}}{|}}{\text{PhCOAc}}} \xrightarrow{\text{H}_2 \quad \text{Pd-C} \quad \text{EtOAc}} \underset{\overset{|}{\text{COOH}}}{\overset{\overset{\text{Me}}{|}}{\text{PhCH}}}$$

JCS <u>C</u> (1967) 136
Org React (1953) <u>7</u> 263

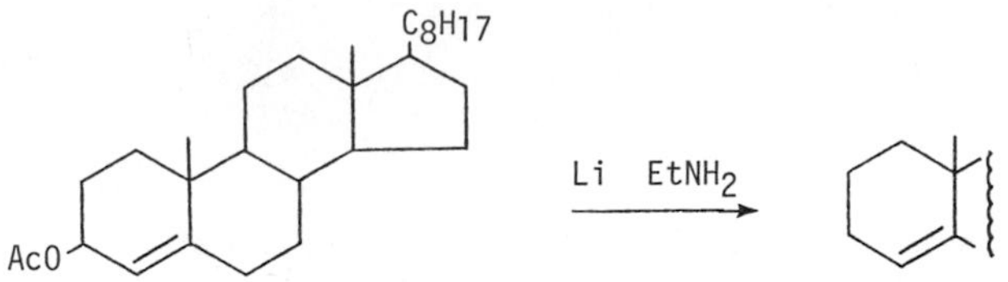
C_8H_{17}
AcO
Li EtNH$_2$

94%

JCS (1957) 1969
JOC (1968) 33 435

Section 159 Hydrides from Ethers

This section lists examples of the hydrogenolysis of ethers, ROR' ⟶ RH. For the conversion ROR' ⟶ RR" (R"=alkyl or aryl) see section 69 (Alkyls and Aryls from Ethers)

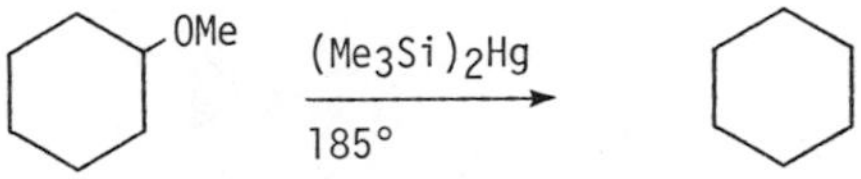
OMe
$(Me_3Si)_2Hg$
185°

58-76%

JCS C (1967) 2188

$PhCH_2OEt \xrightarrow{(i\text{-}Bu)_2AlH} PhMe$ 92%

Izv (1959) 2255
(Chem Abs 54 10837)

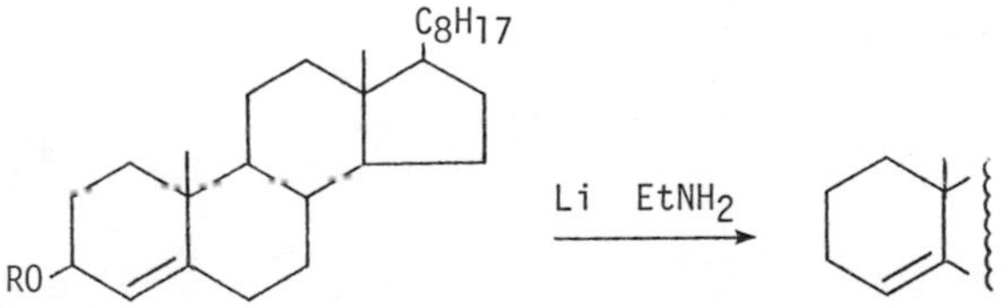
C_8H_{17}
RO
Li EtNH$_2$

R=Me or $PhCH_2$

JCS (1957) 1969

$PhCH=CHCH_2OPh \xrightarrow{LiAlH_4\ \ NiCl_2\ \ THF} PhCH=CHMe$

JACS (1957) 79 5463

PhCH$_2$OBu $\xrightarrow{\text{H}_2 \quad \text{Ni}}$ PhMe

Org React (1953) 7 263
Chem Ind (1963) 1354

~93%

Further examples of the hydrogenolysis of allyl and benzyl ethers are
included in section 39 (Alcohols and Phenols from Ethers and Epoxides)
and section 45A (Protection of Alcohols and Phenols)

Section 160   Hydrides from Halides and Sulfonates

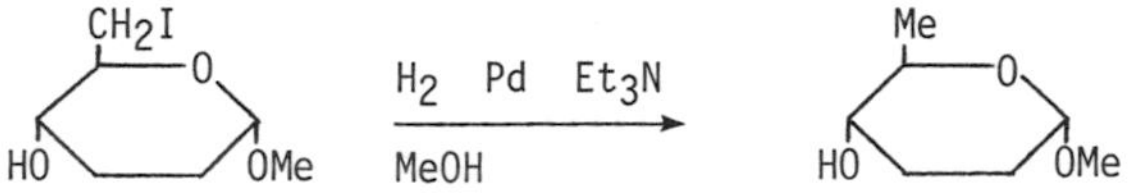

JOC (1969) 34 3519

81%

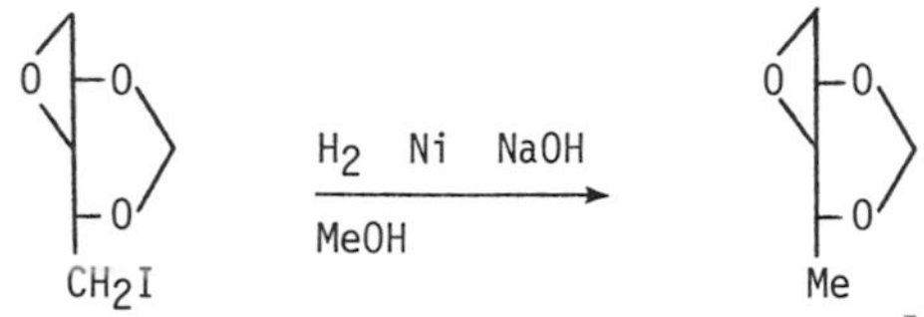

JACS (1950) 72 561
Helv (1946) 29 378 360

37%

PrF $\xrightarrow[190°]{\text{H}_2 \quad \text{Pd-C}}$ PrH

J Phys Chem (1956) 60 1454

100%

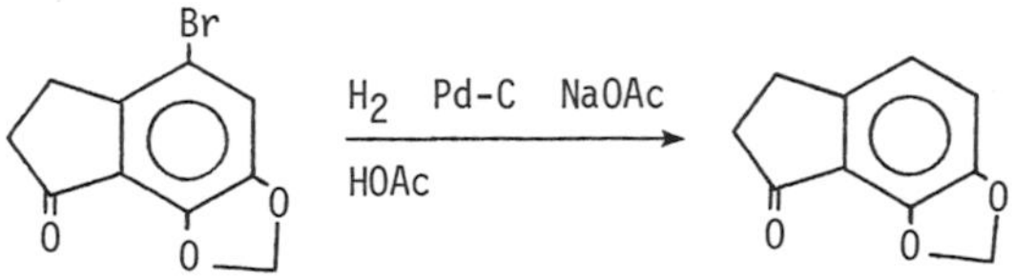

JCS C (1969) 2600

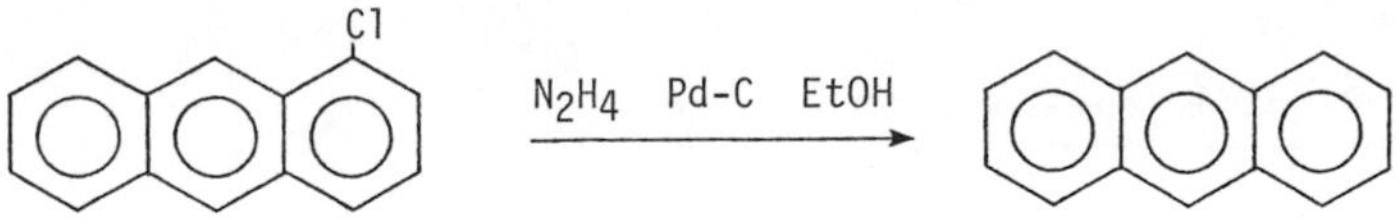
Cl
N_2H_4 Pd-C EtOH

88%

Chem Ind (1959) 1348

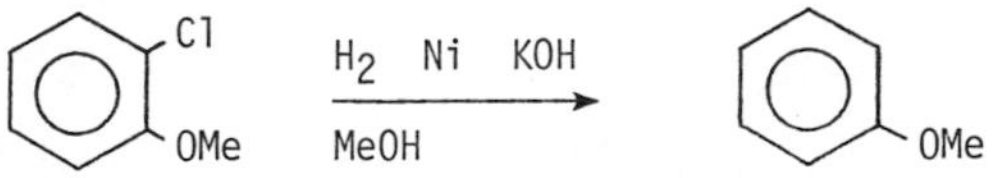
Cl
OMe
H_2 Ni KOH
MeOH
OMe

90%

Ber (1958) 91 1376

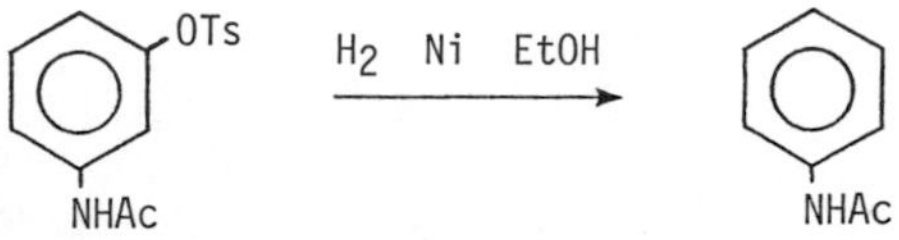
OTs
NHAc
H_2 Ni EtOH
NHAc

90%

JCS (1949) S178
Aust J Chem (1963) 16 647

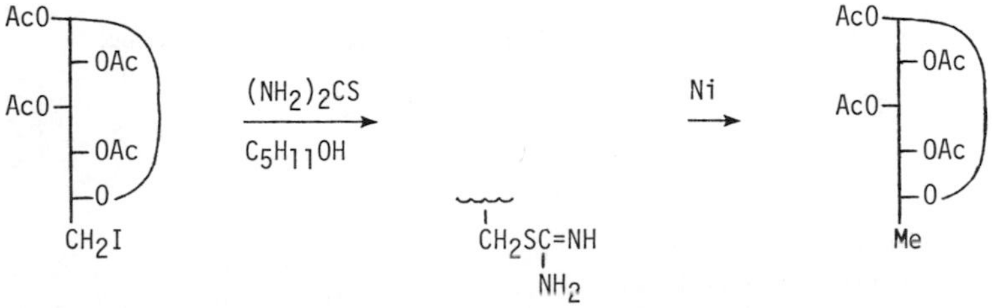
AcO
OAc
AcO
OAc
O
CH_2I
$(NH_2)_2CS$
$C_5H_{11}OH$
$CH_2SC=NH$
NH_2
Ni
AcO
OAc
AcO
OAc
O
Me

55%

Helv (1946) 29 1199
JACS (1969) 91 7342
(1950) 72 561

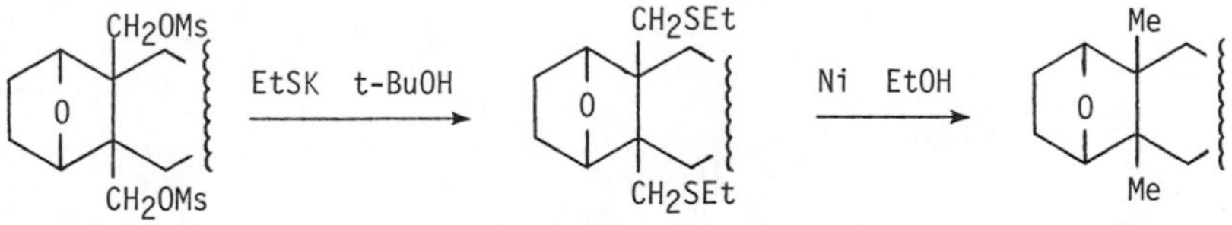
CH_2OMs
O
CH_2OMs
EtSK t-BuOH
CH_2SEt
O
CH_2SEt
Ni EtOH
Me
O
Me

JACS (1953) 75 384

JOC (1944) <u>9</u> 1

$$\underset{(CH_2)_9CH_2Br}{\overset{COOH}{|}} \quad \xrightarrow[\text{or Zn-Cu EtOH}]{\overset{\text{Ni-Al NaOH MeOH}}{\text{or Zn HOAc}}} \quad \underset{(CH_2)_9Me}{\overset{COOH}{|}} \qquad 100\%$$

Bull Soc Chim Fr (1967) 2018

$$C_{16}H_{33}I \quad \xrightarrow{\text{Zn HOAc}} \quad C_{16}H_{34} \qquad\qquad 85\%$$

Org Synth (1943) Coll Vol 2 320

94%

(Aliphatic halides are also reduced)

Tetr Lett (1962) 449
           (1968) 1575
JOC (1964) <u>29</u> 160

80%

JACS (1968) <u>90</u> 3594

$$\underset{\overset{|}{\text{Me}}}{\overset{\overset{\text{Et}}{|}}{\text{i-Pr}(CH_2)_3\text{CCl}}} \quad \xrightarrow{\text{Li NH}_3} \quad \underset{\overset{|}{\text{Me}}}{\overset{\overset{\text{Et}}{|}}{\text{i-Pr}(CH_2)_3\text{CH}}}$$

Rec Trav Chim (1964) <u>83</u> 367
JACS (1951) <u>73</u> 3329
Chem Comm (1969) 138

$$\text{(cyclohexyl-X)} \xrightarrow{\text{Mg \quad i-PrOH \quad decalin}} \text{(cyclohexane)}$$

33% (X=F)
83% (X=Cl)

Proc Chem Soc (1963) 219

$$\text{PhX} \xrightarrow{\text{Mg \quad i-PrOH \quad decalin}} \text{PhH}$$

89% (X=Cl or Br)
95% (X=I)

Proc Chem Soc (1963) 219

$$\underset{\text{Me}}{\text{PrCHBr}} \xrightarrow[\text{2 \quad H}_2\text{SO}_4 \quad \text{H}_2\text{O}]{\text{1 \quad Mg \quad Bu}_2\text{O}} \underset{\text{Me}}{\text{PrCH}_2}$$

50-53%

Org Synth (1943) Coll Vol 2 478

$$\text{C}_8\text{H}_{17}\text{Br} \xrightarrow{\text{BuLi \quad hexane}} \text{C}_8\text{H}_{18} \quad + \quad \text{C}_8\text{H}_{17}\text{Bu}$$

JOC (1963) $\underline{28}$ 280

$$\text{C}_8\text{H}_{17}\text{Cl} \xrightarrow{\text{NaBH}_4 \quad \text{Me}_2\text{SO}} \text{C}_8\text{H}_{18}$$

JOC (1969) $\underline{34}$ 3923
Tetr Lett (1969) 3495

42%

$$\text{PrOTs} \xrightarrow{\text{NaBH}_4 \quad \text{Me}_2\text{SO}} \text{PrH}$$

JOC (1969) $\underline{34}$ 3923

90%

$$\begin{array}{c} \text{C}_8\text{H}_{17}\text{Br} \\ \text{C}_6\text{H}_{13}\underset{\text{Br}}{\text{CHCH}_2\text{Br}} \end{array} \xrightarrow{\text{LiAlH}_4} \text{C}_8\text{H}_{18}$$

~70%

JACS (1948) $\underline{70}$ 3664 3738
Org React (1951) $\underline{6}$ 469

$$\text{PhBr} \xrightarrow{\text{LiAlH}_4 \quad \text{THF}} \text{PhH}$$

95%

JOC (1969) $\underline{34}$ 3918

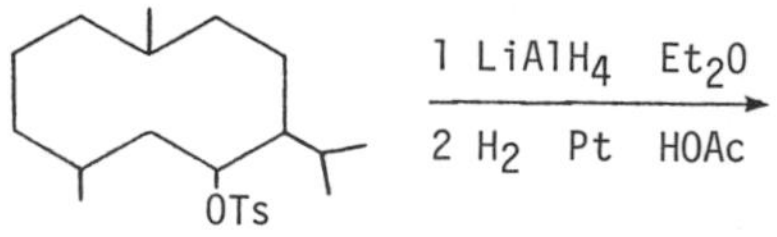

1 LiAlH$_4$ Et$_2$O
2 H$_2$ Pt   HOAc

Tetrahedron (1964) 20 2903
Helv (1949) 32 1371
JACS (1970) 92 553

CN
|
(CH$_2$)$_3$CH$_2$Br

Bu$_3$SnH

CN
|
(CH$_2$)$_3$Me

JOC (1969) 34 2014
Synthesis (1970) 499

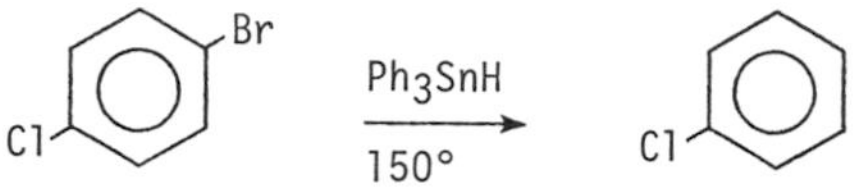

60%

Ph$_3$SnH
150°

JOC (1960) 25 2203
(1969) 34 2014
(1959) 24 294

75%

BuBr

$\xrightarrow{\text{Cr}^{2+}\ \ \text{NH}_2\text{CH}_2\text{CH}_2\text{NH}_2}{\text{Me}_2\text{SO}\ \ \text{H}_2\text{O}}$

BuH

JACS (1966) 88 4094
Angew (1968) 80 271
(Internat Ed 7 247)

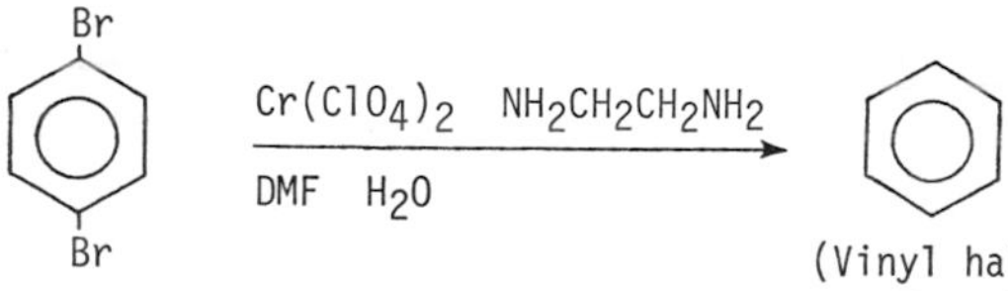

100%

Cr(ClO$_4$)$_2$ NH$_2$CH$_2$CH$_2$NH$_2$

DMF   H$_2$O

100%

(Vinyl halides are also reduced)
Tetrahedron (1968) 24 3503

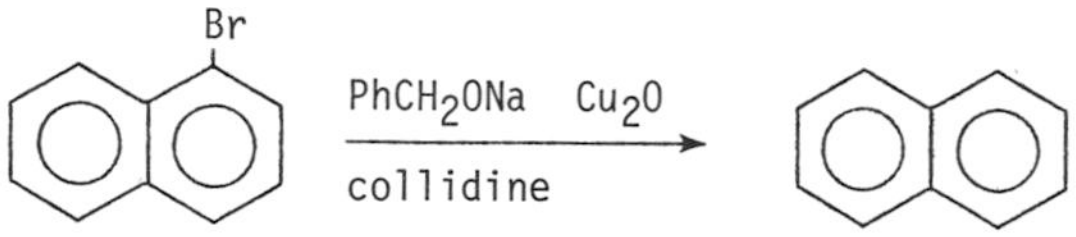

PhCH$_2$ONa   Cu$_2$O
collidine

JCS C (1969) 308

95%

Ac₂O   Cu₂O   Pyr

→

64%

JCS (1964) 1112
Proc Chem Soc (1962) 113

PhCl $\xrightarrow{\text{h}\nu \quad \text{i-PrOH}}$ PhH

72%

Tetr Lett (1969) 1267

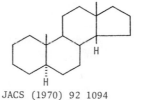

Electrolysis   Et₄NBr
─────────────────────
DMF

94%

(Aromatic halides may also be selectively reduced in the
presence of aliphatic halides)

JOC (1970) 35 1232
Ber (1968) 101 4179

Section 161   Hydrides from Hydrides
               °°°°°°°°°°°°°°°°°°°°°°°°°°

This section lists examples of hydrocarbon epimerization. Many related
reactions are included in section 65 (Alkyls and Aryls from Alkyls and
Aryls)

$\xrightarrow{\text{h}\nu \quad \text{HgBr}_2}$
cyclohexane

70%

JACS (1970) 92 1094

Section 162   Hydrides from Ketones
               °°°°°°°°°°°°°°°°°°°°°°°°°°

This section lists examples of the conversion R₂CO ⟶ RH. For the
conversion R₂CO ⟶ R₂CH₂ or R₂CHR' see section 72 (Alkyls, Methylenes
and Aryls from Ketones)

(t-Bu)₂CO $\xrightarrow{\text{NaNH}_2 \quad \text{C}_6\text{H}_6}$ t-BuH   +   t-BuCONH₂

Compt Rend (1910) 150 661
Org React (1957) 9 1

Bull Acad Sci URSS (1941) 167
(Chem Abs 37 3749)

JACS (1946) 68 2176

$$PhC{\equiv}CCOPh \xrightarrow{\text{NaNH}_2 \quad \text{NH}_3 \quad \text{THF}} PhC{\equiv}CH \qquad\qquad 25\%$$

Bull Chem Soc Jap (1962) 35 1488

Section 163   Hydrides from Nitriles

This section lists examples of the conversion RCN ⟶ RH. For the
conversion RCN ⟶ RMe see section 73 (Alkyls from Nitriles)

$$PhO(CH_2)_3\underset{\underset{CN}{|}}{CHPh} \xrightarrow[\text{toluene}]{\text{Na \quad EtOH}} PhO(CH_2)_3CH_2Ph \qquad\qquad 89\%$$

JACS (1934) 56 1614
Gazz (1963) 93 525
(Chem Abs 59 8825)

$$RCN \xrightarrow{\text{Na \quad NH}_3} RH$$

35% (R=C₁₂H₂₅)
90% (R=PhCH₂)

35% $(R=C_{12}H_{25})$
90% $(R=PhCH_2)$

JACS (1969) 91 2059

$$\underset{\underset{CH_2NMe_2}{|}}{\underset{Me_2NCHCH_2}{\overset{Ph_2CCN}{|}}} \xrightarrow[\text{Et}_2\text{O \quad xylene}]{\text{EtMgBr}} \underset{\underset{CH_2NMe_2}{|}}{\underset{Me_2NCHCH_2}{\overset{Ph_2CH}{|}}}$$

JACS (1952) 74 5793
Chem Comm (1970) 350

Electrolysis LiCl
————————————→
EtNH$_2$

60-80%

Tetr Lett (1968) 1975

Polyphosphoric acid
————————————→
200°

JACS (1953) 75 3600

Section 164   Hydrides from Olefins

This section lists examples of the conversion R$_2$C=CR$_2$ ⟶ R$_2$CH$_2$. For the hydrogenation, dimerization and alkylation of olefins see section 74 (Alkyls, Methylenes and Aryls from Olefins)

Ph$_2$C=CH$_2$ $\xrightarrow{\text{NH}_2\text{NHNa}\quad \text{Et}_2\text{O}}$ Ph$_2$CH$_2$ 97%

Angew (1962) 74 650
(Internat Ed 1 456)

Section 165   Hydrides from Miscellaneous Compounds

This section lists examples of the replacement of miscellaneous functional groups by hydrogen (RX ⟶ RH)

C$_6$H$_{13}$SH $\xrightarrow[300°]{\text{H}_2 \text{ (100 atmos)}\quad \text{MoS}_2}$ C$_6$H$_{14}$ 90%

Coll Czech (1966) 31 2202

C$_8$H$_{17}$SH $\xrightarrow{\text{(EtO)}_3\text{P}}$ C$_8$H$_{18}$ 88%

JACS (1956) 78 6414

$$Ph-X-Ph \xrightarrow{\text{Ni   EtOH}} PhH$$

68% (X=S)
75% (X=SO)
65% (X=SO$_2$)

JACS (1943) $\underline{65}$ 1013
For use of Ni$_2$B see JOC (1965) $\underline{30}$ 1316

$$C_{10}H_{21}-X-C_{10}H_{21} \xrightarrow{\text{Li   MeNH}_2} C_{10}H_{22} \qquad \text{(X=S or SO}_2\text{)}$$

JACS (1960) $\underline{82}$ 2872

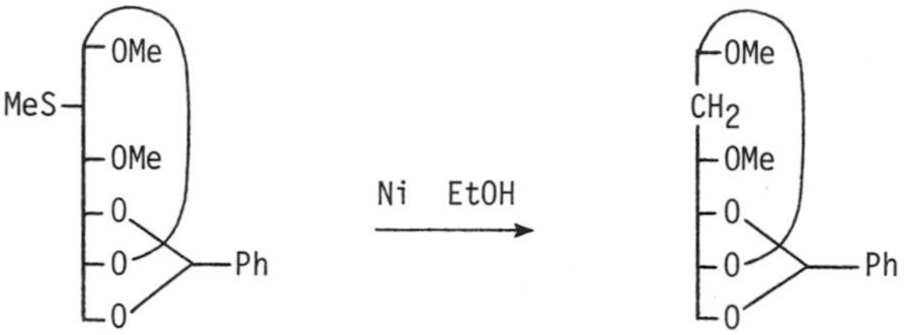

Helv (1946) $\underline{29}$ 371
JACS (1953) $\underline{75}$ 384
(1963) $\underline{85}$ 173

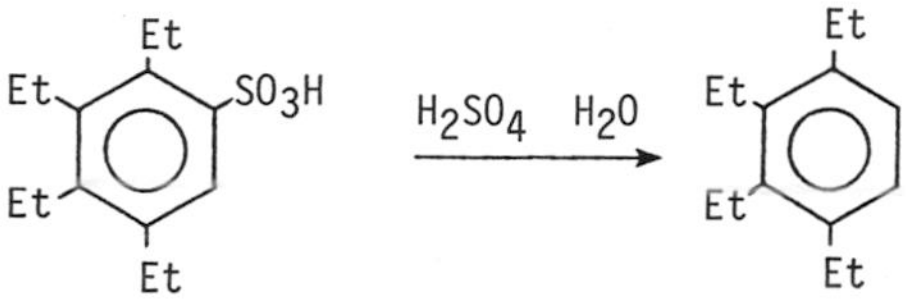

95%

91%

Org React (1942) $\underline{1}$ 370

Chapter 12   PREPARATION
OF
KETONES

Section 166   <u>Ketones from Acetylenes</u>

$$EtC{\equiv}CEt \xrightarrow[\text{2 } H_2O_2 \quad H_2O]{\text{1 } NaBH_4 \quad BF_3{\cdot}Et_2O \quad diglyme} EtCH_2COEt \qquad\qquad 62\%$$

JACS (1961) <u>83</u> 3834
(1967) <u>89</u> 5086
Org React (1963) <u>13</u> 1

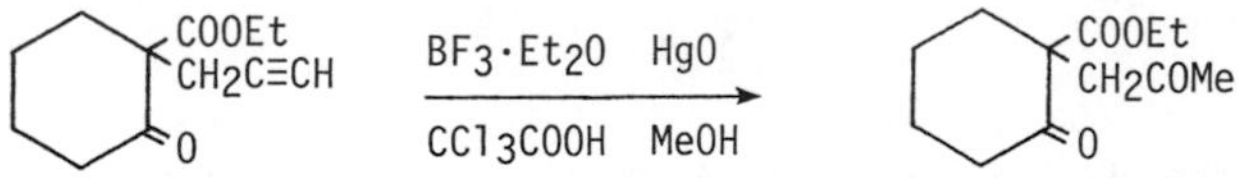

75%

JCS (1952) 4086
Helv (1945) <u>28</u> 1355
(1943) <u>26</u> 680

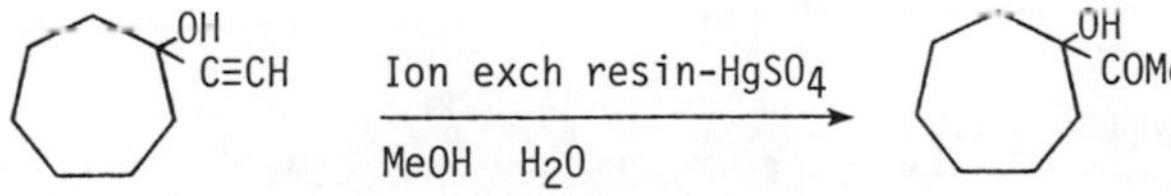

80%

JCS (1954) 3257

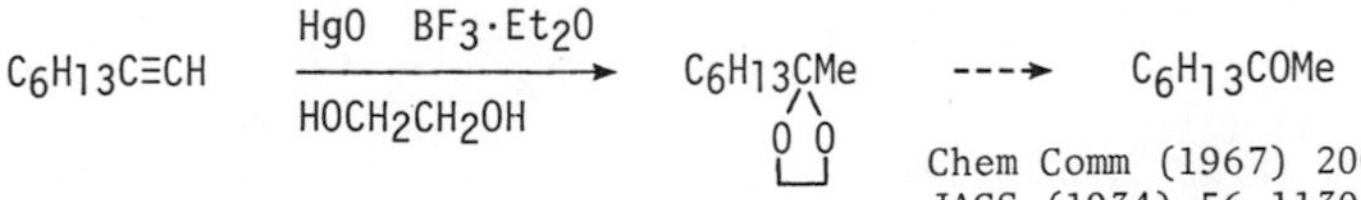

$$C_6H_{13}C{\equiv}CH \xrightarrow[\text{HOCH}_2CH_2OH]{HgO \quad BF_3{\cdot}Et_2O} C_6H_{13}\overset{O\;O}{\underset{\smile}{C}}Me \dashrightarrow C_6H_{13}COMe$$

Chem Comm (1967) 200
JACS (1934) <u>56</u> 1130
(1936) <u>58</u>   80

379

EtC≡CEt $\xrightarrow[\text{2 } H_2O]{\text{1 ClSO}_2\text{CNO CH}_2\text{Cl}_2}$ EtCOCH₂Et   67%

Tetr Lett (1970) 27

Helv (1964) <u>47</u> 194

JACS (1958) <u>80</u> 6118

Section 167   Ketones from Carboxylic Acids, Acid Halides and Anhydrides

C₁₇H₃₄COOH $\xrightarrow[\text{electrolysis}]{\text{HOOCCH}_2\overset{\text{Me}}{\text{CH}}\text{CH}_2\overset{\text{Me}}{\text{CH}}\text{CH}_2\text{COMe}}$ C₁₇H₃₄CH₂C̣HCH₂C̣HCH₂COMe (Me, Me)

Acta Chem Scand (1956) <u>10</u> 478
Advances in Org Chem (1960) <u>1</u> 1

PhCH₂COOH $\xrightarrow{\text{MeLi Et}_2O}$ PhCH₂COMe   76%

Acta Chem Scand (1952) <u>6</u> 782
Org React (1970) <u>18</u> 1

$$\text{(cyclobutane)}\text{COOH, Ph} \xrightarrow{\text{PhLi} \quad \text{Et}_2\text{O}} \text{(cyclobutane)}\text{COPh, Ph}$$

JACS (1969) <u>91</u> 456
      (1970) <u>92</u> 2590

$$\text{EtCOONa} \xrightarrow[]{\text{EtMgBr} \quad \text{Et}_2\text{O}} \text{Et}_2\text{CO}$$

Ber (1909) <u>42</u> 4500
JACS (1933) <u>55</u> 1258
JCS (1950) 2012

Review: The Synthesis of Ketones from Acid Halides and Organometallic
       Compounds of Magnesium, Zinc and Cadmium

Org React (1954) <u>8</u> 28

$$C_{17}H_{35}COCl \xrightarrow[\text{Et}_2\text{O} \quad C_6H_6]{\text{PhCH}_2\text{CH}_2\text{MgBr-CdCl}_2} C_{17}H_{35}COCH_2CH_2Ph \qquad 65\%$$

Org React (1954) <u>8</u> 28
JACS (1950) <u>72</u> 5333

For use of aryl cadmium reagents see   JACS (1945) <u>67</u> 740

$$\underset{(CH_2)_{10}COCl}{\overset{COOEt}{|}} \xrightarrow[\text{Et}_2\text{O} \quad C_6H_6]{C_{18}H_{37}\text{MgBr-ZnCl}_2} \underset{(CH_2)_{10}COC_{18}H_{37}}{\overset{COOEt}{|}} \qquad \sim79\%$$

Org React (1954) <u>8</u> 28

$$\underset{(CH_2)_8COCl}{\overset{COOEt}{|}} \xrightarrow{i\text{-PrCH}_2\text{MgBr} \quad \text{Et}_2\text{O}} \underset{(CH_2)_8COCH_2Pr\text{-}i}{\overset{COOEt}{|}}$$

Org React (1954) <u>8</u> 28
JOC (1961) <u>26</u> 1768
Tetr Lett (1970) 2523

$$C_5H_{11}COCl \xrightarrow{\text{Bu}_2\text{CuLi}} C_5H_{11}COBu \qquad 79\%$$

Tetr Lett (1970) 4647

$(PhCO)_2O$ $\xrightarrow{\text{t-BuMgCl-CdCl}_2 \quad \text{Et}_2\text{O}}$ PhCOBu-t   40%

JOC (1941) 6 462
(1948) 13 592
Ber (1964) 97 1649

$C_5H_{11}COOH$ $\dashrightarrow$ $C_5H_{11}CON$⟨N⟩ $\xrightarrow[\text{Et}_2\text{O}]{\text{EtMgBr}}$ $C_5H_{11}COEt$ ~64%

Annalen (1962) 655 90

PhCOCl $\xrightarrow[\text{HMPA}]{\text{Me}_3\text{SiCH}_2\text{Ph} \quad \text{BuLi}}$ $PhCOCH_2Ph$

Tetr Lett (1970) 1137

PhCOCl $\xrightarrow{\text{Bu}_2\text{Hg} \quad \text{AlBr}_3 \quad \text{CH}_2\text{Cl}_2}$ PhCOBu   73%

J Organometallic Chem (1969) 17 P21

PrCOCl $\xrightarrow{(\text{C}_6\text{H}_{13})_3\text{Al} \quad \text{CH}_2\text{Cl}_2}$ $PrCOC_6H_{13}$ 78%

Tenside (1967) 4 167
(Chem Abs 67 116507)

PhCOCl $\xdashrightarrow{\text{CH}_2\text{N}_2}$ $PhCOCHN_2$ $\xrightarrow{\text{p-Tolylcopper}}$ $PhCOCH_2$-⟨○⟩-Me   ~31%

Chem Comm (1969) 515

$C_{10}H_{21}COCl$ $\xrightarrow[\text{C}_6\text{H}_6 \quad \text{EtOH}]{\overset{\displaystyle R}{|}\text{MgC(COOEt)}_2}$ $C_{10}H_{21}CO\overset{\displaystyle R}{\underset{|}{C}}(COOEt)_2$ $\xrightarrow[\substack{\text{EtCOOH} \\ \text{H}_2\text{O}}]{\text{H}_2\text{SO}_4}$ $C_{10}H_{21}COCH_2R$

JCS (1950) 322
Org Synth (1963) Coll Vol 4 285
For use of malonic tetrahydropyranyl esters see JCS (1952) 3945
"   "   "   "   t-butyl   "   "   JACS (1952) 74 831
"   "   "   "   benzyl   "   "   JCS (1950) 325
and JACS (1963) 85 1409

$$PhCOCl \xrightarrow[\text{Ph}_3\text{CNa} \quad \text{Et}_2\text{O}]{\text{Et}_2\text{CHCOOEt}} PhCOCEt_2 \underset{\overset{|}{COOEt}}{} \xrightarrow[\text{or HI} \quad H_2O]{H_2SO_4 \quad HOAc} PhCOCHEt_2 \qquad 44\%$$

JACS (1941) <u>63</u> 3163

$$PhCOCl \xrightarrow{\text{Naphthalene} \quad AlCl_3 \quad PhNO_2}$$

JCS (1949) S99
Chem Rev (1955) <u>55</u> 229

For further examples of the acylation of aromatic compounds see section
176 (Ketones from Hydrides)

$$PhCOOH \xrightarrow[250°]{\text{C}_8\text{H}_{17}\text{COOH} \quad Fe} PhCOC_8H_{17} \qquad 70\%$$

JOC (1963) <u>28</u> 879
Org Synth (1963) Coll Vol 4 854

$$(C_5H_{11}CO)_2O \xrightarrow{BF_3} (C_5H_{11})_2CO \qquad 64\%$$

JACS (1950) <u>72</u> 3294

$$(PrCO)_2O \xrightarrow{(Me_3Si)_2 \quad AlCl_3} PrCOMe \qquad 75\%$$

Compt Rend (1965) <u>261</u> 1329

$$PhCOCl \xrightarrow{Et_3SiH \quad RhCl_3(PBu_2Ph)_3} Ph_2CO$$

Chem Comm (1970) 1703

$$PhCOCl \xrightarrow[2 \ H_2O]{1 \ MeCH=PPh_3 \quad C_6H_6} PhCOCH_2Me \qquad <85\%$$

Tetr Lett (1960) (4) 7
Annalen (1964) <u>674</u> 11
Ber (1962) <u>95</u> 1513

PrCOOAg $\xrightarrow[\text{2 300°}]{\text{1 Bu}_4\overset{+}{P}\ \overset{-}{Br}\ \text{MeOH}}$ PrCOBu   39%

JOC (1963) 28 1133

1 CH$_2$N$_2$ Zn   NaI
Et$_2$O   HOAc
2 HBr
Et$_2$O

Tetr Lett (1969) 51   75%
JACS (1943) 65 1516

1 CH$_2$=C(OMe)$_2$
2 HCl   H$_2$O   HOAc

JACS (1956) 78 6086

C$_7$H$_{15}$COCl $\xrightarrow[\text{2 H}_2\text{O}]{\text{1 Me}_3\text{N}\ \text{Et}_2\text{O}}$ (C$_7$H$_{15}$)$_2$CO   95%

JACS (1947) 69 2444
Org Synth (1963) Coll Vol 4 555
Aust J Chem (1955) 8 506

For preparation of cyclic ketones from dicarboxylic acids see
JACS (1948) 70 34

(Me$_2$CHCO)$_2$O $\xrightarrow[\text{PhCl}]{\text{Pyridine N-oxide}}$ Me$_2$CO

Tetr Lett (1965) 233

KMnO$_4$ NaBiO$_3$
KOH   H$_2$O   H$_3$PO$_4$ H$_2$O

JOC (1964) 29 2914
JCS (1953) 2129 3580

C_8H_{17}

COOH

1 NaN$_3$ H$_2$SO$_4$
⎯⎯⎯⎯⎯⎯⎯⎯⎯⎯
   CHCl$_3$
2 NaOH   H$_2$O

NH$_2$

1 t-Butyl hypochlorite
⎯⎯⎯⎯⎯⎯⎯⎯⎯⎯⎯⎯⎯⎯
   NaHCO$_3$ Et$_2$O
2 EtONa   EtOH
3 H$_2$SO$_4$ H$_2$O

JOC (1965) 30 3775

O

65%

COOH
COOH
C$_7$H$_{15}$

Ac$_2$O
⎯⎯⎯⎯⎯⎯

O
C$_7$H$_{15}$

~80%

Compt Rend (1955) 240 317
Helv (1945) 28 1651

COOH
COOH

BaO   Fe
⎯⎯⎯⎯⎯⎯⎯
260-280°

O

Helv (1947) 30 2158
Org Synth (1941) Coll Vol 1 192

(CH$_2$)$_6$COOCe$_{1/3}$

(CH$_2$)$_6$COOCe$_{1/3}$

400-450°
⎯⎯⎯⎯⎯⎯

(CH$_2$)$_6$

(CH$_2$)$_6$

CO

Helv (1932) 15 1220 1459
(1928) 11 1174

COOH
COOH

KF   250-280°
⎯⎯⎯⎯⎯⎯⎯⎯⎯⎯⎯
or BaO   320-330°

O

52-72%

JOC (1962) 27 1034

Bu$_2$C(COOH)$_2$

1 Pb(OAc)$_4$ Pyr   C$_6$H$_6$
⎯⎯⎯⎯⎯⎯⎯⎯⎯⎯⎯⎯⎯⎯⎯⎯
2 KOH   MeOH   H$_2$O

Bu$_2$CO

70%

Tetr Lett (1966) 6145
JOC (1964) 29 2914

$$Ph-\underset{}{\square}\underset{COOH}{\overset{COOH}{<}} \xrightarrow[\begin{array}{c}Me_2CO\\ 2\ NaN_3\end{array}]{1\ ClCOOEt\ \ Et_3N} Ph-\underset{}{\square}\underset{CON_3}{\overset{CON_3}{<}} \xrightarrow[H_2O]{H_2SO_4} Ph-\underset{}{\square}\!\!=\!O$$

$$\begin{array}{c}JOC\ (1962)\ \underline{27}\ 1647\\ (1964)\ \underline{29}\ 2914\end{array}$$

45%

Section 168   Ketones from Alcohols and Phenols

$$Me_2\underset{OH}{\overset{}{C}}CH=CH_2 \xrightarrow[\text{}]{\overset{\overset{OEt}{|}}{CH_2=CMe}\ \ TsOH\ \ MeOH} Me_2C=CHCH_2CH_2COMe$$

41%

Helv (1967) $\underline{50}$ 2091

$$HO\overset{\overset{Me}{|}}{C}HCH=CH_2 \dashrightarrow MeCOCH_2COO\overset{\overset{Me}{|}}{C}HCH=CH_2 \xrightarrow[220°]{Ph_2O} \underset{trans}{MeCH=CHCH_2CH_2COMe}$$

76%

Tetr Lett (1969) 3253

$$i\text{-}PrOH \xrightarrow[BF_3]{MeCOCH_2COOEt} MeCO\overset{\overset{}{|}}{C}H\underset{i\text{-}Pr}{COOEt} \dashrightarrow MeCOCH_2\underset{i\text{-}Pr}{}$$

60-67%

Org Synth (1955) Coll Vol 3 405

$$AcO-\underset{}{\bigcirc}\!-OH \xrightarrow{CrO_3\ HOAc\ H_2O} AcO-\underset{}{\bigcirc}\!=O$$

69%

JCS (1949) 615
(1940) 10

For use of CrO_3-$Mn(NO_3)_2$ see JOC (1967) $\underline{32}$ 1098

$$\text{MeO}-\langle\rangle-(CH_2)_3\underset{\overset{|}{OH}}{CH}CH=CH_2 \xrightarrow[\text{(2 phases)}]{CrO_3 \quad Et_2O \quad H_2O} \text{MeO}-\langle\rangle-(CH_2)_3COCH=CH_2$$

63%

JCS C (1966) 1972
Chem Pharm Bull (1964) 12 1184
JOC (1971) 36 387

$$\xrightarrow[\text{HOAc}]{Na_2Cr_2O_7}$$

96%

Org Synth (1963) Coll Vol 4 195
Annalen (1967) 707 203

$$\text{PhCHC}\equiv\text{CH} \xrightarrow[\text{(Jones' reagent)}]{CrO_3 \quad H_2SO_4 \quad Me_2CO \quad H_2O} \text{PhCOC}\equiv\text{CH}$$
$$\overset{|}{OH}$$

77%

JCS (1946) 39
      (1970) 2631
JOC (1960) 25 1434

$$\xrightarrow[\text{DMF}]{CrO_3 \quad H_2SO_4}$$

79%

Ber (1961) 94 729

$$C_6H_{13}\underset{\overset{|}{OH}}{CHMe} \xrightarrow[]{CrO_3\cdot Pyr \quad CH_2Cl_2} C_6H_{13}COMe$$

97%

Tetr Lett (1968) 3363
For use of CrO$_3$ in Pyr-H$_2$O see Tetrahedron (1962) 18 1351
For in situ preparation of CrO$_3$-Pyr see JOC (1970) 35 4000

$$\xrightarrow[\text{CCl}_4]{\text{t-Butyl chromate}}$$

Biochem J (1962) 84 195
Helv (1957) 40 487

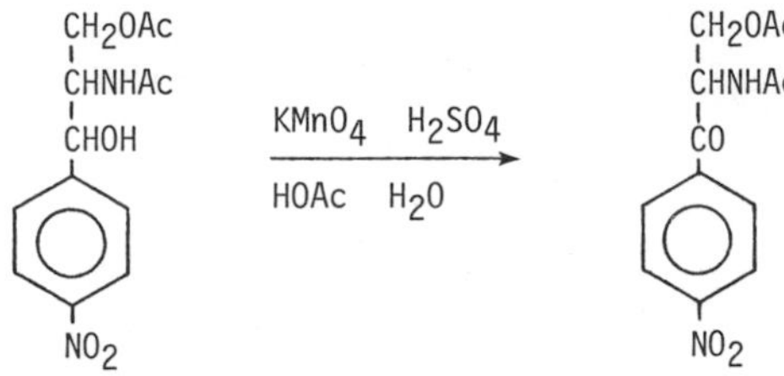
CH2OAc
CHNHAc
CHOH
KMnO4 H2SO4
HOAc H2O
CH2OAc
CHNHAc
CO
NO2
NO2

86%

Ber (1960) 93 387
(1961) 94 169

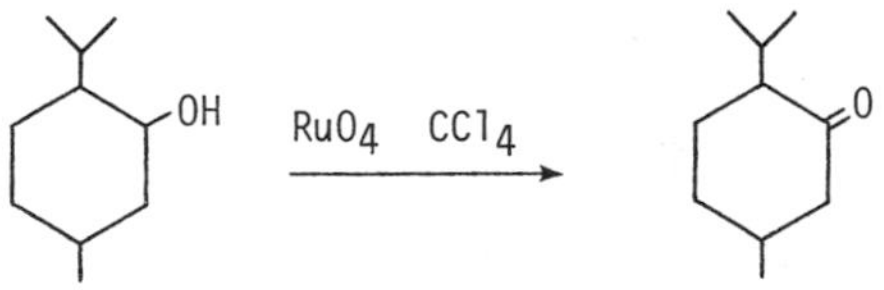
OH
RuO4 CCl4
O

96%

JACS (1958) 80 6682

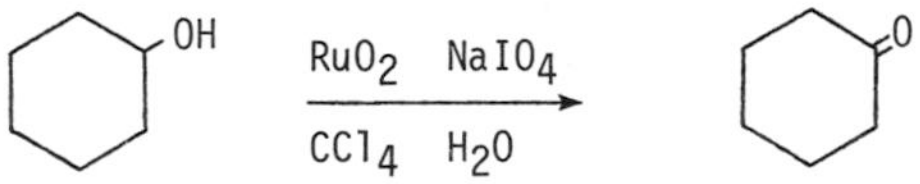
OH
RuO2 NaIO4
CCl4 H2O
O

79%

Tetr Lett (1967) 4729
(1970) 4233
Tetrahedron (1963) 19 1959

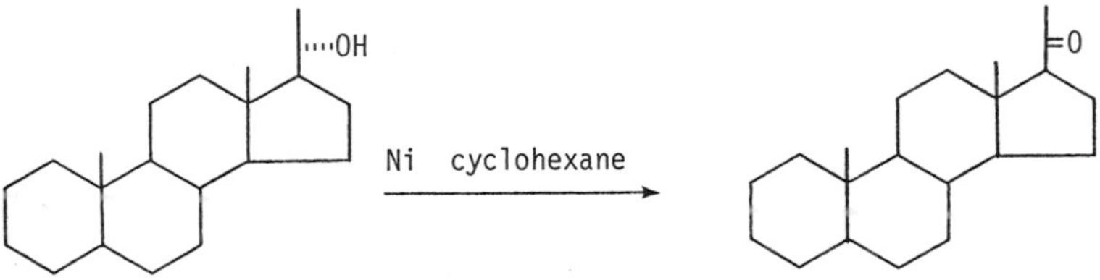
OH
Ni cyclohexane
=O

JCS C (1969) 968
JOC (1968) 33 175 2814
(1958) 23 899

BuCH(OH)Me —[Li $NH_2CH_2CH_2NH_2$]→ BuCOMe 45%

JOC (1962) 27 2662

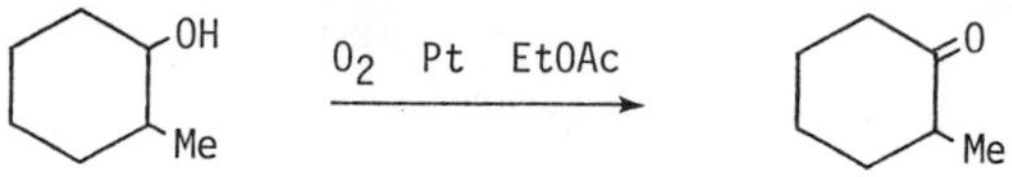

O_2 Pt EtOAc

50%

JACS (1955) 77 190
Tetrahedron (1968) 24 6583
Angew (1957) 69 600
Advances in Carbohydrate Chem
                    (1962) 17 169

PrCHMe O_2 hν Ph_2CO PrCOMe
|
OH

Ber (1963) 96 509

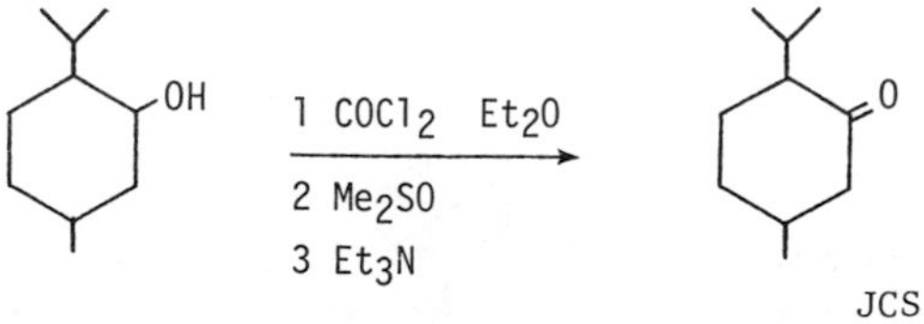

1 $COCl_2$ Et_2O

2 Me_2SO

3 Et_3N

20%

JCS (1964) 1855

DCC CF_3COOH

Pyr Me_2SO C_6H_6

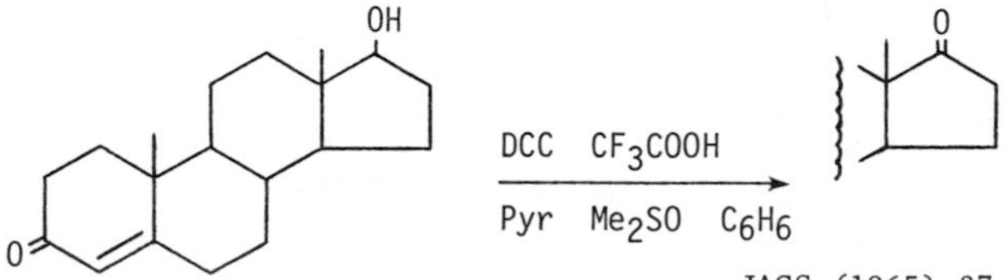

92%

JACS (1965) 87 5670
Chem Rev (1967) 67 247
Can J Chem (1966) 44 2517

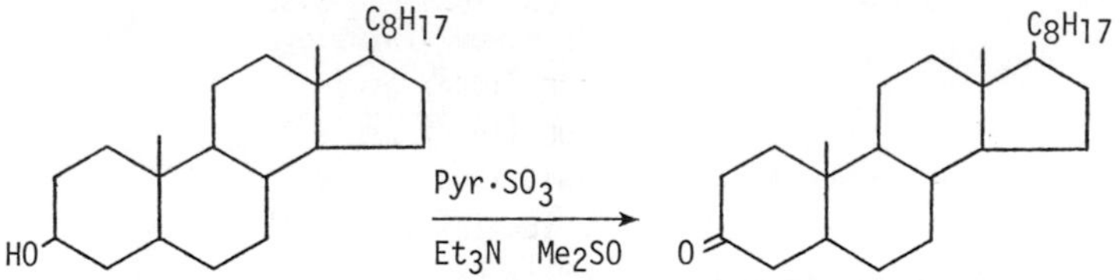

$Pyr \cdot SO_3$

Et_3N Me_2SO

83%

JACS (1967) 89 5505

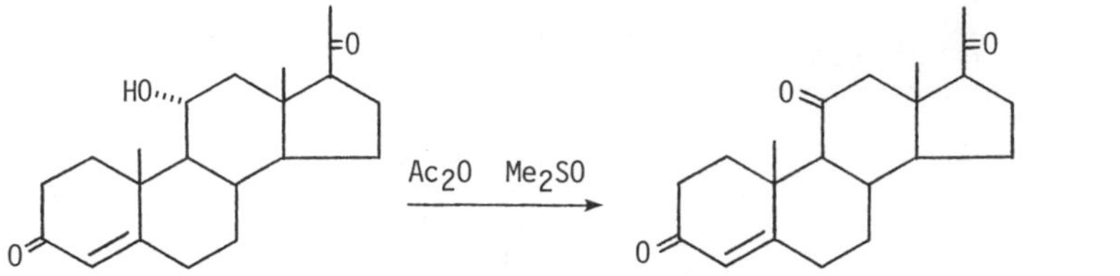

JACS (1967) 89 2416
Carbohydrate Res (1966) 2 251

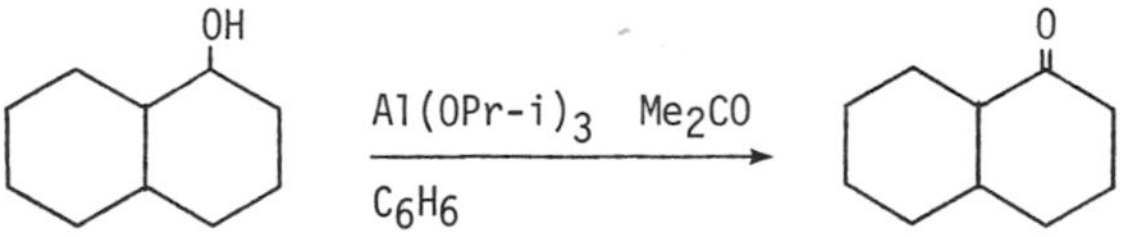

30%

Org React (1951) 6 207

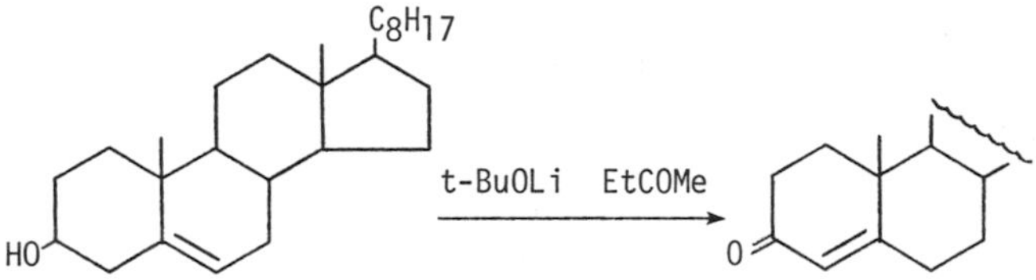

80%

JCS C (1969) 804

68%

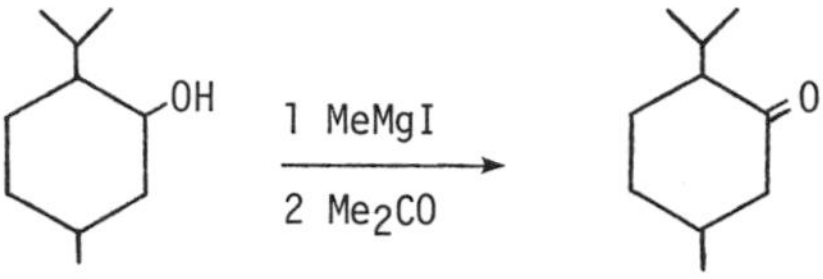

Tetr Lett (1964) 3481

$Ph(CH_2)_3\overset{CHMe}{\underset{OH}{}}$ $\xrightarrow{\text{NBA} \quad Me_2CO \quad H_2O}$ $Ph(CH_2)_3COMe$ 59%

Chem Comm (1965) 5
JACS (1954) 76 3682
Chem Rev (1963) 63 21

For oxidation with NBS   see JCS (1959) 2594
                                and JOC (1957) 22 1678

"   "   "   N-bromocaprolactam see JOC (1960) 25 263

"   "   "   NCS   "   Helv (1953) 36 1763

"   "   "   Tribromoisocyanuric acid see
                                Bull Chem Soc Jap (1958) 31 450

"   "   "   1-chlorobenzotriazole see JCS C (1969) 1474

Br$_2$ NaHCO$_3$
H$_2$O

44%

JACS (1945) $\underline{67}$ 312

MeCH$_2$CMe$_2$ $\xrightarrow{\text{Br}_2}$ MeCHCMe$_2$ $\xrightarrow{\text{H}_2\text{O}}$ MeCOCHMe$_2$ 55%
     | | |
     OH   BrBr

JACS (1933) $\underline{55}$ 1136

PhCHMe $\xrightarrow[\text{Pyr CCl}_4]{\text{t-Butyl hypochlorite}}$ PhCOMe   83%
   |
   OH
                                    Helv (1953) $\underline{36}$ 1763
                                    JACS (1955) $\underline{77}$ 172

C$_5$H$_{11}$CHMe $\xrightarrow[]{\text{INO}_3 \ \ \text{Pyr} \ \ \text{CHCl}_3}$ C$_5$H$_{11}$COMe   65%
        |
        OH
                                    JCS $\underline{C}$ (1970) 676

MeCHCH$_2$CH$_2$CHMe $\xrightarrow[]{\text{Pb(OAc)}_4 \ \ \text{Pyr}}$ MeCOCH$_2$CH$_2$COMe   89%
   | |
   OH   OH
                                    Tetr Lett (1964) 3071
                                    JACS (1960) $\underline{82}$ 4956

Ag$_2$CO$_3$-celite
C$_6$H$_6$

99%

Compt Rend (1968) $\underline{C}$ $\underline{267}$ 900
JACS (1955) $\underline{77}$ 490

$$\text{cyclohexanol} \xrightarrow{\text{AgO HOAc H}_3\text{PO}_4} \text{cyclohexanone}$$

55%

Tetr Lett (1967) 4193
         (1968) 5685

$$\xrightarrow[\text{Me}_2\text{SO}]{\text{Argentic picolinate}}$$

Tetr Lett (1967) 415
Can J Chem (1969) 47 1649

$$\underset{\overset{|}{\text{OH}}}{\text{EtCHMe}} \xrightarrow[\text{O}_2]{\text{PdCl}_2 \quad \text{Cu(NO}_3)_2} \text{EtCOMe}$$

JOC (1967) 32 2816

$$\xrightarrow[\text{3,5-dione C}_6\text{H}_6]{\text{4-Phenyl-1,2,4-triazoline-}}$$

75%

Chem Comm (1966) 744

$$\underset{\overset{|}{\text{OH}}}{\text{PhCHMe}} \xrightarrow[\text{C}_6\text{H}_6]{\overset{\text{NCOOEt}}{\underset{\text{NCOOEt}}{\|}}} \text{PhCOMe}$$

87%

JOC (1967) 32 727
Bull Chem Soc Jap (1968) 41 1491

$$\text{Me}_2\text{CHOH} \xrightarrow{\text{XeO}_3 \quad \text{H}_2\text{O}} \text{Me}_2\text{CO}$$

JACS (1964) 86 2078

$$\xrightarrow[\text{MeCN}]{\text{MnO}_2}$$

100%

Proc Chem Soc (1964) 110
Chem Comm (1966) 121

$$\xrightarrow[\text{Pet ether}]{\text{MnO}_2}$$

(Applicable to allylic and benzylic alcohols only)
JCS (1952) 1094
JACS (1955) 77 4330
JOC (1959) 24 1051

$$\xrightarrow[\text{HMPA}]{\text{CrO}_3}$$

(Applicable to allylic and benzylic alcohols only)
Bull Soc Chim Fr (1969) 335

PhCHCH=CH$_2$ $\xrightarrow{\text{Chloranil\quad CCl}_4}$ PhCOCH=CH$_2$ 66%
  |
  OH
                            (Applicable to allylic and benzylic alcohols only)
                                      JCS (1956) 3070

$$\xrightarrow{\text{DDQ\quad dioxane}}$$

(Applicable to allylic and benzylic alcohols only)
Acta Chem Scand (1961) 15 218
Tetr Lett (1960) (9) 14
Chem Pharm Bull (1970) 18 2343
Chem Rev (1967) 67 153

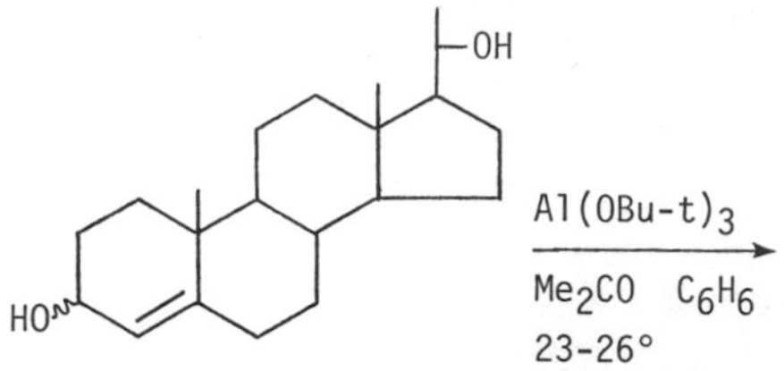

$$\xrightarrow[\begin{array}{c}Me_2CO \quad C_6H_6 \\ 23\text{-}26°\end{array}]{Al(OBu\text{-}t)_3}$$

> 65%

(Method for selective oxidation of allylic alcohols)

Helv (1961) <u>44</u> 179

$$PhCHC_8H_{17} \xrightarrow{N_2O_4 \quad CHCl_3} PhCOC_8H_{17}$$
$$\overset{|}{OH}$$

92%

(Applicable to benzylic alcohols only)

JCS (1957) 5087

$$PhCHMe \xrightarrow{O_2 \quad h\nu \quad Me_2SO} PhCOMe$$
$$\overset{|}{OH}$$

53%

(Applicable to benzylic alcohols only)

Bull Chem Soc Jap (1965) <u>38</u> 1225

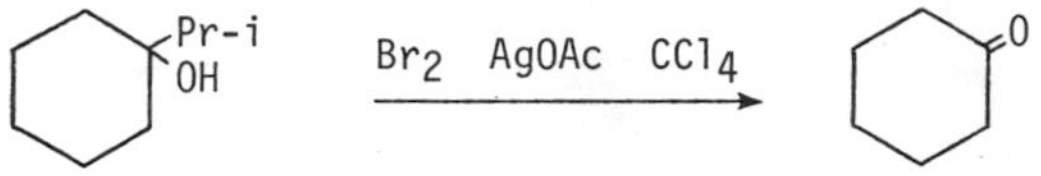

$$\xrightarrow{Br_2 \quad AgOAc \quad CCl_4}$$

38%

JACS (1965) <u>87</u> 1807

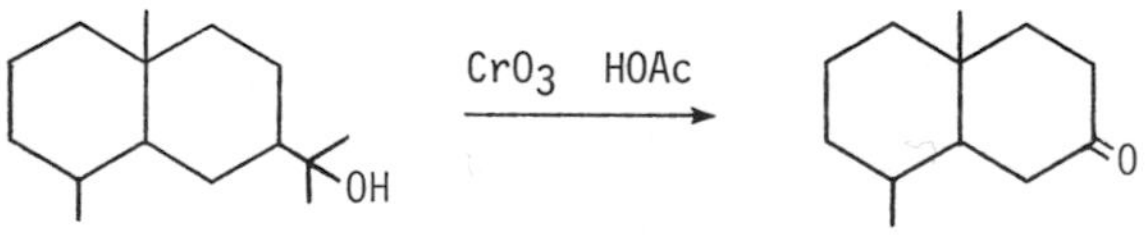

$$\xrightarrow{CrO_3 \quad HOAc}$$

Helv (1942) <u>25</u> 1364

$$\xrightarrow[155°]{(COOH)_2}$$

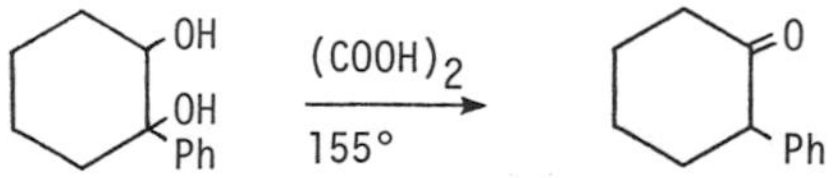

82%

Rec Trav Chim (1948) <u>67</u> 489
JACS (1957) <u>79</u> 147
Helv (1948) <u>31</u> 1077

OHOH   Ac₂O   OHOAc   Zn

$$\text{PhC-CHMe} \xrightarrow[\text{Pyr}]{\text{Ac}_2\text{O}} \text{PhC-CHMe} \xrightarrow{\text{170°}} \text{PhCHCOMe} \qquad \sim 67\%$$

Me   Me   Me

JOC (1970) <u>35</u> 660

$$\text{Me}_2\text{C-CMe}_2 \xrightarrow{\text{H}_2\text{SO}_4 \quad \text{H}_2\text{O}} \text{Me}_3\text{CCOMe} \qquad 65\text{-}72\%$$

OHOH

Org Synth (1941) Coll Vol 1 462
JACS (1960) <u>82</u> 4965

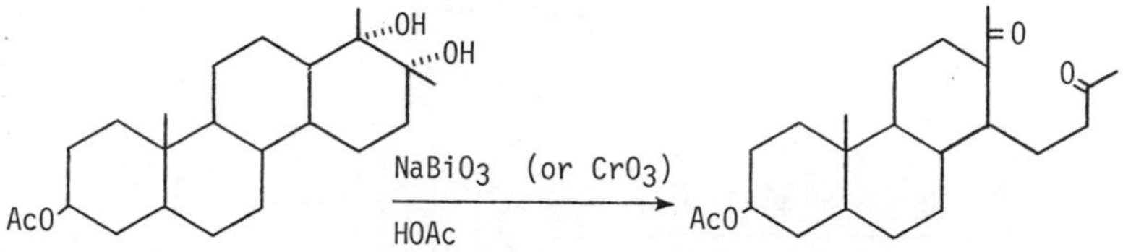

NaBiO₃ (or CrO₃) / HOAc

JACS (1960) <u>82</u> 4965   100%

Pb(OAc)₄ (or CrO₃) / HOAc

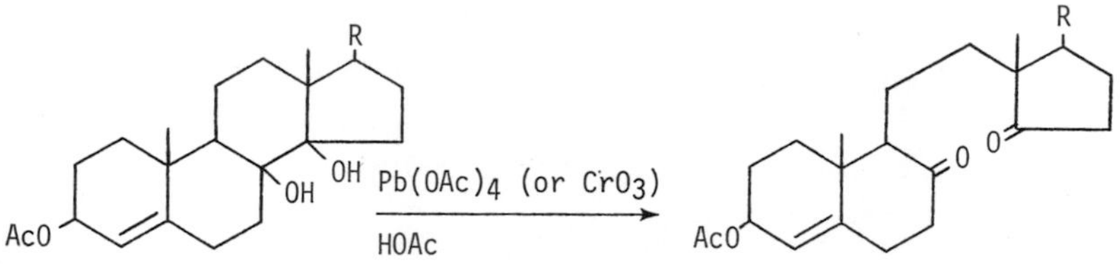

Helv (1959) <u>42</u> 1620

CrO₃ / HOAc

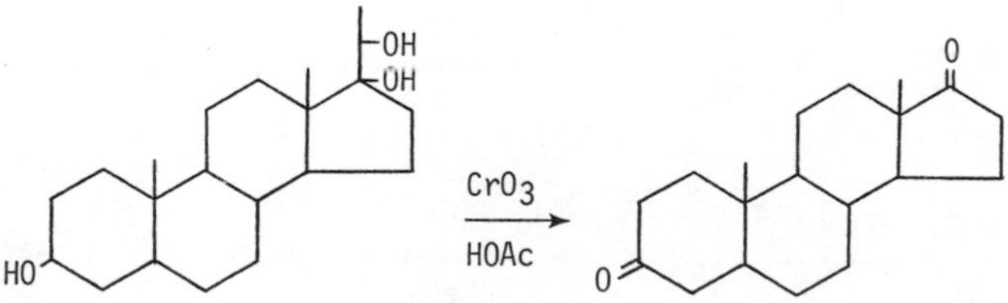

Helv (1938) <u>21</u> 546
JACS (1962) <u>84</u> 2241

$$\text{Me}_2\text{C-CMe}_2 \xrightarrow{\text{Nickel peroxide}} \text{Me}_2\text{CO} \qquad 61\%$$

OHOH

Chem Pharm Bull (1964) <u>12</u> 403

Helv (1941) 24 828
JACS (1956) 78 3087
Org React (1944) 2 341

Me_2C-CMe_2
   | |
   OH OH

$\xrightarrow{K_2S_2O_8 \quad AgNO_3 \quad H_2O}$

Me_2CO 100%

JACS (1954) 76 6345

$\xrightarrow[\text{N-ethylmorpholine}]{\text{H}_2 \text{ (2,500 psi)} \quad \text{Pd}}$
EtOH

40%

JACS (1946) 68 2172

$\xrightarrow{\text{Li} \quad \text{EtNH}_2}$

96%

JACS (1955) 77 6042

Section 169   Ketones from Aldehydes
               ∘∘∘∘∘∘∘∘∘∘∘∘∘∘∘∘∘∘∘∘∘∘∘∘∘∘∘

$C_6H_{13}CHO$
$\xrightarrow[\text{glyme}]{\overset{+ \; OMe}{Ph_3PCHMe} \; Cl^- \quad t\text{-BuOK}}$
$C_6H_{13}CH=\overset{OMe}{C}Me$
$\xrightarrow[\text{MeOH}]{\text{HCl}}$
$C_6H_{13}CH_2COMe$ 57%

Tetr Lett (1964) 3323

$C_6H_{13}CHO$
$\xrightarrow[\text{2 } I_2]{\text{1 } Ph_3P=CHMe \quad THF}$
$C_6H_{13}CH_2COMe$

Tetr Lett (1970) 447

$$C_5H_{11}CHO \xrightarrow[\text{THF}]{\overset{-SMe}{\underset{}{MeCPO(OEt)_2}} \ \overset{+}{Li}} C_5H_{11}CH=\overset{SMe}{\underset{}{CMe}} \xrightarrow[\text{H}_2\text{O}]{HgCl_2 \ \ MeCN} C_5H_{11}CH_2COMe \quad 50\%$$

JOC (1970) <u>35</u> 777

$$PhCHO \xrightarrow[\text{EtOH} \ \ H_2O]{CH_2(CN)_2 \ \ glycine} PhCH=C(CN)_2 \xrightarrow[\text{2 NaOH} \ \ H_2O]{1 \ \ CH_2N_2 \ \ Et_2O} PhCOMe \quad 99\%$$

Tetr Lett (1963) 955

$$PhCHO \ - \!\!-\!\!\rightarrow \ PhCH\!\!\left\langle \begin{array}{c} O \\ O \end{array} \right. \xrightarrow[\text{2 H}_2\text{O}]{1 \ \ BuLi \ \ Et_2O \ \ cyclohexane} PhCOBu \quad 87\%$$

JOC (1965) <u>30</u> 226

$$PhCHO \xrightarrow{Cyclohexylamine} PhCH=N\!\!\bigcirc \xrightarrow[\text{Me}_2\text{SO}]{NaH} \overset{Me}{\underset{}{PhC}}=N\!\!\bigcirc \xrightarrow{Acid} PhCOMe$$

Can J Chem (1970) <u>48</u> 570

$$EtCHO \xrightarrow[\text{KCN}]{Me_2NH \cdot HCl} \underset{NMe_2}{EtCHCN} \xrightarrow[\text{2 H}_2\text{O}]{1 \ \ EtBr \ \ KNH_2 \ \ NH_3} EtCOEt \quad <75\%$$

Bull Soc Chim Fr (1961) 1653
JACS (1960) 82 1960
JOC (1961) <u>26</u> 4740

JOC (1951) <u>16</u> 221
Chem Comm (1967) 1258

$$PhCHCHO \xrightarrow{Br_2} PhCCHO \xrightarrow{\overset{-}{C}N} PhC\overset{O}{\overset{\|}{C}}-CHCN \xrightarrow[\text{2 MeMgBr}]{\text{1 ZnBr}_2 \ \text{Et}_2\text{O}} PhCHCOMe \quad 50\text{-}70\%$$

with Me substituents below:

PhCH(Me)CHO → PhC(Br)(Me)CHO → PhCO-CH(Me)CN → PhCH(Me)COMe   50-70%

Tetr Lett (1970) 2947

$$PhCHO \xrightarrow{BuMgBr} PhCHOMgBr \xrightarrow[\overset{\|}{NCOOEt}]{NCOOEt} PhCOBu \quad 50\%$$

(Bu substituent on PhCHOMgBr)

Bull Chem Soc Jap (1968) <u>41</u> 1491
JCS <u>C</u> (1966) 313

$$\xrightarrow[\text{EtONa}]{\overset{Me}{\overset{|}{BrCHCOOEt}}} \qquad \xrightarrow[\text{2 Cu 180}°]{\text{1 NaOH EtOH}} \qquad 21\%$$

JCS (1943) 15
JACS (1963) <u>85</u> 955
Org React (1949) <u>5</u> 413
Chem Rev (1955) <u>55</u> 283

$$\xrightarrow[\text{EtOH CHCl}_3]{\text{CH}_2\text{N}_2 \ \text{Et}_2\text{O}} \qquad 62\%$$

Org React (1954) <u>8</u> 364

$$\xrightarrow{\text{MeCHN}_2 \ \text{Et}_2\text{O}} \qquad 100\%$$

JACS (1950) <u>72</u> 2737

$$C_6H_{13}CHO \xrightarrow[\text{peroxide}]{C_6H_{13}CH=CH_2 \ \text{diacetyl}} C_6H_{13}COC_8H_{17} \quad 75\%$$

JOC (1949) <u>14</u> 248
Aust J Chem (1967) <u>20</u> 2033

PhCHO $\xrightarrow[\text{MeOH} \quad H_2O]{\text{HOAc} \quad \text{KOH} \quad \text{electrolysis}}$ PhCOMe   18%

Tetr Lett (1968) 1781

Me_3CCHO $\xrightarrow{AlCl_3}$ $Me_2CHCOMe$ 100%

Ber (1936) 69B 2244

$\underset{\text{Me}}{\text{PhCHCHO}}$ $\xrightarrow[\text{2 } H_2SO_4 \quad H_2O]{\text{1 } NH_2CONHNH_2}$ $PhCH_2COMe$ 65%

Zh Obshch Khim (1948) 18 2000
(Chem Abs 43 4632)

$\xrightarrow[\text{NaOAc}]{Ac_2O}$ $\xrightarrow[\substack{\text{2 } O_3 \\ \text{3 Zn HOAc}}]{\text{1 } Br_2 \quad CHCl_3}$ 57%

JACS (1950) 72 2617

$\xrightarrow{\text{Morpholine}}$ $\xrightarrow[\text{h}\nu \quad \text{MeOH}]{O_2 \quad \text{rose bengal}}$

Tetr Lett (1968) 3271 3267
Chem Comm (1969) 314
For oxidation of enamines with $Na_2Cr_2O_7$ see JACS (1955) 77 1212 1216
"   "   "   "   " O_3 "   JACS (1952) 74 3627

$\xrightarrow[\substack{\text{octane} \quad \text{2,2'-bipyridyl} \\ Cu(OAc)_2 \quad \text{DMF}}]{O_2 \quad \text{1,4-diazabicyclo[2.2.2.]-}}$ 90%

Tetr Lett (1969) 985
Chem Ind (1970) 1144
Ber (1967) 100 259

Section 170   Ketones from Alkyls and Methylenes
°°°

This section lists examples of the oxidation of methylenes (RCH_2R' ⟶
$RCOR'$), the replacement of alkyl groups by ketonic groups and the
degradation of alkyl groups to ketones (R_2CHR' ⟶ R_2CO). For the
acylation of hydrocarbons (RH ⟶ RCOR') see section 176 (Ketones from
Hydrides)

$$Cl_2 \quad NO \quad HCl$$
$$\overline{\qquad\qquad\qquad}$$
$$h\nu \quad CCl_4$$

Ber (1965) <u>98</u> 3493 3501

$$NOCl \quad h\nu \quad C_6H_6$$

71%

JOC (1953) <u>18</u> 115
Compt Rend (1965) <u>260</u> 4514

$$h\nu \quad PhNO_2$$

Chem Comm (1970) 1390

$C_{15}H_{31}COOMe$

$$CrO_3 \quad HOAc \quad CH_2Cl_2$$
$$\overline{\qquad\qquad\qquad\qquad}$$

$Me(CH_2)_nCO(CH_2)_{13-n}COOMe$

n=7 to 11

Chem Ind (1966) 2168
JACS (1952) <u>74</u> 3910

$$CrO_3{\cdot}Pyr \quad CH_2Cl_2$$
$$\overline{\qquad\qquad\qquad\qquad}$$

71%

JOC (1969) <u>34</u> 3587

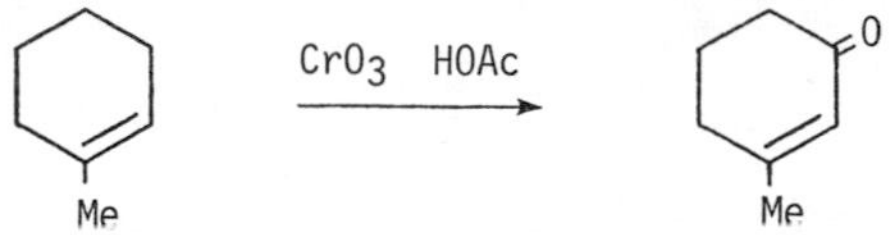
CrO_3 HOAc
O
Me
Me

< 20%

JACS (1941) 63 758
(1949) 71 2226

For oxidation with Na_2CrO_4 in HOAc see JOC (1970) 35 192

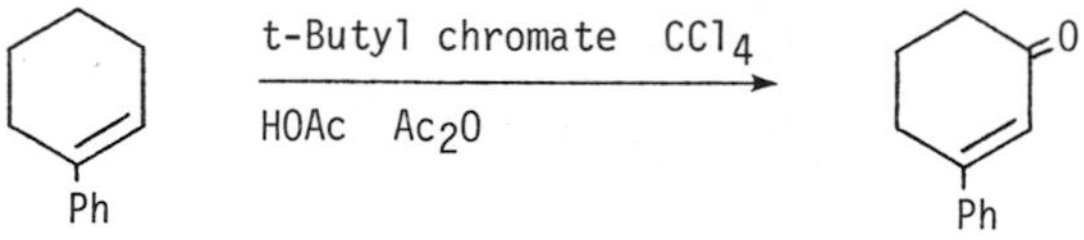
t-Butyl chromate CCl_4
HOAc Ac_2O
O
Ph
Ph

22%

JCS (1951) 516
Helv (1952) 35 284

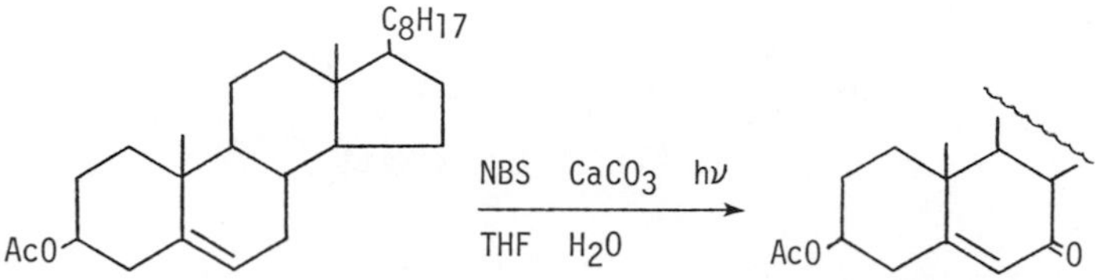
C_8H_{17}
AcO
NBS $CaCO_3$ hν
THF H_2O
AcO
O

81%

Chem Comm (1969) 1220

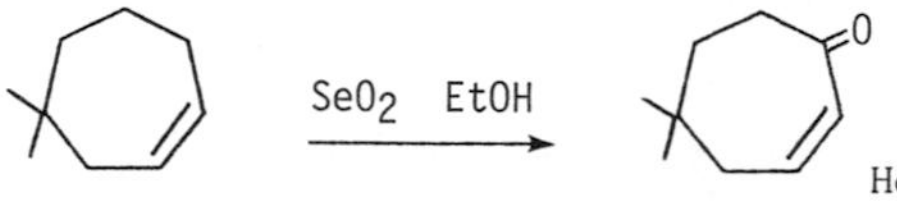
SeO_2 EtOH
O

Helv (1940) 23 524 1477
Org React (1949) 5 331

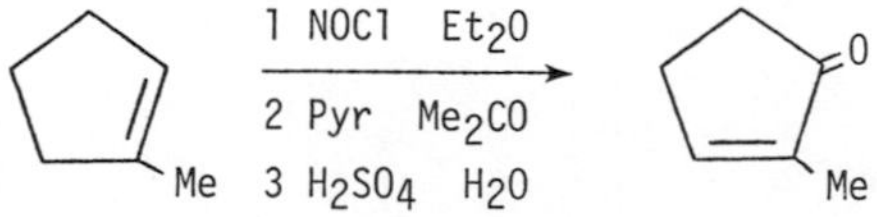
Me
1 NOCl Et_2O
2 Pyr Me_2CO
3 H_2SO_4 H_2O
O
Me

Zh Org Khim (1965) 1 865
(Chem Abs 63 6873)
JCS (1951) 516

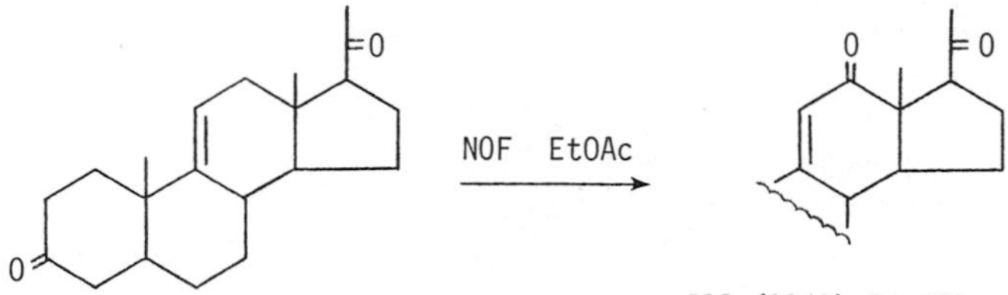
O
O
NOF EtOAc
O
O

30%

JOC (1969) 34 409

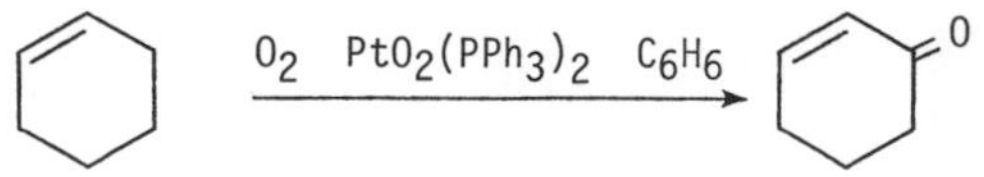
O2 PtO2(PPh3)2 C6H6
O

JACS (1967) 89 4809

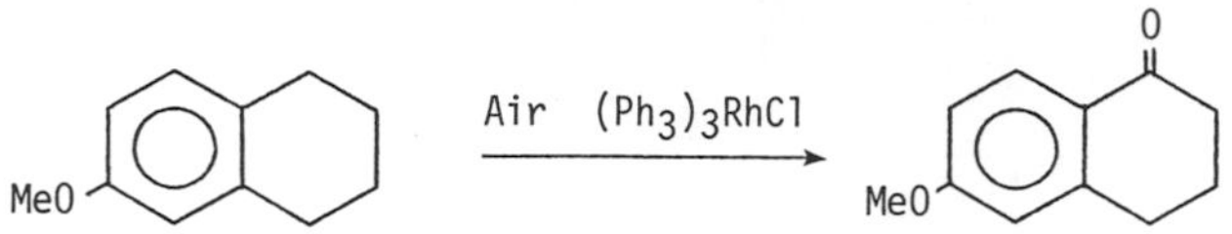
MeO
Air (Ph3)3RhCl
O
MeO

40%

Tetr Lett (1968) 2917
(1967) 3665

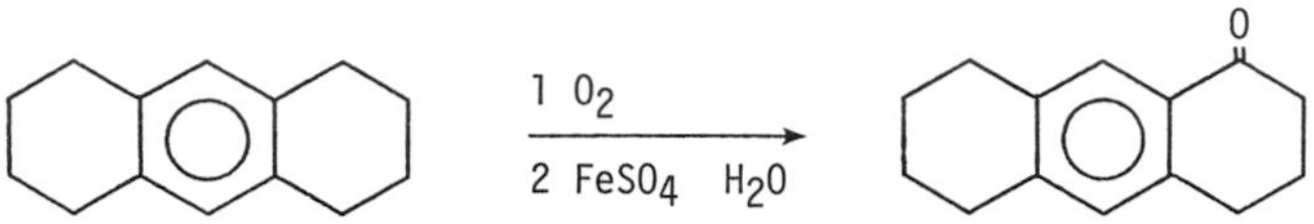
1 O2
2 FeSO4 H2O
O

Ber (1943) 76B 1130

$PhCH_2Me$ —— Argentic picolinate / Me_2SO ——→ PhCOMe

Tetr Lett (1967) 415

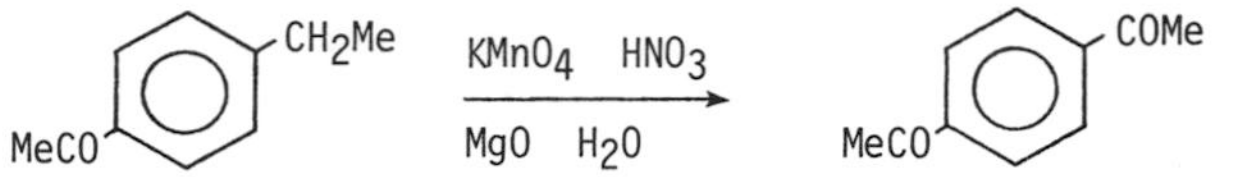
CH2Me
MeCO
KMnO4 HNO3
MgO H2O
COMe
MeCO

82%

JOC (1961) 26 4151

$PhCH_2Me$ —— $(NH_4)_2S_2O_8$ $AgNO_3$ H_2O ——→ PhCOMe <73%

Org Prep and Procedures (1970) 2 207

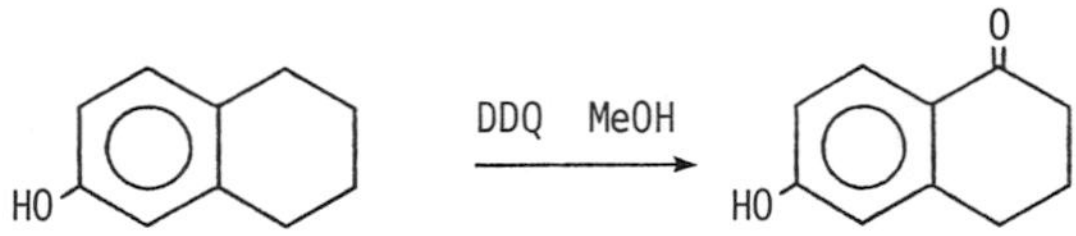
HO
DDQ MeOH
O
HO

78%

Chem Ind (1970) 158

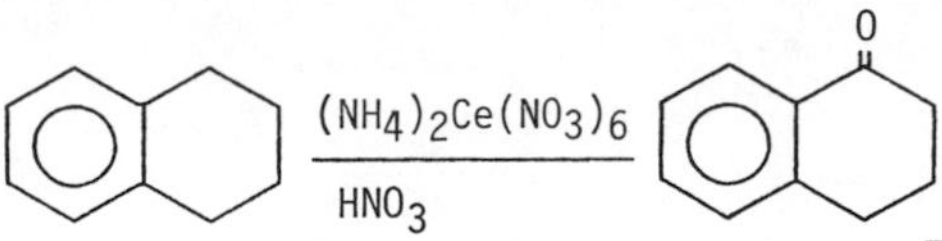

76%

Tetr Lett (1966) 4493

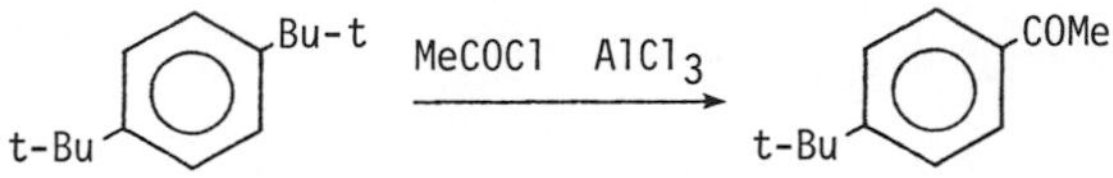

72%

JACS (1942) 64 2421
Helv (1960) 43 1473

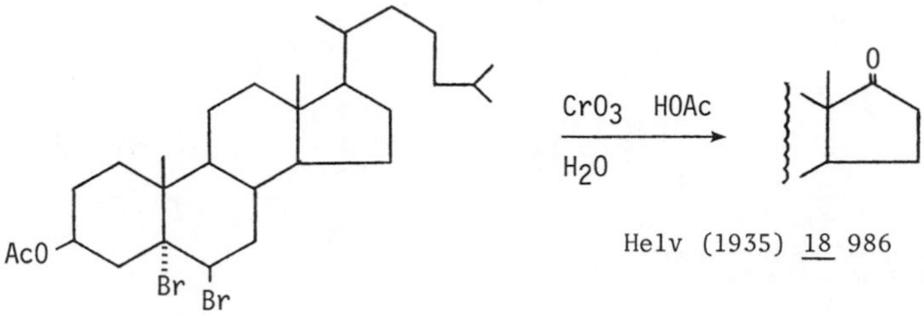

Helv (1935) 18 986

Section 171   Ketones from Amides
               °°°°°°°°°°°°°°°°°°°°°°°°

PrCHCONMe₂ $\xrightarrow{\text{EtLi} \quad \text{Et}_2\text{O}}$ PrCHCOEt   80%
  | |
  Me   Me   JOC (1959) 24 701
                                                         JACS (1939) 61 232
                                                         Ber (1959) 92 2555

PhCHMe   Ni   200°
  | $\xrightarrow{\hspace{2cm}}$ PhCOMe
HCONH
                          Compt Rend (1947) 225 457

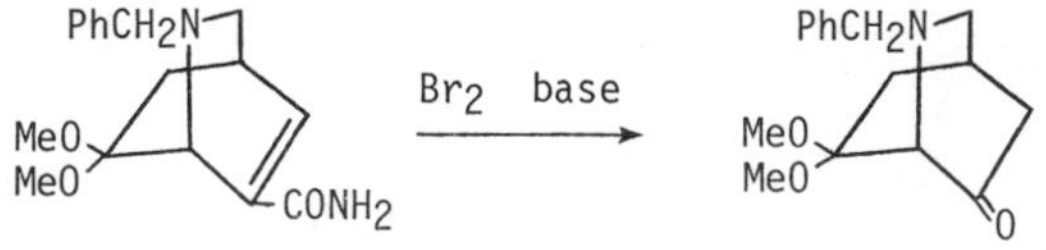
PhCH2N
MeO
MeO
CONH2
Br2 base
PhCH2N
MeO
MeO
O

JACS (1968) 90 2448
Org React (1946) 3 267

$(Me_2CH)_2NAc \xrightarrow{K_2S_2O_8 \quad K_2HPO_4 \quad H_2O} Me_2CO$

JOC (1964) 29 3632

Amides may also be converted into ketones via intermediate amines. See section 96 (Amines from Amides) and section 172 (Ketones from Amines)

Section 172 Ketones from Amines

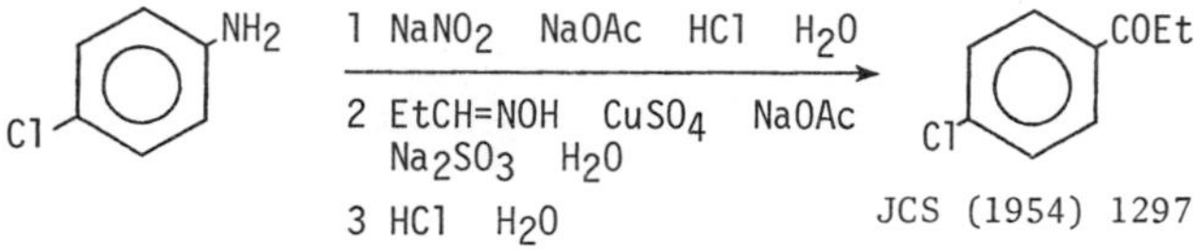
NH2
Cl
1 NaNO2 NaOAc HCl H2O
2 EtCH=NOH CuSO4 NaOAc
Na2SO3 H2O
3 HCl H2O
COEt
Cl

30%

JCS (1954) 1297

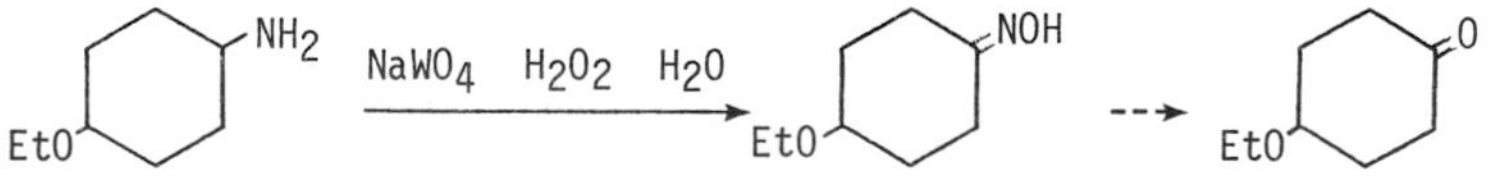
NH2
EtO
NaWO4 H2O2 H2O
NOH
EtO
O
EtO

87%

Ber (1960) 93 132

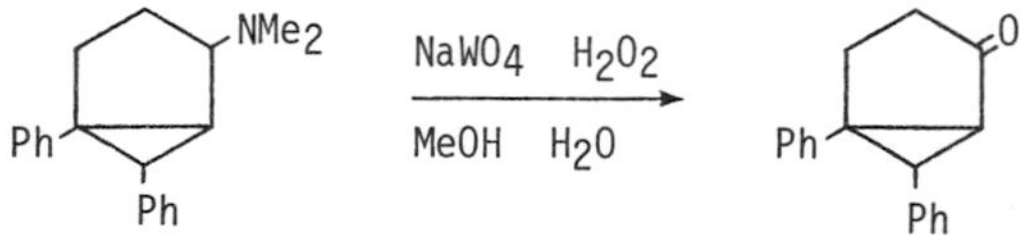
NMe2
Ph
Ph
NaWO4 H2O2
MeOH H2O
O
Ph
Ph

54%

JACS (1968) 90 4892

(R=H or Me)

JCS (1960) 3559
JCS C (1966) 995

$(Me_2CH)_2NH$ $\xrightarrow{Ph_2CO \quad h\nu \quad C_6H_6}$ $Me_2C=NCHMe_2$ $\xrightarrow[H_2O]{HCl}$ Me_2CO ~95%

JACS (1965) 87 2996

EtCHNHR $\xrightarrow[\text{t-BuOH} \quad H_2O]{KMnO_4 \quad CaSO_4}$ EtCOMe   91% (R=H)
| 96% (R=EtCH)
Me |
                                                                               Me
JOC (1967) 32 3129

1 NCS MeOH THF

2 EtONa EtOH

$\xrightarrow[MeOH]{H_2SO_4}$

31%

Tetrahedron (1969) 25 455]
Ber (1955) 88 883
JACS (1968) 90 3245

t-Butyl hypochlorite

$\xrightarrow{}$

$NaHCO_3$ Et_2O

73%

JACS (1954) 76 5554

Argentic picolinate

$\xrightarrow{}$

H_2O

41%

JCS (1965) 4962

OF_2 Freon 11

$\xrightarrow{}$

66%

JACS (1964) 86 1392

Me$_2$CHNH$_2$ $\xrightarrow[\text{2 HCl H}_2\text{O}]{\text{1 2,4-Dinitrobenzaldehyde Pyr}}$ Me$_2$CO   46%

JCS (1954) 209

(CH$_2$)$_{11}$CHNH$_2$ $\xrightarrow[\text{2 (COOH)}_2]{\substack{\text{1 3,5-Di-t-butyl-1,2-benzoquinone}\\ \text{(or 3,5-dinitromesitylglyoxal)}}}$ (CH$_2$)$_{11}$CO   86-97%

JACS (1969) 91 1429

$\overset{\text{Me}}{C_6H_{13}\text{CHNH}_2}$ $\xrightarrow{\text{2,4-Dinitrochlorobenzene}}$ $\overset{\text{Me}}{C_6H_{13}\text{CHNH}}$ (NO$_2$, NO$_2$ ring) $\xrightarrow[\text{H}_2\text{O}]{\text{CrO}_3\ \text{H}_2\text{SO}_4}$ C$_6$H$_{13}$COMe   21%

JOC (1962) 27 452

Me$_2$N (steroid lactone) $\xrightarrow{\text{H}_2\text{O}_2\ \text{H}_2\text{O}}$ Me$_2$N$^+$–O (cyclohexane) $\xrightarrow[\text{2 NaOH H}_2\text{O}]{\text{1 (CF}_3\text{CO)}_2\text{O}}$ O=(cyclohexane)   57%

Tetrahedron (1967) 23 4681
JACS (1968) 90 5622

Section 173   Ketones from Esters

C$_8$H$_{17}$CH=CH(CH$_2$)$_7$COOMe $\xrightarrow[\text{xylene}]{\text{EtONa}}$ C$_8$H$_{17}$CH=CH(CH$_2$)$_6$CHCOOMe / C$_8$H$_{17}$CH=CH(CH$_2$)$_7$CO

$\downarrow$ KOH EtOH

[C$_8$H$_{17}$CH=CH(CH$_2$)$_7$]$_2$CO

Aust J Chem (1955) 8 506

$$i\text{-PrCH}_2\text{COOEt} \xrightarrow{\text{NaH}} i\text{-PrCH}_2\overset{\underset{\displaystyle i\text{-Pr}}{\big|}}{\text{COCHCOOEt}} \dashrightarrow (i\text{-PrCH}_2)_2\text{CO}$$

$$52\%$$ JACS (1946) $\underline{68}$ 2647

$$\text{Me}_2\text{CHCOOEt} \xrightarrow[\text{2 PhCOCl}]{\text{1 Ph}_3\text{CNa Et}_2\text{O}} \text{Me}_2\overset{\underset{\displaystyle \text{COPh}}{\big|}}{\text{CCOOEt}} \xrightarrow[\text{H}_2\text{O}]{\text{H}_2\text{SO}_4 \quad \text{HOAc}} \text{Me}_2\text{CHCOPh}$$

JACS (1941) $\underline{63}$ 3163 3156

$$\text{C}_7\text{H}_{15}\text{COOEt} \xrightarrow[\text{hexane}]{\text{t-BuLi THF}} \text{C}_7\text{H}_{15}\overset{\underset{\displaystyle \text{C}_6\text{H}_{13}}{\big|}}{\text{COCHCOOEt}} \xrightarrow[\text{H}_2\text{O}]{\text{LiOH}} (\text{C}_7\text{H}_{15})_2\text{CO}$$

Dokl (1964) $\underline{155}$ 1352
(Chem Abs $\underline{61}$ 1750)

For the preparation of ketones by alkylation of β-ketoesters with halides
followed by hydrolysis see section 175 (Ketones from Halides and Sulfonates)

$$\text{C}_5\text{H}_{11}\text{COOEt} \xrightarrow[\text{Me}_2\text{SO}]{\overset{+ \quad -}{\text{MeSOCH}_2}} \text{C}_5\text{H}_{11}\text{COCH}_2\text{SOMe}$$

1 MeI   NaH   Me$_2$SO
2 Al-Hg   H$_2$O   THF → C$_5$H$_{11}$COCH$_2$Me   62%

Al-Hg   H$_2$O   THF → C$_5$H$_{11}$COMe   100%

JOC (1966) $\underline{31}$ 2355
JACS (1965) $\underline{87}$ 1345

$$\text{C}_{17}\text{H}_{35}\text{COOMe} \xrightarrow[\text{Me}_2\text{SO THF}]{\overset{- \quad +}{\text{MeSO}_2\text{CH}_2 \text{ Na}}} \text{C}_{17}\text{H}_{35}\text{COCH}_2\text{SO}_2\text{Me} \xrightarrow[\text{H}_2\text{O THF}]{\text{Al-Hg}} \text{C}_{17}\text{H}_{35}\text{COMe} \quad 67\%$$

JOC (1968) $\underline{33}$ 61

$$\xrightarrow[\text{THF}]{\overset{- \quad +}{[\text{LiCH}_2\text{SONAr}] \text{ Li}}}$$

84%

JACS (1968) $\underline{90}$ 5548

PhCH$_2$COOEt $\xrightarrow[\text{2 HCl}\quad\text{H}_2\text{O}]{\overset{\overset{\text{OEt}}{|}}{\underset{}{\text{1 CH}_2=\text{CNMe}_2}}}$ PhCH$_2$COMe   54%

Chem Pharm Bull (1969) 17 2314

PhCOOEt $\xrightarrow[\text{2 NaOH}\quad\text{MeOH}]{\text{1 Ph}_3\text{P=CH}_2\quad\text{Et}_2\text{O}}$ PhCOMe   41%

Ber (1962) 95 1513

Me$_2$CHOAc $\xrightarrow{\text{Br}_2\quad\text{H}_2\text{O}\quad\text{pH 4.6}}$ Me$_2$CO

JACS (1967) 89 3555

$\xrightarrow[\text{Me}_2\text{CO}]{\text{CrO}_3\quad\text{H}_2\text{SO}_4}$ 83-87%

Org Synth (1962) 42 79

PhCH$_2$CH$_2$C(COOEt)$_2$ $\xrightarrow[\text{2 NaNO}_2\quad\text{HOAc}]{\text{1 N}_2\text{H}_4}$ PhCH$_2$CH$_2$C(CON$_3$)$_2$ $\xrightarrow[\text{2 Acid}]{\text{1 ROH}}$ PhCH$_2$CH$_2$COMe
$\quad\quad\quad|$ $\quad\quad\quad\quad\quad\quad\quad\quad\quad\quad\quad\quad\quad\quad\quad|$
$\quad\quad$ Me $\quad\quad\quad\quad\quad\quad\quad\quad\quad\quad\quad\quad\quad\quad\quad$ Me   28%

Org React (1946) 3 337
JOC (1962) 27 1647

Section 174   Ketones from Ethers and Epoxides
               ∘∘∘∘∘∘∘∘∘∘∘∘∘∘∘∘∘∘∘∘∘∘∘∘∘∘∘∘∘∘∘∘∘∘∘

(Me$_2$CH)$_2$O $\xrightarrow[\text{BuOH}\quad\text{CuCl}]{\text{t-Butyl perbenzoate}}$ Me$_2$C(OBu)$_2$ $\xrightarrow{\text{Acid}}$ Me$_2$CO

Angew (1961) 73 65
Tetrahedron (1961) 13 241

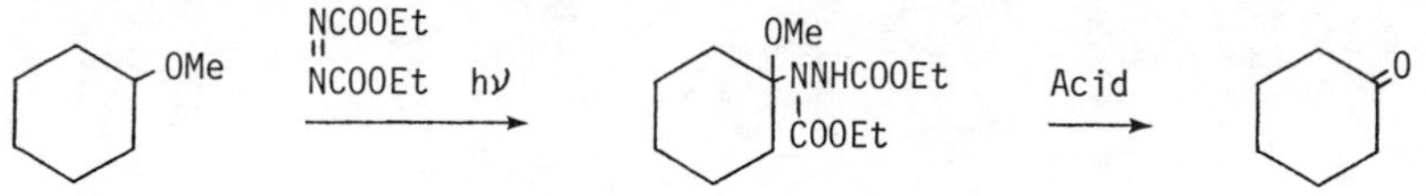

Chem Comm (1965) 259

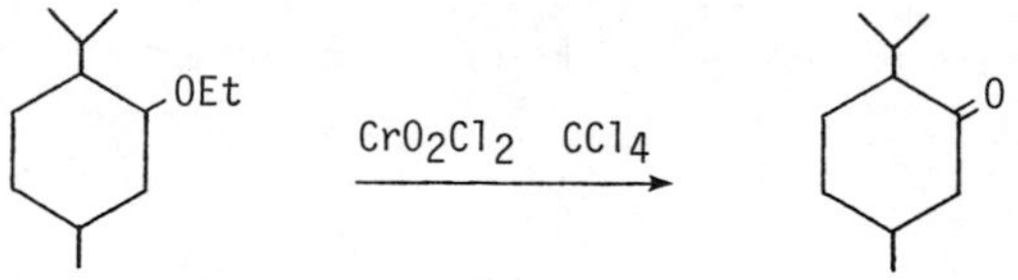

J Prakt Chem (1938) 151 61

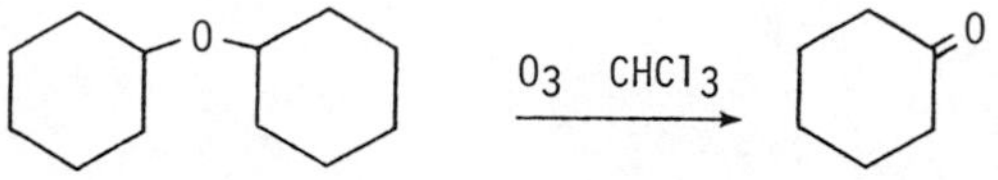

87%

JOC (1962) 27 2392

25%

Me$_2$CHOCH$_2$Ph $\xrightarrow{\text{Br}_2 \quad \text{H}_2\text{O}}$ Me$_2$CO + PhCHO

                                    45%   55%

JACS (1967) 89 3550

EtCHOMe $\xrightarrow[\text{MeOH}]{\text{Electrolysis} \quad \text{MeONa}}$ EtC(OMe)$_2$ $\dashrightarrow$ EtCOMe
  | |
  Me   Me

JACS (1969) 91 2803

Me$_2$CHOCPh$_3$ $\xrightarrow{\sim 228°}$ Me$_2$CO

JACS (1930) 52 753
       (1924) 46 2580

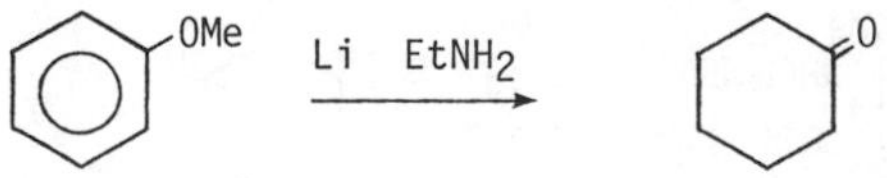

40%

JACS (1955) 77 6042
JCS (1942) 689
Org Synth (1963) Coll Vol 4 903

Some of the methods listed in section 54 (Aldehydes from Ethers and
Epoxides) may also be applied to the preparation of ketones from ethers

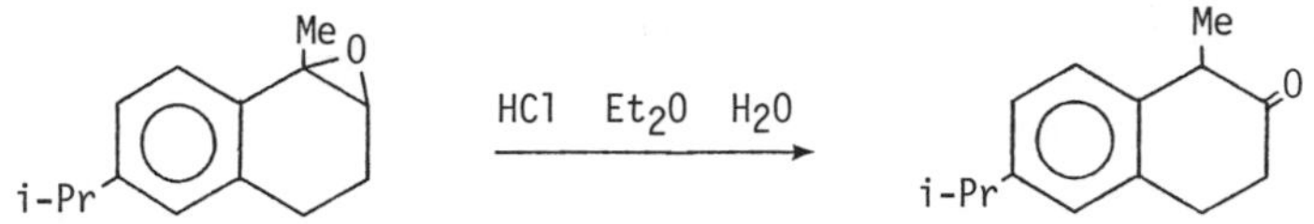

79%

JACS (1962) 84 284

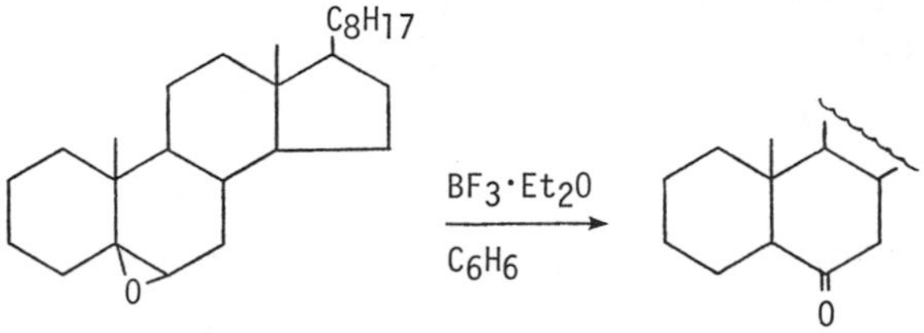

JCS (1957) 4596 4765
Helv (1953) 36 398

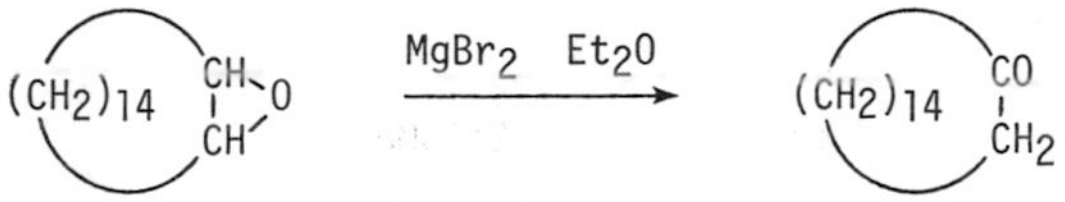

75%

Helv (1948) 31 1077
Chem Rev (1959) 59 737

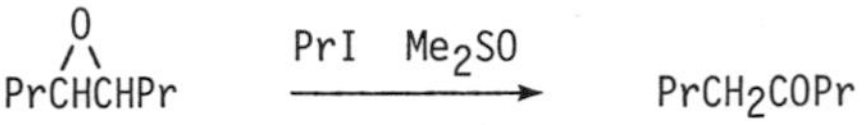

100%

Chem Comm (1968) 227

JACS (1968) <u>90</u> 4193
       (1965) <u>87</u> 1405

80%

$$\underset{Me}{\overset{O}{\underset{\diagup\diagdown}{MeCHCHMe}}} \quad \xrightarrow{\text{Co}_2(\text{CO})_8 \quad \text{MeOH}} \quad MeCH_2COMe$$

JOC (1962) <u>27</u> 2706

77%

$$\underset{\underset{Me}{|}}{\overset{O}{\underset{\diagup\diagdown}{PhC-CH_2}}} \quad \xrightarrow[\text{MeOH} \quad H_2O]{H_2O_2 \quad KOH} \quad PhCOMe$$

JACS (1957) <u>79</u> 503

85-90%

Section 175   <u>Ketones from Halides and Sulfonates</u>
○○

$$BuBr \quad \xrightarrow[\text{EtOH}]{MeCOCH_2COOEt \quad EtONa} \quad \underset{\underset{Bu}{|}}{MeCOCHCOOEt} \quad \xrightarrow{NaOH \quad H_2O} \quad MeCOCH_2Bu \qquad \sim 40\%$$

Org Synth (1941) Coll Vol 1 248 351

For cleavage of β-ketoesters with CaI$_2$ see   JOC (1966) <u>31</u> 3267
  "   "   "   "   "   LiI "   Org Synth (1965) <u>45</u> 7
  "   "   "   "   by pyrolysis see JOC (1957) <u>22</u> 1189

$$BuBr \quad \xrightarrow[\text{THF}]{\overset{-\quad\;\; -\quad\;\; +}{CH_2COCHCOOMe} \;\; 2Na} \quad BuCH_2COCH_2COOMe \quad \xrightarrow[\substack{2 \; BuLi \\ 3 \; PhCH_2Cl}]{1 \; NaH} \quad \underset{\underset{CH_2Ph}{|}}{BuCHCOCH_2COOMe}$$

| | Hydrolysis | | Hydrolysis

BuCH$_2$COMe $\underset{\underset{CH_2Ph}{|}}{BuCHCOMe}$

JACS (1970) <u>92</u> 6702

$C_8H_{17}Br$ $\xrightarrow[\text{NaH} \quad \text{t-BuOH}]{CH_2(COOBu\text{-}t)_2}$ $C_8H_{17}CH(COOBu\text{-}t)_2$ $\xrightarrow[2 \ C_7H_{15}COCl]{1 \ \text{NaH} \quad C_6H_6}$ $C_8H_{17}\underset{\underset{\displaystyle COC_7H_{15}}{|}}{C}(COOBu\text{-}t)_2$

$\downarrow$ TsOH   HOAc

$C_8H_{17}CH_2COC_7H_{15}$

JACS (1952) <u>74</u> 831   46%
JCS (1952) 3945
(1950) 325

BuI $\xrightarrow[\text{EtOH}]{MeCOCH_2COMe \quad K_2CO_3}$ $BuCH_2COMe$ 60%

JOC (1965) <u>30</u> 3321

BuBr $\xrightarrow[2 \quad \text{Cl}\diagdown\text{O}]{1 \ \text{Mg} \quad Et_2O}$ Bu$\diagup$O $\xrightarrow[340\text{-}350°]{\text{Pt-C}}$ $BuCO(CH_2)_3Me$ 71%

Izv (1964) 747
(Chem Abs <u>61</u> 3057)

BuBr $\xrightarrow[2 \ PhCN]{1 \ \text{Mg} \quad Et_2O}$ PhCOBu

JACS (1957) <u>79</u> 881

Further examples of the reaction RMgX + R'CN $\longrightarrow$ RCOR' are included in section 178 (Ketones from Nitriles)

$\xrightarrow[\begin{array}{l}2 \ CdCl_2 \\ 3 \ MeCOCl \quad C_6H_6\end{array}]{1 \ \text{Mg} \quad EtBr \quad Et_2O}$

JCS (1955) 3986
Org React (1954) <u>8</u> 28

BuBr $\xrightarrow[2 \ Ac_2O]{1 \ \text{Mg} \quad Et_2O}$ BuCOMe   79%

JACS (1945) <u>67</u> 154
JOC (1948) <u>13</u> 592

Further examples of the preparation of ketones by reaction of organometallic derivatives of Li, Mg, Cd and Zn with carboxylic acids, acid halides and anhydrides are included in section 167 (Ketones from Carboxylic Acids, Acid Halides and Anhydrides)

i-PrCl $\xrightarrow[\substack{\text{2 Reflux} \\ \text{3 } CO_2}]{\text{1 Li } Et_2O}$ $(i\text{-}PrCH_2CH_2)_2CO$

JACS (1953) <u>75</u> 1771

PhBr $\xrightarrow[\substack{\text{2 } CO_2}]{\text{1 Li } Et_2O}$ Ph_2CO

JACS (1933) <u>55</u> 1258   70%
(1955) <u>77</u> 2806

i-PrCH$_2$CH$_2$Br $\xrightarrow[\substack{\text{2 } EtCONMe_2}]{\text{1 Li } Et_2O}$ $i\text{-}PrCH_2CH_2COEt$ 75%

JOC (1959) <u>24</u> 701

$\xrightarrow[\substack{\text{2 } Me_2NCOOEt}]{\text{1 EtLi } Et_2O}$ 75%

Tetr Lett (1970) 5219

BuBr $\xrightarrow[\substack{\text{2 CNCMe}_2 \\ | \\ \text{CH}_2\text{Bu-t}}]{\text{1 Li}}$ $\underset{\substack{| \\ CH_2Bu\text{-}t}}{\overset{\overset{Li}{|}}{BuC}}=NCMe_2$ $\xrightarrow[\substack{\text{2 } (COOH)_2}]{\text{1 EtBr}}$ BuCOEt   87%

JACS (1970) <u>92</u> 6675

BuX $\xrightarrow[\substack{\text{2 PrCHO}}]{\text{1 Mg}}$ $\underset{\substack{| \\ Pr}}{BuCHOMgX}$ $\xrightarrow{\substack{NCOOEt \\ \| \\ NCOOEt}}$ BuCOPr   66%

Bull Chem Soc Jap (1968) <u>41</u> 1491

i-PrBr $\xrightarrow[\text{2 MeCH=CHCOMe}]{\text{1 Mg   Et}_2\text{O}}$ MeCHCH$_2$COMe
                                                        |
                                                      i-Pr

JACS (1951) <u>73</u> 2721

For promotion of 1:4 addition of Grignard
reagents by CuX$_2$ see JACS (1941) <u>63</u> 2308
and JACS (1965) <u>87</u> 82
and JOC (1966) <u>31</u> 3128

$\xrightarrow[\text{2 2-Chlorocyclohexanone}]{\text{1 Mg   Et}_2\text{O}}$

48%

JOC (1959) <u>24</u> 843
(1947) <u>12</u> 737

PhBr $\xrightarrow[\begin{array}{l}\text{2 CuBr}\\\text{3 N}_2\text{CHCOPh}\end{array}]{\text{1 Li   Et}_2\text{O}}$ PhCH$_2$COPh

35%

Chem Comm (1969) 515

$\xrightarrow[2]{\text{1 Mg   THF}}$

3 MeLi
4 (COOH)$_2$

CH$_2$CHCOMe (Me substituent)

82%

JACS (1969) <u>91</u> 5887

BuBr $\xrightarrow[2]{\text{1 Mg   Et}_2\text{O}}$ BuCOCH$_2$CH$_2$Ph

3 (COOH)$_2$ H$_2$O

71%

JACS (1970) <u>92</u> 1084

PhCl $\dashrightarrow$ PhMgCl $\dashrightarrow$ PhHgCl $\xrightarrow{\text{CO   RhCl}_3\text{   MeCN}}$ Ph$_2$CO

<48%

JACS (1968) <u>90</u> 5546

BuBr --→ BuHgBr $\xrightarrow[\text{2 Ph}_3\text{P C}_6\text{H}_6]{\text{1 Co}_2\text{(CO)}_8 \text{ THF}}$ Bu$_2$CO   <42%

JACS (1969) <u>91</u> 3037

$\xrightarrow{\begin{array}{l}\text{1 Li Et}_2\text{O}\\ \text{2 9-Borabicyclo[3.3.1]-}\\ \text{   nonane}\\ \text{3 MeSO}_3\text{H}\\ \text{4 PhCOCH}_2\text{Br t-BuOK THF}\end{array}}$ 95%

JACS (1969) <u>91</u> 4304 6852

PhCH$_2$Br $\xrightarrow{\text{Li[PhCOFe(CO)}_4\text{] C}_6\text{H}_6}$ PhCH$_2$COPh   67%

JOC (1970) <u>35</u> 4183
Trans New York Acad Sci (1965) <u>27</u> 724

C$_6$H$_{13}$Br $\xrightarrow[\substack{\text{2 Ph}_3\overset{+}{\text{P}}\text{CMe}_2 \overset{-}{\text{Cl}} \text{THF} \\ \quad\quad |\\ \quad\quad\text{COOEt}}]{\text{1 Mg}}$ C$_6$H$_{13}$COPr-i   45%

Tetr Lett (1969) 23

PhCH$_2$Br —

$\xrightarrow[\text{2 Zn HOAc H}_2\text{O}]{\text{1 } \text{NCH=CMe} \text{DMF}}$ PhCH$_2$CH$_2$COMe   85%

$\xrightarrow[\text{2 Zn HOAc H}_2\text{O}]{\text{1 } \text{NClCOCHN} \text{DMA}}$ (PhCH$_2$CH$_2$)$_2$CO   40%

Aust J Chem (1967) <u>20</u> 2441

BuBr --→ BuC(COOH)$_2$ $\xrightarrow[\text{Pyr C}_6\text{H}_6]{\text{Pb(OAc)}_4}$ BuCOEt
         |
         Et

Tetr Lett (1966) 6145

BuBr --→ BuOCHCH=CHMe $\xrightarrow[\text{Me}_2\text{NCH}_2\text{CH}_2\text{NMe}_2]{\text{PrLI}}$ BuCHCH$_2$COMe   ~30%
                | |
                Me   Me

Tetr Lett (1969) 821

i-PrI ─┬─ 1 MeCH$\overset{S}{\underset{S}{\diagdown}}$ BuLi THF ─────→ i-PrCOMe   84%
       │   2 HgCl$_2$
       │
       └─ 1 CH$_2\overset{S}{\underset{S}{\diagdown}}$ BuLi THF ─────→ (i-Pr)$_2$CO   70%
           2 BuLi i-PrI
           3 HgCl$_2$

Angew (1965) 77 1134
(Internat Ed 4 1075)
Tetr Lett (1969) 173
JOC (1968) 33 298
Synthesis (1969) 17

$\xrightarrow[\text{2 H}_2\text{SO}_4 \text{ H}_2\text{O}]{\text{1 Na}_2\text{Cr}_2\text{O}_7 \text{ H}_2\text{O}}$

Ber (1956) 89 1732

C$_6$H$_{13}$CHMe $\xrightarrow{\text{MgO Me}_2\text{SO}}$ C$_6$H$_{13}$COMe   32%
          |
          I

JCS (1964) 520
JOC (1959) 24 1792
    (1960) 25 670
Chem Rev (1967) 67 247

$\xrightarrow{\overset{+-}{\text{Me}_3\text{NO}}\ \ \text{CHCl}_3}$

50%

Ber (1961) 94 1360

C$_6$H$_{13}$CHMe $\xrightarrow[\text{2 O}_2 \text{ Et}_2\text{O}]{\text{1 Mg Et}_2\text{O}}$ C$_6$H$_{13}$CHMe $\xrightarrow{\text{NaOH}}$ C$_6$H$_{13}$COMe
          | |
          Cl   OOH
                                                    91%

JACS (1955) 77 6032
Ber (1960) 93 2151

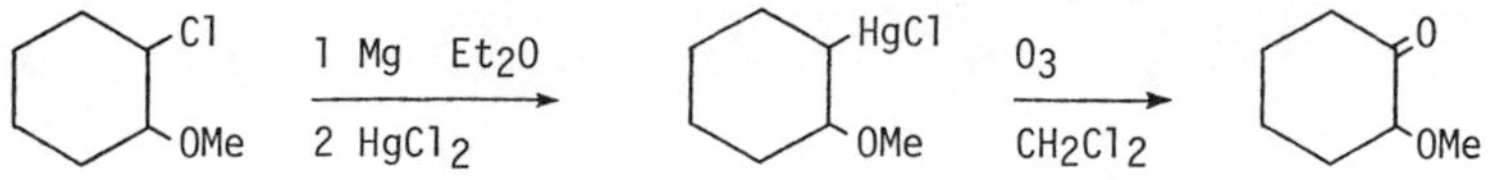

Tetr Lett (1970) 2679

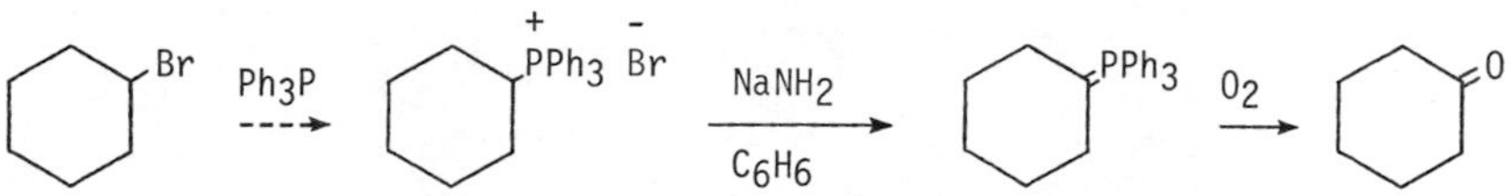

Ber (1963) 96 1899
(1961) 94 1987

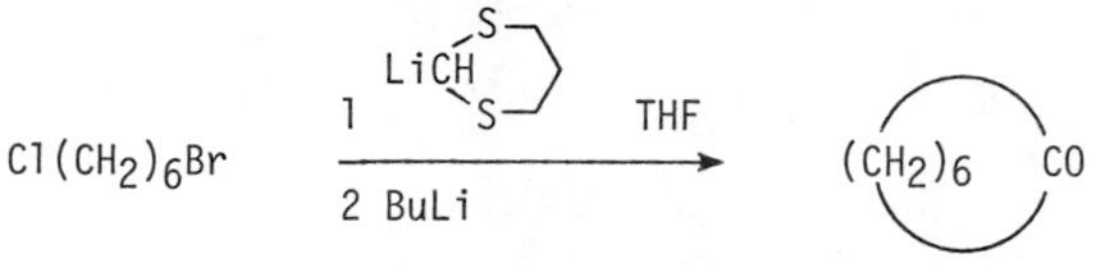

$Cl(CH_2)_6Br$ $\xrightarrow[\text{2 BuLi}]{\text{1 LiĊH}\begin{smallmatrix}S\\S\end{smallmatrix}\quad THF}$ $(CH_2)_6\quad CO$ 68%

JOC (1968) 33 300

Ph_2CBr_2 $\xrightarrow[\text{2 HCl H}_2\text{O}]{\text{1 Morpholine}}$ Ph_2CO 63%

Bull Soc Chim Fr (1965) 3544
Org Synth (1941) Coll Vol 1 95
JACS (1966) 88 3515

Section 176   Ketones from Hydrides (RH)

This section lists examples of the replacement of hydrogen by ketonic
groups e.g. RH → RCOMe (R=alkyl, aryl, vinyl etc.). For the oxidation
of methylenes R_2CH_2 → R_2CO, and the degradation of alkyls R_2CHR' →
R_2CO, see section 170 (Ketones from Alkyls and Methylenes)

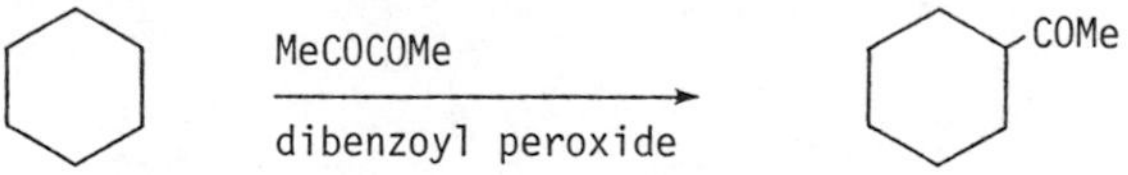

JACS (1968) 90 3588

$$Me_2CHCH_3 \xrightarrow{\text{MeCOCl \quad AlCl}_3} Me_2CHCH_2COMe$$

Ber (1936) 69B 2244

$$BuCH=CH_2 \xrightarrow{\text{MeCOOH \quad (CF}_3CO)_2O} BuCH=CHCOMe$$ 22%

JCS (1953) 3628

Review: The Friedel-Crafts Acylation Reaction and its Application to
Polycyclic Aromatic Hydrocarbons Chem Rev (1955) 55 229

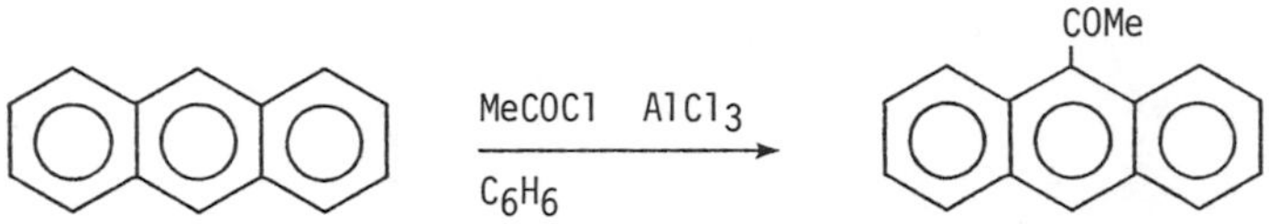

57-60%

Org Synth (1963) Coll Vol 4 8
                (1955) Coll Vol 3 23

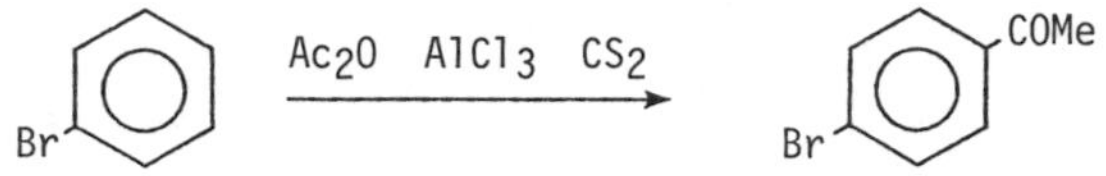

69-79%

Org Synth (1941) Coll Vol 1 109

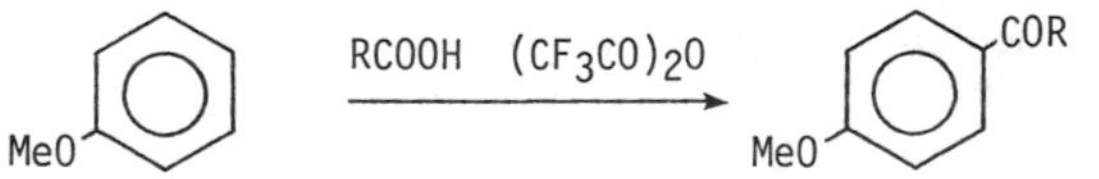

78% (R=Me)
                              56% (R=Ph)

JCS (1951) 718

$$PhH \xrightarrow{\overset{+}{RCO} \ \overset{-}{SbF}_6} PhCOR$$ 81-93% (R=Et)
                                                                         86-93% (R=Ph)

JACS (1962) 84 2733

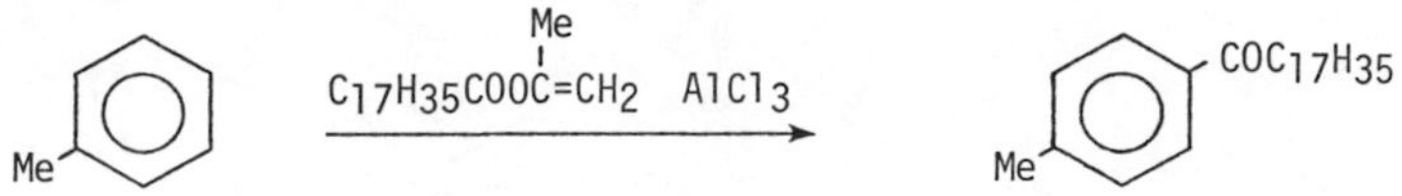

83%

JOC (1970) 35 2351

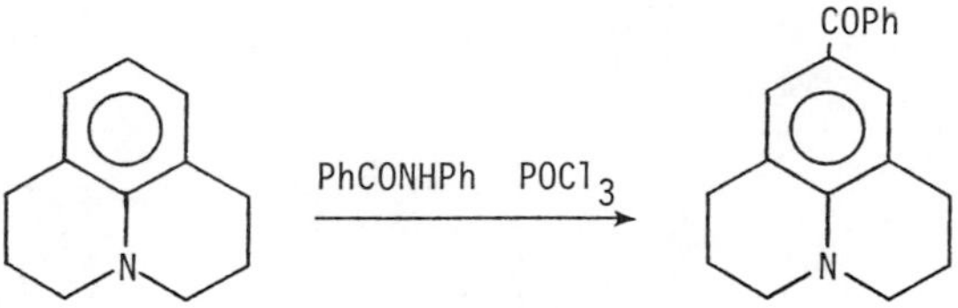

70%

JOC (1952) 17 1281
JCS (1932) 642

Section 177   Ketones from Ketones

All reactions of enones forming <u>saturated</u> ketones (e.g. RCH=CHCOR $\longrightarrow$
RCH$_2$-CHCOR) are listed in section 180 (Ketones from Miscellaneous Compounds)
    |
    R'

t-BuCOCH$_2$Me $\xrightarrow[\text{2 BuI}]{\text{1 NaH THF}}$ t-BuCOCHMe   57%
                                                Bu

Bull Soc Chim Fr (1968) 4990
Tetrahedron (1968) 24 6583

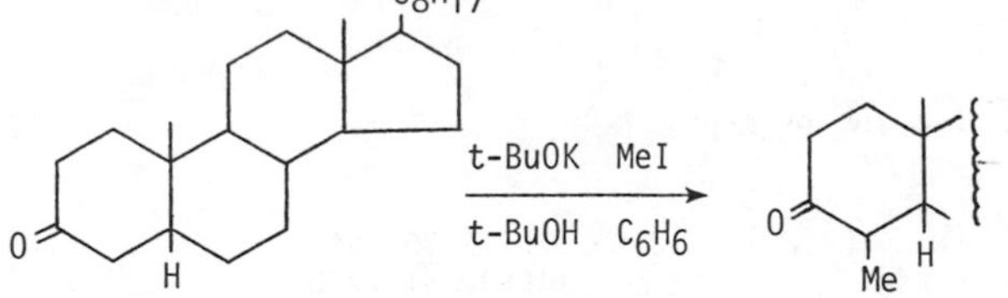

40%

JACS (1958) 80 5220

JCS (1956) 4490
    (1960) 67

44%

JACS (1960) 82 2847
     (1962) 84 3402

Et
|
Me$_2$CONa   PhCH$_2$Br
MeCOCH$_3$ ——————————————→ MeCOC(CH$_2$Ph)$_3$
           Et$_2$O   C$_6$H$_6$ Bull Soc Chim Fr (1961) 836
                                                       (1956) 1392

45%

JACS (1969) 91 1264
     (1953) 75 369
Org Synth (1955) Coll Vol 3 44

            NaNH$_2$ PhBr   NH$_3$
MeCOCH$_2$Et ——————————————→ MeCOCHEt   65%
                                    |
                                    Ph
                             JACS (1959) 81 1169

            1 Ph$_3$CNa   Et$_2$O
PhCOCH$_2$Et ——————————————→ PhCOCHEt   62%
            2 EtBr |
                                   Et
                             JACS (1957) 79 881
                                  (1959) 81 1745

i-PrCOCHMe₂ $\xrightarrow[\text{2 Me}_2\text{SO}_4 \text{ or MeI}]{\text{1 BuMgBr   HMPA}}$ i-PrCOCMe₂
                                                                          |
                                                                          Me

Compt Rend (1966) C 263 488

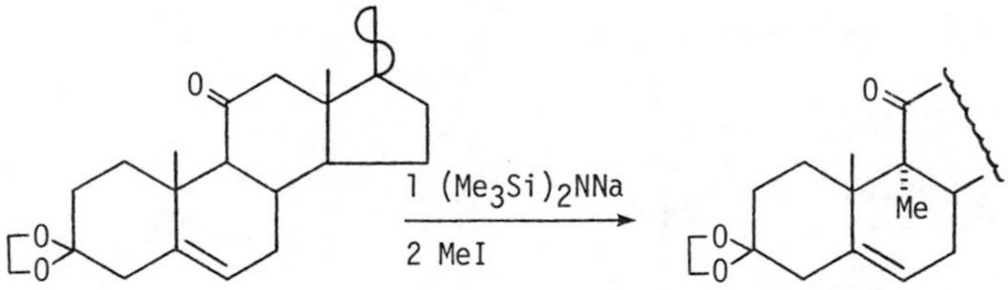

Chem Comm (1969) 1498

PhCOCH₃ $\xrightarrow[\text{di-t-butyl peroxide}]{\text{C}_6\text{H}_{13}\text{CH=CH}_2}$ PhCOCH₂C₈H₁₇   10%

JCS (1965) 1918

PhCOCH₃ $\xrightarrow[\text{xylene}]{\text{PhCH}_2\text{OH   PhCH}_2\text{OLi}}$ PhCOCH₂CH₂Ph   70%

JACS (1956) 78 4950

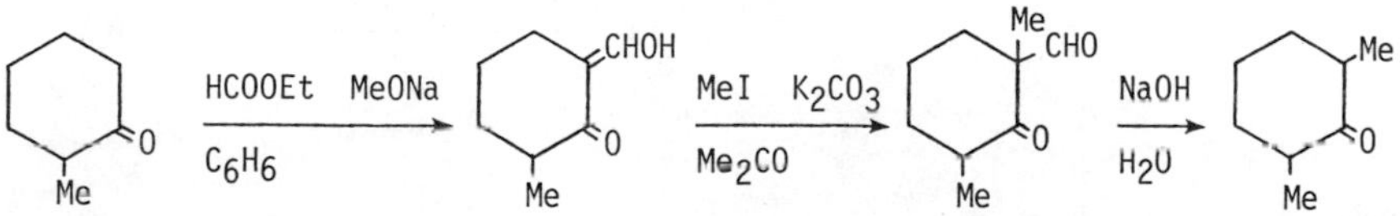

JCS (1954) 1373
JACS (1947) 69 1361
     (1957) 79 6313

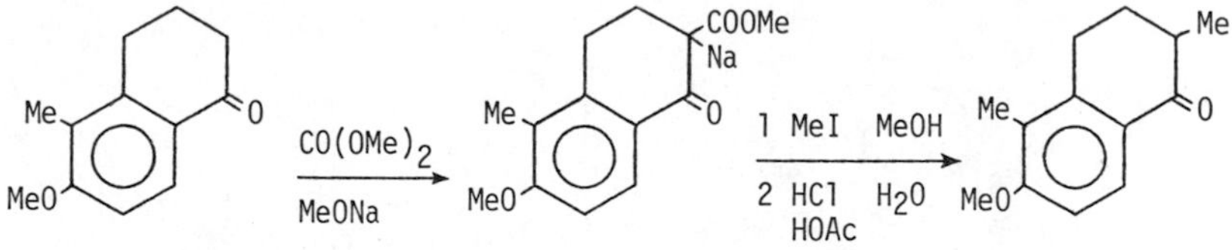

32%

Tetrahedron (1957) 1 49

83%

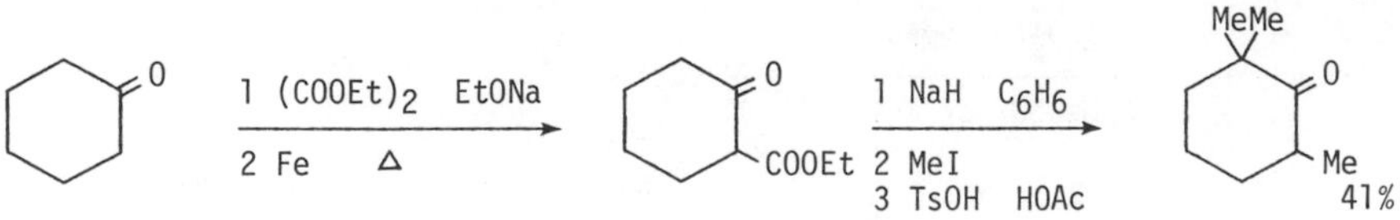

JACS (1958) 80 4072
Org Synth (1943) Coll Vol 2 531

For the preparation of β-ketoesters from ketones via enamines see
JACS (1963) 85 207

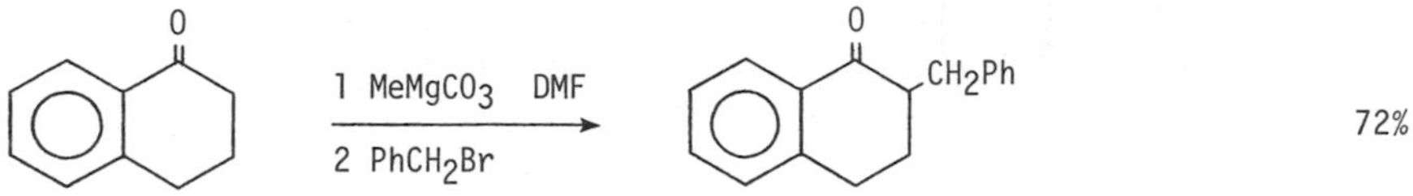

JACS (1959) 81 2598

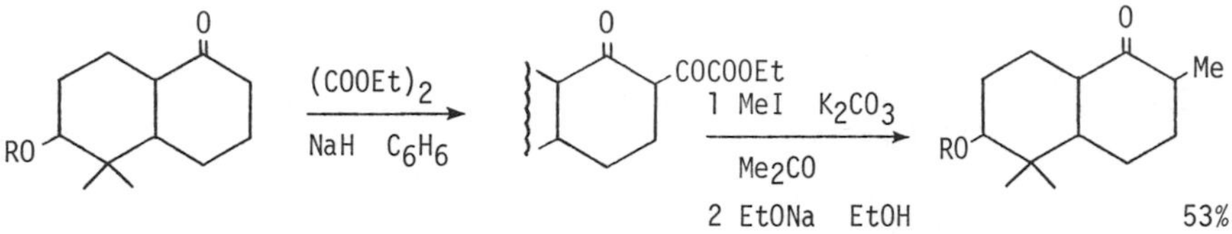

JACS (1958) 80 1967 5220

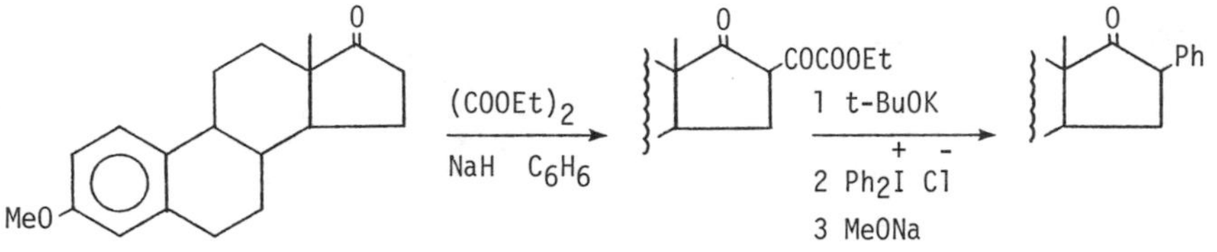

J Med Chem (1967) 10 106

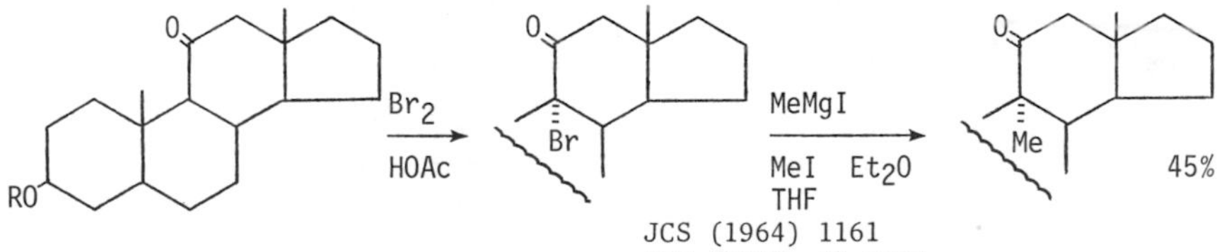

JCS (1964) 1161
JOC (1961) 26 2426
Tetr Lett (1969) 2269

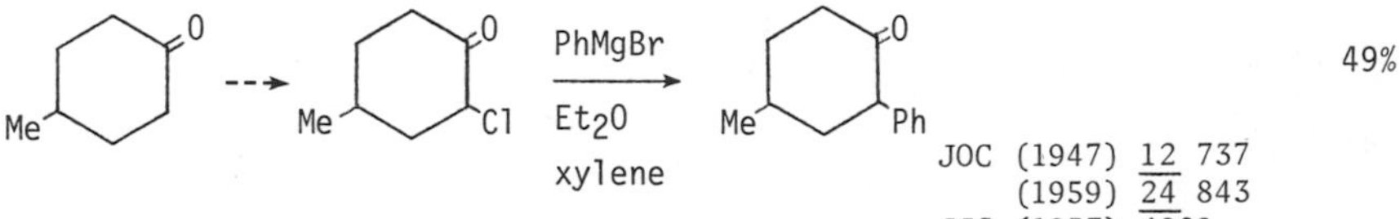

JOC (1947) 12 737
      (1959) 24 843
JCS (1957) 4089

i-PrCOCHMe$_2$ $\dashrightarrow$ i-PrCOCMe$_2$ $\xrightarrow{\text{i-Pr}_2\text{CuLi}}$ i-PrCOCMe$_2$
                             | |
                             Br   i-Pr

Tetr Lett (1971) 177

PhCOCH$_3$ $\dashrightarrow$ PhCOCH$_2$Br $\xrightarrow[\text{THF}]{\text{Et}_3\text{B}\quad\text{t-BuOK}}$ PhCOCH$_2$Et

JACS (1968) 90 6218
     (1969) 91 4304 6852

C$_6$H$_{13}$COCH$_2$C$_5$H$_{11}$ $\dashrightarrow$ C$_6$H$_{13}$COCHC$_5$H$_{11}$ $\xrightarrow[\text{2 MeI}]{\text{1 Zn}\quad\text{Me}_2\text{SO}\quad\text{C}_6\text{H}_6}$ C$_6$H$_{13}$COCHC$_5$H$_{11}$
                                                            | |
                                                            Br   Me

JACS (1967) 89 5727

$\xrightarrow[\text{CCl}_4]{\text{Ac}_2\text{O}}$

$\xrightarrow[\begin{array}{c}\text{MeOCH}_2\text{CH}_2\text{OMe}\\\text{2 MeI}\end{array}]{\text{1 MeLi}\quad\text{Et}_2\text{O}}$

JOC (1965) 30 2502
Tetr Lett (1971) 105

$\xrightarrow[\text{HClO}_4]{(\text{PhCO})_2\text{O}}$

$\xrightarrow[\text{2 CH}_2\text{I}_2\quad\text{Zn-Cu}]{\text{1 MeLi}}$

65-75%

JOC (1969) 34 1962

$\xrightarrow[\text{C}_6\text{H}_6]{\text{Pyrrolidine}}$

$\xrightarrow[\text{2 H}_2\text{SO}_4\quad\text{H}_2\text{O}]{\text{1 BuI}\quad\text{toluene}}$

44%

JACS (1963) 85 207
Advances in Org Chem (1963) 4 1

For preparation of enamines from unreactive ketones
                        see JCS (1965) 5142
                        and JOC (1967) 32 213

$C_5H_{11}COCH_3$ $\xrightarrow{\text{Cyclohexylamine}}$ $C_5H_{11}CCH_3$ $\xrightarrow[\begin{array}{c}2\ BuI\\3\ Acid\end{array}]{1\ EtMgBr}$ $C_5H_{11}COCH_2Bu$

(with $\underset{N-\text{cyclohexyl}}{\overset{\|}{}}$)

JACS (1963) 85 2178

(cyclohexanone) $\xrightarrow[\text{2 MeI \ K}_2\text{CO}_3]{\text{1 HCHO \ Me}_2\text{NH·HCl}}$ $\underset{CH_2NMe_3}{}$ $^+$ I$^-$ $\xrightarrow[\substack{\text{pentyl}\\\text{borane}}]{\text{Tricyclo-}}$ (product CH_2–cyclopentyl)   85%

JACS (1968) 90 4166

$PhCH_2COMe$ $\xrightarrow[\text{Me}_2\text{NH·HCl}]{\text{HCHO}}$ $\underset{CH_2NMe_2·HCl}{PhCHCOMe}$ $\xrightarrow[\text{Ni \ EtOH}]{\text{H}_2\ (60\text{-}100\ \text{atmos})}$ $\underset{Me}{PhCHCOMe}$ 56%

JACS (1953) 75 1128

(tetralone) $\xrightarrow[\text{EtOH \ H}_2\text{O}]{\text{PhCHO \ NaOH}}$ (=CHPh) $\xrightarrow[\text{EtOH}]{\text{H}_2\ \text{Pd-C}}$ (CH$_2$Ph)   91%

Org Prep and Procedures (1970) 2 37

(decalone) $\xrightarrow[\text{C}_6\text{H}_6]{\text{HCOOEt \ MeONa}}$ (=CHOH) $\xrightarrow[\text{HCl \ MeOH}]{\text{H}_2\ \text{Pd-C}}$ (Me)   70%

JOC (1965) 30 2502

(tetralone) $\xrightarrow[\substack{\text{C}_6\text{H}_6\\ \text{2 PhCOCl \ Pyr}}]{\text{1 HCOOEt \ MeONa}}$ (=CHOCOPh) $\xrightarrow[\substack{\text{Et}_3\text{N}\\ \text{i-PrOH}}]{\text{H}_2\ \text{PtO}_2}$ (Me)   60-78%

Synthesis (1970) 476
Org React (1954) 8 119

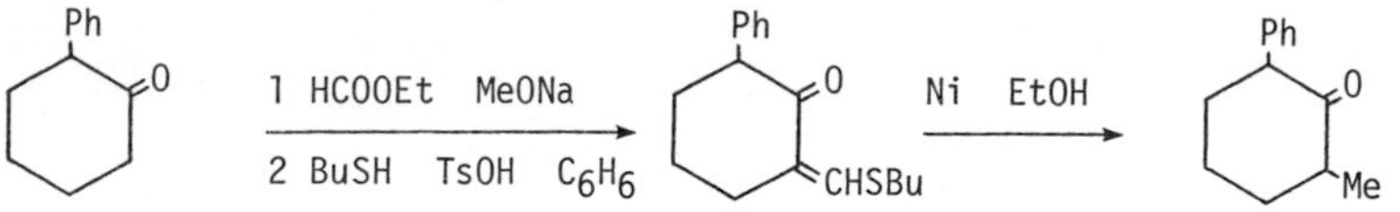
Ph
O
1 HCOOEt MeONa
2 BuSH TsOH C_6H_6
CHSBu
Ni EtOH
Me

79%

Chem Ind (1960) 1534

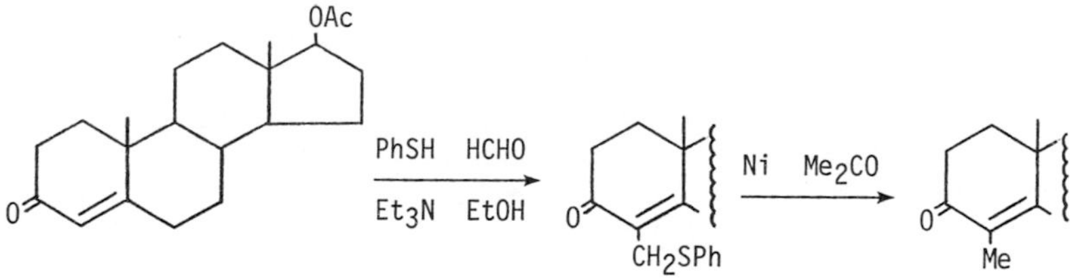
OAc
O
PhSH HCHO
Et_3N EtOH
CH_2SPh
Ni Me_2CO
Me

~80%

JCS (1962) 1091

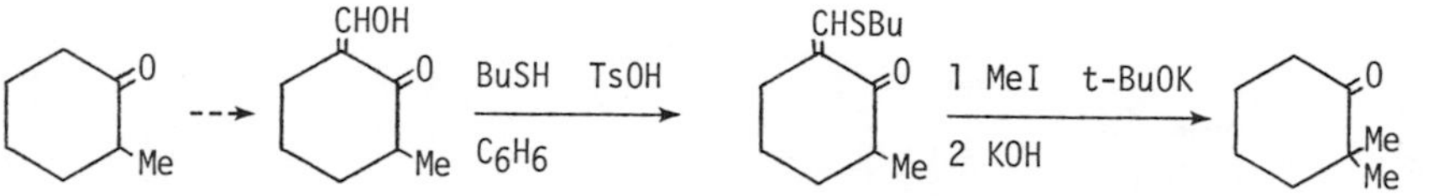
O
Me
CHOH
BuSH TsOH
C_6H_6
CHSBu
1 MeI t-BuOK
2 KOH
Me
Me

JOC (1962) 27 1615 1620
Tetrahedron (1968) 24 3095

74%

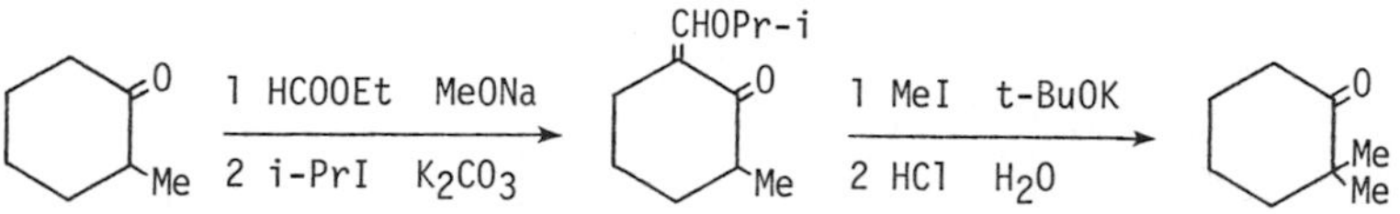
O
Me
1 HCOOEt MeONa
2 i-PrI K_2CO_3
CHOPr-i
1 MeI t-BuOK
2 HCl H_2O
Me
Me

32%

JACS (1947) 69 1361

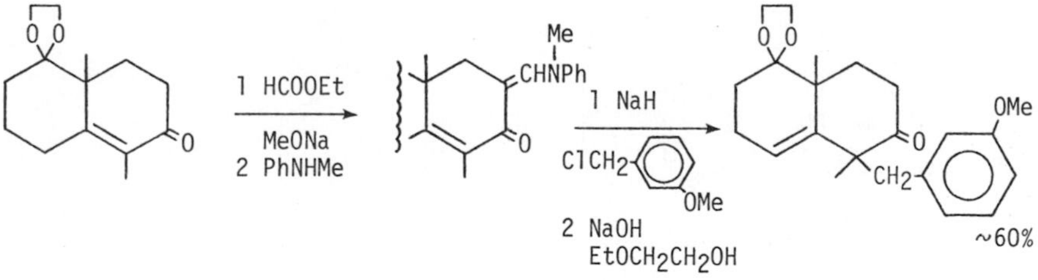
O O
O
1 HCOOEt
MeONa
2 PhNHMe
Me
CHNPh
1 NaH
$ClCH_2$
OMe
2 NaOH
$EtOCH_2CH_2OH$
CH_2
OMe

~60%

Tetrahedron (1969) 25 4011
JOC (1970) 35 468

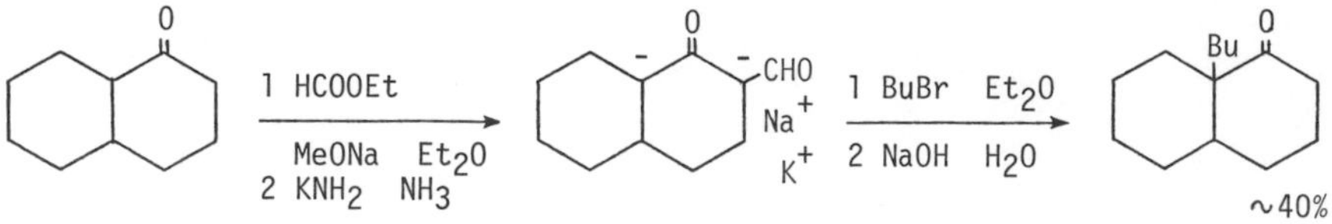

JACS (1965) 87 82

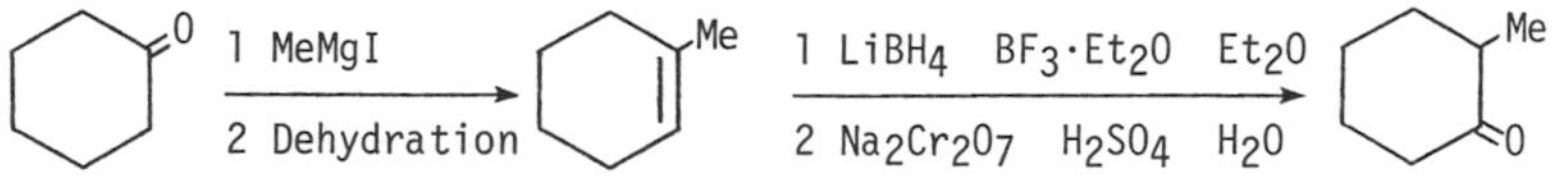

JACS (1961) 83 2951

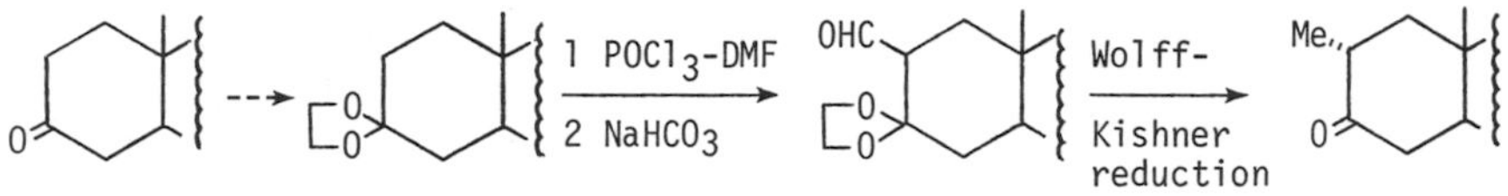

Tetr Lett (1964) 2161

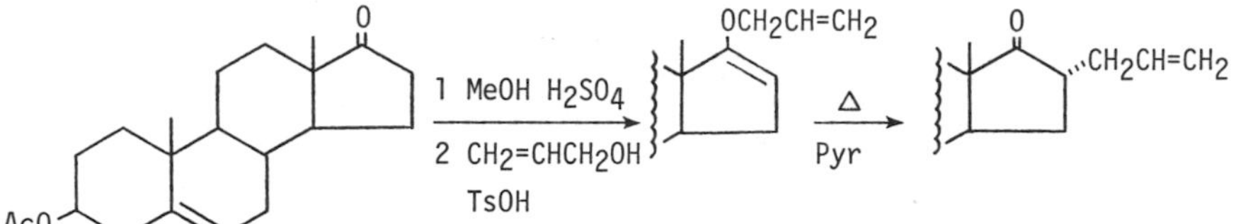

Gazz (1965) 95 351
(Chem Abs 63 11647)

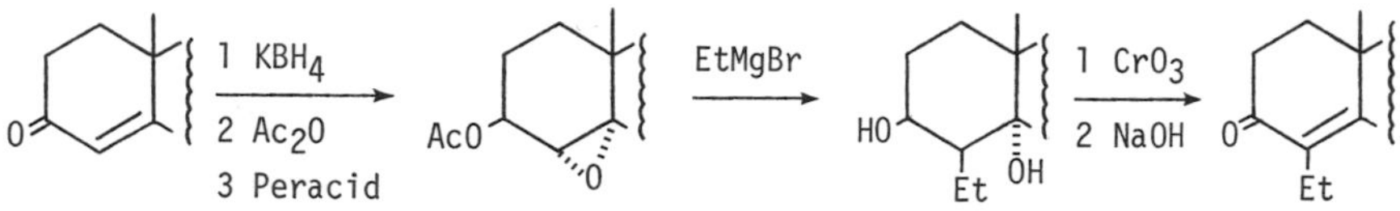

Bull Soc Chim Fr (1964) 321

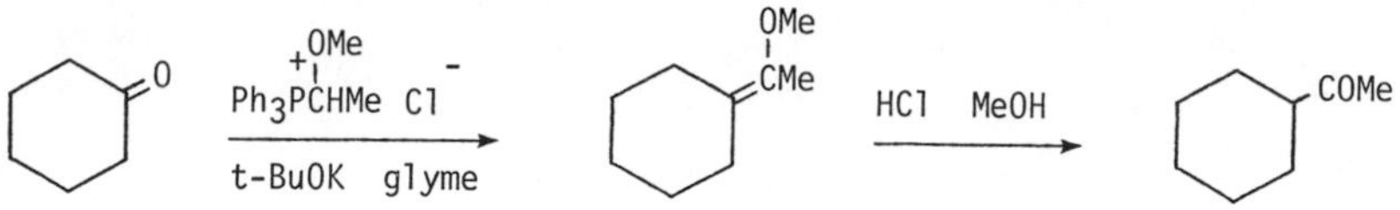

Tetr Lett (1964) 3323

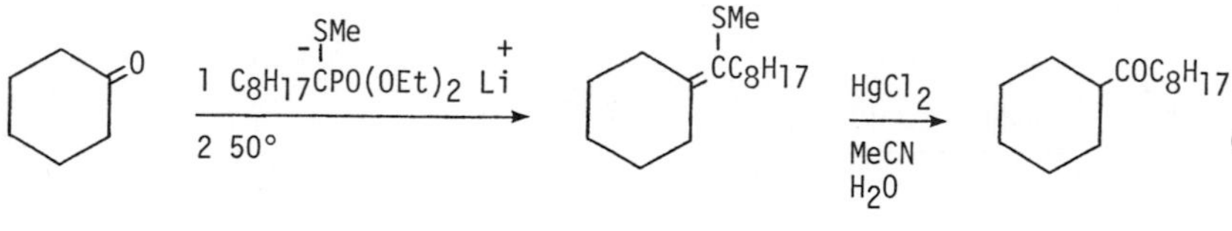

JOC (1970) 35 777

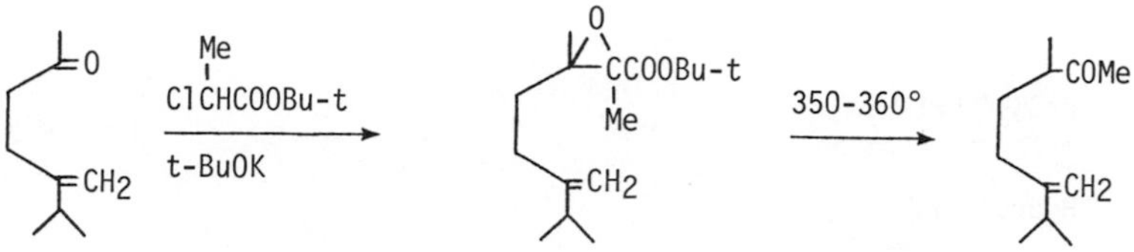

JACS (1963) 85 955
Org React (1949) 5 413
Chem Rev (1955) 55 283

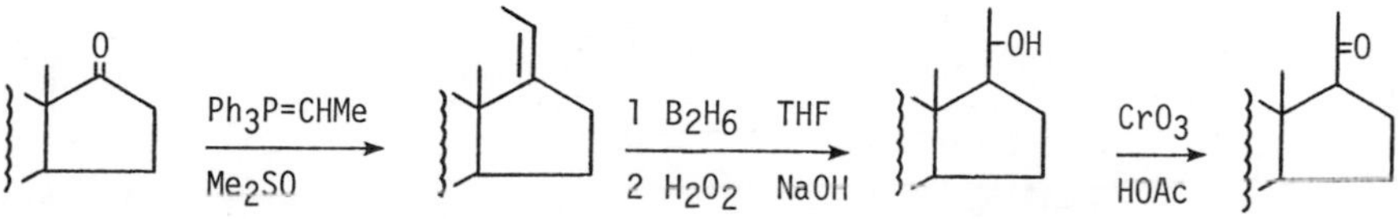

JOC (1966) 31 24

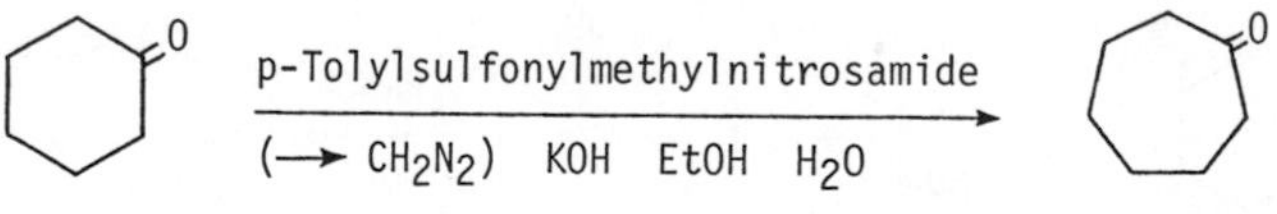

p-Tolylsulfonylmethylnitrosamide
(⟶ CH₂N₂) KOH EtOH H₂O

33-36%

Org Synth (1963) Coll Vol 4 225
Org React (1954) 8 364
JACS (1966) 88 3515

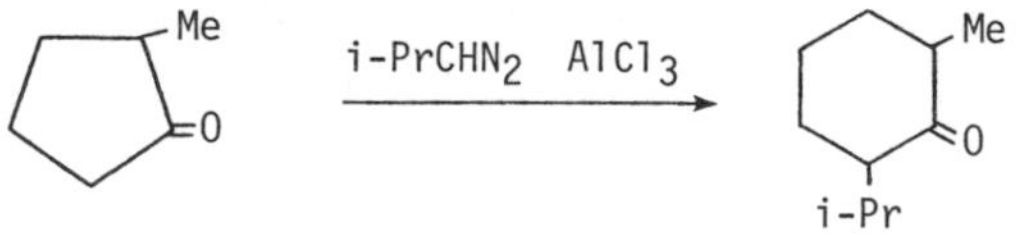
Me
i-PrCHN2 AlCl3
Me
i-Pr

Tetr Lett (1962) 775

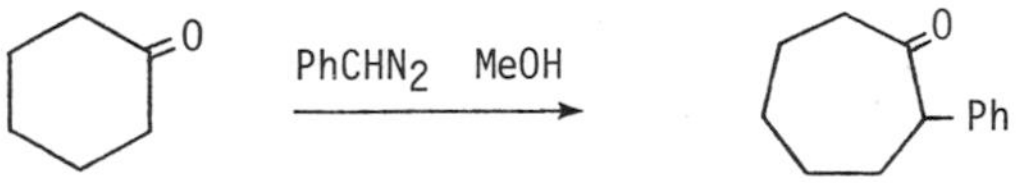
PhCHN2 MeOH
Ph

76%

Org React (1954) 8 364

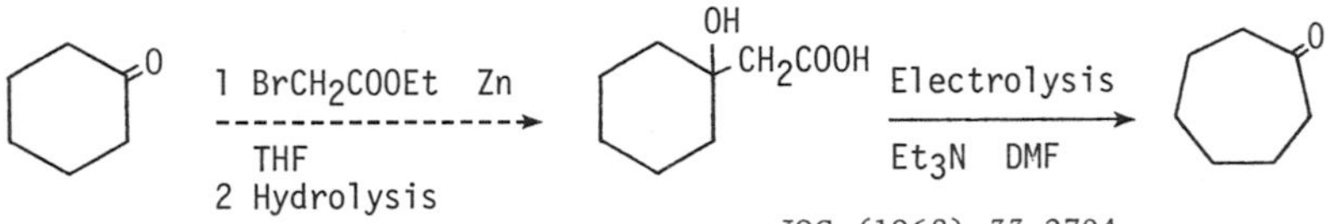
1 BrCH2COOEt Zn
THF
2 Hydrolysis
OH
CH2COOH
Electrolysis
Et3N DMF

<55%

JOC (1968) 33 2704

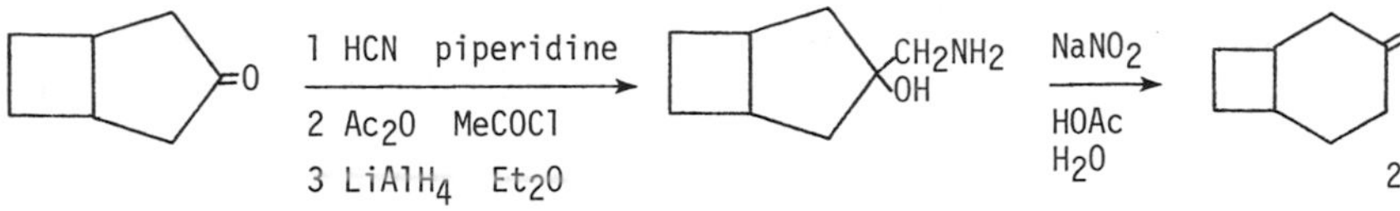
1 HCN piperidine
2 Ac2O MeCOCl
3 LiAlH4 Et2O
CH2NH2
OH
NaNO2
HOAc
H2O

22%

JOC (1964) 29 2914
JACS (1952) 74 2278
JCS C (1970) 1454

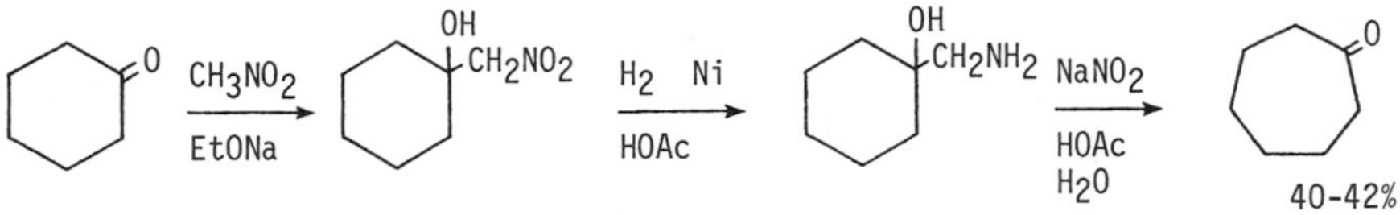
CH3NO2
EtONa
OH
CH2NO2
H2 Ni
HOAc
OH
CH2NH2
NaNO2
HOAc
H2O

40-42%

Org Synth (1963) Coll Vol 4 221

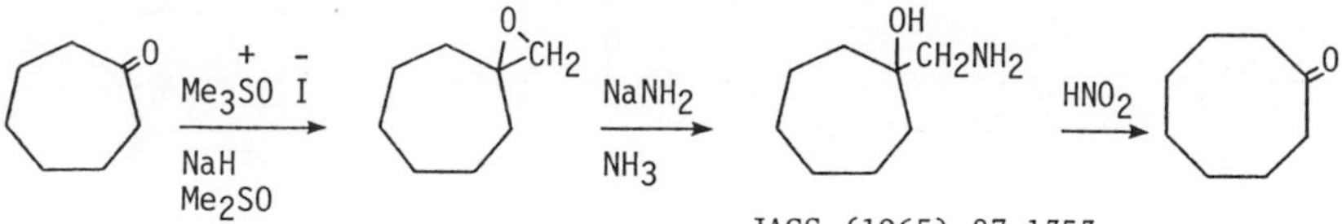

JACS (1965) 87 1353

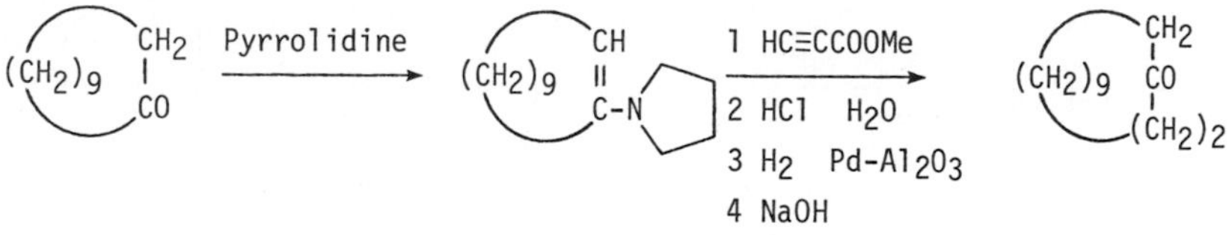

30%

JOC (1967) 32 926

$$EtCOCH_2Me \xrightarrow{\quad HClO_4 \quad H_2O \quad} EtCH_2COMe \qquad \sim 63\%$$

JOC (1960) 25 1252
      (1969) 34 806

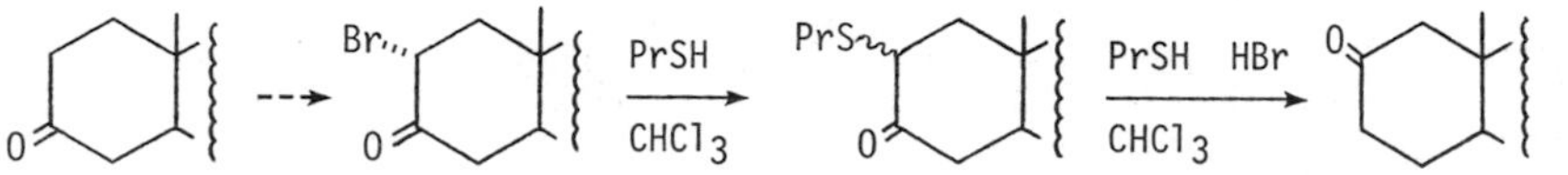

JOC (1963) 28 2626

11%

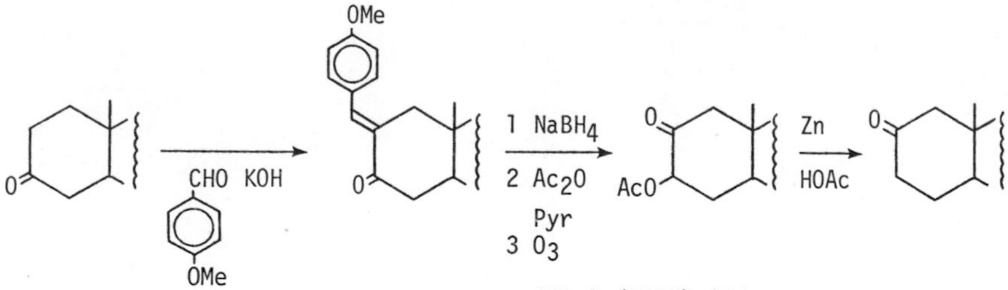

JCS C (1970) 244

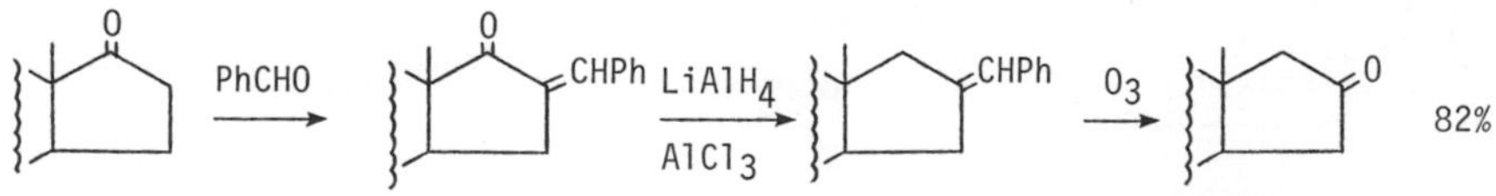

Chem Comm (1967) 898
Compt Rend (1967) C 265 929

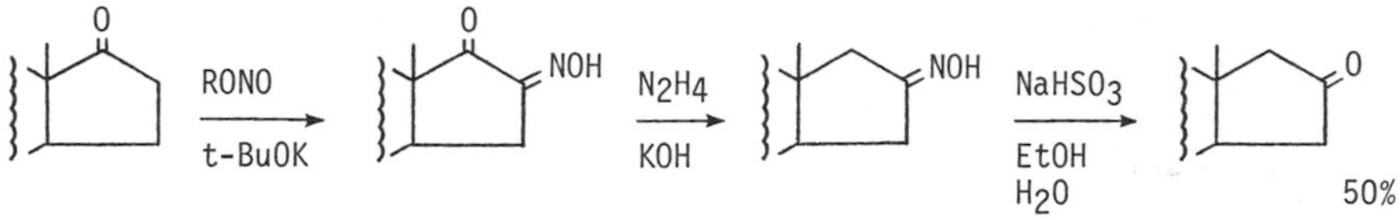

82%

Chem Comm (1968) 1350

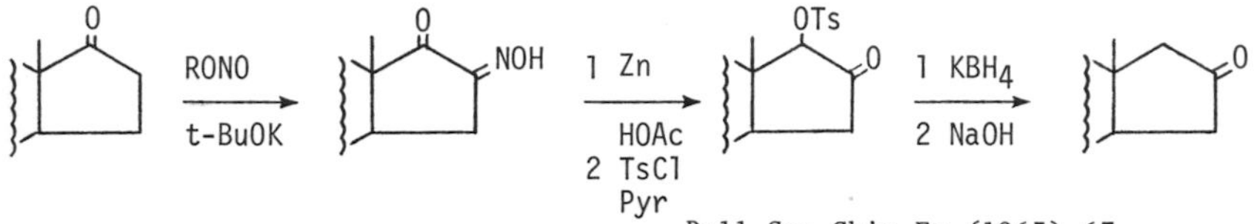

50%

Bull Soc Chim Fr (1965) 67

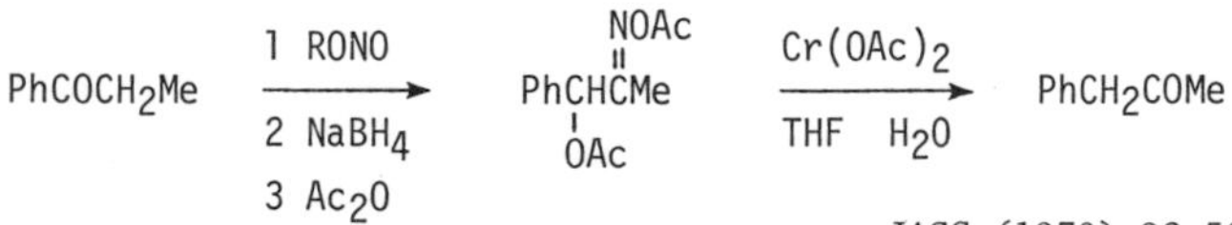

$$PhCOCH_2Me \xrightarrow[\substack{2\ NaBH_4 \\ 3\ Ac_2O}]{1\ RONO} \underset{OAc}{\overset{NOAc}{PhCHCMe}} \xrightarrow[\substack{THF\ H_2O}]{Cr(OAc)_2} PhCH_2COMe$$

JACS (1970) 92 5276

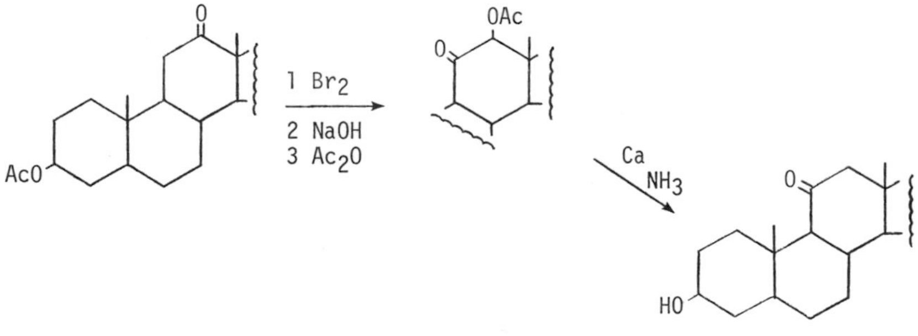

JCS (1956) 4330 4344

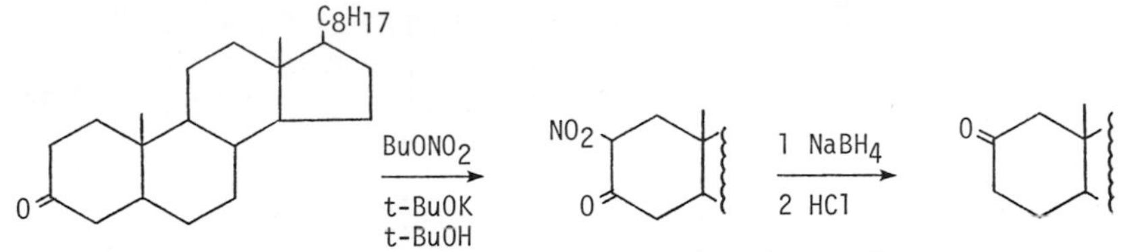

JOC (1968) 33 1733

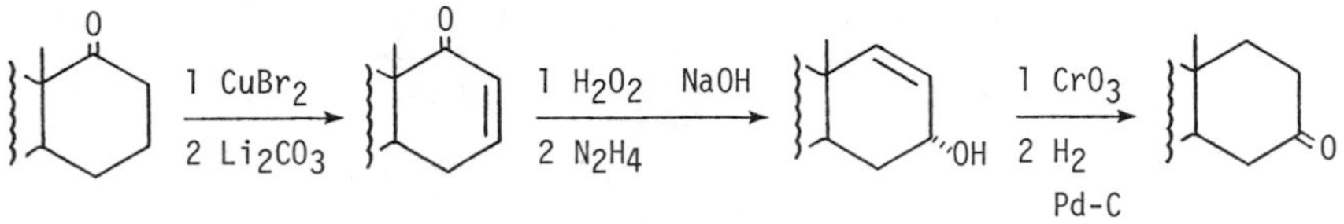

JCS C (1970) 1454

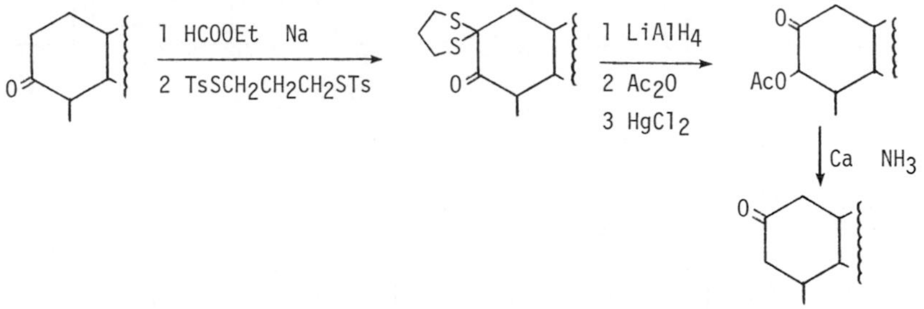

JOC (1969) 34 4188

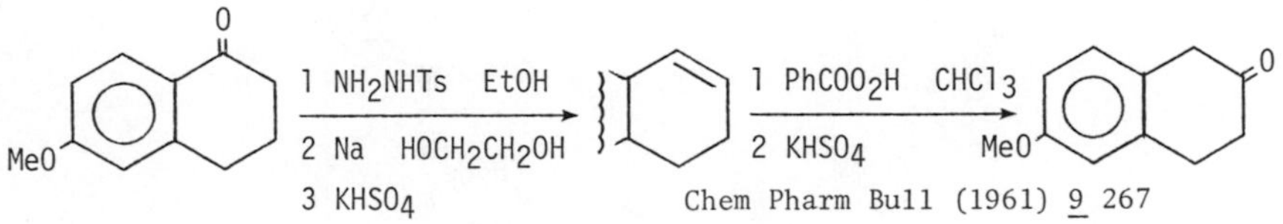

Chem Pharm Bull (1961) 9 267

PrCOCH$_2$Pr $\xrightarrow{\text{h}\nu \quad \text{t-BuOH}}$ PrCOMe

Tetr Lett (1968) 5385
Chem Comm (1969) 204

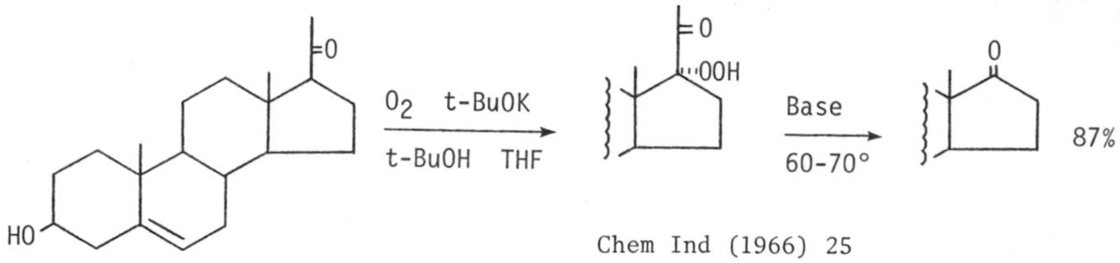

Chem Ind (1966) 25

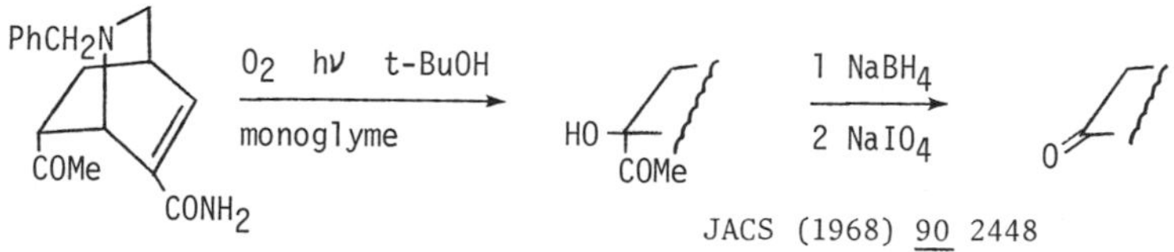

JACS (1968) <u>90</u> 2448

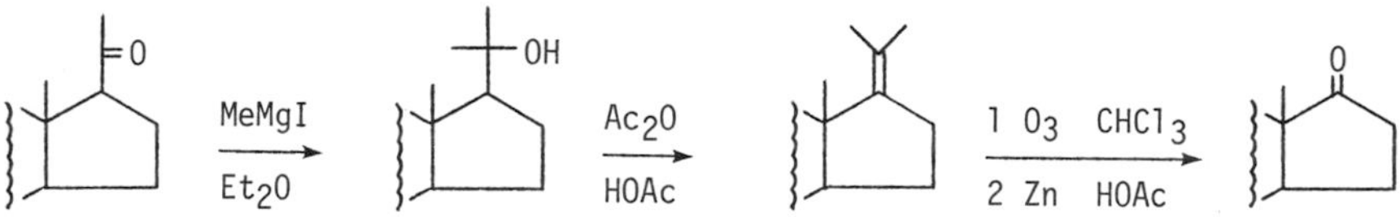

JACS (1942) <u>64</u> 1276

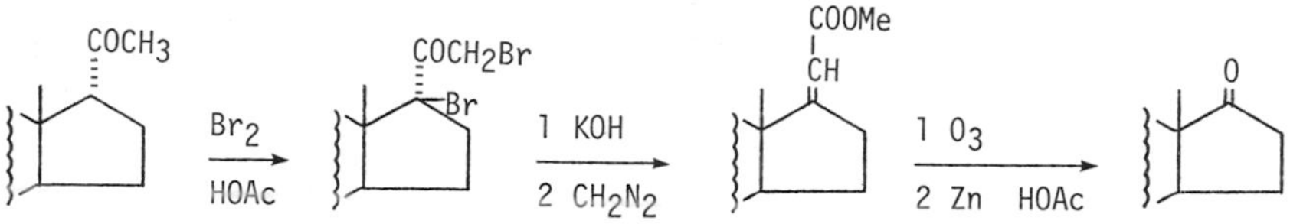

Helv (1944) <u>27</u> 549

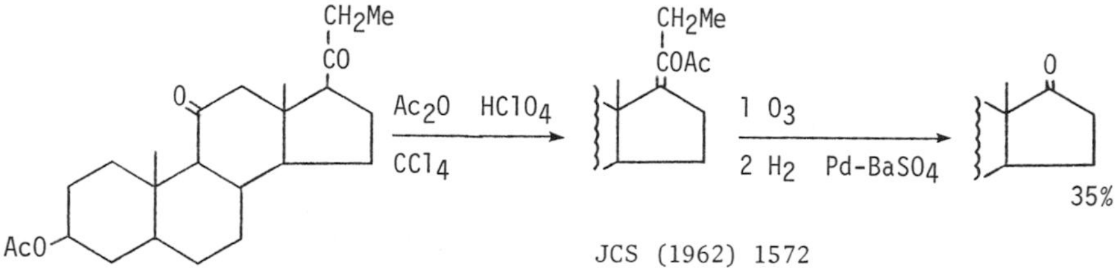

JCS (1962) 1572

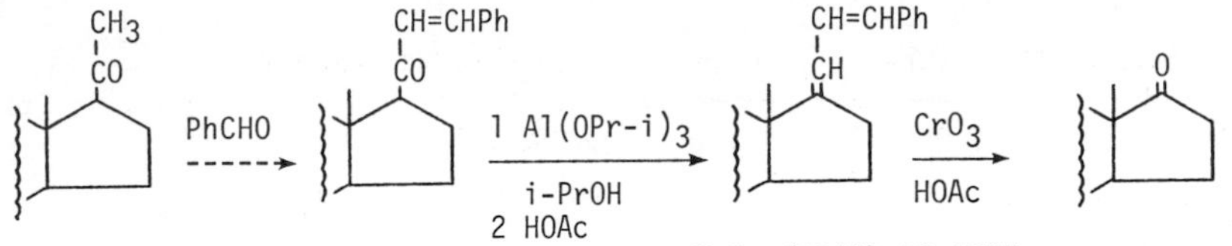

Helv (1949) 32 1795

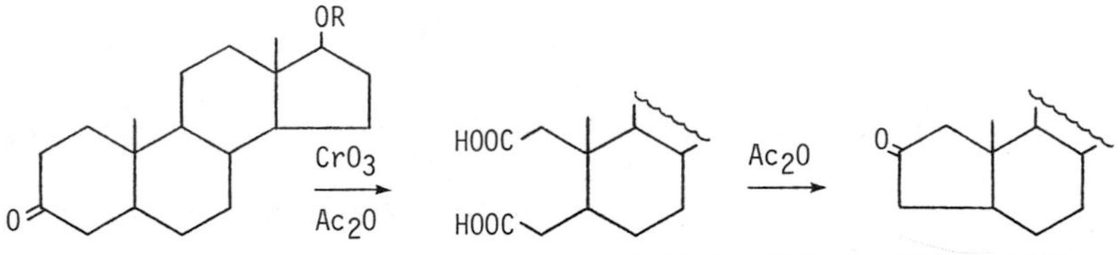

Bull Soc Chim Fr (1958) 1573

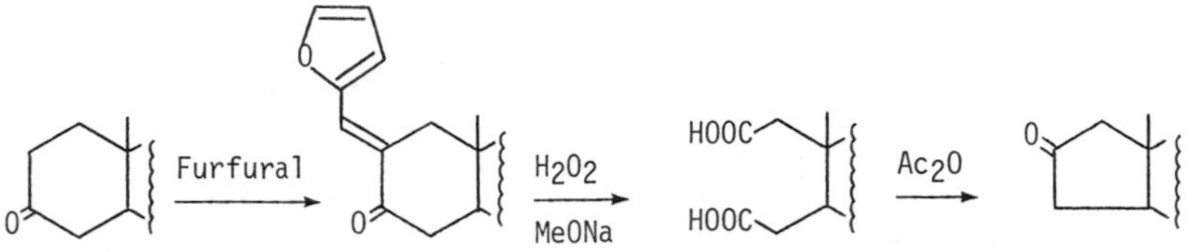

Bull Soc Chim Fr (1958) 1573

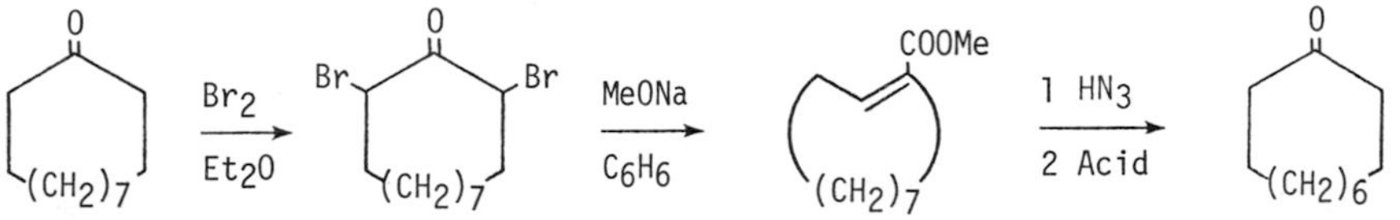

JOC (1968) 33 2157

85-90%

Section 178   Ketones from Nitriles
             ○○○○○○○○○○○○○○○○○○○○○○○

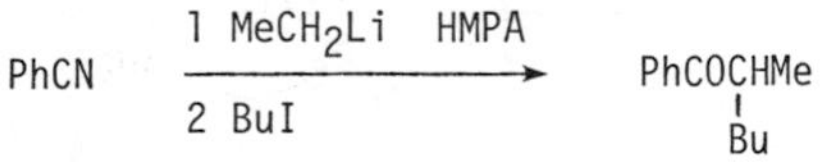

PhCN → PhCOCHMe \
                    Bu   70%

1 MeCH₂Li HMPA
2 BuI

Compt Rend (1967) C 265 245
Bull Soc Chim Fr (1968) 4990

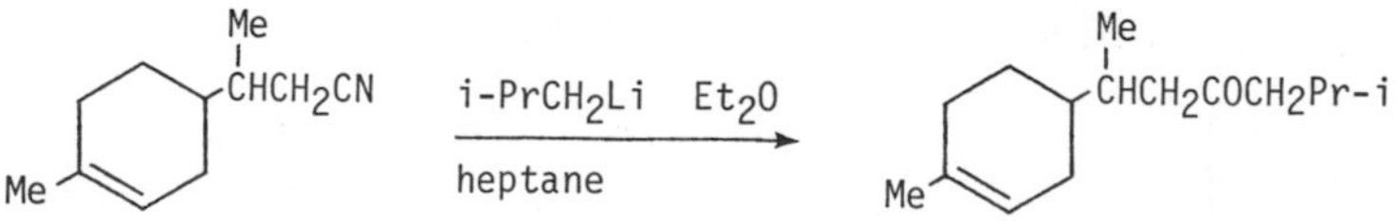
Me
CHCH2CN
Me
i-PrCH2Li Et2O
heptane
Me
CHCH2COCH2Pr-i
Me

76%

JACS (1970) 92 336

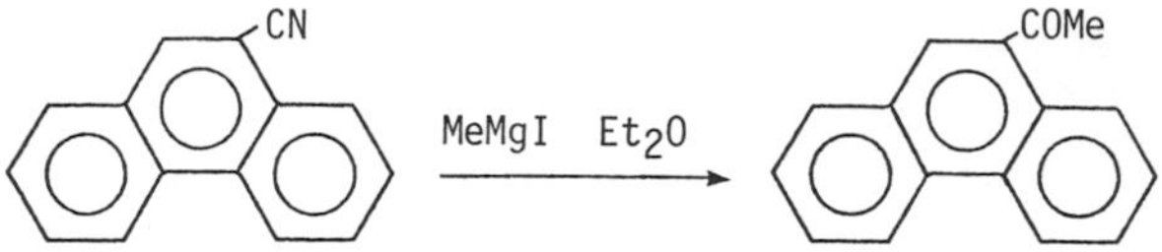
CN
MeMgI Et2O
COMe

52-59%

Org Synth (1955) Coll Vol 3 26

$BuCN \xrightarrow{Bu_3P=CHPh \quad Et_2O} BuCOCH_2Ph$ 73%

JACS (1967) 89 7009

$C_{11}H_{23}CN \xrightarrow{Et_3Al \quad C_6H_6} C_{11}H_{23}COEt$ 89%

Ber (1964) 97 2661

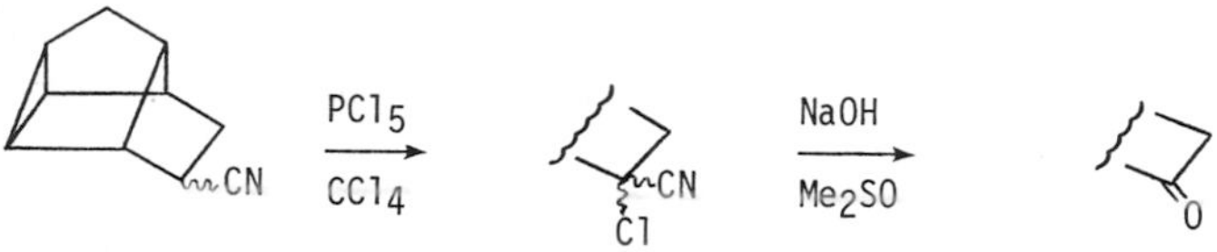
CN
PCl5
CCl4
CN
Cl
NaOH
Me2SO
O

47%

Tetr Lett (1967) 437

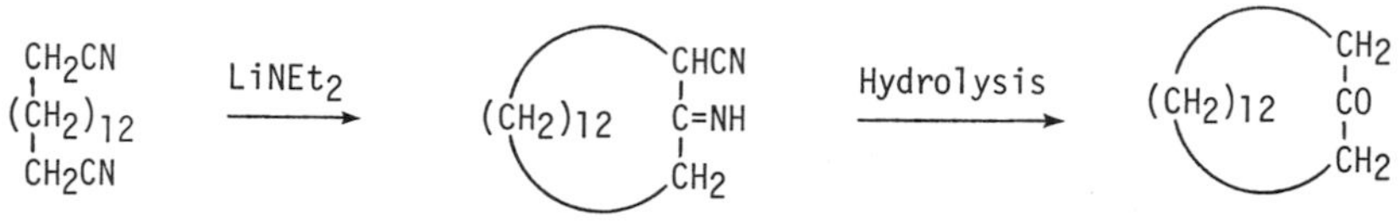
CH2CN
(CH2)12
CH2CN
LiNEt2
(CH2)12
CHCN
C=NH
CH2
Hydrolysis
(CH2)12
CH2
CO
CH2

Annalen (1933) 504 94
(1934) 513 43

Section 179   Ketones from Olefins

$$EtCH=CH_2 \xrightarrow[\substack{2\ BrCH_2COMe\ \ potassium\ 2,6- \\ di\text{-}t\text{-}butylphenoxide}]{1\ B_2H_6\quad THF} Et(CH_2)_3COMe \qquad 84\%$$

JACS (1969) <u>91</u> 6852 2147

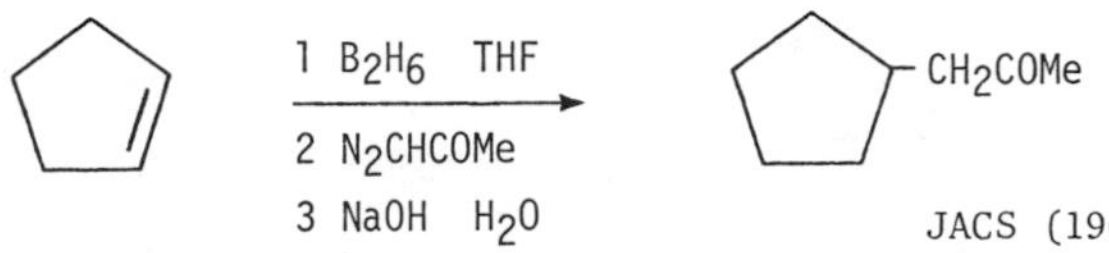

1 B₂H₆   THF
2 N₂CHCOMe
3 NaOH H₂O

→ CH₂COMe   67%

JACS (1968) <u>90</u> 5936

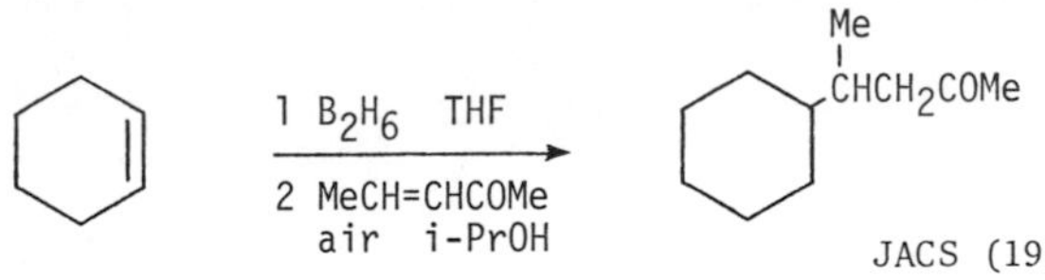

1 B₂H₆   THF
2 MeCH=CHCOMe
  air   i-PrOH

Me
|
CHCH₂COMe   98%

JACS (1970) <u>92</u> 714

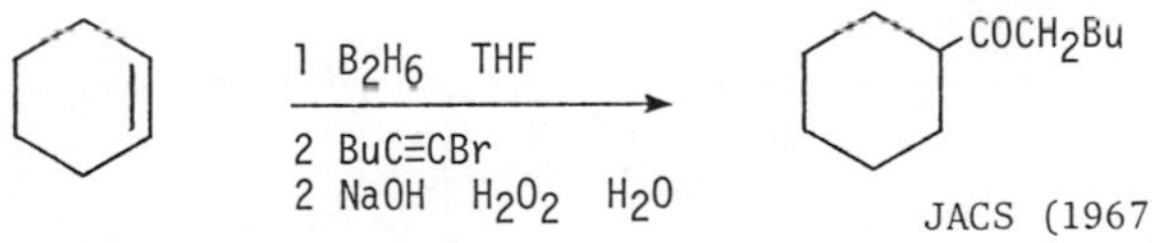

1 B₂H₆   THF
2 BuC≡CBr
2 NaOH H₂O₂ H₂O

COCH₂Bu   79%

JACS (1967) <u>89</u> 5086

$$C_6H_{13}CH=CH_2 \xdashrightarrow{B_2H_6} (C_6H_{13}CH_2CH_2)_3B \xrightarrow[\substack{2\ (CF_3CO)_2O \\ 3\ NaOH\ \ H_2O_2\ \ H_2O}]{1\ NaCN\quad diglyme} (C_6H_{13}CH_2CH_2)_2CO$$

~95%

Chem Comm (1970) 1529

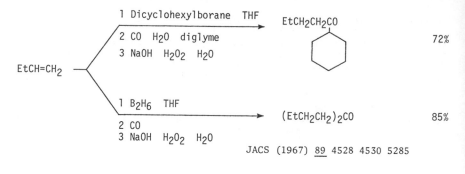

EtCH=CH$_2$

1 Dicyclohexylborane THF

2 CO H$_2$O diglyme
3 NaOH H$_2$O$_2$ H$_2$O

EtCH$_2$CH$_2$CO — (cyclohexyl)   72%

1 B$_2$H$_6$ THF

2 CO
3 NaOH H$_2$O$_2$ H$_2$O

(EtCH$_2$CH$_2$)$_2$CO   85%

JACS (1967) <u>89</u> 4528 4530 5285

$$\text{cyclopentene} \xrightarrow[\text{cyclohexane}]{\text{MeCOCl AlCl}_3} \text{cyclopentyl–COMe}$$

49%

Ber (1936) <u>69B</u> 1820

$$C_6H_{13}CH=CH_2 \xrightarrow[\text{di-t-butyl peroxide}]{\text{MeCH}_2\text{COMe}} C_6H_{13}CH_2CH_2\underset{\underset{Me}{|}}{CH}COMe$$

~30%

JCS (1965) 1918

$$\text{(bicyclic)=CH}_2 \xrightarrow[\text{dibenzoyl peroxide}]{\text{MeCHO}} \text{(bicyclic)CH}_2\text{COMe}$$

30%

Aust J Chem (1967) <u>20</u> 2033
JOC (1964) <u>29</u> 245
(1949) <u>14</u> 248

$$\text{(cyclohexenyl)Me} \xrightarrow[\text{2 Na}_2\text{Cr}_2\text{O}_7 \text{ H}_2\text{SO}_4 \text{ H}_2\text{O}]{\text{1 LiBH}_4 \text{ BF}_3 \cdot \text{Et}_2\text{O}} \text{(cyclohexanone)Me}$$

78%

JACS (1961) <u>83</u> 2951
JOC (1962) <u>27</u> 2938
Org React (1963) <u>13</u> 1

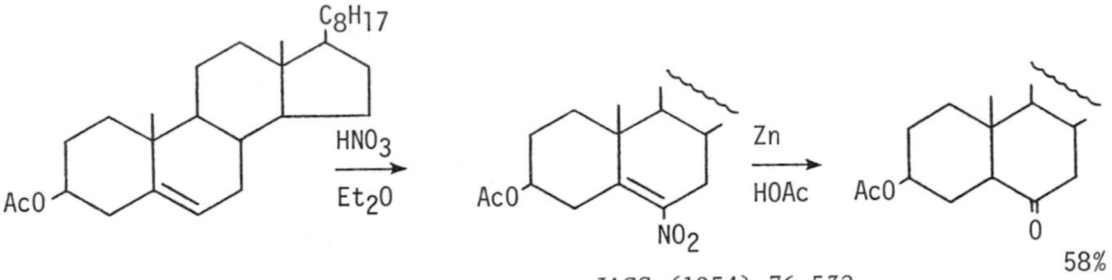

JACS (1954) 76 532

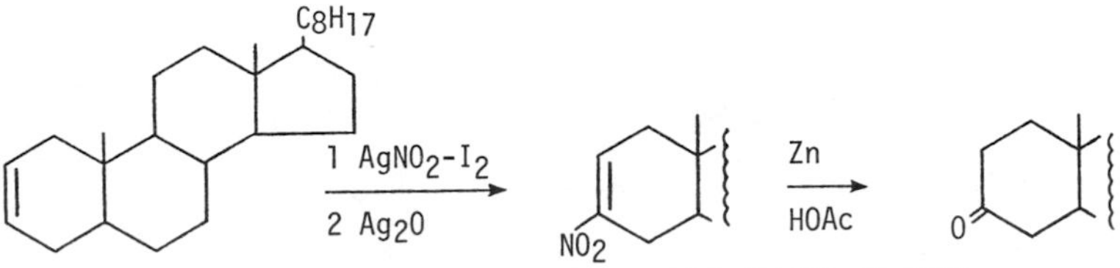

JOC (1969) 34 2628

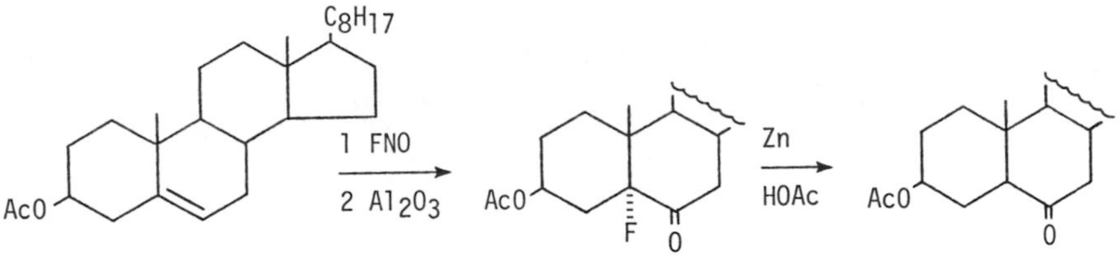

Chem Ind (1965) 1929

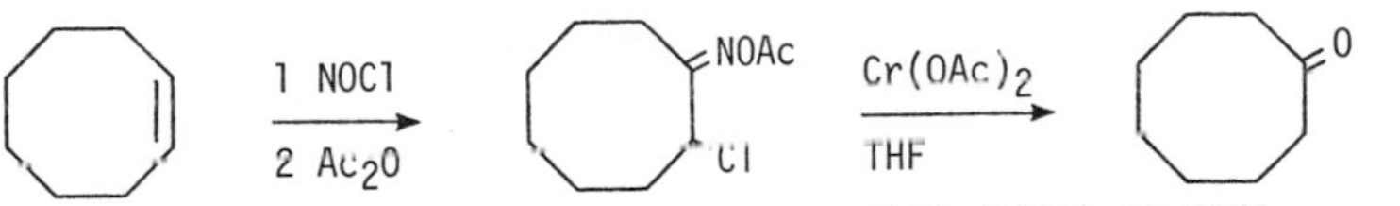

JACS (1970) 92 5276
Zh Org Khim (1966) 2 2178
(Chem Abs 66 75730)

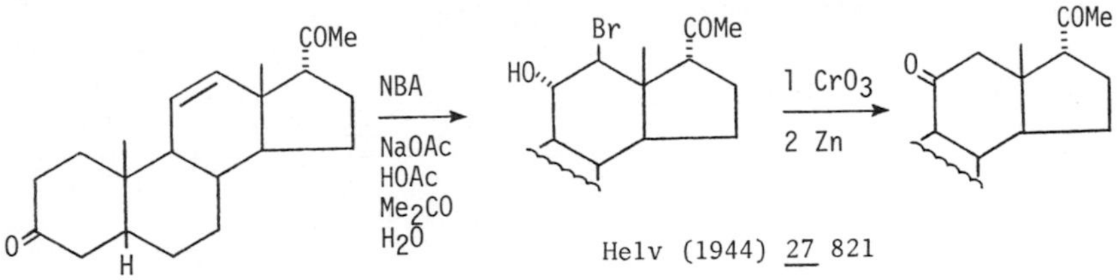

Helv (1944) 27 821

$$\xrightarrow[\text{CHCl}_3]{\text{CrO}_2\text{Cl}_2}$$

$$\xrightarrow[\text{2 Zn}]{\text{1 CrO}_3}$$

JACS (1956) 78 3749
JOC (1968) 33 3970

$$\xrightarrow[\text{MeCN}]{\text{N}_3\text{CN}}$$

=NCN

$$\xrightarrow[\text{Ag}^+ \ \text{H}_2\text{O}]{\text{Acid}}$$

91%

JACS (1964) 86 4506

PhC=CH$_2$
|
Me

$$\xrightarrow[\text{2 H}_2\text{SO}_4 \ \text{H}_2\text{O}]{\text{1 Tl(NO}_3)_3 \ \text{MeOH}}$$

PhCH$_2$COMe

81%

Tetr Lett (1970) 5275

C$_6$H$_{13}$CH=CH$_2$

$$\xrightarrow[\text{PrOH}]{\text{O}_2 \ \text{PdCl}_2 \ \text{CuCl}_2 \cdot 2\text{H}_2\text{O}}$$

C$_6$H$_{13}$COMe

68%

JOC (1969) 34 3949
Synthesis (1970) 225

$$\xrightarrow[\text{BF}_3 \cdot \text{Et}_2\text{O} \ \text{CH}_2\text{Cl}_2]{\text{(CF}_3\text{CO)}_2\text{O} \ \text{H}_2\text{O}_2}$$

41%

JOC (1967) 32 2669
(1962) 27 666

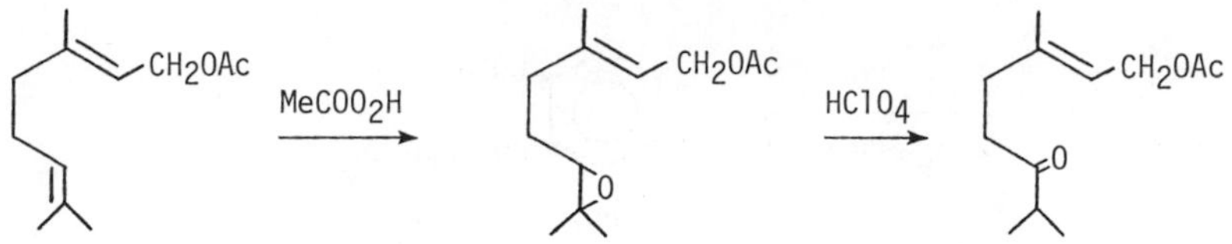
CH2OAc
MeCOO2H
CH2OAc
O
HClO4
CH2OAc
O

60%

Tetr Lett (1964) 3481

Further examples of the epoxidation of olefins and the conversion of epoxides into ketones are included in section 134 (Ethers and Epoxides from Olefins) and section 174 (Ketones from Ethers and Epoxides)

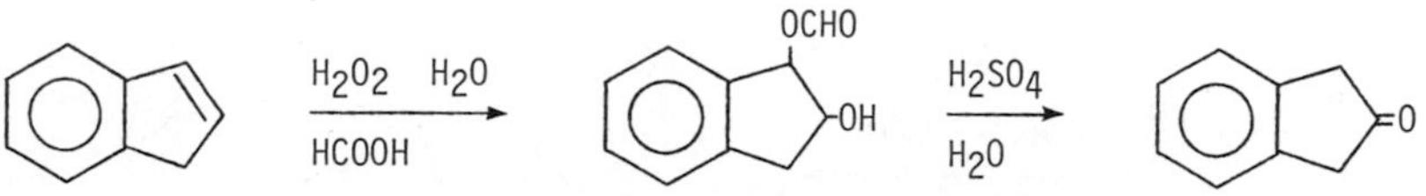
H2O2 H2O
HCOOH
OCHO
OH
H2SO4
H2O
O

69-81%

Org Synth (1961) 41 53

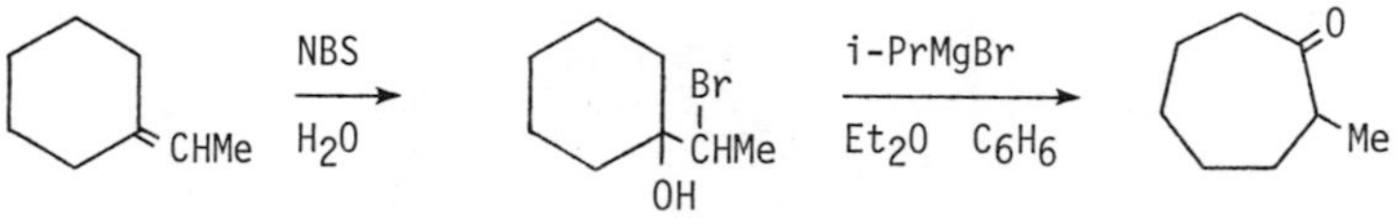
CHMe
NBS
H2O
Br
CHMe
OH
i-PrMgBr
Et2O C6H6
O
Me

40%

JOC (1970) 35 2670

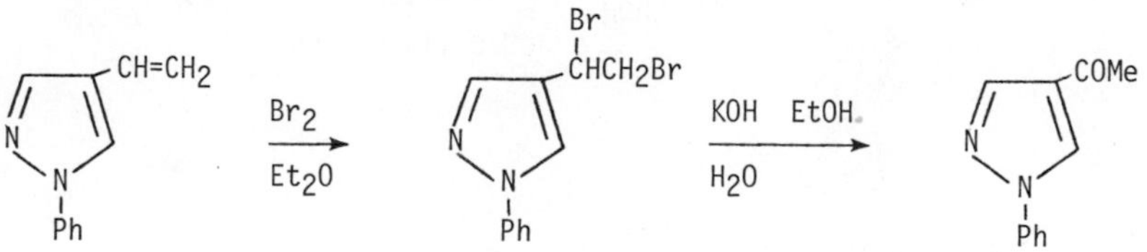
CH=CH2
N
N
Ph
Br2
Et2O
Br
CHCH2Br
N
N
Ph
KOH EtOH
H2O
COMe
N
N
Ph

71%

JCS (1963) 3967

$$\xrightarrow[\text{2 } H_2 \quad Pd\text{-}CaCO_3]{\text{1 } O_3 \quad HOAc}$$

Ber (1954) <u>87</u> 993

Ozonides may also be reduced by the following reagents:

Zn-HOAc	JACS (1943) <u>65</u> 752
Ni	JACS (1941) <u>63</u> 3540
HCOOH	JACS (1948) <u>70</u> 4069
Tetracyanoethylene	Ber (1963) <u>96</u> 1564
	Bull Soc Chim Fr (1964) 729
Ph_3P	Angew (1956) <u>68</u> 473
$P(OMe)_3$-MeOH	JOC (1960) <u>25</u> 1031
NaI-HOAc	JOC (1968) <u>33</u> 1656
	Ber (1955) <u>88</u> 795
$PhNHNH_2$	Ber (1954) <u>87</u> 993

$$\xrightarrow[\substack{K_2CO_3 \quad H_2O \\ \text{dioxane}}]{KMnO_4 \quad NaIO_4}$$

JOC (1966) <u>31</u> 3028

$$\underset{\overset{|}{Bu}}{PhC}=CHPh \xrightarrow[\text{dioxane} \quad H_2O]{OsO_4 \quad NaIO_4} PhCOBu \quad + \quad PhCHO$$

71%

JACS (1966) <u>88</u> 476
(1957) <u>79</u> 6313
JCS <u>C</u> (1966) 655

$$\xrightarrow[Pyr]{OsO_4 \quad C_6H_6} \quad \xrightarrow[\text{or NaIO}_4]{Pb(OAc)_4}$$

JCS <u>C</u> (1966) 655

$$\xrightarrow[CCl_4]{RuO_4}$$

27%

Bull Soc Chim Fr (1964) 729
JCS (1962) 4745

$$\underset{\underset{\text{t-BuCH}_2\text{C=CH}_2}{|}}{\overset{\text{Me}}{}} \xrightarrow{\text{K}_2\text{Cr}_2\text{O}_7 \quad \text{H}_2\text{SO}_4 \quad \text{H}_2\text{O}} \text{t-BuCH}_2\text{COMe} \qquad 56\%$$

JACS (1950) $\underline{72}$ 3701

MeCHCH$_2$CH$_2$CH=CMe$_2$ $\xrightarrow[\text{2 CH}_2\text{N}_2]{\text{1 CrO}_3 \quad \text{HOAc}}$ MeCHCH$_2$CH$_2$COOMe $\xrightarrow[\text{2 Ac}_2\text{O}]{\text{1 PhMgBr}}$ MeCHCH$_2$CH=CPh$_2$

1 NBS
2 Ac$_2$O　HOAc

MeCO $\xleftarrow{\text{CrO}_3 \quad \text{HOAc}}$ MeCH=CH-CH=CPh$_2$

JACS (1970) $\underline{92}$ 2059

$$\underset{\underset{\text{Me}}{|}}{\text{PhC=CH}_2} \xrightarrow{\text{N-Nitrosopiperidine} \quad h\nu} \underset{\underset{\underset{91\%}{\text{Me}}}{|}}{\text{PhC=NOH}} \dashrightarrow \text{PhCOMe}$$

JACS (1965) $\underline{87}$ 4642

$\xrightarrow[300°]{\text{N}_2\text{O}}$

JCS (1951) 2999 3009

$$\text{PhCH=CMe}_2 \xrightarrow{\text{NH}_2\text{NHNa} \quad \text{Et}_2\text{O}} \underset{53\%}{\text{Me}_2\text{C=NNH}_2} \dashrightarrow \text{Me}_2\text{CO}$$

Angew (1962) $\underline{74}$ 650
(Internat Ed $\underline{1}$ 456)

Ph$_2$C=CHEt $\xrightarrow{\text{NaN}_3 \quad \text{H}_2\text{SO}_4}$ PhCOCH$_2$Et   64%

JACS (1950) 72 5777

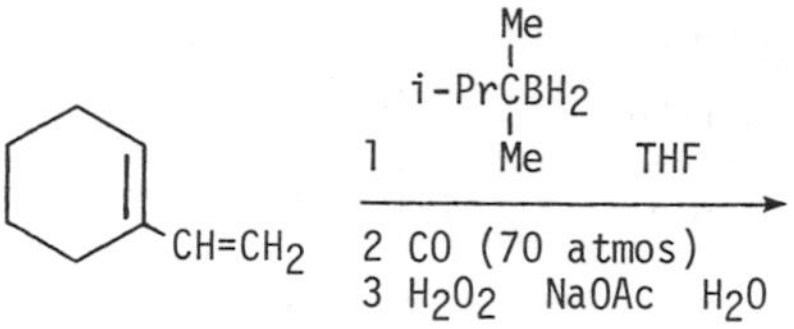

60%

Chem Comm (1968) 594

Section 180   Ketones from Miscellaneous Compounds

$\xrightarrow[\text{2 MeI}]{\text{1 Li \quad NH}_3}$

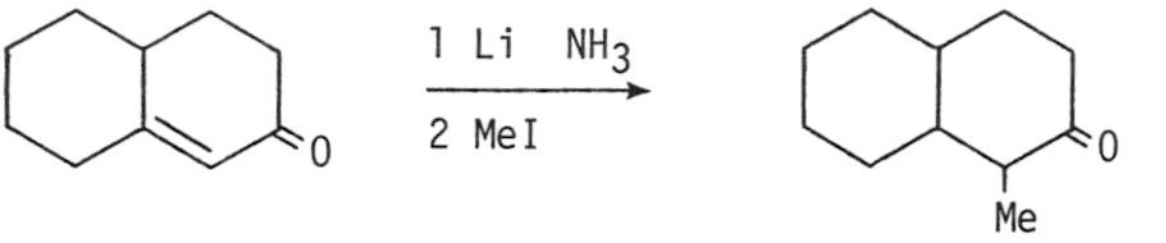

< 83%

JACS (1965) 87 275
Tetrahedron (1964) 20 357
JOC (1967) 32 2851
Bull Soc Chim Fr (1969) 4356 4348

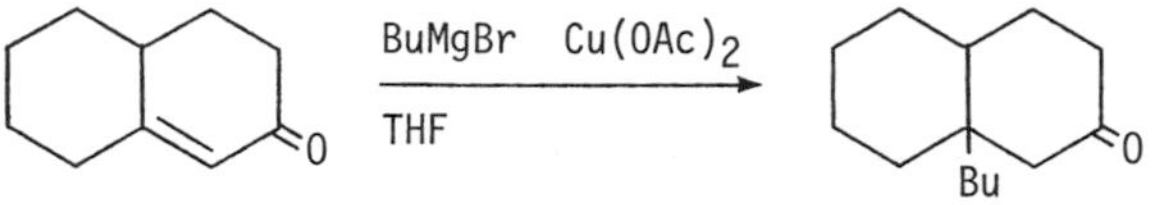

$\xrightarrow[\text{THF}]{\text{BuMgBr \quad Cu(OAc)}_2}$

71%

JACS (1965) 87 82
JOC (1966) 31 3128
JCS (1943) 501

MeCH=CHCOMe $\xrightarrow[\text{Et}_2\text{O}]{\text{MeMgBr-Bu}_3\text{PCuI}}$ Me$_2$CHCH$_2$COMe   99%

JOC (1966) <u>31</u> 3128

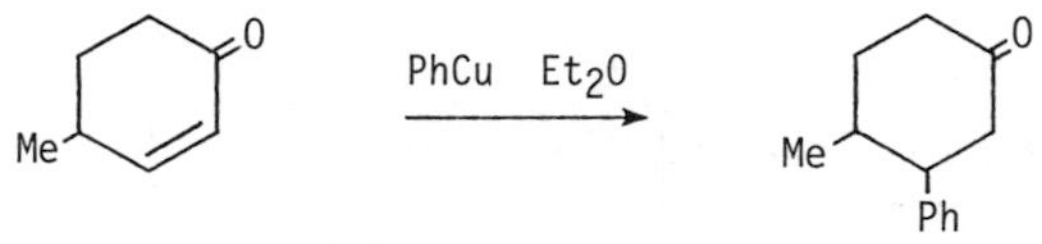

PhCu Et$_2$O

Tetr Lett (1970) 1579

H$_2$ Pd-C

HCl EtOH   cis

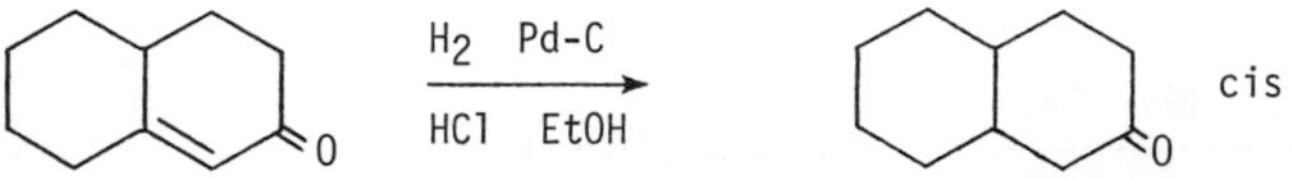

JOC (1958) <u>23</u> 1853
Chem Ind (1966) 1796

For reduction of enones with H$_2$ and Rh see JOC (1968) <u>33</u> 3695

PhCH=CHCOMe $\xrightarrow{\text{H}_2 \quad \text{Ni} \quad \text{CH}_2\text{Cl}_2}$ PhCH$_2$CH$_2$COMe   93-96%

Bull Soc Chim Fr (1954) 522

Me$_2$C=CHCOMe $\xrightarrow{\text{Ni} \quad \text{H}_2\text{O}}$ Me$_2$CHCH$_2$COMe   95%

Bull Chem Soc Jap (1967) <u>40</u> 1548

$(\text{Ph}_3\text{P})_3\text{RhCl}$

MeOH

40%

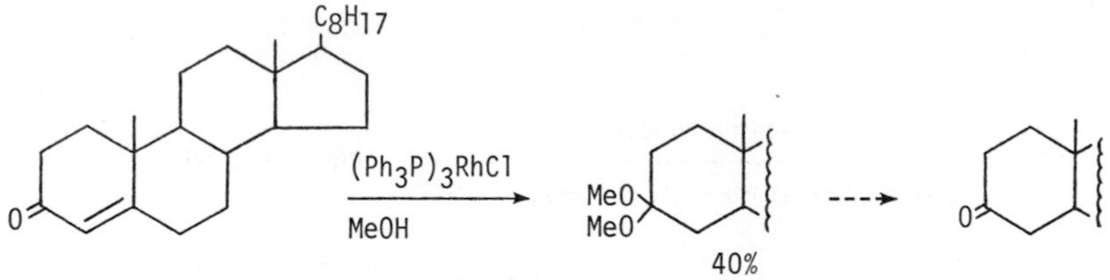

Ber (1968) <u>101</u> 1154

H_2 $(Ph_3P)_3RhCl$ NaI

EtOH C_6H_6

JACS (1966) <u>88</u> 4537

1 Li NH_3 Et_2O

2 MeOH

3 CrO_3 H_2SO_4 Me_2CO

trans   63%

JOC (1970) <u>35</u> 753
JACS (1965) <u>87</u> 275

t-BuCH=CHCOBu-t $\xrightarrow{\text{Na HMPA THF}}$ t-BuCH$_2$CH$_2$COBu-t

JACS (1970) <u>92</u> 2783
Bull Soc Chim Fr (1968) 595

1 Na trimesitylborane

$MeOCH_2CH_2OMe$

2 MeOH

JACS (1970) <u>92</u> 696

85-95%

$Me_2C=CHCOMe$ $\xrightarrow{\text{CrCl}_2 \ \text{NH}_3 \ \text{H}_2\text{O}}$ Me_2CHCH_2COMe

Angew (1968) <u>80</u> 271
(Internat Ed <u>7</u> 247)

PhCH=CHCOPh $\xrightarrow{\substack{\text{Diethyl 1,4-dihydro-2,6-} \\ \text{dimethylpyridine-3,5-dicarboxylate} \\ 156°}}$ PhCH$_2$CH$_2$COPh

JCS (1960) 3257

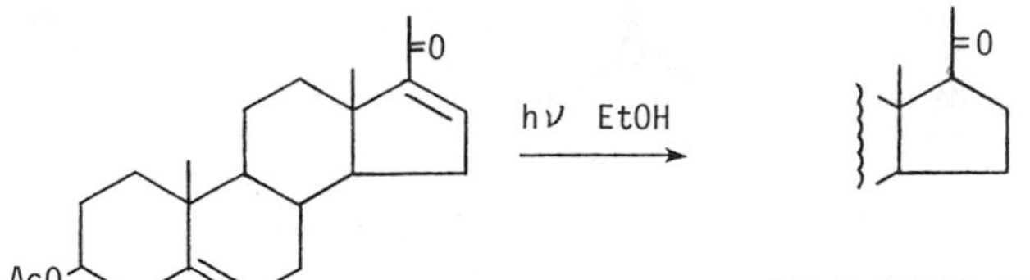

$h\nu$ EtOH

40%

JCS <u>C</u> (1967) 2032

$Me_2C=CHCOMe$ $\xrightarrow{Ph_3SnH}$ Me_2CHCH_2COMe

Compt Rend (1965) <u>260</u> 581

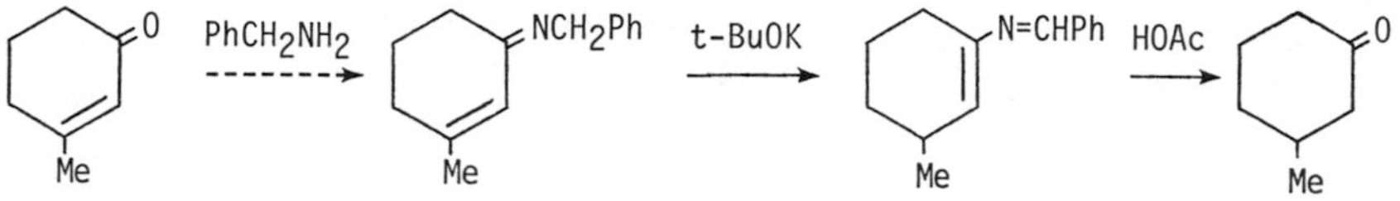

$PhCH_2NH_2$ t-BuOK   HOAc

40-70%

JACS (1967) <u>89</u> 2794

Zn   HOAc

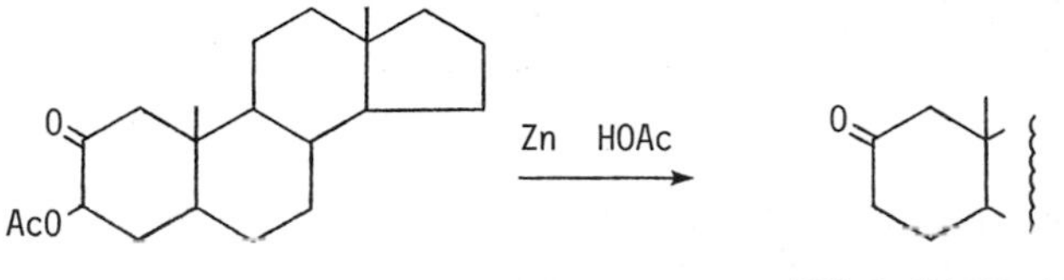

JCS <u>C</u> (1970) 244
JACS (1955) <u>77</u> 4367

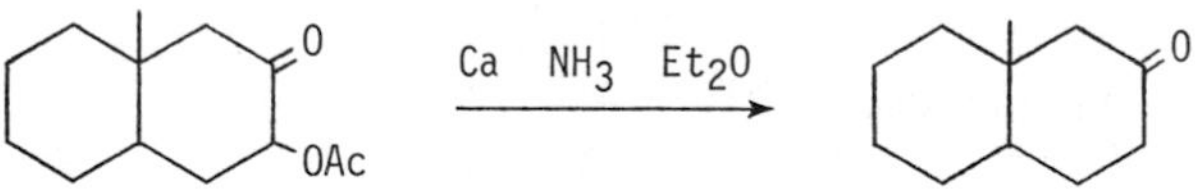

Ca NH_3 Et_2O

80%

JOC (1969) <u>34</u> 4188
JCS (1956) 4344

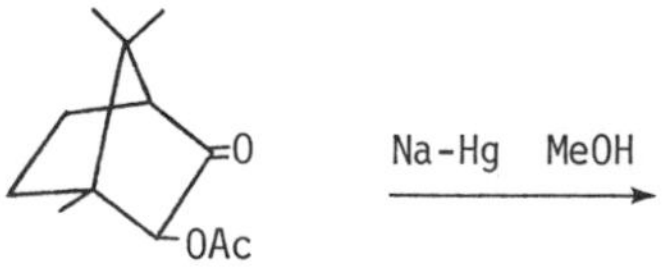

Na-Hg   MeOH

95%

Bull Chem Soc Jap (1969) 42 2068

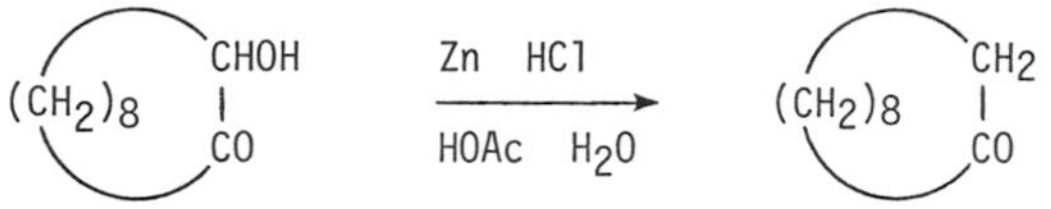

$(CH_2)_8$ with CHOH, CO

Zn   HCl
HOAc H_2O

$(CH_2)_8$ with CH_2, CO

75-78%

Org Synth (1963) Coll Vol 4 218
JACS (1964) 86 3068

$\overset{OH}{EtCOCEt_2}$

HI   HOAc H_2O

$EtCOCHEt_2$

80%

JACS (1964) 86 3068
JOC (1938) 3 456

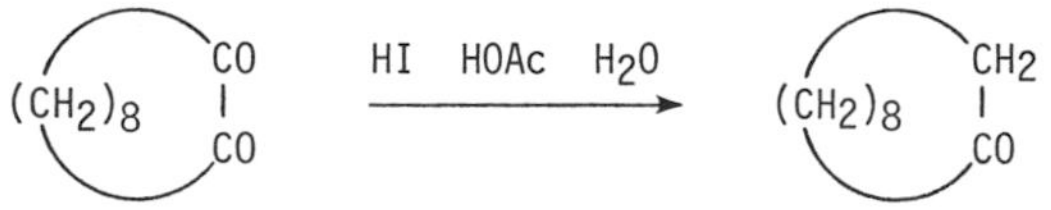

$(CH_2)_8$ with CO, CO

HI   HOAc H_2O

$(CH_2)_8$ with CH_2, CO

80%

JACS (1964) 86 3068

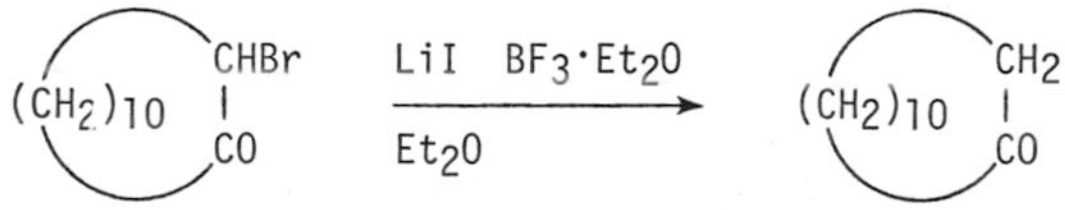

$(CH_2)_{10}$ with CHBr, CO

LiI $BF_3 \cdot Et_2O$
Et_2O

$(CH_2)_{10}$ with CH_2, CO

100%

Tetr Lett (1971) 137

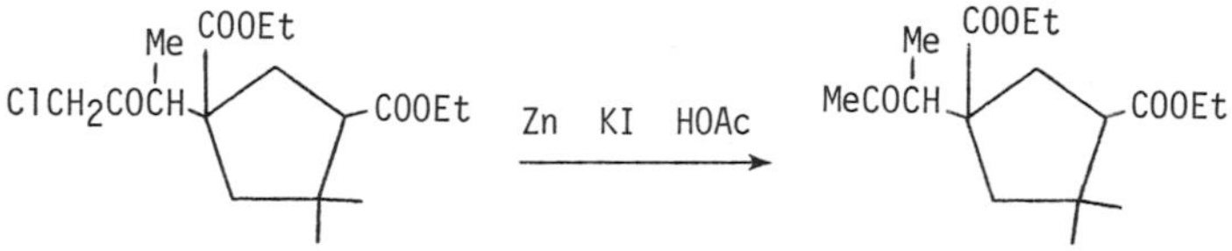

$ClCH_2COCH$, Me, COOEt, COOEt

Zn   KI   HOAc

$MeCOCH$, Me, COOEt, COOEt

88%

JACS (1961) 83 3114
Helv (1944) 27 821
Tetr Lett (1968) 1575

$PhCOCH_2Cl$ $\xrightarrow{\text{CrSO}_4 \quad \text{DMF} \quad \text{H}_2\text{O}}$ PhCOMe   79%

JACS (1963) 85 2768
(1945) 67 1728
Angew (1968) 80 271
(Internat Ed 7 247)

$\xrightarrow[\text{MeOH} \quad \text{Me}_2\text{CO}]{\text{H}_2 \quad \text{Pd} \quad \text{KOAc}}$

JCS (1964) 5535

$PhCOCH_2Cl$ $\xrightarrow{\text{Bu}_2\text{SnH}_2 \quad \text{Et}_2\text{O}}$ PhCOMe   95%

JOC (1963) 28 2165

$\xrightarrow[\text{or 2,6-dimethylphenol}]{\text{PhNMe}_2}$

70-76%

Chimia (1967) 21 464
JCS C (1969) 301

$i\text{-PrCH}_2\text{COCHPr-}i$
$\underset{\text{Br}}{|}$ $\xrightarrow[\text{2 EtOH}]{\text{1 Ph}_3\text{P} \quad \text{C}_6\text{H}_6}$ $i\text{-PrCH}_2\text{COCH}_2\text{Pr-}i$

JCS (1962) 2337
Tetr Lett (1962) 471

$PhCOCHMe$
$\underset{\text{Br}}{|}$ $\xrightarrow{\text{Ph}_2\text{PH} \quad \text{CCl}_4}$ $PhCOCH_2Me$ 100%

JOC (1969) 34 2687

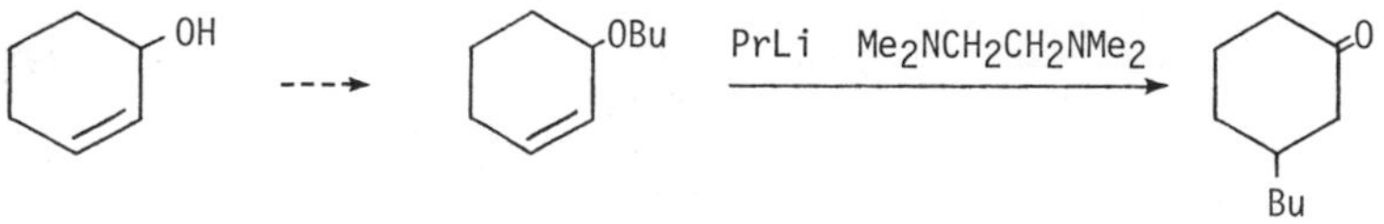

OH
OBu
PrLi Me2NCH2CH2NMe2
O
Bu

Tetr Lett (1969) 821

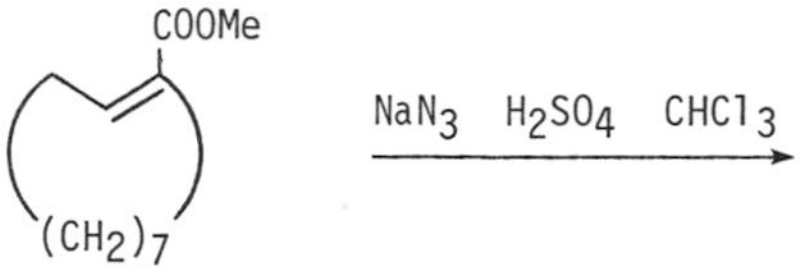

COOMe
(CH2)7
NaN3 H2SO4 CHCl3
O
(CH2)6

JOC (1968) 33 2157

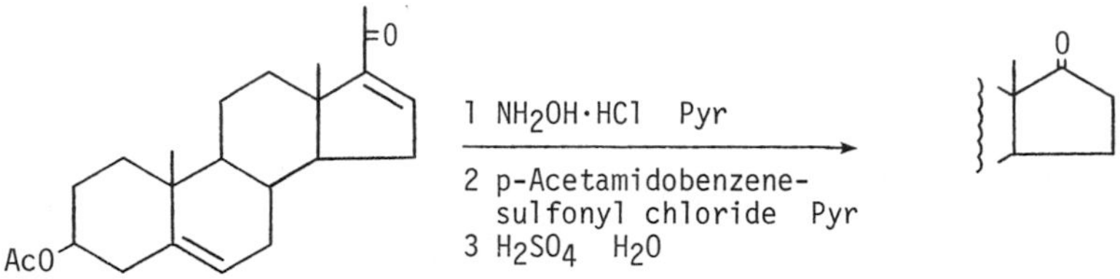

O
AcO
1 NH2OH·HCl Pyr
2 p-Acetamidobenzene-sulfonyl chloride Pyr
3 H2SO4 H2O
O

74%

JOC (1956) 21 520
Org React (1960) 11 1

$Me_2C(OH)CH(OAc)Bu \xrightarrow{Zn\ \ 154°} Me_2CHCOBu$ 68%

Chem Comm (1968) 1639
JOC (1970) 35 660

$Me_2C{=}CHCH_2CH_2CH(OH)Me \xrightarrow{\text{Polyphosphoric acid}} Me_2CHCH_2CH_2CH_2COMe$ 80%

Bull Soc Chim Fr (1963) 1799
JACS (1965) 87 2772

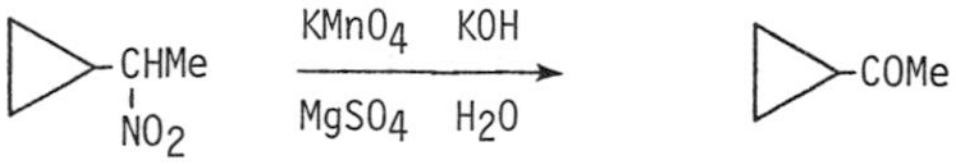

CHMe
NO2
KMnO4 KOH
MgSO4 H2O
COMe

77%

JOC (1962) 27 3699

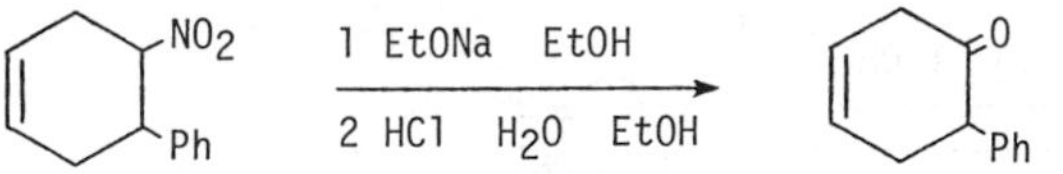

1 EtONa EtOH
2 HCl H₂O EtOH

78%

JOC (1952) 17 581
Chem Rev (1955) 55 137

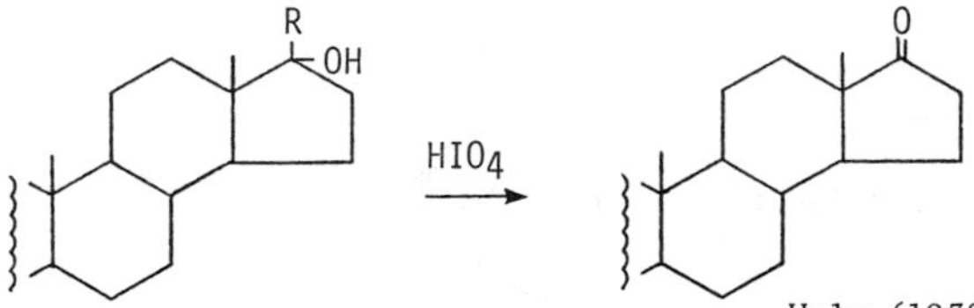

HIO₄

R = CHOH or CHO
        |
        Me

Helv (1938) 21 546
     (1940) 23 170
     (1941) 24 945
Org React (1944) 2 341

Section 180A   Protection of Ketones

Ketals are stable to the following reagents: NaOH, NaH, Na-NH₃, RMgX, NaBH₄, LiAlH₄, H₂-Pd, CrO₃ and SOCl₂-Pyr

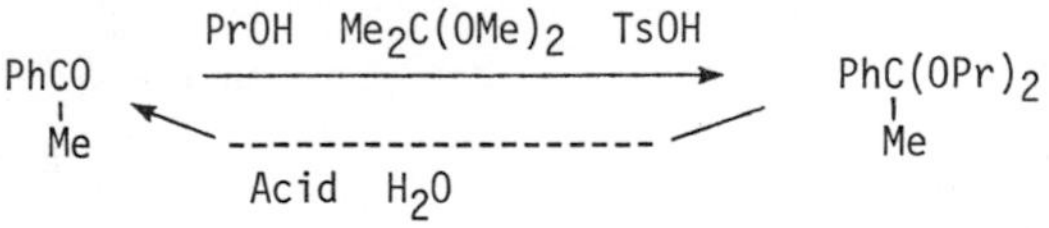

PhCO   PrOH Me₂C(OMe)₂ TsOH   PhC(OPr)₂
 | ←─────────────────────────→ |
 Me   Acid H₂O   Me

JOC (1960) 25 521 525

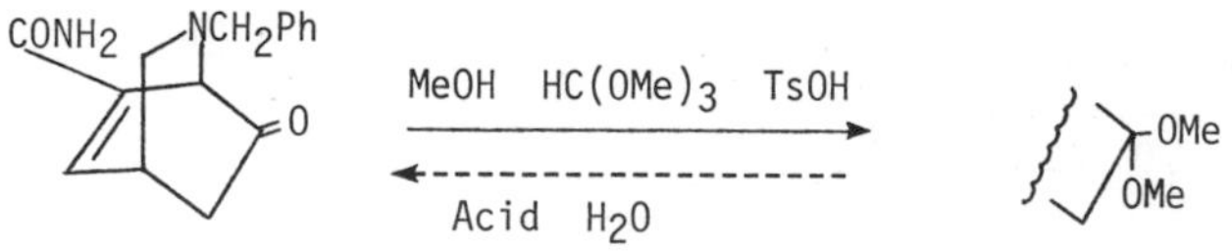

JACS (1968) 90 2448
Annalen (1962) 656 97

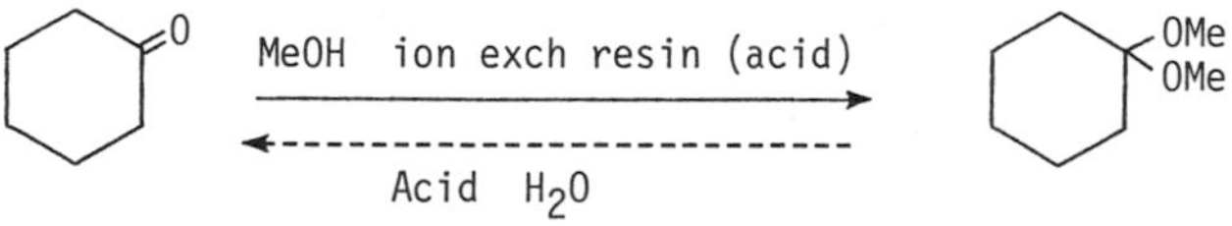

JOC (1959) 24 1731

The following catalysts may also be used for the preparation of ketals:

$(Ph_3P)_3RhCl$ . .Ber (1968) 101 1154

SeO_2 . . . . .JACS (1954) 76 6113

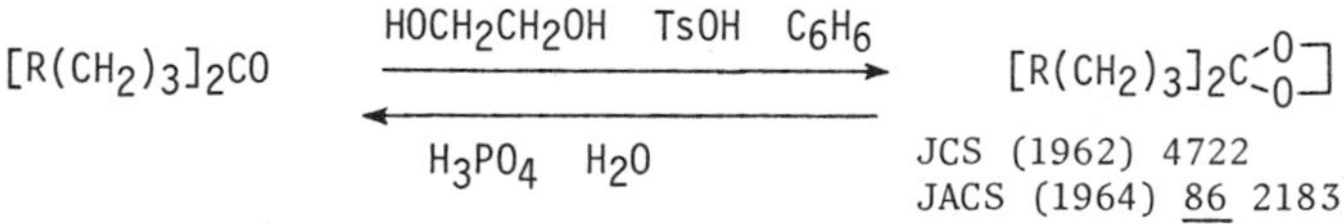

JCS (1962) 4722
JACS (1964) 86 2183

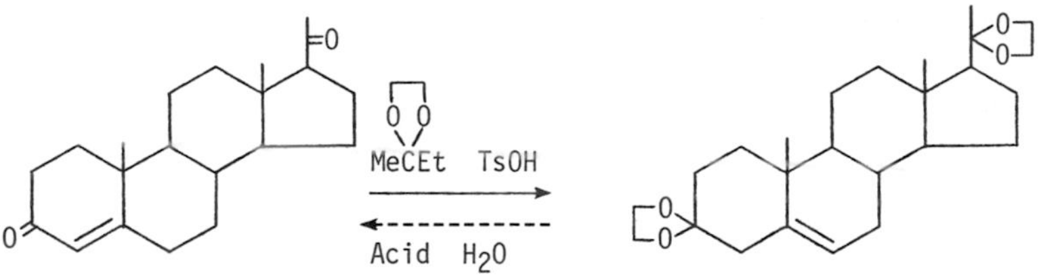

JACS (1954) 76 1359
(1953) 75 1716

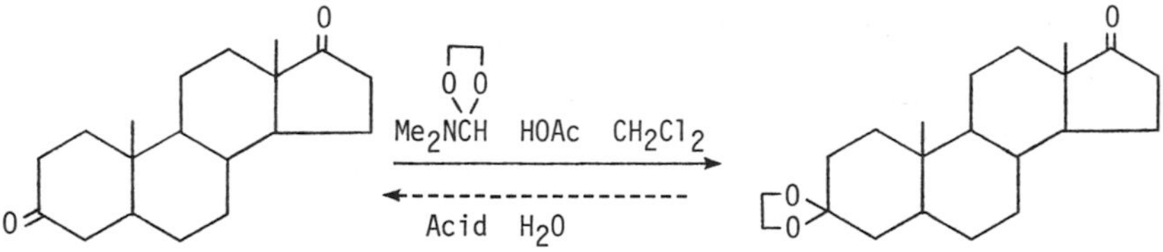

Steroids (1963) 1 45

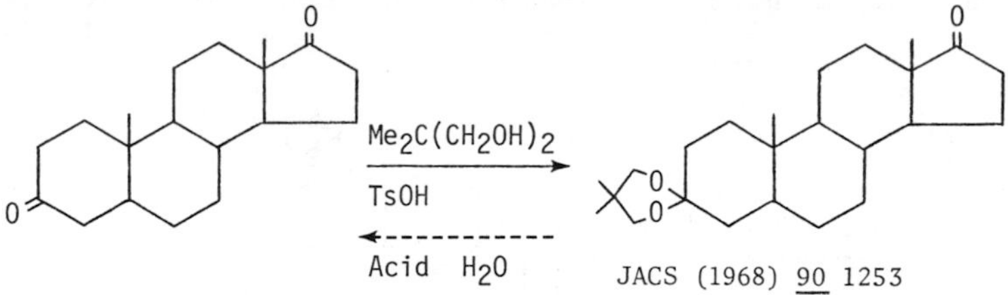
O
O
$Me_2C(CH_2OH)_2$
TsOH
Acid H_2O
O
O
O

JACS (1968) 90 1253

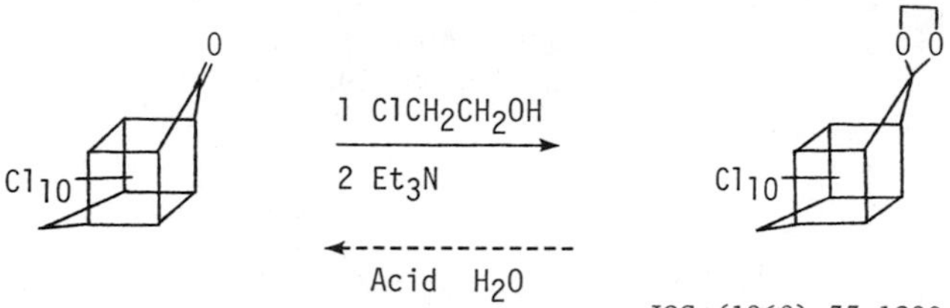
O
Cl_{10}
1 $ClCH_2CH_2OH$
2 Et_3N
Acid H_2O
O O
Cl_{10}

JOC (1968) 33 1280

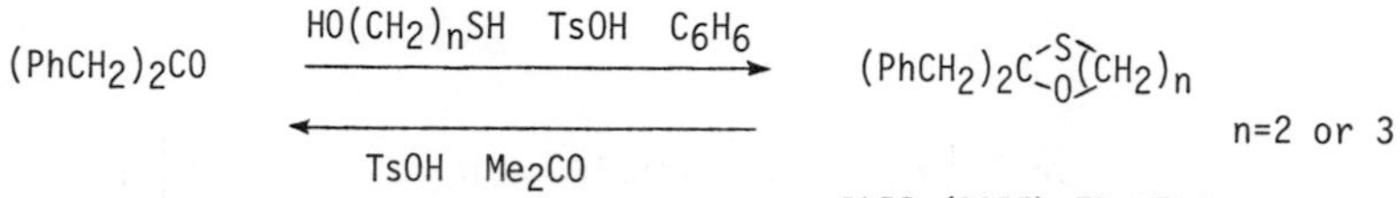
$(PhCH_2)_2CO$
$HO(CH_2)_nSH$ TsOH C_6H_6
TsOH Me_2CO
$(PhCH_2)_2C$ S O $(CH_2)_n$
n=2 or 3

JACS (1953) 75 3704
(1954) 76 1945
For cleavage with Ni see JACS (1958) 80 4723

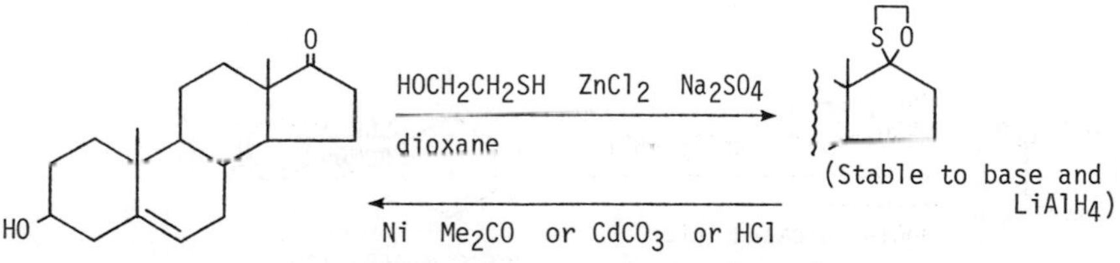
O
HO
$HOCH_2CH_2SH$ $ZnCl_2$ Na_2SO_4
dioxane
Ni Me_2CO or $CdCO_3$ or HCl
S O
(Stable to base and $LiAlH_4$)

JACS (1951) 73 4961

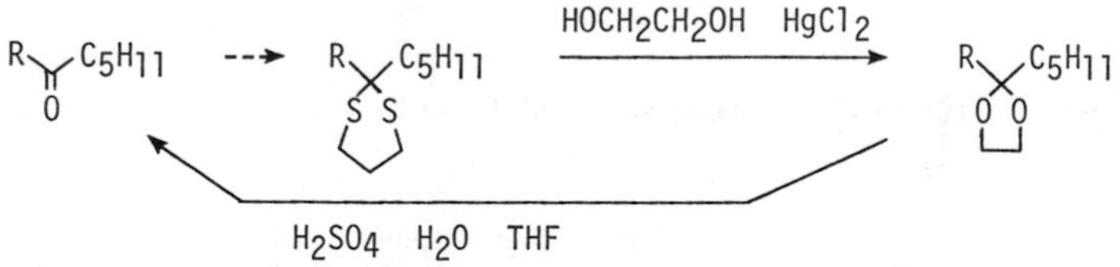
R C_5H_{11}
O
R C_5H_{11}
S S
$HOCH_2CH_2OH$ $HgCl_2$
R C_5H_{11}
O O
H_2SO_4 H_2O THF

JACS (1968) 90 3245

Thioketals are stable to the following reagents: NaOH, LiAlH$_4$, RMgX, CrO$_3$-Pyr

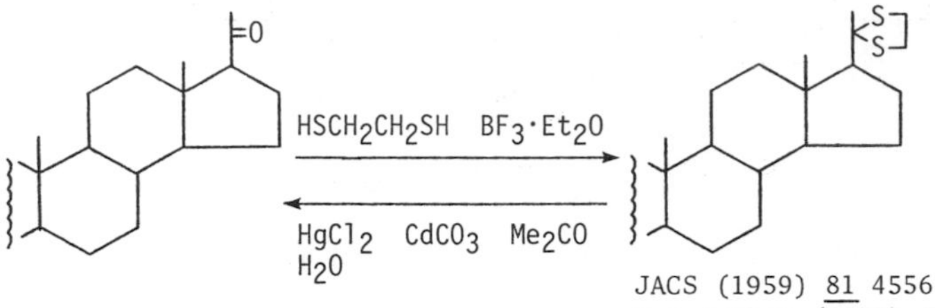

JACS (1959) 81 4556
Can J Chem (1955) 33 716

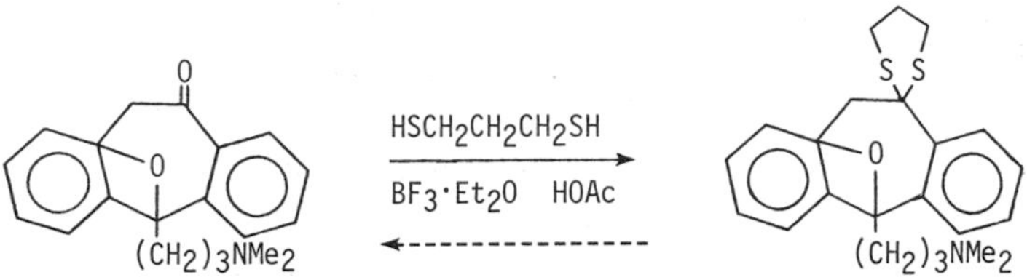

J Med Chem (1970) 13 191

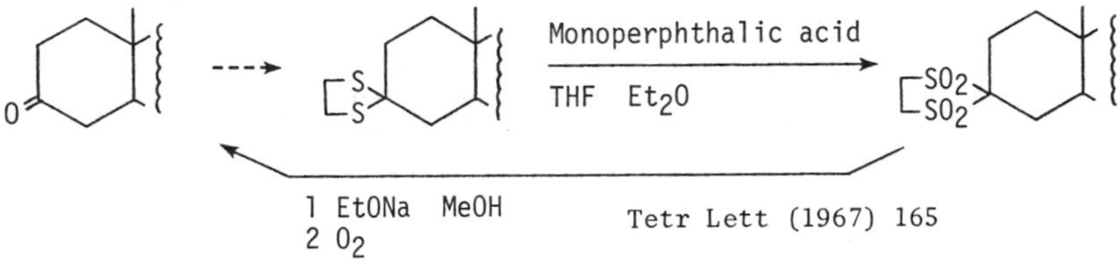

Tetr Lett (1967) 165

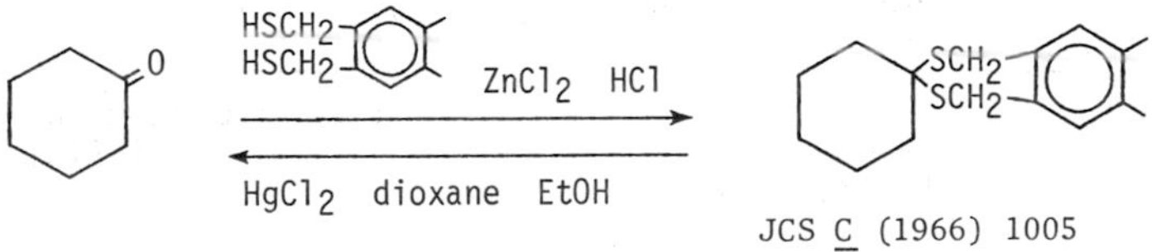

JCS C (1966) 1005

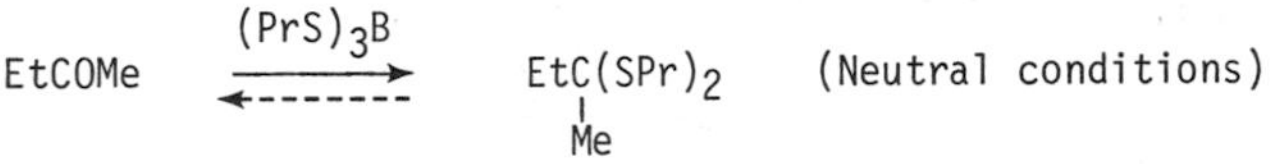

EtCOMe $\xrightarrow{\text{(PrS)}_3\text{B}}$ EtC(SPr)$_2$ (Neutral conditions)
                                  |
                                  Me

Can J Chem (1969) 47 859
            (1965) 43 307

$$(RCOOCH_2)_2CO \underset{HgCl_2 \quad Me_2CO \quad H_2O}{\overset{EtSH \quad ZnCl_2}{\rightleftharpoons}} (RCOOCH_2)_2C(SEt)_2$$

Can J Chem (1955) 33 716

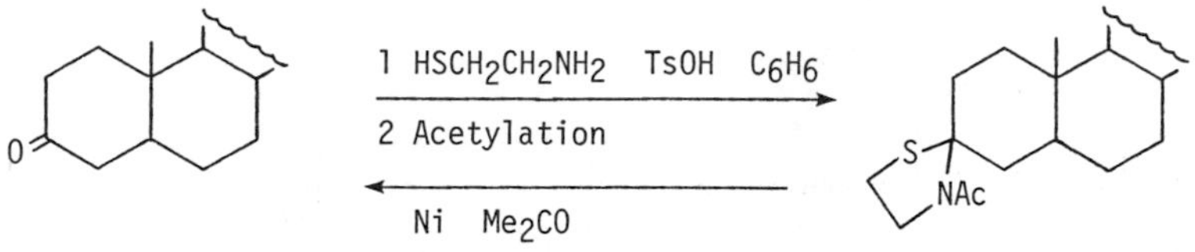
1 HSCH2CH2NH2 TsOH C6H6
2 Acetylation
Ni Me2CO
O
S
NAc

JOC (1962) 27 1112

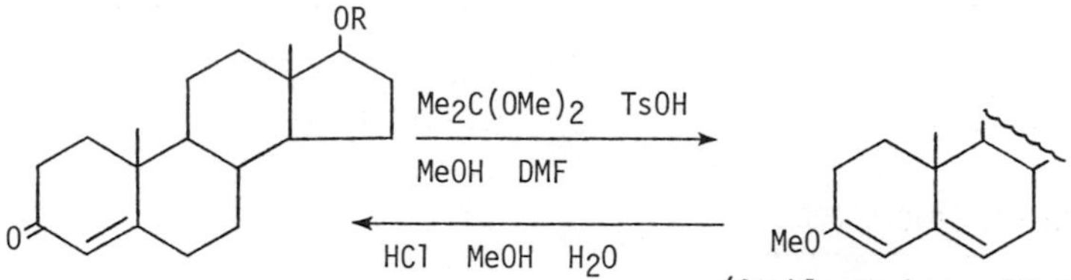
OR
Me2C(OMe)2 TsOH
MeOH DMF
HCl MeOH H2O
O
MeO

(Stable to base, RMgX and $LiAlH_4$)

JOC (1961) 26 3925

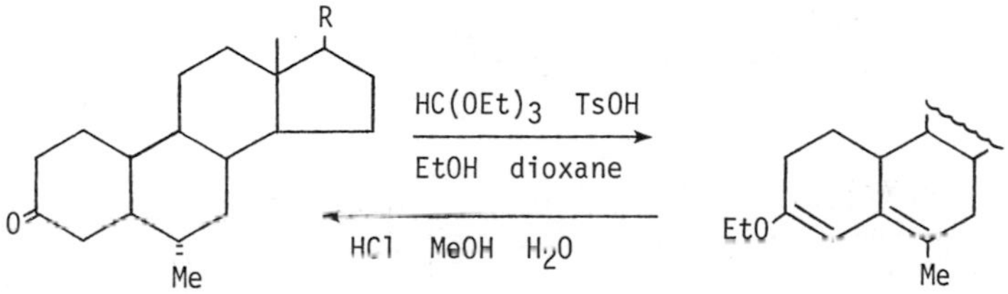
R
HC(OEt)3 TsOH
EtOH dioxane
HCl MeOH H2O
O
Me
EtO
Me

JACS (1959) 81 4566

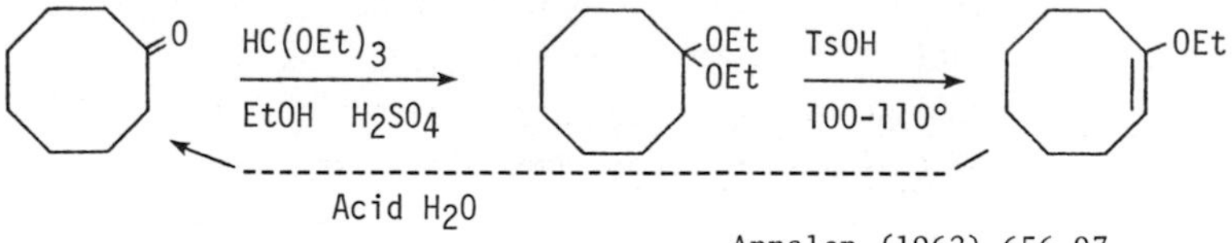
O
HC(OEt)3
EtOH H2SO4
OEt
OEt
TsOH
100-110°
OEt
Acid H2O

Annalen (1962) 656 97
Gazz (1962) 92 309
(Chem Abs 57 12572)

(Stable to LiAlH$_4$)   (Stable to acid)

JACS (1951) $\underline{73}$ 1528

(Stable to RMgX and LiAlH$_4$)

Coll Czech (1961) $\underline{26}$ 1646
JACS (1956) $\underline{78}$ 430
       (1952) $\underline{74}$ 3627

JCS $\underline{C}$ (1966) 531

JACS (1970) $\underline{92}$ 5276

Oximes may also be cleaved by the following reagents:

NaHSO$_3$	JOC (1966) $\underline{31}$ 3446
MeCOCH$_2$CH$_2$COOH-HCl	JACS (1959) $\underline{81}$ 4629
Zn-HOAc	J Indian Chem Soc (1969) $\underline{46}$ 44
TiCl$_3$	Tetr Lett (1971) 195
(NH$_4$)$_2$Ce(NO$_3$)$_6$	Can J Chem (1969) $\underline{47}$ 145
Fe(CO)$_5$-BF$_3$·Et$_2$O	JOC (1967) $\underline{32}$ 2938

For preparation of oximes of unreactive and acid or base sensitive ketones
    (using Me$_2$SO-NH$_2$OH·HCl) see Chem Ind (1969) 240

(Stable to acid and base)

HOAc H_2O

JOC (1962) <u>27</u> 914
For cleavage of methoximes with O_3 see JOC (1969) <u>34</u> 2961

Semicarbazones are stable to $NaBH_4$ and Oppenauer oxidation

$(COOH)_2$ H_2O

Helv (1932) <u>15</u> 1220

Semicarbazones may also be cleaved by the following reagents:

$H_2SO_4-H_2O$	JCS (1946) 27
$HCl-H_2O$	JACS (1950) <u>72</u> 1751
$MeCOCOOH$	JCS (1953) 3864
	JACS (1955) <u>77</u> 1221
$MeCOCH_2COMe-HCl$	Annalen (1962) <u>656</u> 119
Ac_2O-Pyr	JOC (1956) <u>21</u> 795
$(NH_4)_2Ce(NO_3)_6$	Can J Chem (1969) <u>47</u> 145
$MeCOO_2H$	Ber (1961) <u>94</u> 712
$NaNO_2-HOAc$	Rec Trav Chim (1946) <u>65</u> 796
	JACS (1955) <u>77</u> 4781
	(1961) <u>83</u> 4249

(Stable to base, B_2H_6, $LiAlH_4$ and CrO_3)

Chem Comm (1969) 445

For cleavage of dimethylhydrazones with O_3 see   JOC (1969) <u>34</u> 2961
    "   "   "   "   " peracid see Ber (1961) <u>94</u> 712

Annalen (1962) <u>656</u> 119

Phenylhydrazones may also be cleaved by the following reagents:

MnO_2 JOC (1967) <u>32</u> 2252

$MeCOO_2H$ JOC (1967) <u>32</u> 2865

2,4-Dinitrophenylhydrazones are stable to CrO_3 and B_2H_6

$$C_6H_{13}\underset{\underset{Me}{|}}{CO} \xrightarrow[\text{KHCO}_3 \quad \text{HOCH}_2\text{CH}_2\text{OH} \quad \text{H}_2\text{O}]{\text{2,4-Dinitrophenylhydrazine}} C_6H_{13}\underset{\underset{Me}{|}}{C}=NN$$

Aust J Chem (1968) <u>21</u> 271

2,4-Dinitrophenylhydrazones may also be cleaved by the following reagents:

$MeCOCH_2CH_2COOH-HCl$	JACS (1959) <u>81</u> 4629
$SnCl_2-HCl-Me_2CO$	Nature (1954) <u>173</u> 266
$CrCl_2-HCl$	JCS (1962) 4729
O_3	Chem Comm (1968) 433
	JOC (1951) <u>16</u> 556
$CuCO_3\cdot Cu(OH)_2-HCOOH$	Nature (1954) <u>173</u> 541

(Stable to acids, peracids, CrO_3

and Br_2)

Gazz (1961) <u>91</u> 1250
(Chem Abs <u>56</u> 10211)
Steroids (1967) <u>10</u> 411
JACS (1953) <u>75</u> 650

Chem Comm (1970) 1420

Chapter 13   PREPARATION
OF
NITRILES

Nitriles from Acetylenes
°°°°°°°°°°°°°°°°°°°°°°°°°°°°°°

$$BuC\equiv CH \xrightarrow[\substack{2\ MeLi\ \ Et_2O \\ 3\ (CN)_2}]{1\ (i\text{-}Bu)_2AlH} BuCH=CHCN \dashrightarrow BuCH_2CH_2CN$$

87%

JACS (1968) <u>90</u> 7139

$$C_9H_{19}C\equiv CH \xrightarrow[2\ ClCN]{1\ Mg\ \ EtMgBr} C_9H_{19}C\equiv CCN \dashrightarrow C_9H_{19}CH_2CH_2CN$$

Annales de Chimie (1926) <u>5</u> 5

$$MeC\equiv CH \xrightarrow[NO]{COCl_2\ \ h\nu} MeCN$$

JACS (1963) <u>85</u> 3506

Section 182   Nitriles from Carboxylic Acids
°°°°°°°°°°°°°°°°°°°°°°°°°°°°°°°°°°

$$\underset{(CH_2)_3COOH}{\overset{CN}{|}} \xrightarrow{\text{Electrolysis}\ \ MeONa\ \ MeOH} \underset{(CH_2)_6CN}{\overset{CN}{|}} \qquad \sim 41\%$$

Z. Naturforsch (1947) 2b 185
Advances in Org Chem (1960) <u>1</u> 1

i-PrCOOH $\xrightarrow{PhCH_2CN \quad H_2SO_4}$ i-PrCN 40%

JCS (1956) 1686

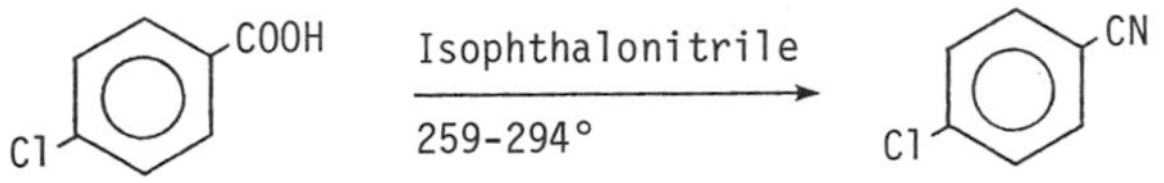

COOH
Cl
Isophthalonitrile
259-294°
CN
Cl

93%

JOC (1958) 23 1350

$PhCH{=}CHCOOZn_{1/2}$ $\xrightarrow{Pb(CNS)_2 \quad \Delta}$ PhCH=CHCN 27-36%

JACS (1916) 38 2120

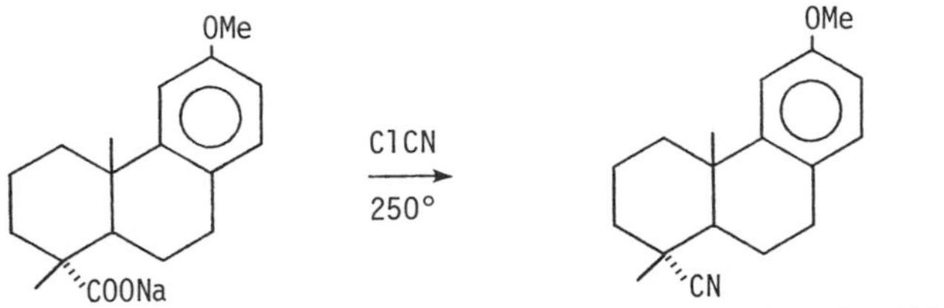

OMe
COONa
ClCN
250°
OMe
CN

26%

JCS (1961) 3185

t-BuCOOH $\xrightarrow[2\ DMF]{1\ ClSO_2NCO \quad pentane}$ t-BuCN 68%

Ber (1967) 100 2719
Tetr Lett (1968) 1631

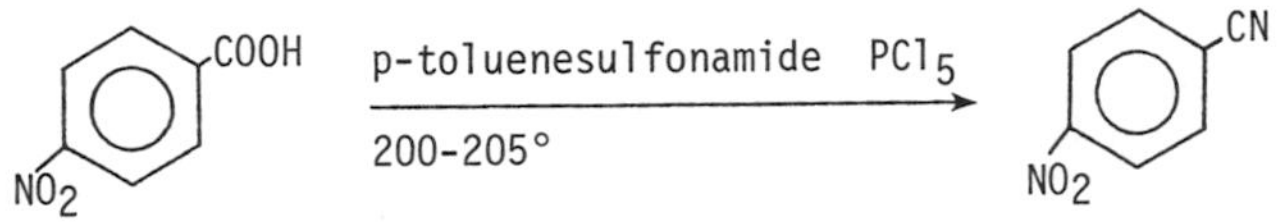

COOH
NO2
p-toluenesulfonamide PCl5
200-205°
CN
NO2

85-90%

Org Synth (1955) Coll Vol 3 646
JCS (1946) 763

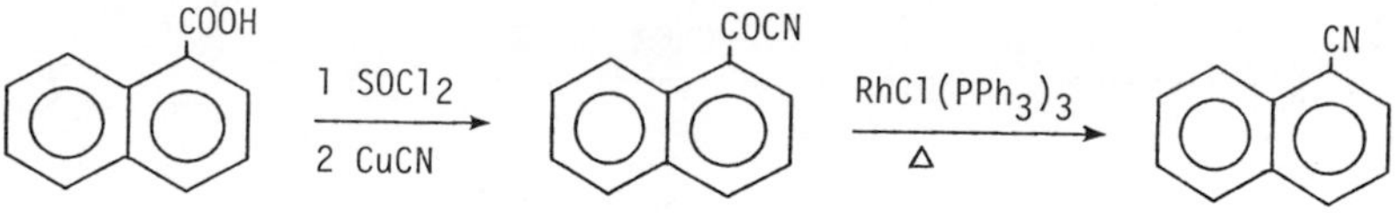

JACS (1967) **89** 2338

$$(t\text{-Bu})_2CHCOOH \xrightarrow[\text{2 NaNH}_2 \ \ NH_3]{\text{1 SOCl}_2} (t\text{-Bu})_2CHCONH_2 \xrightarrow[C_6H_6]{\text{SOCl}_2} (t\text{-Bu})_2CHCN \qquad 61\%$$

JACS (1960) **82** 2498

Further examples of the preparation of nitriles from amides are included in section 186 (Nitriles from Amides)

JACS (1953) **75** 2347
JOC (1961) **26** 3507

Section 183   <u>Nitriles from Alcohols</u>
ooooooooooooooooooooo

Annalen (1963) **661** 157

$$C_6H_{13}\underset{\underset{Me}{|}}{CHOH} \xrightarrow{\quad CCl_4 \quad Ph_3P \quad} C_6H_{13}\underset{\underset{Me}{|}}{CHCl} \xrightarrow{\quad NaCN \quad Me_2SO \quad} C_6H_{13}\underset{\underset{Me}{|}}{CHCN}$$

JOC (1967) <u>32</u> 855

JACS (1970) <u>92</u> 336

Further examples of the preparation of nitriles from halides and sulfonates
are included in section 190 (Nitriles from Halides and Sulfonates)

$$PhCH_2OH \xrightarrow[\text{MeONa \quad MeOH}]{\quad NH_3 \quad O_2 \quad CuCl_2 \quad} PhCN \qquad\qquad 30\%$$

Rec Trav Chim (1963) <u>82</u> 757

Section 184   <u>Nitriles from Aldehydes</u>

$$PhCHO \xrightarrow[\text{H}_2\text{O}]{\quad TsCl \quad KCN \quad} \underset{\underset{OTs}{|}}{PhCHCN} \xrightarrow{\quad PhH \quad AlCl_3 \quad} \underset{\underset{Ph}{|}}{PhCHCN} \qquad 57\%$$

JOC (1954) <u>19</u> 1699

$$PhCHO \xrightarrow[\text{MeOCH}_2\text{CH}_2\text{OMe}]{\quad (EtO)_2OPCH_2CN \quad NaH \quad} \underset{66\%}{PhCH=CHCN} \dashrightarrow PhCH_2CH_2CN$$

JACS (1961) <u>83</u> 1733

PrCHO $\xrightarrow[\text{HOAc}]{\overset{\displaystyle \text{COOEt}}{\underset{}{\text{CH}_2\text{CN}}}\ \ \text{H}_2\ \ \text{Pd}}$ PrCH$_2$CH$\overset{\text{COOEt}}{\underset{}{|}}$CN $\xrightarrow[\text{2}\quad \Delta]{\text{1 Base}}$ PrCH$_2$CH$_2$CN

Org Synth (1955) Coll Vol 3 385
JACS (1959) 81 5397

Arch Pharm (1957) 290 218

PhCHO $\xrightarrow{\text{KCN}}$ PhCH$\overset{}{\underset{\text{OH}}{|}}$CN $\xrightarrow[\text{CHCl}_3]{\text{SOCl}_2}$ PhCH$\overset{}{\underset{\text{Cl}}{|}}$CN $\xrightarrow[\text{HOAc}]{\text{Zn}}$ PhCH$_2$CN $\sim$56%

J Soc Chem Ind (1935) 54 98T

JACS (1940) 62 1512
JCS (1946) 958
Org React (1942) 1 210

PhCHO $\xrightarrow[\substack{\text{NaOAc}\quad \text{Ac}_2\text{O}\\ \text{2 NaOH}\ \ \text{H}_2\text{O}\\ \text{3 HCl}\ \ \text{H}_2\text{O}\\ \text{4 NH}_2\text{OH}\cdot\text{HCl}\\ \text{NaOH}\ \ \text{H}_2\text{O}}]{\text{1 AcNHCH}_2\text{COOH}}$ PhCH$_2$C$\overset{\text{COOH}}{\underset{\text{NOH}}{\|}}$ $\xrightarrow{\text{Ac}_2\text{O}}$ PhCH$_2$CN 43-60%

JACS (1942) 64 885
Org React (1946) 3 198

Chem Comm (1966) 17

$C_6H_{13}CHO$ $\xrightarrow{\text{NH}_3 \quad \text{Pb(OAc)}_4 \quad C_6H_6}$ $C_6H_{13}CN$ 59%

Chem Ind (1965) 988

$C_6H_{13}CHO$ $\xrightarrow[\text{MeONa} \quad \text{MeOH}]{\text{NH}_3 \quad O_2 \quad \text{CuCl}_2}$ $C_6H_{13}CN$ 63%

Rec Trav Chim (1963) <u>82</u> 757

PhCHO $\xrightarrow[\text{2 } I_2]{\text{1 } NH_3 \quad \text{MeONa} \quad \text{MeOH}}$ PhCN   50%

Bull Chem Soc Jap (1966) <u>39</u> 854

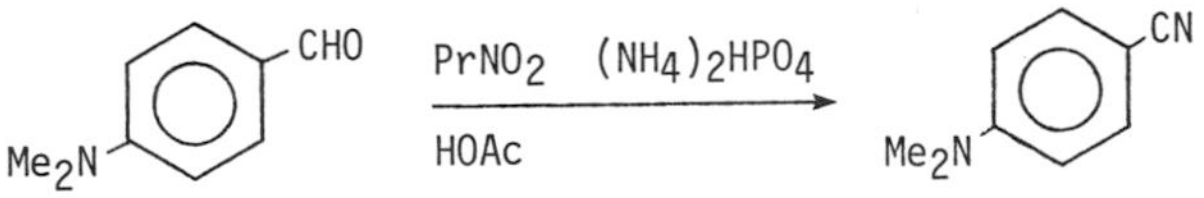

JACS (1961) <u>83</u> 2203   77%

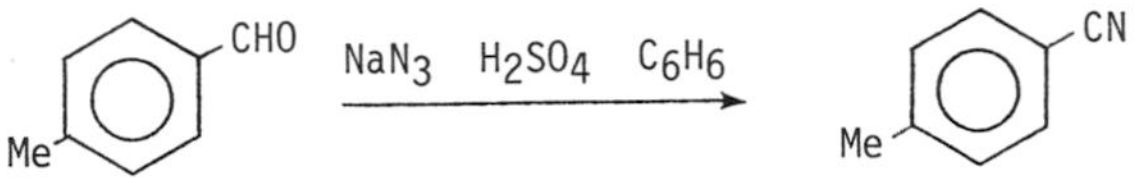

JACS (1952) <u>74</u> 1168
Org React (1946) <u>3</u> 307   44%

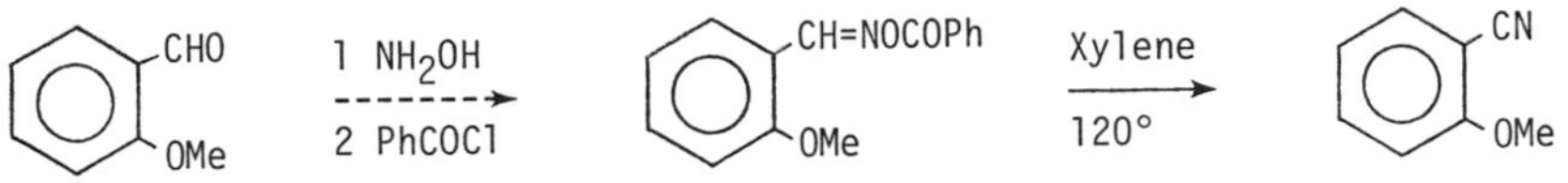

JCS (1950) 1243

PhCH=CHCHO $\xrightarrow[\text{Pyr}\quad C_6H_6]{CF_3COONHCOCF_3}$ PhCH=CHCN   88%

JACS (1959) <u>81</u> 6340

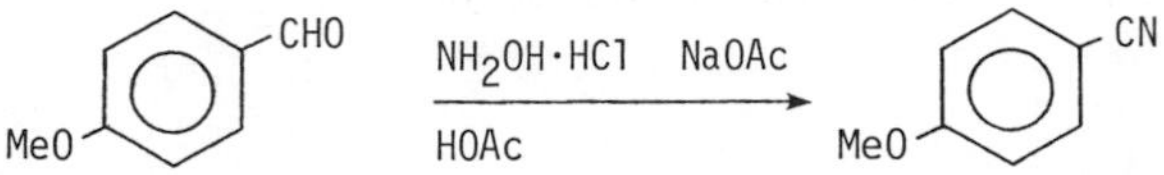

$\xrightarrow[\text{HOAc}]{NH_2OH \cdot HCl \quad NaOAc}$

67%

Chem Ind (1961) 1873
JCS (1965) 1564

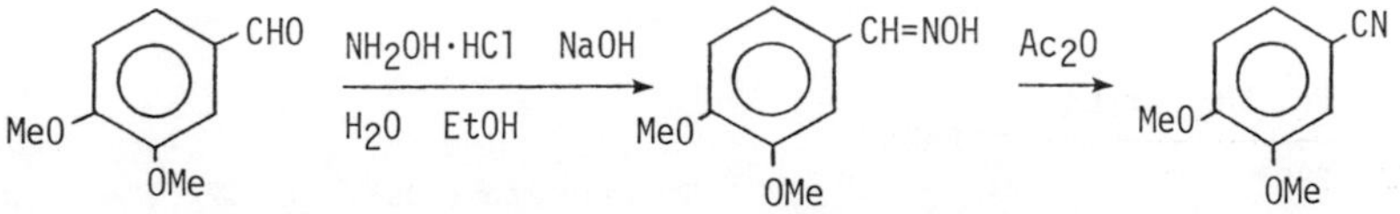

$\xrightarrow[\text{H}_2\text{O}\quad\text{EtOH}]{NH_2OH \cdot HCl \quad NaOH}$ $\xrightarrow{Ac_2O}$

70-76%

Org Synth (1943) Coll Vol 2 622

The following reagents may also be used for dehydration of oximes to
nitriles:

HCl-EtOH   Can J Chem (1967) <u>45</u> 1014
MsCl-Pyr-collidine Tetr Lett (1965) 2497
PhNCO-Et$_3$N   JOC (1961) <u>26</u> 782
PhN=CCl$_2$ Bull Chem Soc Jap (1962) <u>35</u> 1104
(PhO)$_2$POH-CCl$_4$-Et$_3$N JOC (1969) <u>34</u> 2805
p-chlorophenyl chlorothionoformate Chem Comm (1970) 1014

BuCHO $\xrightarrow[\text{C}_6\text{H}_6]{NH_2NMe_2}$ BuCH=NNMe$_2$ $\xrightarrow[\substack{\text{2 MeONa}\quad\text{MeOH}\\ \text{or H}_2\text{O}_2\quad\text{H}_2\text{O}}]{\text{1 MeI}}$ BuCN   42-47%

JOC (1962) <u>27</u> 4372
(1966) <u>31</u> 4100

PhCH$_2$CHO $\xrightarrow{Ph_2NNH_2}$ PhCH$_2$CH=NNPh$_2$ $\xrightarrow{O_2 \quad h\nu}$ PhCH$_2$CN   41%

Tetr Lett (1970) 2085

PhCHO $\xrightarrow[\text{2 190-200°}]{\text{1 } \text{NH}_2\text{N} \diagup \text{ TsOH } \text{C}_6\text{H}_6}$ PhCN   69%

Z Chem (1964) 4 304

Some of the methods listed in section 192 (Nitriles from Ketones) may also be used for the preparation of nitriles from aldehydes

Section 185   Nitriles from Alkyls

This section lists examples of the replacement of alkyl groups by nitriles. For the replacement of hydrogen by nitrile, RH $\longrightarrow$ RCN, see section 191 (Nitriles from Hydrides)

PhEt $\xrightarrow[389°]{\text{O}_2 \text{ NH}_3 \text{ Cr}_2\text{O}_3}$ PhCN   8%

Tetrahedron (1968) 24 6277

Section 186   Nitriles from Amides

25% (R=Et)
71% (R=H)

Tetr Lett (1970) 1963

PhCH$_2$CONH$_2$ --► PhCH$_2$CONHBr $\xrightarrow[\text{C}_6\text{H}_6]{\text{Ph}_3\text{P}}$ PhCH$_2$CN   60%

JCS (1960) 2976

PhCONH$_2$ $\xrightarrow{\text{Ph}_3\text{P} \quad \text{CCl}_4 \quad \text{THF}}$ PhCN   82%

Tetr Lett (1970) 4383

JACS (1949) 71 2650

The following reagents may also be used for the conversion of amides into nitriles:

SOCl$_2$	Org Synth (1963) Coll Vol 4 436
SOCl$_2$-DMF	Chem Ind (1964) 752
TsCl-Pyr	JACS (1955) 77 1701
COCl$_2$-Pyr	JCS (1954) 3730
PhSO$_3$H (235°)	JCS (1946) 763
P$_2$O$_5$	Org Synth (1963) Coll Vol 4 144
POCl$_3$-NaHSO$_3$	JOC (1957) 22 1142
AlCl$_3$	JACS (1940) 62 1432
Et$_3$SiH-ZnCl$_2$	Compt Rend (1962) 254 2357
Catechyl phosphorus trichloride	Ber (1963) 96 1387
Me$_2$NCHF$_2$	Coll Czech (1963) 28 2047
DCC-Pyr	JOC (1961) 26 3356
NaBH$_4$-Pyr	Chem Pharm Bull (1969) 17 98
	JOC (1967) 32 846
BuLi	JOC (1967) 32 3640
	(1966) 31 3873

Section 187   Nitriles from Amines
°°°°°°°°°°°°°°°°°°°°°°°

Org Synth (1932) Coll Vol 1 514
Rec Trav Chim (1961) 80 1075

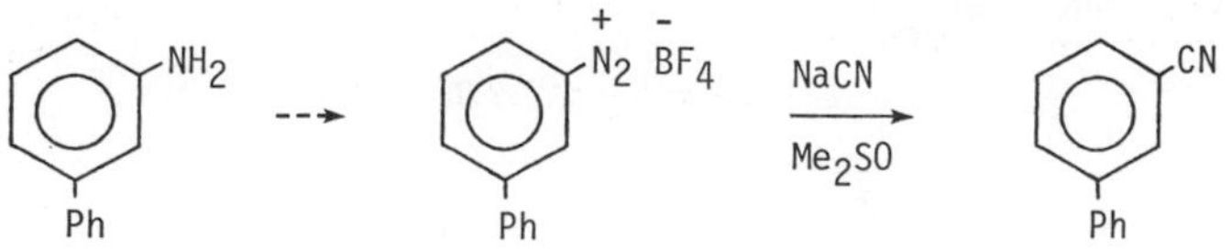

Org Prep and Procedures (1969) <u>1</u> 221

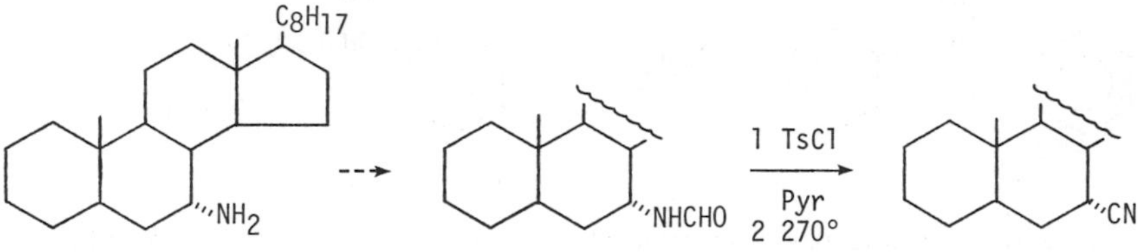

Tetr Lett (1966) 5087
JOC (1958) <u>23</u> 1221

$C_7H_{15}CH_2NH_2$ $\xrightarrow[C_6H_6]{\text{Nickel peroxide}}$ $C_7H_{15}CN$ 96%

Chem Pharm Bull (1963) <u>11</u> 296

$PhCH_2NH_2$ $\xrightarrow{Pb(OAc)_4 \quad C_6H_6}$ $PhCN$ 59%

Tetrahedron (1967) <u>23</u> 721

$BuCH_2NH_2$ $\xrightarrow{AgO \quad H_2O}$ $BuCN$ 87%

Tetr Lett (1968) 5685

$PrCH_2NH_2$ $\xrightarrow[2 \text{ CsF \quad MeCN}]{1 \text{ Cl}_2 \quad NaHCO_3 \quad H_2O}$ $PrCN$ > 81%

JOC (1968) <u>33</u> 1008

$C_5H_{11}CH_2NH_2$ $\xrightarrow{\text{IF}_5 \quad \text{Pyr} \quad CH_2Cl_2}$ $C_5H_{11}CN$ 36%

JOC (1961) 26 2531

Section 188   Nitriles from Esters

The reaction ROTs $\longrightarrow$ RCN is included in section 190 (Nitriles from Halides and Sulfonates)

$EtCH_2COOMe$ $\xrightarrow{\text{MeONO} \quad \text{MeONa} \quad \text{DMF}}$ EtCN   18%

J Prakt Chem (1969) 311 370

$PhCH=CHCH_2CH(COOEt)_2$ $\xrightarrow[\text{NaOEt}]{C_5H_{11}ONO}$ $PhCH=CHCH_2\underset{\underset{NOH}{\|}}{C}COOH$ $\xrightarrow{Ac_2O}$ $PhCH=CHCH_2CN$

JCS (1950) 926

Section 189   Nitriles from Ethers

NaCN hν
t-BuOH H₂O   30%

JACS (1966) 88 2884

Electrolysis Et₄NCN
MeCN   45%

JACS (1969) 4181

$$ArOMe \xrightarrow{KCN} MeCN$$

Ber (1933) 66B 1623

Section 190   Nitriles from Halides and Sulfonates
○○○

1 NaNH$_2$ NH$_3$
2 PrCH$_2$CN

30%

J Med Chem (1971) 14 72

$$EtI \xrightarrow[EtONa]{\overset{COOEt}{\overset{|}{CH_2CN}}} Et_2\overset{COOEt}{\underset{|}{C}}CN \xrightarrow{NaOH \quad H_2O} Et_2\overset{COOH}{\underset{|}{C}}CN \xrightarrow{\Delta} Et_2CHCN$$

Δ

JACS (1959) 81 5397
(1938) 60 131
Org React (1957) 9 107

$$EtBr \xrightarrow{CH_3CN \quad NaNH_2 \quad NH_3} EtCH_2CN$$

58%

JACS (1945) 67 2152
Ber (1968) 101 3113
Org React (1957) 9 107

Further examples of the alkylation of nitriles with halides are included in section 193 (Nitriles from Nitriles)

PhBr ---->   PhB

ClCH$_2$CN   potassium 2,6-
―――――――――――――――――――
di-t-butylphenoxide   THF

PhCH$_2$CN

<75%

JACS (1969) 91 6854

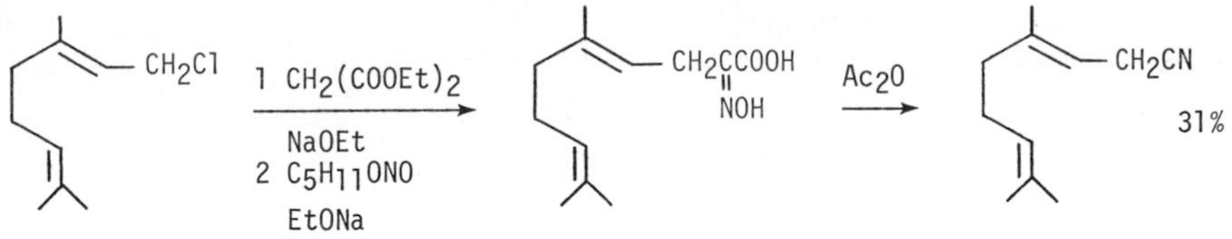

JCS (1950) 926

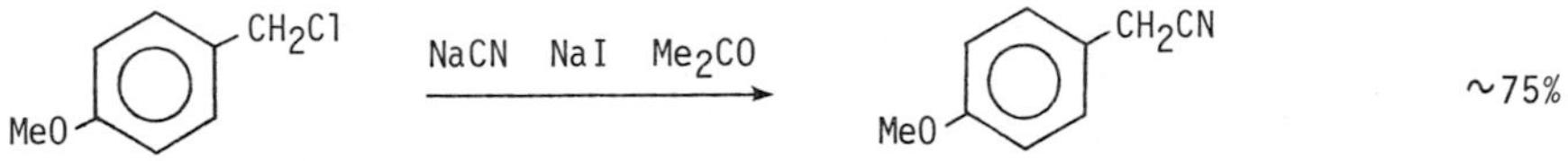

~75%

Org Synth (1963) Coll Vol 4 576

BuBr $\xrightarrow[]{\text{NaCN} \quad \text{Me}_2\text{SO}}$ BuCN

JOC (1960) 25 257 877

92%

92%

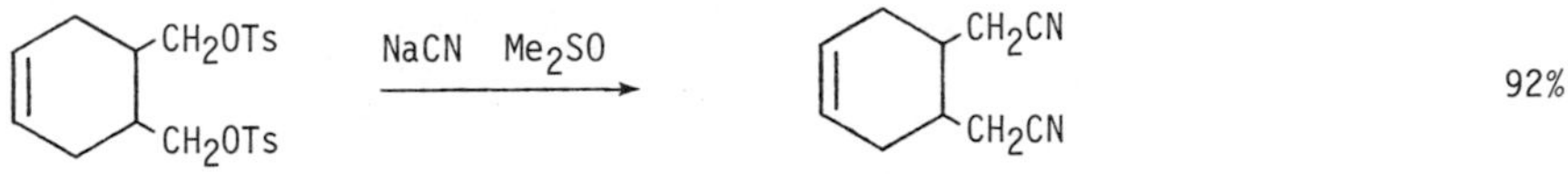

Tetr Lett (1964) 2273
JACS (1970) 92 336
JOC (1958) 23 797

BuBr $\xrightarrow[\text{H}_2\text{O}]{\text{NaCN} \quad \overset{+}{\text{PhCH}_2\text{NMe}_3} \ \overset{-}{\text{CN}}}$ BuCN

69%

Chem Pharm Bull (1962) 10 427

PhCH$_2$Br $\xrightarrow[\text{C}_6\text{H}_6]{\text{Ion exch resin (CN form)}}$ PhCH$_2$CN

72%

JOC (1963) 28 698

$$C_7H_{15}C\equiv CCH=CHCH_2Br \xrightarrow[\text{xylene}]{\text{CuCN \quad NaI}} C_7H_{15}C\equiv CCH=CHCH_2CN \qquad 31\%$$

JACS (1953) <u>75</u> 3430
Ber (1954) <u>87</u> 712

$$RBr \xrightarrow[\text{2 ClCN or (CN)}_2]{\text{1 Mg \quad Et}_2\text{O}} RCN$$

63% (R=PhCH$_2$CH$_2$)
80% (R=Ph)
Annales de Chimie (1915) <u>4</u> 28
Compt Rend (1911) <u>152</u> 388

$$PhCH=CHBr \xrightarrow[\text{MeOH}]{\text{K}_4\text{Ni}_2\text{(CN)}_6 \quad \text{KCN}} PhCH=CHCN \qquad 78\%$$

JACS (1969) <u>91</u> 1233
JOC (1969) <u>34</u> 3626

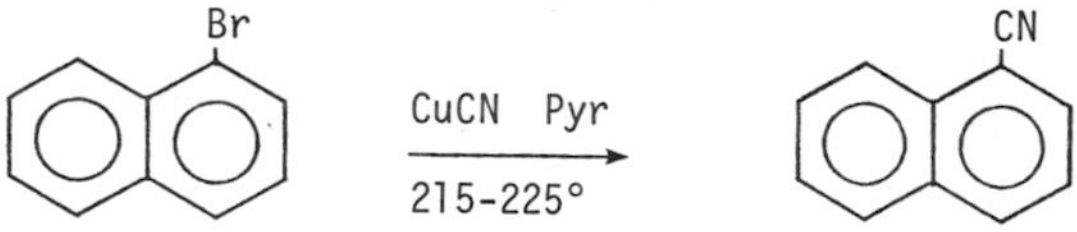

82-90%

Org Synth (1955) Coll Vol 3 631 212
JACS (1966) <u>88</u> 3318
JOC (1952) <u>17</u> 298

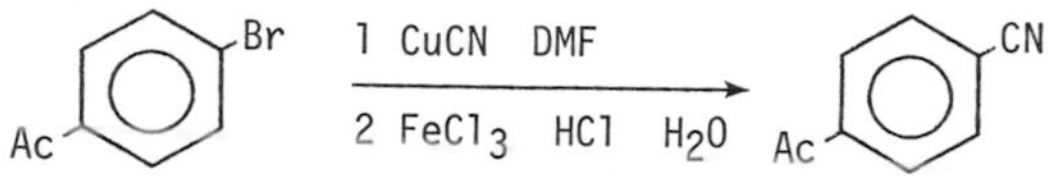

88%

JOC (1961) <u>26</u> 2522

The following solvents may also be used in the above reaction:

N-Methylpyrrolidone JOC (1961) <u>26</u> 2525
HMPA   JOC (1969) <u>34</u> 3626
Me$_2$SO   Proc Chem Soc (1962) 113

$$PhCH_2Br \xrightarrow[\text{2 NaNH}_2 \quad \text{NH}_3]{\text{1 Ph}_3\text{P}} PhCH=PPh_3 \xrightarrow{\text{NO}} PhCN \qquad 24\%$$

Chem Comm (1969) 166
Annalen (1969) <u>721</u> 34
Bull Chem Soc Jap (1967) <u>40</u> 2983

Section 191   Nitriles from Hydrides (RH)

This section lists examples of the replacement of hydrogen by CN,
RH → RCN. For the replacement of alkyl groups by CN, RR → RCN see
section 185(Nitriles from Alkyls)

$$
\underset{\underset{Me}{|}}{\overset{\overset{Me}{|}}{i\text{-PrCH}}} \xrightarrow{\text{ClCN \quad benzoyl peroxide}} \underset{\underset{Me}{|}}{\overset{\overset{Me}{|}}{i\text{-PrCCN}}}
$$

JACS (1969) 91 3028
Ber (1963) 96 670

$$
MeCH=CH_2 \xrightarrow{\text{Pd(CN)}_2 \quad \text{DMF}} MeCH=CHCN
$$
                                                                    20%

JACS (1966) 88 4105

$$
C_5H_{11}C\equiv CH \xrightarrow[\text{2 ClCN}]{\text{1 EtMgBr \quad Et}_2O} C_5H_{11}C\equiv CCN
$$
                                                                    67%

Bull Soc Chim Fr (1915) 17 228

~60%

JOC (1954) 19 1699
    (1949) 14 839

< 35%

JACS (1921) 43 898
JCS (1954) 678
Chem Met Eng (1921) 24 638
(Chem Abs 15 1703)

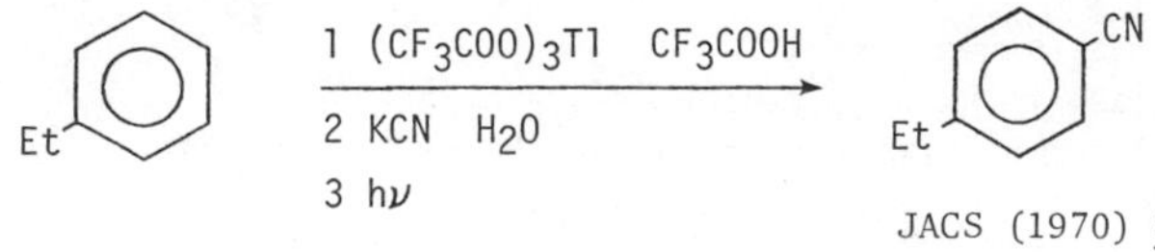

Et
1 $(CF_3COO)_3Tl$ CF_3COOH
2 KCN H_2O
3 hν
CN
Et

80%

JACS (1970) 92 3520

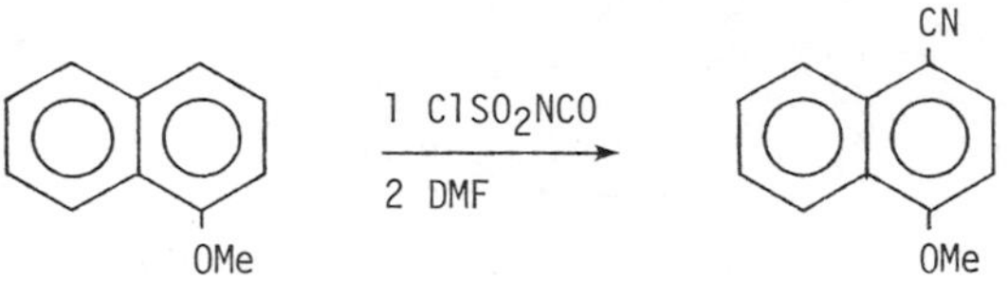

OMe
1 $ClSO_2NCO$
2 DMF
CN
OMe

81%

Ber (1967) 100 2719

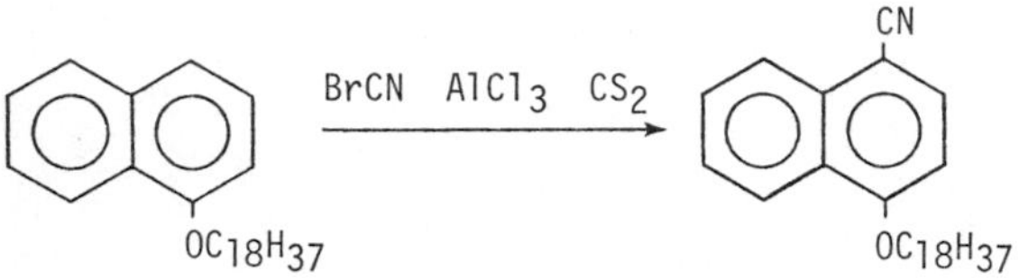

$OC_{18}H_{37}$
BrCN $AlCl_3$ CS_2
CN
$OC_{18}H_{37}$

80%

JCS (1954) 678

1 CCl_3CN $AlCl_3$
2 KOH
CN

65%

Ber (1933) 66B 339

NO_2
OMe
KCN O_2 hν
t-BuOH H_2O
NO_2
CN
OMe

53%

JACS (1966) 88 2884

PhC≡C
NaCN electrolysis
MeOH
CN
PhC≡C

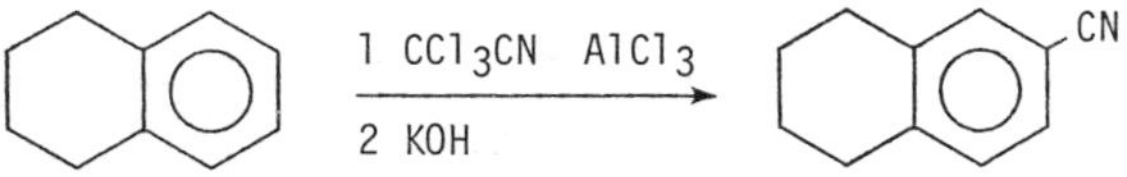

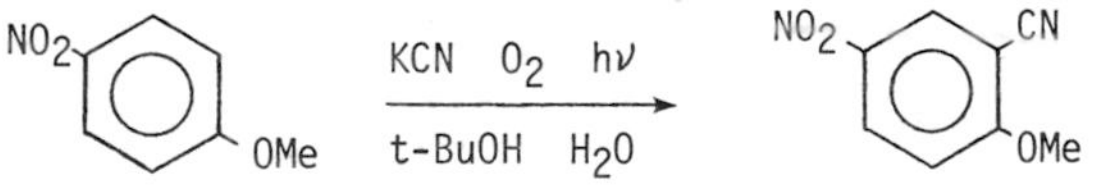

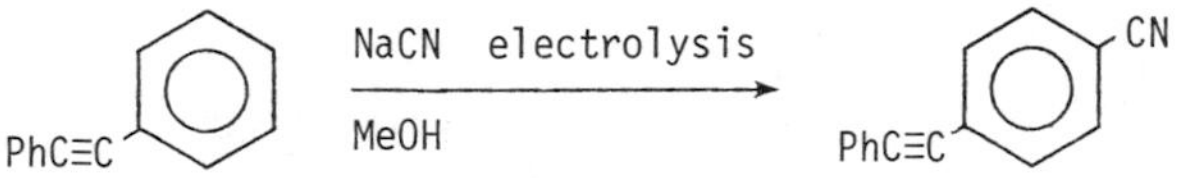

35%

Chem Comm (1970) 711
JACS (1969) 91 4181

Section 192   <u>Nitriles from Ketones</u>

$$\underset{\overset{|}{Me}}{\overset{EtCO}{}} \quad \xrightarrow[\text{HOAc } C_6H_6]{\overset{COOEt}{\underset{|}{CH_2CN}} \text{ alanine}} \quad \underset{\overset{|}{Me}}{\overset{COOEt}{EtC=CCN}} \quad \xrightarrow[Et_2O]{PhCH_2MgCl} \quad \underset{\overset{|}{Me} \; \overset{|}{COOEt}}{\overset{PhCH_2}{EtC\!-\!CHCN}}$$

$$\downarrow KOH$$

$$\underset{\overset{|}{Me}}{\overset{PhCH_2}{EtCCH_2CN}}$$

~70%

Org Synth (1963) Coll Vol 4 93
JCS <u>C</u> (1969) 121

$$\text{(cyclohexanone)} \quad \xrightarrow[\text{2 165-175°}]{\overset{COOH}{1 \; \underset{|}{CH_2CN}} \; NH_4OAc} \quad \text{(cyclohexene-CH}_2CN\text{)} \quad \dashrightarrow \quad \text{(cyclohexane-CH}_2CN\text{)}$$

76-91%

Org Synth (1963) Coll Vol 4 234

$$\underset{\overset{|}{Me}}{\overset{EtCO}{}} \quad \xrightarrow[\text{HOAc EtOH}]{\overset{COOEt}{\underset{|}{CH_2CN}} \; H_2 \; Pd\text{-}C \; NH_4OAc} \quad \underset{\overset{|}{Me}}{\overset{COOEt}{EtCHCHCN}} \quad \xrightarrow{\Delta} \quad \underset{\overset{|}{Me}}{EtCHCH_2CN} \qquad 64\%$$

JACS (1959) <u>81</u> 5397

$$\xrightarrow[\text{NaOEt DMF}]{(EtO)_2OPCH_2CN} \quad \overset{CHCN}{\text{(=CHCN)}} \quad \dashrightarrow \quad \overset{CH_2CN}{\text{(CH}_2CN)}$$

84%

JOC (1965) <u>30</u> 505

$$Ph_2CO \quad \xrightarrow[\text{2 } H_3PO_4]{\text{1 } LiCH_2CN \; THF \; hexane} \quad Ph_2C=CHCN \quad \dashrightarrow \quad Ph_2CHCH_2CN$$

58%

JOC (1968) <u>33</u> 3402

1 KCN HOAc

2 POCl₃ Pyr

H₂ Pd-C

EtOAc HOAc

35%

JCS C (1968) 2283

$C_{14}H_{29}CH_2COPh$ $\xrightarrow[\text{HCl H}_2\text{O dioxane}]{\text{i-PrCH}_2\text{CH}_2\text{ONO}}$ $C_{14}H_{29}\underset{\underset{NOH}{\|}}{C}COPh$ $\xrightarrow[\text{NaOH H}_2\text{O}]{\text{TsCl}}$ $C_{14}H_{29}CN$ >63%

JACS (1953) 75 2347
JOC (1961) 26 3507

Some of the methods listed in section 184 (Nitriles from Aldehydes) may also be applied to the preparation of nitriles from ketones

Section 193   Nitriles from Nitriles

Review: The Alkylation of Esters and Nitriles   Org React (1957) 9 107

$BuCH_2CN$ $\xrightarrow[\text{toluene}]{\text{BuBr NaNH}_2}$ Bu_3CCN 80%

JACS (1948) 70 3091

$PhCH_2CN$ $\xrightarrow{\text{Br}_2}$ $\underset{\underset{Br}{|}}{PhCHCN}$ $\xrightarrow{\text{PhH AlCl}_3}$ Ph_2CHCN 74%

JOC (1949) 14 839

PhCH$_2$CN $\xrightarrow[\text{Me}_2\text{SO} \quad \text{H}_2\text{O}]{\text{MeBr} \quad \text{NaOH}}$ PhCHCN   79%
                                             | Me

JOC (1969) 34 226

PhCH$_2$CN $\xrightarrow[\text{2 Cyclohexyl bromide}]{\text{1 NaNH}_2 \quad \text{NH}_3}$ PhCHCN   65-77%

Org Synth (1955) Coll Vol 3 219
JCS (1946) 25
JOC (1958) 23 1346

PhCH$_2$CN $\xrightarrow[\text{2 CO}_2]{\text{1 BuLi}}$ PhCHCN $\xrightarrow[\substack{\text{2 BuBr} \\ \text{3 } \Delta \text{ (decarbox)}}]{\text{1 NaNH}_2 \quad \text{NH}_3}$ PhCHCN   < 70%
                            | COOH | Bu

JOC (1966) 31 3873

For further examples of the alkylation of nitriles with halides see section 190 (Nitriles from Halides and Sulfonates)

Section 194   Nitriles from Olefins
ooooooooooooooooooooo

EtCH=CH$_2$ $\xrightarrow{\text{B}_2\text{H}_6}$ (EtCH$_2$CH$_2$)$_3$B $\xrightarrow[\text{t-butylphenoxide} \quad \text{THF}]{\text{Cl}_2\text{CHCN} \quad \text{potassium 2,6-di-}}$ (EtCH$_2$CH$_2$)$_2$CHCN

                                                               ~85%

JACS (1970) 92 5790

EtCH=CH$_2$ $\xrightarrow{\text{B}_2\text{H}_6}$ (EtCH$_2$CH$_2$)$_3$B $\xrightarrow[\text{t-butylphenoxide} \quad \text{THF}]{\text{ClCH}_2\text{CN} \quad \text{potassium 2,6-di-}}$ Et(CH$_2$)$_3$CN

JACS (1969) 91 6854

$$BuCH=CH_2 \xrightarrow[\text{THF}]{B_2H_6} (BuCH_2CH_2)_3B \xrightarrow[\text{2 KOH } H_2O]{\text{1 } N_2CHCN} Bu(CH_2)_3CN \qquad 95\%$$

JACS (1968) <u>90</u> 6891

$$C_6H_{13}CH=CH_2 \xrightarrow{CH_3CN \quad \text{dibenzoyl peroxide}} C_6H_{13}(CH_2)_3CN \qquad 17\%$$

JCS (1965) 1918 1932

$$\xrightarrow[\text{(PhO)}_3P]{HCN \quad Pd[P(OPh)_3]_4} \qquad 83\%$$

Chem Comm (1969) 112

$$\begin{array}{c} COOMe \\ | \\ (CH_2)_3CH=CH_2 \end{array} \xrightarrow{HCN \text{ (28 atmos)} \quad Co_2(CO)_8} \begin{array}{c} COOMe \\ | \\ (CH_2)_3CHMe \\ | \\ CN \end{array} \qquad \sim19\%$$

JACS (1954) <u>76</u> 5364

$$PhCH=CH_2 \longrightarrow \begin{array}{c} PhCCH_2N \\ \| \\ NOH \end{array} \xrightarrow[\text{Et}_3N]{TsCl} PhCN$$

JACS (1965) <u>87</u> 4642

Section 195   Nitriles from Miscellaneous Compounds

$$MeCH=CHCN \xrightarrow{Bu_3SnH \quad MeOH} MeCH_2CH_2CN \qquad 90\%$$

Tetr Lett (1967) 4805
(1969) 489

H$_2$ Pd-C
─────────────
EtOAc   HOAc

92%

JCS C (1968) 2283

PhCHCN $\xrightarrow{\text{Zn   HOAc}}$ PhCH$_2$CN
 |
 Cl

J Soc Chem Ind (1935) 54 98T

H$_2$ Pd-BaSO$_4$
─────────────
EtOH

51%

Arch Pharm (1957) 290 218

C$_7$H$_{15}$CH$_2$NO$_2$ $\xrightarrow[\text{2 Br}_2 \ \text{CCl}_4]{\text{1 Base}}$ C$_7$H$_{15}$CHNO$_2$ $\xrightarrow[\text{C}_6\text{H}_6]{\text{Ph}_3\text{P}}$ C$_7$H$_{15}$CN   50%
                                                                 |
                                                                 Br

JCS (1960) 2976

PhC≡CCONH$_2$ $\xrightarrow{\text{NaOCl   H}_2\text{O}}$ PhCH$_2$CN

Rec Trav Chim (1920) 39 704
                      (1927) 46 268

PhCH$_2$COCOOH $\xrightarrow{\text{NH}_2\text{OH·HCl   H}_2\text{O}}$ PhCH$_2$CN   90%

Can J Chem (1961) 39 1340
JCS (1950) 926

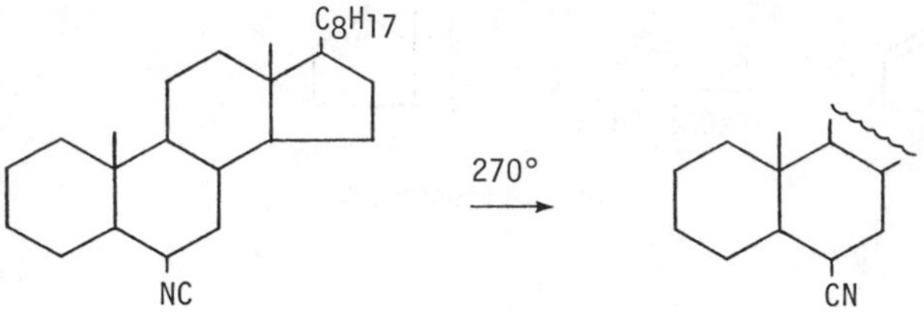

270°

<70%

Tetr Lett (1966) 5087

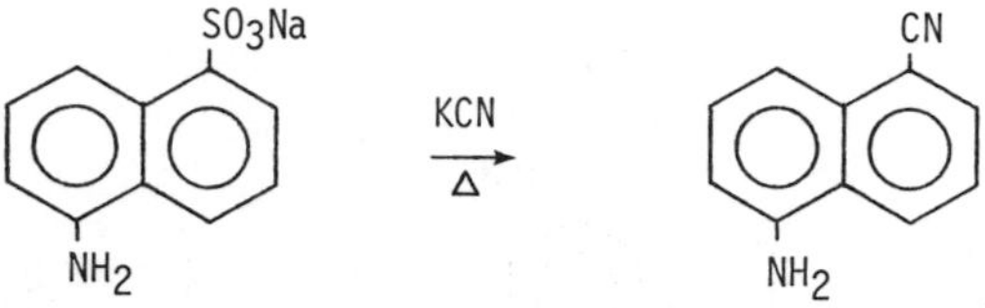

KCN
Δ

JCS (1954) 678

Chapter 14 PREPARATION OF OLEFINS

Olefins from Acetylenes

$BuC{\equiv}CH \xrightarrow[\text{2 BuLi}]{\text{1 Dicyclohexylborane}} BuCH_2CHLi_2 \xrightarrow{(C_5H_{11})_2CO} BuCH_2CH=C(C_5H_{11})_2$

30-35%

Tetr Lett (1966) 4315

$BuC{\equiv}CH \xrightarrow[\text{2 NaOH } I_2 \text{ THF } H_2O]{\text{1 (i-PrCH)}_2BH \text{ THF}}$ BuCH=CHCHPr-i (Me)

cis

75%

JACS (1967) $\underline{89}$ 3652

$BuC{\equiv}CH \xrightarrow[\text{THF}]{\text{(i-Bu)}_2AlH} BuCH_2CH[Al(Bu-i)_2]_2 \xrightarrow[\text{2 HCHO}]{\text{1 MeONa}} BuCH_2CH=CH_2$

Tetr Lett (1966) 6021

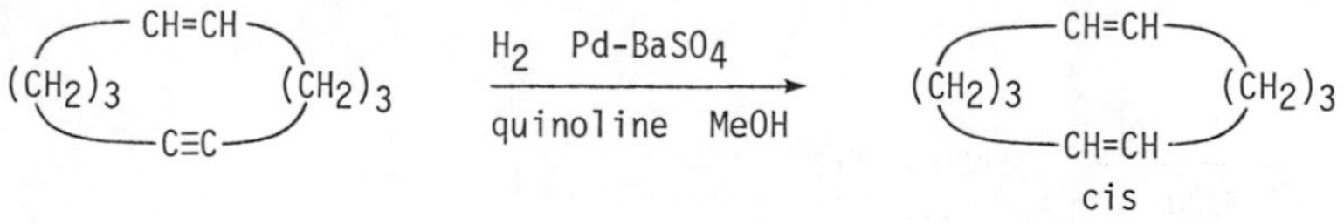

$\xrightarrow[\text{quinoline MeOH}]{H_2 \quad Pd\text{-}BaSO_4}$

cis

JACS (1956) $\underline{78}$ 2518

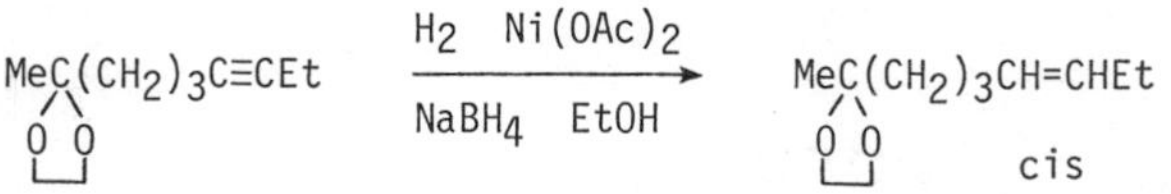

MeC(CH$_2$)$_3$C≡CEt $\xrightarrow[\text{NaBH}_4 \quad \text{EtOH}]{\text{H}_2 \quad \text{Ni(OAc)}_2}$ MeC(CH$_2$)$_3$CH=CHEt   80%

(with O O bridge on left reactant and product)

cis

Tetrahedron (1969) <u>25</u> 5149

BuC≡CEt $\xrightarrow[\text{LiCl} \quad \text{MeNH}_2]{\text{Electrolysis}}$ BuCH=CHEt   58%

trans

JOC (1968) <u>33</u> 2727

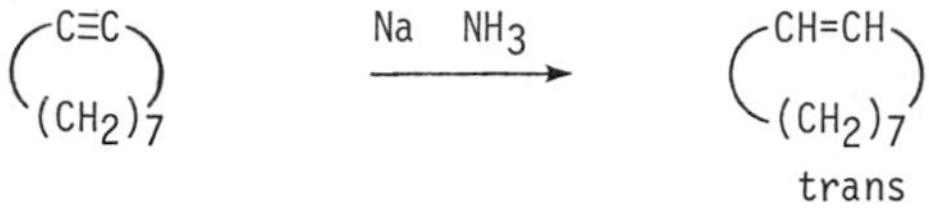

CH=CH
(CH$_2$)$_7$ <86%

trans

JACS (1952) <u>74</u> 3643
(1963) <u>85</u> 622

PrC≡C(CH$_2$)$_4$C≡CH $\xrightarrow[\text{NH}_3 \quad \text{Et}_2\text{O}]{\text{NaNH}_2 \quad \text{Na}}$ PrCH=CH(CH$_2$)$_4$C≡CH   75%

trans

JCS (1955) 3558

Li   MeNH$_2$

trans

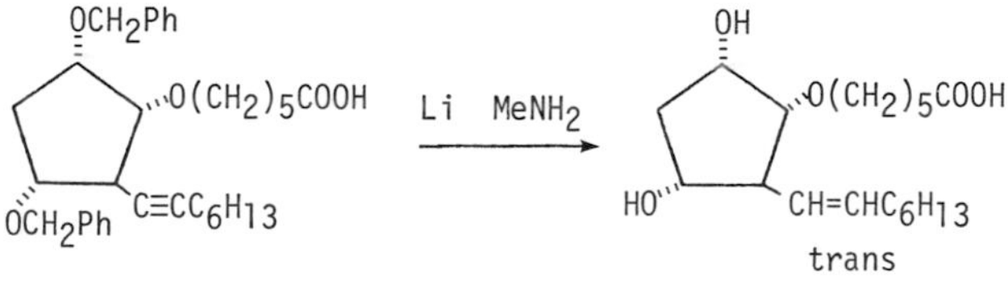

Chem Comm (1968) 634

PhC≡CPh $\xrightarrow[\text{2 MeOH}]{\text{1 Li} \quad \text{THF}}$ PhCH=CHPh

cis

JOC (1970) <u>35</u> 1702

$$\underset{\text{COOH}}{\overset{\text{C}\equiv\text{CPh}}{\bigcirc}} \quad \xrightarrow{\text{CrSO}_4 \quad \text{DMF} \quad \text{H}_2\text{O}} \quad \underset{\text{COOH}}{\overset{\text{CH}=\text{CHPh}}{\bigcirc}} \quad \text{trans} \qquad 85\%$$

JACS (1964) <u>86</u> 4358

$$\text{EtC}\equiv\text{CEt} \quad \xrightarrow[\text{2 HOAc}]{\text{1 NaBH}_4 \quad \text{BF}_3\cdot\text{Et}_2\text{O} \quad \text{diglyme}} \quad \begin{array}{c}\text{EtCH}=\text{CHEt}\\ \text{cis}\end{array} \qquad 68\%$$

JACS (1959) <u>81</u> 1512

$$\underset{}{\overset{\text{OH}}{\underset{\text{C}\equiv\text{CH}}{\bigcirc}}} \quad \xrightarrow{(\text{i-Bu})_2\text{AlH} \quad \text{C}_6\text{H}_6} \quad \underset{}{\overset{\text{OH}}{\underset{\text{CH}=\text{CH}_2}{\bigcirc}}} \qquad 48\%$$

JOC (1959) <u>24</u> 627
(1963) <u>28</u> 1254
Ber (1956) <u>89</u> 444

$$\text{EtC}\equiv\text{CEt} \underset{\text{2 100-130°}}{\overset{\text{1 (i-Bu)}_2\text{AlH}}{\bigg\langle}} \begin{array}{l} \xrightarrow{\text{2 MeLi}} \begin{array}{c}\text{EtCH}=\text{CHEt}\\ \text{cis}\end{array} \qquad 90\% \\ \xrightarrow{\text{1 (i-Bu)}_2\text{AlH-MeLi}} \begin{array}{c}\text{EtCH}=\text{CHEt}\\ \text{trans}\end{array} \qquad 88\% \end{array}$$

JACS (1967) <u>89</u> 5085 5086

$$\text{EtC}\equiv\text{CEt} \quad \xrightarrow[\text{117-138°}]{\text{LiAlH}_4 \quad \text{diglyme}} \quad \begin{array}{c}\text{EtCH}=\text{CHEt}\\ \text{trans}\end{array} \qquad 97\%$$

Tetrahedron (1967) <u>23</u> 4509

$$\text{C}_{10}\text{H}_{21}\text{C}\equiv\text{C(CH}_2)_5\text{COOH} \xrightarrow{\text{HI}} \underset{\underset{\text{I}}{|}}{\text{C}_{10}\text{H}_{21}\text{CH}=\text{C(CH}_2)_5\text{COOH}} \xrightarrow[\text{HOAc}]{\text{Zn}} \text{C}_{10}\text{H}_{21}\text{CH}=\text{CH(CH}_2)_5\text{COOH}$$

Compt Rend (1916) <u>162</u> 944

MeC≡CMe $\xrightarrow[\text{2 HCl H}_2\text{O}]{\text{1 B}_2\text{Cl}_4}$ MeCH=CHMe

JACS (1967) <u>89</u> 4217

Section 197   Olefins from Carboxylic Acids, Acid Halides and Anhydrides

The decarboxylation of αβ- and βγ-unsaturated acids, RCH=CHCOOH ⟶
RCH=CH₂ and RCH=CHCH₂COOH ⟶ RCH₂CH=CH₂, is included in section 152
(Hydrides from Carboxylic Acids)

$\underset{\text{(CH}_2)_4\text{COOH}}{\text{COOMe}}$ $\xrightarrow[\text{electrolysis MeOH}]{\text{CH}_2=\text{CH(CH}_2)_8\text{COOH}}$ $\underset{\text{(CH}_2)_{12}\text{CH=CH}_2}{\text{COOMe}}$

JCS (1953) 2393
    (1954) 448 4219

For further examples of the Kolbe electrolysis see section 62 (Alkyls
from Carboxylic Acids)

BuCH₂CH₂COOH $\xrightarrow{\text{Pb(OAc)}_4 \quad \text{Cu(OAc)}_2 \cdot \text{H}_2\text{O} \quad \text{Pyr}}$ BuCH=CH₂   72%

Tetrahedron (1968) <u>24</u> 2215
JCS <u>C</u> (1969) 1047

Tetr Lett (1968) 2471   60%

PhCH₂CH₂COCl $\xrightarrow[\text{or RhCl(PPh}_3)_3 \quad \text{toluene}]{\text{PdCl}_2 \quad 200°}$ PhCH=CH₂   53-71%

JACS (1968) <u>90</u> 94 99

$\underset{\text{Bu}_2\text{CCH}_2\text{Pr}}{\overset{\text{COOH}}{|}} \xrightarrow[\text{220-240}°]{\text{CuCN}} \text{Bu}_2\text{C=CHPr}$ 73%

JACS (1950) $\underline{72}$ 2792

$\text{C}_{13}\text{H}_{27}\text{CH}_2\text{CH}_2\text{COOH} \xrightarrow{\text{SeO}_2 \quad \Delta} \text{C}_{13}\text{H}_{27}\text{CH=CH}_2$

J Chem Soc Jap (1938) $\underline{59}$ 262 271

Electrolysis
Et$_3$N Pyr H$_2$O   35%

Tetr Lett (1968) 5117 5123

Pb(OAc)$_4$ O$_2$ Pyr   76%

JACS (1968) $\underline{90}$ 113
Helv (1958) $\underline{41}$ 1191

PbO$_2$ 170 190°   30-37%

Helv (1958) $\underline{41}$ 1191
JACS (1952) $\underline{74}$ 4370

t-BuOOH $\xrightarrow[\text{or } h\nu]{\Delta}$ 38%

Chem Comm (1969) 98

Section 198   Olefins from Alcohols

BuOH --→ BuOCH₂CH=CH₂ $\xrightarrow{\text{PrLi pentane}}$ BuCH=CH₂   ~73%

Tetr Lett (1969) 821

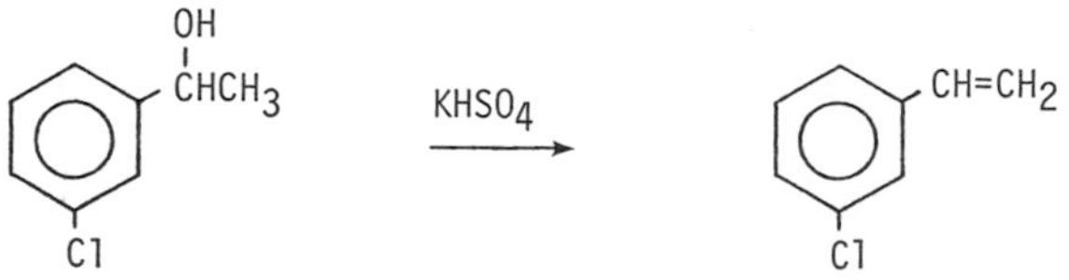

KHSO₄   80-82%

Org Synth (1955) Coll Vol 3 204
        (1943) Coll Vol 2 606
        (1941) Coll Vol 1 430

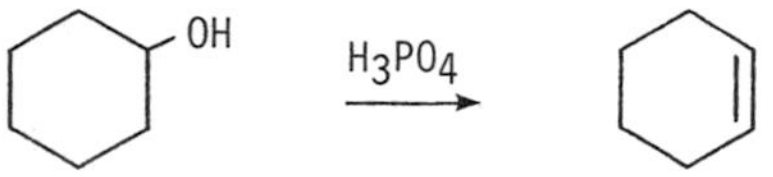

H₃PO₄   79-84%

Org Synth (1943) Coll Vol 2 151

OH
Me₂CCH₂Me $\xrightarrow{\text{I}_2}$ Me₂C=CHMe

JACS (1915) 37 1748

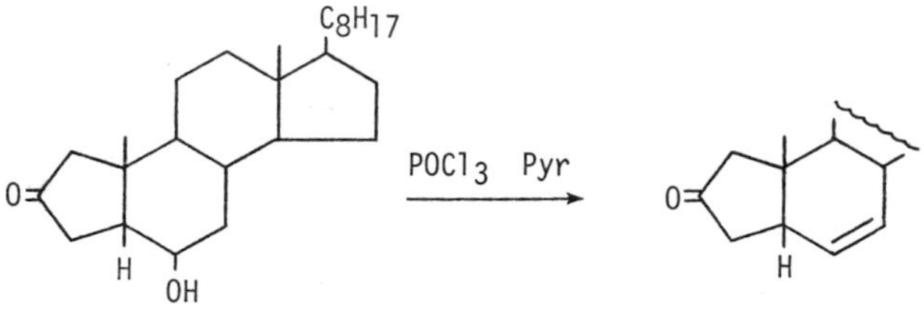

POCl₃ Pyr   83%

JACS (1961) 83 5003
Chem Pharm Bull (1961) 9 854
JCS C (1967) 1115

$(t-Bu)_2CCH_3$ | $\overset{SOCl_2 \quad Pyr}{\longrightarrow}$ | $(t-Bu)_2C=CH_2$ ~70%
$\overset{|}{OH}$

JACS (1960) 82 2498

$\overset{Me_2SO \quad 160°}{\longrightarrow}$ 88%

JOC (1964) 29 123

$\overset{MsCl}{\underset{Pyr}{\longrightarrow}}$ $\overset{t-BuOK}{\underset{\underset{C_6H_6}{Me_2SO}}{\longrightarrow}}$ 57%

Steroids (1964) 4 55
JOC (1964) 29 742
(1961) 26 2883

Further examples of the preparation of olefins from sulfonates are included in section 205 (Olefins from Halides and Sulfonates)

$\underset{\overset{|}{OH}}{CH_2-CHCMe_2}$ $\overset{1 \quad Ft_3NSO_2\overset{+}{N}\overset{-}{C}OOMc}{\underset{2 \quad NaH \quad THF}{\longrightarrow}}$ $\underset{\overset{|}{Me}}{CH_2-CHC=CH_2}$ 55%

JACS (1970) 92 5224
JOC (1970) 35 2594

$\overset{MeOSOCl}{\underset{Pyr \quad Et_2O}{\longrightarrow}}$ $\overset{215-275°}{\longrightarrow}$

JACS (1954) 76 1213

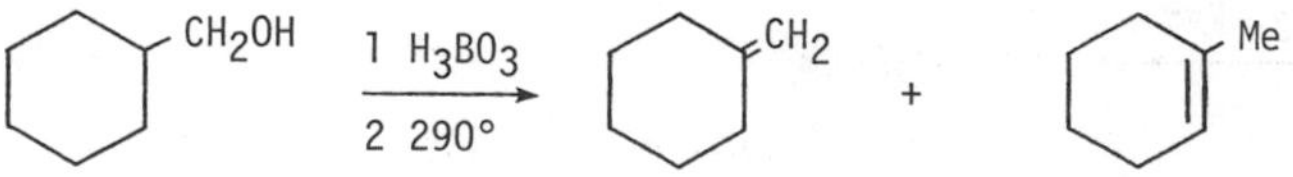

JOC (1961) 26 4193
Chem Rev (1960) 60 431

$C_{16}H_{33}CH_2CH_2OH$ $\xrightarrow[\text{(vapor phase)}]{Ac_2O\ (600°)}$ $C_{16}H_{33}CH=CH_2$ 68%

Chem Ind (1965) 681

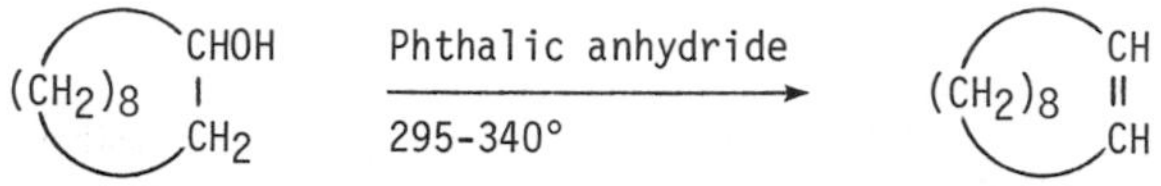

80%

JACS (1952) 74 3636

For examples of the preparation of olefins from esters, by pyrolysis and by other methods, see section 203 (Olefins from Esters)

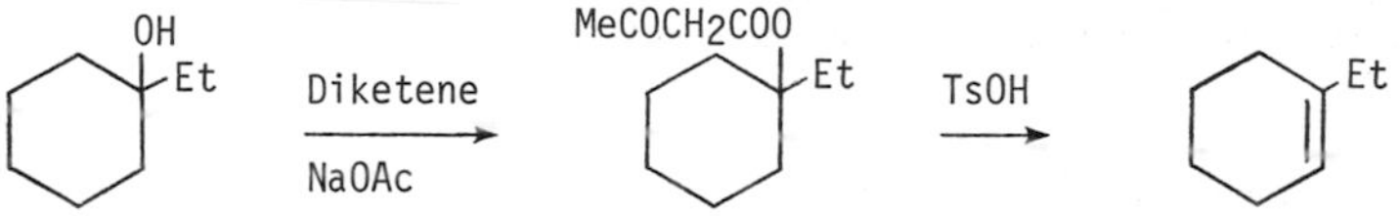

69%

Arkiv Kemi (1961) 17 401

$C_6H_{13}CH_2CH_2OH$ $\xrightarrow[\text{2 Bu}_3\text{P}]{\text{1 BrCH}_2\text{COCl}\ \ \text{Pyr}}$ $C_6H_{13}CH_2CH_2OCOCH_2\overset{+}{P}Bu_3\ \ \overset{-}{Br}$ $\xrightarrow{170°}$ $C_6H_{13}CH=CH_2$

~72%

JOC (1964) 29 1003
JACS (1961) 83 3336

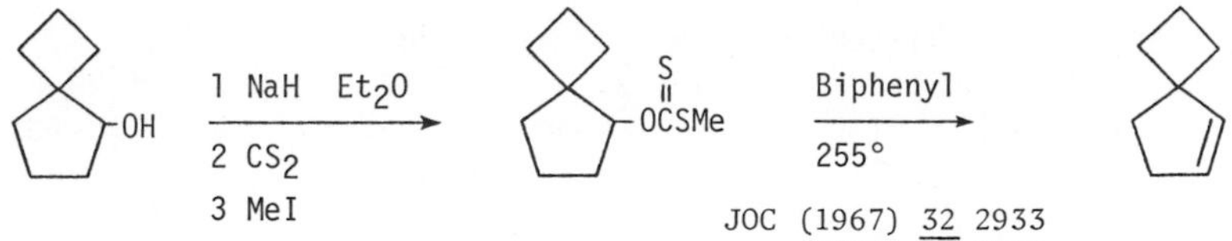

JOC (1967) 32 2933
(1968) 33 2214
Can J Chem (1970) 48 970
Org React (1962) 12 1

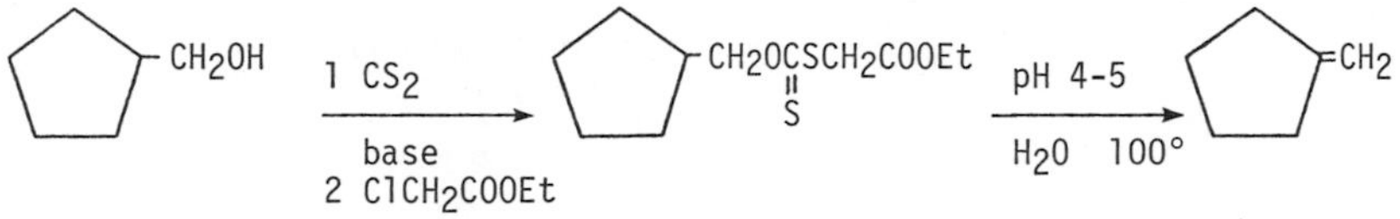

Ind Eng Chem Prod Res Dev (1970) 9 230

$C_6H_{13}CH_2CH_2OH \xrightarrow[\text{DMF}]{\text{NaH} \quad Me_2NCSCl} C_6H_{13}CH_2CH_2OC(=S)NMe_2 \xrightarrow{180\text{-}200°} C_6H_{13}CH{=}CH_2$ 86%

JOC (1969) 34 3604

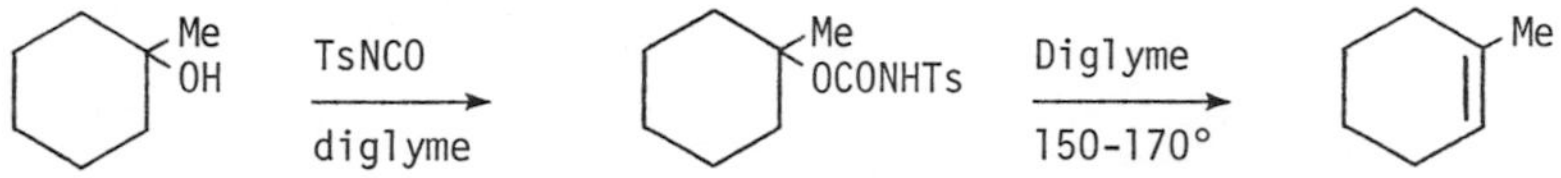

Chem Comm (1970) 606

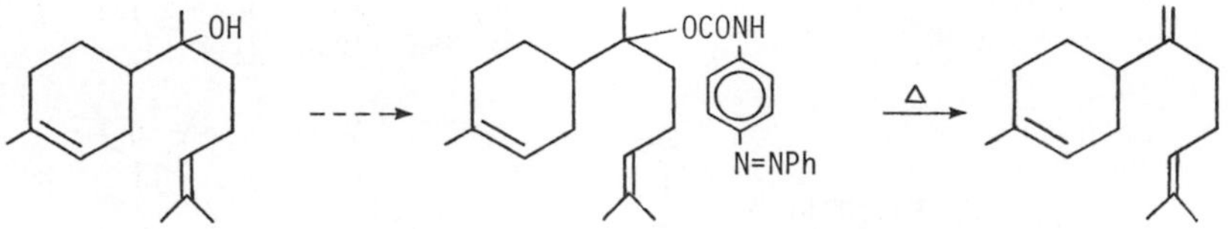

Aust J Chem (1954) 7 298
JACS (1953) 75 2118

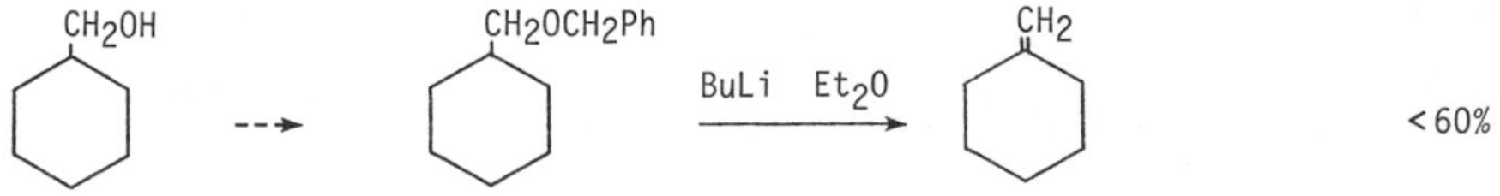

JCS (1958) 843

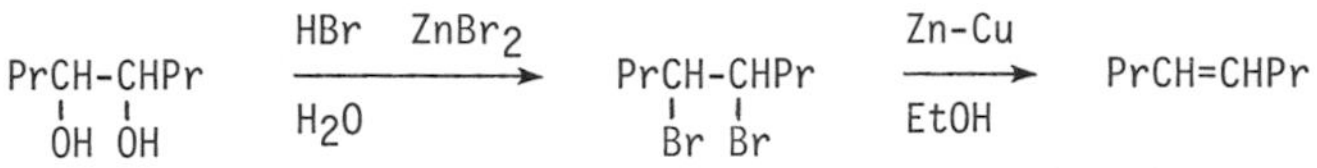

$$PrCH\text{-}CHPr \xrightarrow[\text{H}_2\text{O}]{\text{HBr} \quad \text{ZnBr}_2} PrCH\text{-}CHPr \xrightarrow[\text{EtOH}]{\text{Zn-Cu}} PrCH\text{=}CHPr$$

JACS (1937) 59 403
JCS (1951) 1079

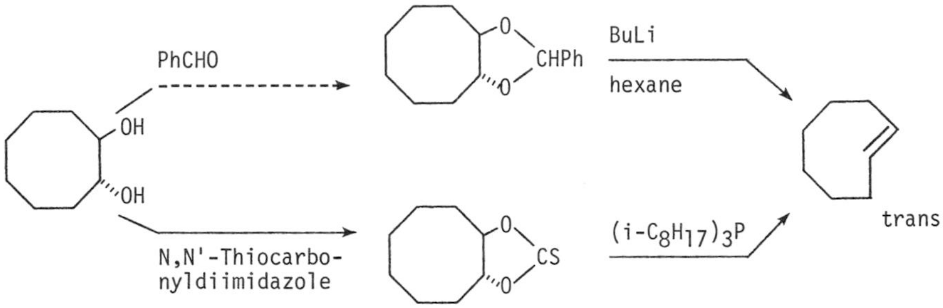

Chem Comm (1968) 1593
Tetr Lett (1968) 3655
JACS (1965) 87 934
(1963) 85 2677

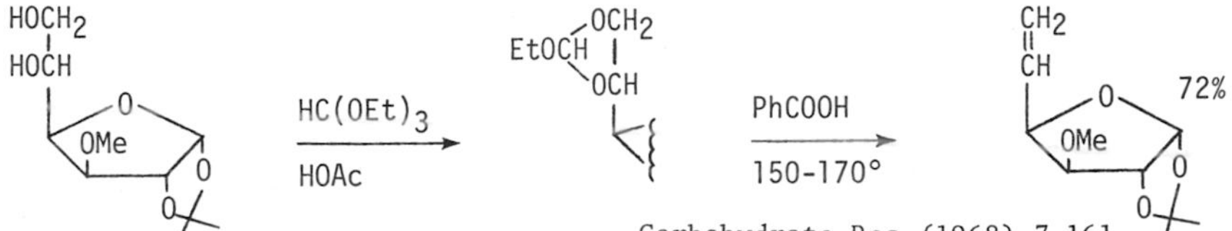

72%

Carbohydrate Res (1968) 7 161
Aust J Chem (1968) 21 2013

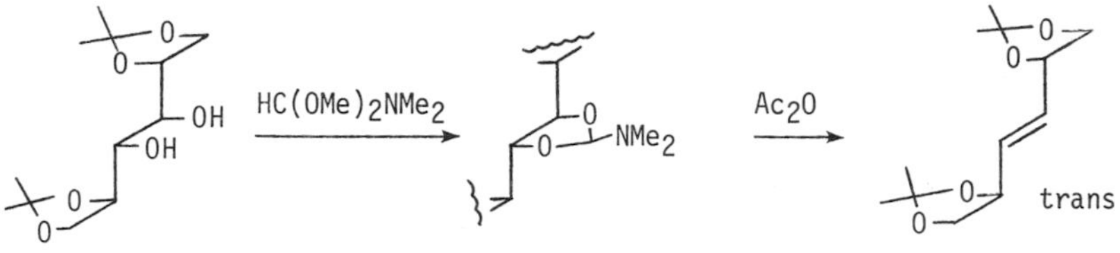

trans

Tetr Lett (1970) 5223

For the preparation of olefins from diols via bis sulfonates see section
205 (Olefins from Halides and Sulfonates)

Section 199   <u>Olefins from Aldehydes</u>

Examples of the decarbonylation of $\alpha\beta$-unsaturated aldehydes, RCH=CHCHO $\longrightarrow$ RCH=CH$_2$, are included in section 154 (Hydrides from Aldehydes)

PhCH=CHCHO $\xrightarrow{\text{PhCH}_2\text{COOH Ac}_2\text{O PbO}}$ PhCH=CHCH=CHPh   23-25%

Org Synth (1943) Coll Vol 2 229

PhCH$_2$CHO $\xrightarrow{\text{KOH   EtOH}}$ PhCH=CHCH$_2$Ph   91%

JOC (1966) <u>31</u> 396

$\xrightarrow{\text{(EtO)}_3\text{P   160°}}$

Izv (1960) 1030
(Chem Abs <u>54</u> 24627)

EtCH$_2$CHO $\xrightarrow[\text{MeOH}]{\text{Br}_2}$ EtCH-CHOMe $\xrightarrow{\text{EtMgBr}}$ EtCH-CHOMe $\xrightarrow[\text{PrOH}]{\text{Zn}}$ EtCH=CHEt $\sim$ 20%
                                      | | | |
                                     Br Br   Br Et

JACS (1932) <u>54</u> 751

$\xrightarrow[\text{Et}_2\text{O}]{\text{CH}_3\text{MgI}}$ $\xrightarrow{\text{KHSO}_4}$ 70%

JACS (1946) <u>68</u> 1085

Reviews: The Wittig Reaction   Org React (1965) <u>14</u> 270
                                          Quart Rev (1963) <u>17</u> 406

New Reactions of Alkylidenephosphoranes and their Preparative Uses
                                          Angew (1965) <u>77</u> 850
                                          (Internat Ed <u>4</u> 830)

PhCHO
$\xrightarrow{\begin{array}{c}1 \ Ph_3P=CH_2 \ \ Et_2O\\ \hline 2 \ 65°\end{array}}$
PhCH=CH$_2$ 67%

Ber (1954) <u>87</u> 1318

PhCHO
$\xrightarrow{Ph_3P=CHBu-t \ \ Et_2O}$
PhCH=CHBu-t   84%

JACS (1965) <u>87</u> 4156

PhCHO
$\xrightarrow[EtOLi \ \ EtOH]{2Br^- \ \begin{array}{c}Ph_3PCH_2\\ Ph_3PCH_2\end{array}}$

PhCH=CH / PhCH=CH   84%

Org React (1965) <u>14</u> 270

—CHO
$\xrightarrow[MeOH \ \ MeCN]{Ph_3PCH_2Pr^+ \ F^-}$
—CH=CHPr   41%

Ber (1970) <u>103</u> 2077

PhCH=CHCHO
$\xrightarrow[DMF \ \ MeONa]{PhCH_2Br-(Me_2N)_3P}$
PhCH=CHCH=CHPh   89%

Annalen (1965) <u>682</u> 58

PhCHO
$\xrightarrow{Ph_2PO(CH_2Ph) \ \ t\text{-}BuOK \ \ C_6H_6}$
PhCH=CHPh   70%

Ber (1959) <u>92</u> 2499

$$\text{PhCHO} \quad \xrightarrow[150°]{\overset{\displaystyle +}{\text{Ph}_3\text{PCH}_2\text{Et}} \; \overset{\displaystyle -}{\text{Br}} \;\; \overset{\displaystyle O}{\overset{\displaystyle /\!\!\backslash}{\text{CH}_2\text{-CH}_2}}} \quad \text{PhCH=CHEt} \qquad 74\%$$

Angew (1968) 80 535
(Internat Ed 7 536)

$$\text{PhCHO} \quad \xrightarrow[\text{2 Crystallization}]{\overset{\displaystyle \text{Me}}{\underset{}{\text{1 LiCHPO(NMe}_2)_2}}} \quad \underset{\text{diastereomer A}}{\overset{\displaystyle \text{OH Me}}{\text{PhCH-CHPO(NMe}_2)_2}} \quad \xrightarrow{\Delta} \quad \begin{array}{l}\text{PhCH=CHMe} \\ \text{cis}\end{array}$$

diastereomer B $\xrightarrow{\Delta}$ trans

$$\overset{\text{MnO}_2}{\swarrow} \qquad \overset{\text{NaBH}_4}{\nearrow}$$

$$\underset{\text{Me}}{\text{PhCOCHPO(NMe}_2)_2}$$

JACS (1966) 88 5653 5652
(1968) 90 6816
JOC (1969) 34 3053

$$\text{PrCHO} \quad \xrightarrow{\text{Ph}_3\text{P=CHPr}} \quad \underset{\underset{+ \quad -}{\text{Ph}_3\text{P} \quad O}}{\text{PrCHCHPr}} \quad \xrightarrow[\substack{\text{2 HCl} \\ \text{3 t-BuOK}}]{\text{1 PhLi}} \quad \begin{array}{l}\text{PrCH=CHPr} \\ \text{trans}\end{array} \qquad 72\%$$

Angew (1966) 78 115
(Internat Ed 5 126)
JACS (1970) 92 226

$$\xrightarrow[\text{2 65°}]{\overset{\displaystyle \text{Li}}{\text{1 Me}_2\text{CPS(OMe)}_2} \;\; \text{THF}}$$

68%

JACS (1966) 88 5654

$$\text{C}_{11}\text{H}_{23}\text{CHO} \quad \xrightarrow[\text{2 Toluene reflux}]{\overset{\displaystyle \text{Li}}{\text{1 LiCH}_2\text{SON-}\langle\rangle\text{-Me}} \;\; \text{THF}} \quad \text{C}_{11}\text{H}_{23}\text{CH=CH}_2 \qquad 56\%$$

JACS (1966) 88 5656
(1968) 90 5548

Further examples of the Wittig Reaction and a list of bases which may be
used for the generation of Wittig reagents are included in section 207
(Olefins from Ketones)

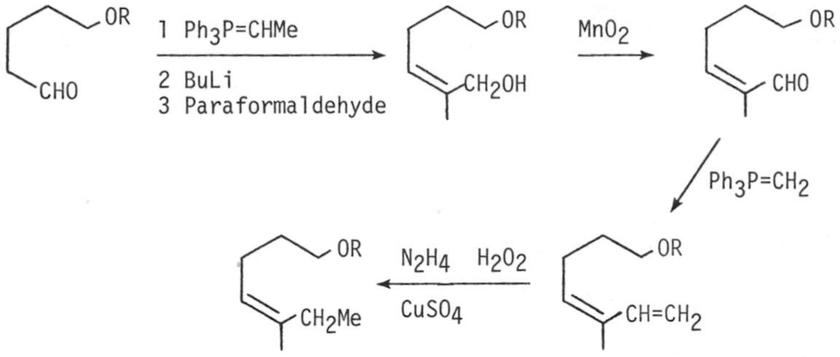

1 Ph₃P=CHMe

2 BuLi
3 Paraformaldehyde

MnO₂

Ph₃P=CH₂

N₂H₄   H₂O₂

CuSO₄

32%

JACS (1970) 92 6635

$C_6H_{13}CHO$
1 Ph₃P=CHMe   THF

2 BuLi

$C_6H_{13}CH-C=PPh_3$ Me I

OLi   Me

$C_6H_{13}CH=CMe_2$ <50%

JACS (1970) 92 226

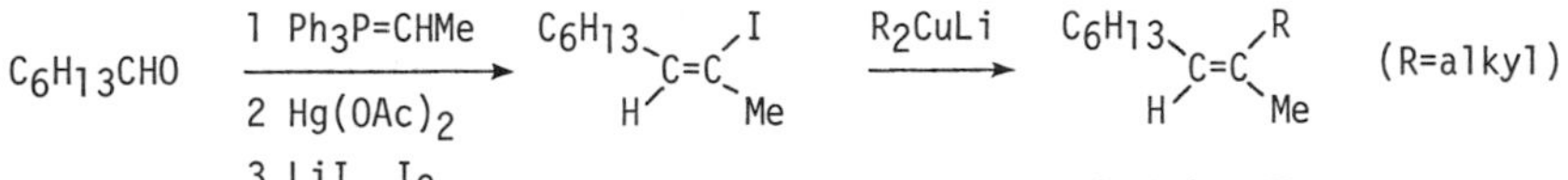

$C_6H_{13}CHO$
1 Ph₃P=CHMe

2 Hg(OAc)₂

3 LiI   I₂

R₂CuLi

(R=alkyl)

Tetr Lett (1970) 447

PhCHO
Me₃SiCH₂Ph   BuLi

HMPA

PhCH=CHPh   >50%

Tetr Lett (1970) 1137

$C_{11}H_{23}CHO$
CH₂I₂   Mg-Hg   Et₂O

$C_{11}H_{23}CH=CH_2$ 65%

Tetrahedron (1970) 26 1281
J Organometallic Chem (1968) 12 263

PhCHO $\xrightarrow{\text{C}_5\text{H}_{11}\text{CHLi}_2 \quad \text{THF}}$ PhCH=CHC$_5$H$_{11}$ 45-50%

Tetr Lett (1966) 4315

PhCH=CHCHO $\xrightarrow{\text{LiCH}_2\text{NC} \quad \text{THF}}$ PhCH=CHCH=CH$_2$ 28%

Angew (1968) 80 842
(Internat Ed 7 805)

(cyclohexyl)–CHO $\xrightarrow[\text{2 SOCl}_2]{\text{1 HCN}}$ (cyclohexyl)–CHCN(Cl) $\xrightarrow{\text{LiAlH}_4}$ (cyclohexyl)–CHCH$_2$ (N–H aziridine) $\xrightarrow{\text{BuONO}}$ (cyclohexyl)–CH=CH$_2$ 83%

Tetr Lett (1969) 4001

MeCH$_2$CHO - - → MeCH$_2$CH(OEt)$_2$ $\xrightarrow{\text{TsOH}}$ MeCH=CHOEt $\xrightarrow{\text{(i-Bu)}_2\text{AlH}_4}$ MeCH=CH$_2$ 44%

JOC (1966) 31 329

Some of the methods included in section 207 (Olefins from Ketones) may
also be applied to the preparation of olefins from aldehydes

Section 200   Olefins from Alkyls, Methylenes and Aryls
ooooooooooooooooooooooooooooooooooooooooooooo

PhCH$_2$CH$_2$Ph $\xrightarrow[260°]{\text{RhCl(PPh}_3)_3}$ PhCH=CHPh   37%

Tetr Lett (1970) 1825

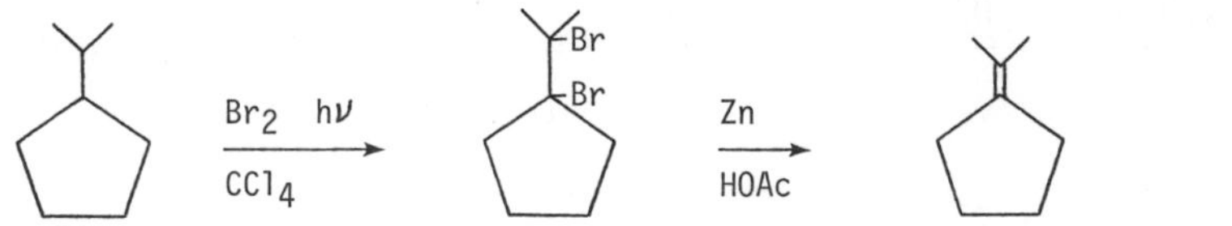

Rec Trav Chim (1964) **83** 67

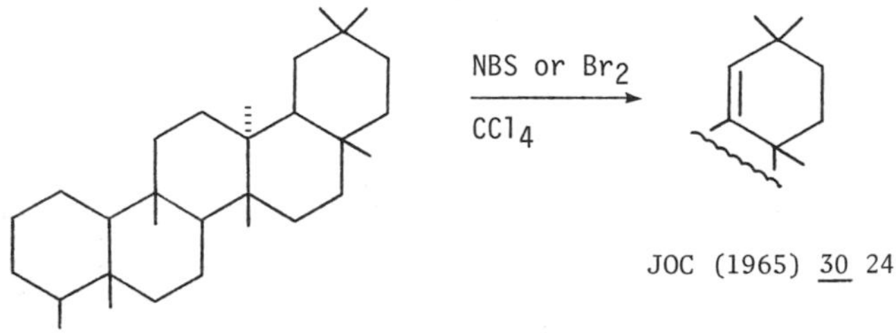

~37%

JOC (1965) **30** 2479

40%

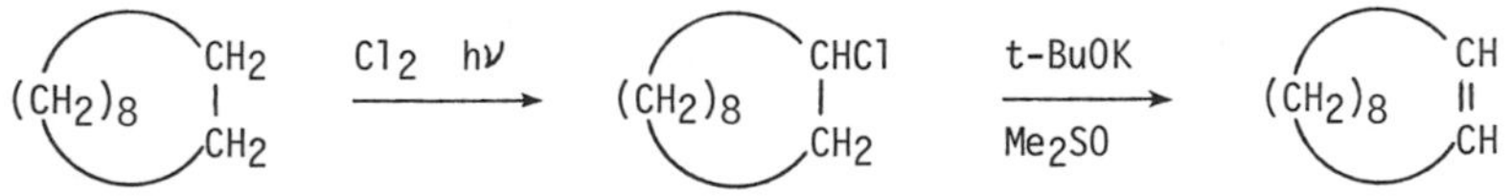

JOC (1967) **32** 510

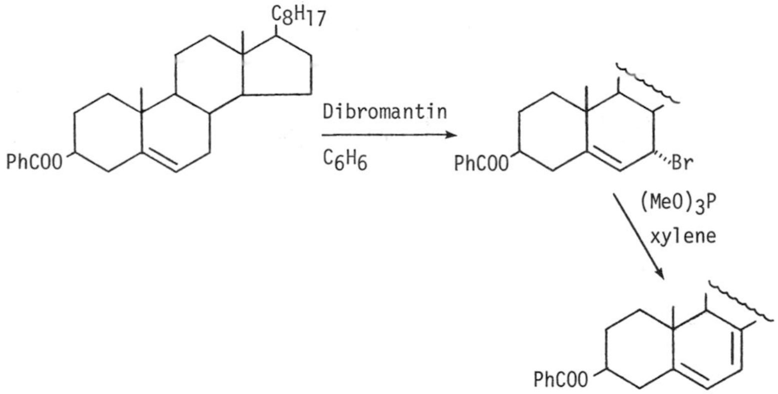

Helv (1958) **41** 70

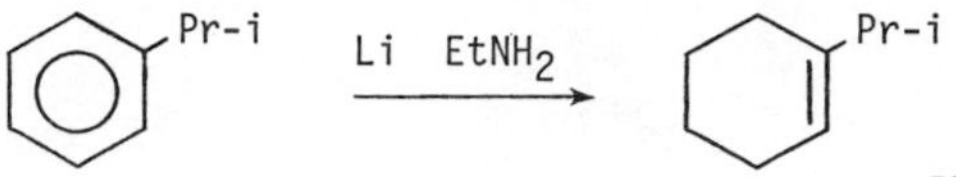

JOC (1963) 28 1094

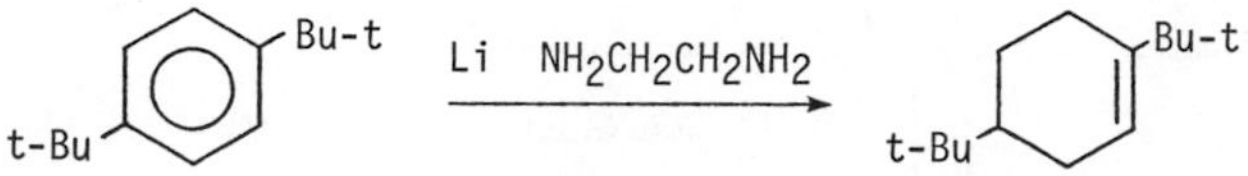

69%

JOC (1966) 31 965
Aust J Chem (1964) 17 55

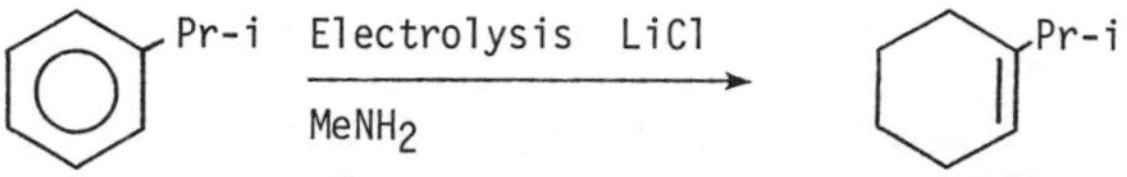

44%

JACS (1964) 86 5272

<81%

Section 201   Olefins from Amides

$$t\text{-BuCH}_2\text{CMe}_2 \xrightarrow[\text{(vapor phase)}]{510°} t\text{-BuCH=CMe}_2 \quad + \quad t\text{-BuCH}_2\text{C=CH}_2$$
|
NHAc   Me

JOC (1958) 23 996
JACS (1959) 81 651
     (1958) 80 4588

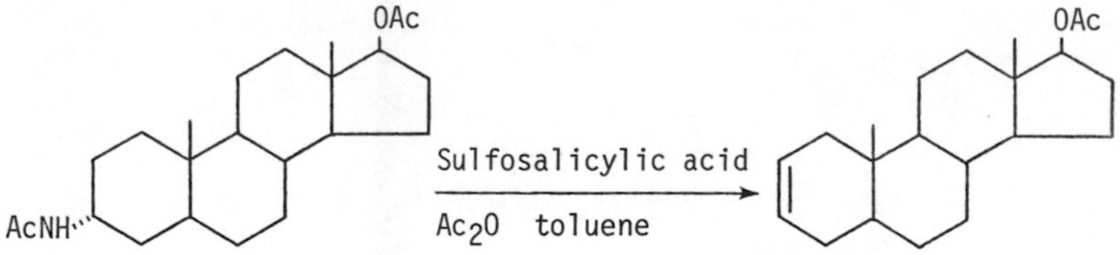

Sulfosalicylic acid

Ac₂O   toluene

Tetr Lett (1965) 2369

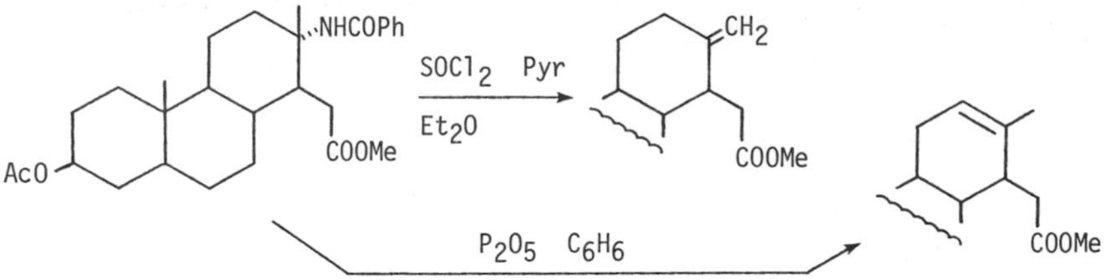

Bull Soc Chim Fr (1966) 2404

Section 202   Olefins from Amines

Review: The Hofmann Elimination Reaction and Amine Oxide Pyrolysis
                         Org React (1960) 11 317

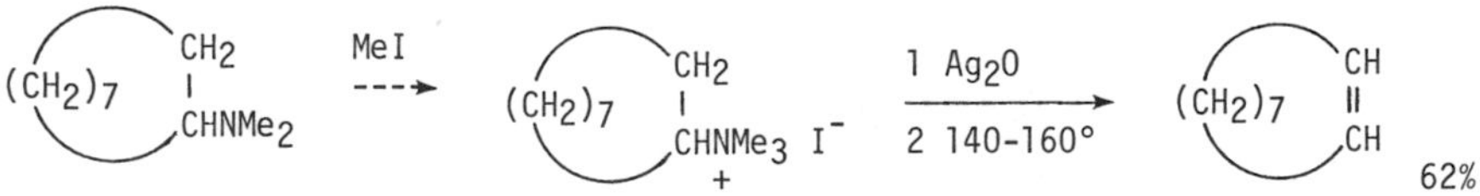

JACS (1952) 74 3643
JCS (1955) 4016

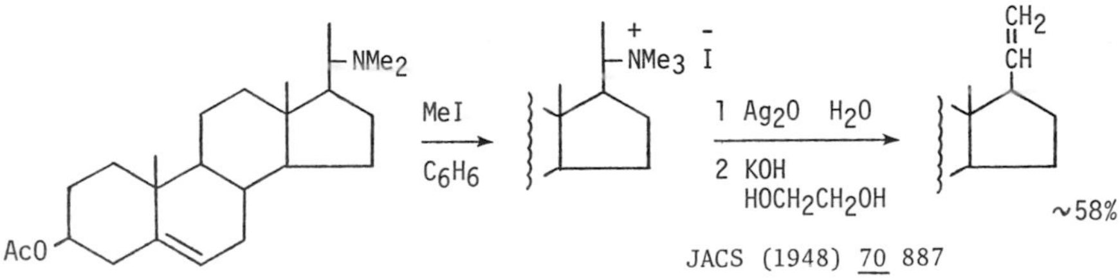

62%

JACS (1948) 70 887

~58%

The following bases may also be used in place of Ag$_2$O in the Hofmann
degradation of amines:
Ag$_2$SO$_4$-Ba(OH)$_2$, NaOH, TlOH, ion exchange resin   Org React (1960) 11 317
                                        Ag$_2$CO$_3$ JCS (1935) 1685

$$BuCH_2CH_2NH_2 \xrightarrow[H_2O]{Me_2SO_4 \quad NaOH} BuCH_2CH_2NMe_3^+ (SO_4)_{1/2}^- \xrightarrow[reflux]{H_2SO_4 \quad H_2O} BuCH=CH_2 \quad 60\%$$

Org React (1960) <u>11</u> 317

(A) PhLi
or (B) KNH$_2$

(A) cis
(B) trans

66-68%

Annalen (1958) <u>612</u> 102

$$\xrightarrow[MeOH]{H_2O_2 \quad H_2O} \qquad \xrightarrow{160°} \qquad 79\text{-}88\%$$

Org Synth (1963) Coll Vol 4 612

Me
|
PhCHCHCH$_3$
|
$^+$NMe$_2$
|
$^-$O

$$\xrightarrow[\text{or } H_2O \quad 132\text{-}138°]{Me_2SO \quad THF \quad 25°}$$

Me
|
PhC=CHMe   +

Me
|
PhCHCH=CH$_2$

JACS (1962) <u>84</u> 1734

$$\xrightarrow[Ac_2O]{\text{Sulfosalicylic acid}} \qquad > 70\%$$

Tetr Lett (1965) 2369

$$BuCH_2CH_2NH_2 \xrightarrow[\text{2 NaH \quad TsCl}]{1 \text{ TsCl \quad DMF}} BuCH_2CH_2NTs_2 \xrightarrow{KI \quad DMF} BuCH=CH_2 \quad 31\%$$

JACS (1969) <u>91</u> 2384

$PhCH_2CH_2NH_2$ $\xrightarrow{\text{(benzene ring with COCl and SO}_2\text{Cl)}}$ $PhCH_2CH_2N\langle \begin{smallmatrix} CO \\ SO_2 \end{smallmatrix} \rangle$(benzene ring) $\xrightarrow[\Delta]{KOH}$ $PhCH=CH_2$ < 65%

Tetr Lett (1967) 3027
JACS (1969) 91 2384

$Me_2CHCH_2NH_2$ $\xrightarrow{C_8H_{17}ONO \quad CHCl_3 \quad HOAc}$ $Me_2C=CH_2$ 27%

JACS (1969) 91 1790

$\xrightarrow{BuONO}$

Tetr Lett (1969) 4001

Section 203   Olefins from Esters

For the preparation of olefins from sulfonic esters see section 205
(Olefins from Halides and Sulfonates)

$HC\equiv C(CH_2)_7CH_2COOEt$ $\xrightarrow[Et_2O]{PhMgBr}$ $HC\equiv C(CH_2)_7CH_2\underset{OH}{\overset{}{C}}Ph_2$ $\xrightarrow{220\text{-}230°}$ $HC\equiv C(CH_2)_7CH=CPh_2$ 92%

JCS (1957) 1622

$C_{15}H_{31}COOMe$ $\xrightarrow[Et_2O]{EtMgBr}$ $C_{15}H_{31}\underset{OH}{\overset{}{C}}Et_2$ $\xrightarrow{HCOOH}$ $C_{15}H_{31}\underset{Et}{\overset{}{C}}=CHMe$ 96%

JACS (1945) 67 2239

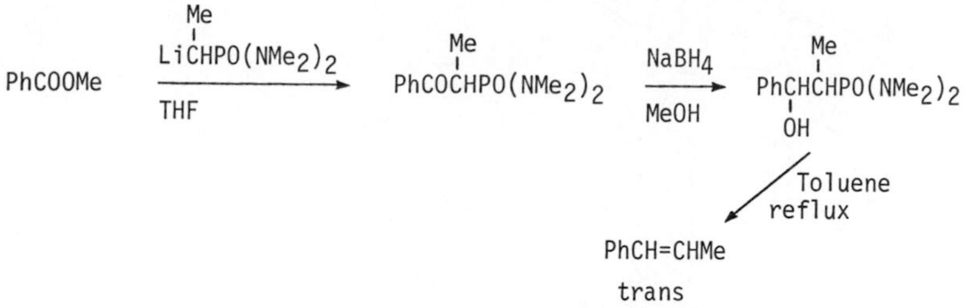
PhCOOMe
Me
LiCHPO(NMe2)2
THF
Me
PhCOCHPO(NMe2)2
NaBH4
MeOH
Me
PhCHCHPO(NMe2)2
OH
Toluene
reflux
PhCH=CHMe
trans

JACS (1966) 88 5653

Review: Pyrolytic Cis Eliminations Chem Rev (1960) 60 431

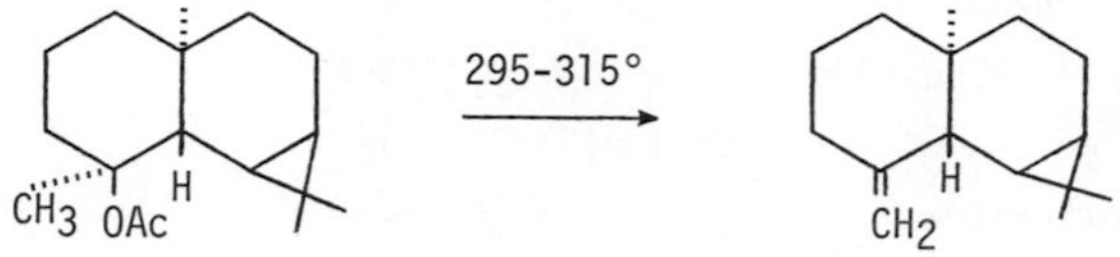
295-315°
CH3 OAc
H
H
CH2

JACS (1959) 81 1968

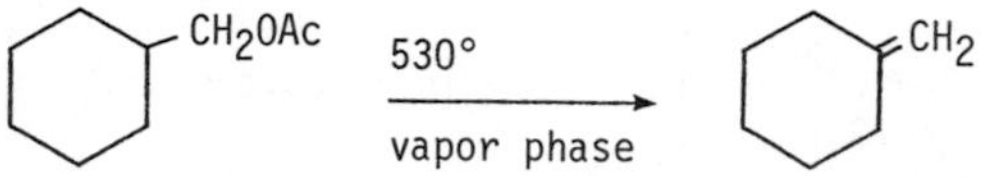
CH2OAc
530°
vapor phase
CH2

JACS (1959) 81 647 651

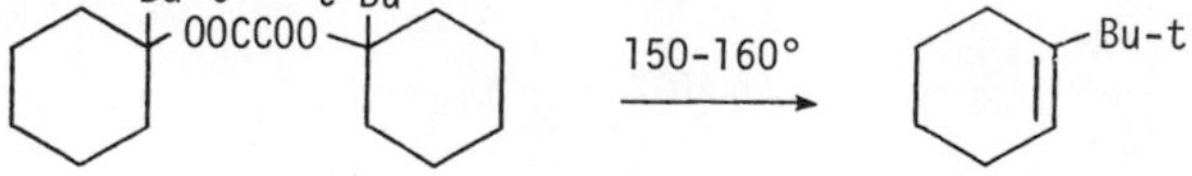
Bu-t
t-Bu
OOCCOO
150-160°
Bu-t

<80%

JACS (1969) 91 3324
JOC (1965) 30 689

330°
——→

71%

JCS (1965) 4379

Esters of the following acids have also been used for the preparation of olefins by liquid-phase pyrolysis:

Stearic	JACS (1948) 70 2690
	JCS (1957) 1998
d-Camphoric	JACS (1952) 74 3944
2,4,6-Triethylbenzoic	JACS (1953) 75 6011
2-Naphthoic	JACS (1954) 76 5692
3,5-Dinitrobenzoic	Chem Ind (1954) 1426
Anthraquinone-2-carboxylic	Helv (1944) 27 713 821
Ethyl carbonic	JACS (1952) 74 5454
2-Naphthyl carbonic	JACS (1954) 76 6108
Phenylcarbamic	JACS (1953) 75 2118
	Can J Chem (1953) 31 688

Further examples of the preparation of olefins by pyrolysis of esters are included in section 198 (Olefins from Alcohols)

PhCH₂CHPh
|
COO

Et   Et

NaNH₂ NH₃ Et₂O
——————————————→

PhCH=CHPh

Et

JACS (1953) 75 6011

Section 204 Olefins from Ethers and Epoxides

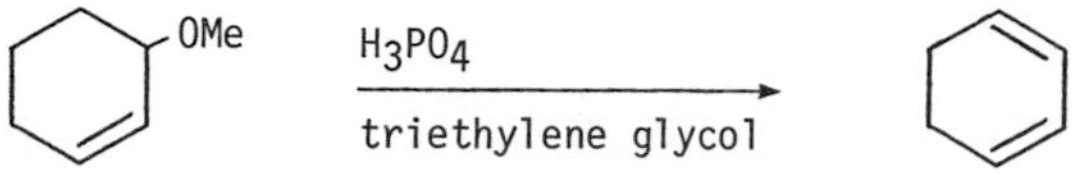
OMe
H_3PO_4
triethylene glycol

82%

Ber (1967) <u>100</u> 1764

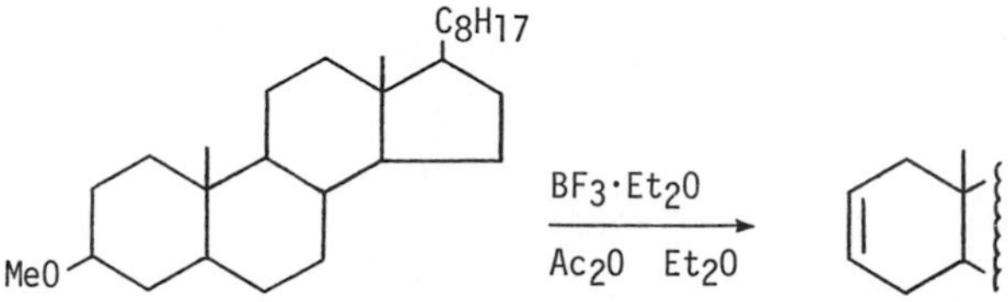
C_8H_{17}
MeO
$BF_3 \cdot Et_2O$
Ac_2O Et_2O

24%

Tetr Lett (1964) 759

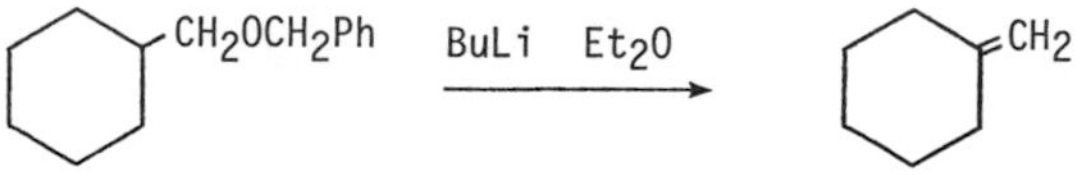
CH_2OCH_2Ph
BuLi Et_2O
CH_2

~60%

JCS (1958) 843

$$C_6H_{13}\overset{OMe}{\overset{|}{C}}HCH_3 \xrightarrow{\text{i-PrLi pentane}} C_6H_{13}CH{=}CH_2$$

JACS (1951) <u>73</u> 5708 1263

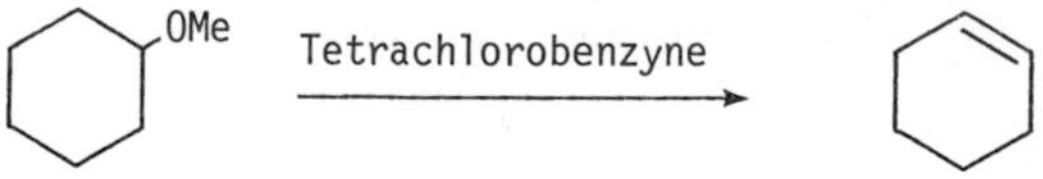
OMe
Tetrachlorobenzyne

Tetr Lett (1968) 4455

$EtCH_2CH_2OBu \xrightarrow[140-150°]{N_2CHCOOEt} EtCH{=}CH_2$ 6%

Rec Trav Chim (1955) 74 143

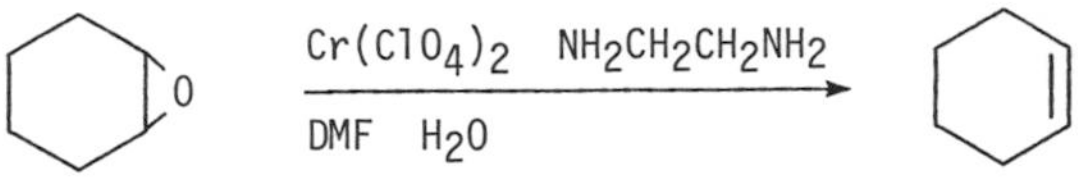
O
Cr(ClO4)2 NH2CH2CH2NH2
DMF H2O

92%

Tetrahedron (1968) 24 3503

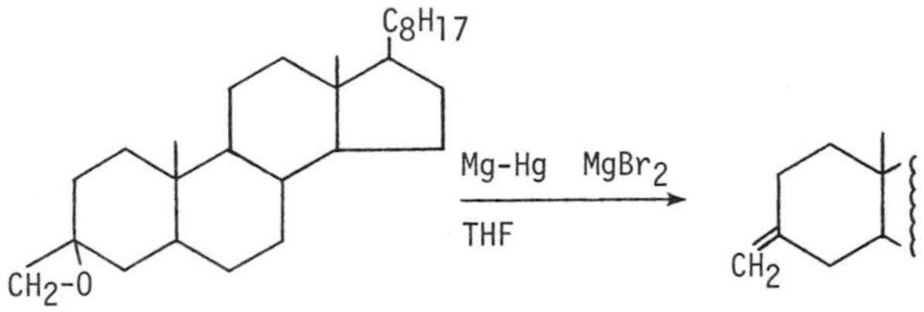
C8H17
CH2-O
Mg-Hg MgBr2
THF
CH2

80%

Chem Comm (1970) 144

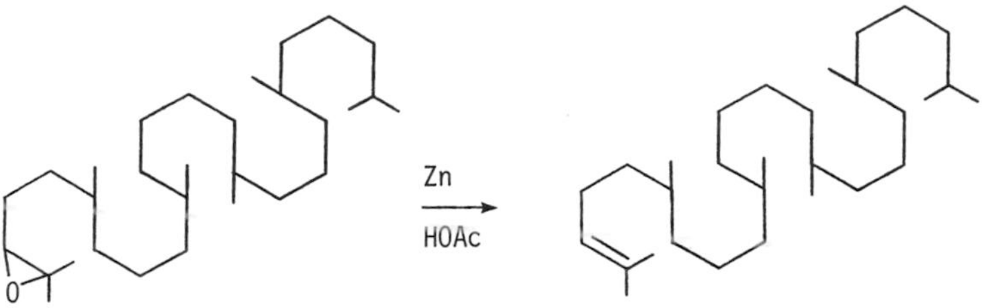
O
Zn
HOAc

Chem Comm (1970) 1450

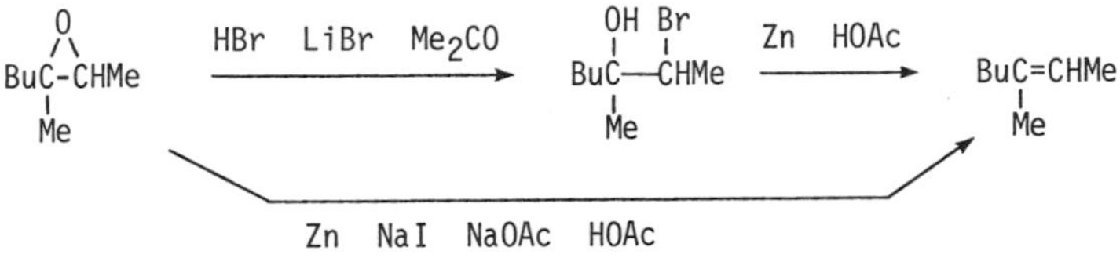
O
BuC-CHMe
Me
HBr LiBr Me2CO
OH Br
BuC—CHMe
Me
Zn HOAc
BuC=CHMe
Me
Zn NaI NaOAc HOAc

JCS (1959) 112
(1955) 1370

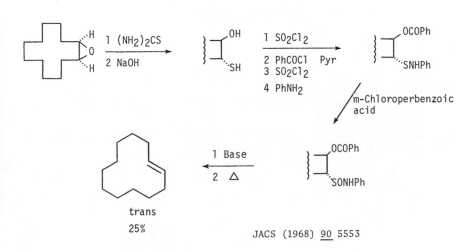

Aust J Chem (1966) 19 1265
JCS (1959) 112

MeCHCHMe
  O
 cis

1 BuLi-HPO(NMe2)2   THF
————————————————————————→
2 CaCO3   silica gel
       toluene

MeCH=CHMe   20%
  trans
JOC (1969) 34 3053

PhCHCH2
  O

Ph3P   Ph3B   Et2O   70°
————————————————————————→
      Ph3P   180°

PhCH=CH2

Ber (1955) 88 1654
Chem Ind (1959) 330

1 (NH2)2CS
——————————→
2 NaOH

1 SO2Cl2
——————————→
2 PhCOCl   Pyr
3 SO2Cl2
4 PhNH2

                    m-Chloroperbenzoic
                    acid

1 Base
←——————
2 Δ

trans
25%

JACS (1968) 90 5553

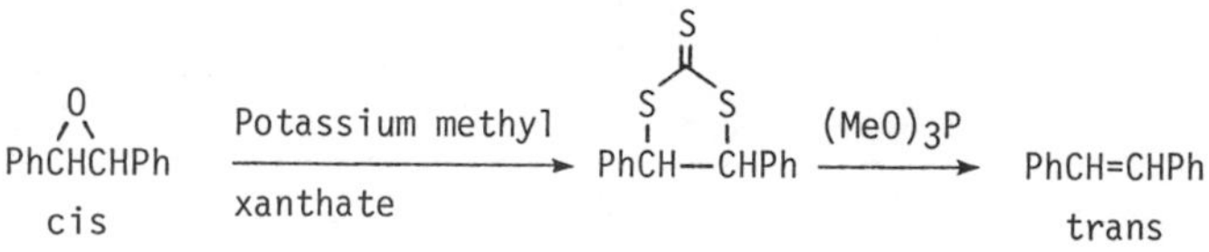

JACS (1965) <u>87</u> 934

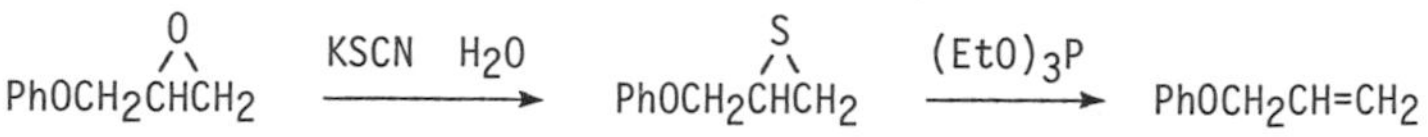

JOC (1961) <u>26</u> 3467

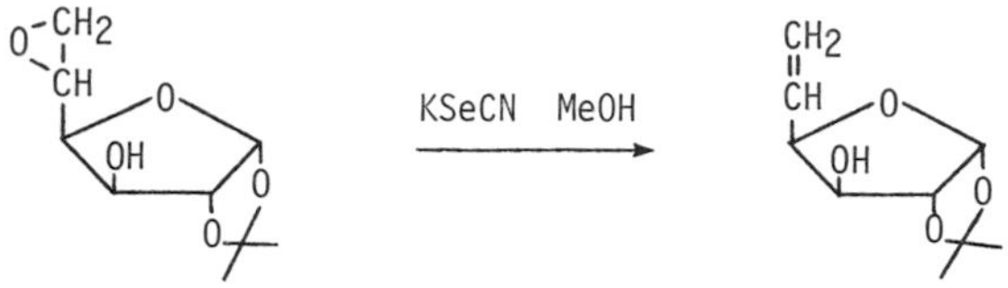

64%

Carbohydrate Res (1967) <u>5</u> 282

Section 205   Olefins from Halides and Sulfonates

The reduction of vinyl halides, $R_2C=\overset{R}{C}X \longrightarrow R_2C=\overset{R}{C}H$, is included in section 160 (Hydrides from Halides and Sulfonates)

$$PrCH_2Br \xrightarrow[\substack{2 \ NaNH_2 \\ THF}]{1 \ Ph_3P} PrCH=PPh_3 \xrightarrow{O_2} PrCH=CHPr$$

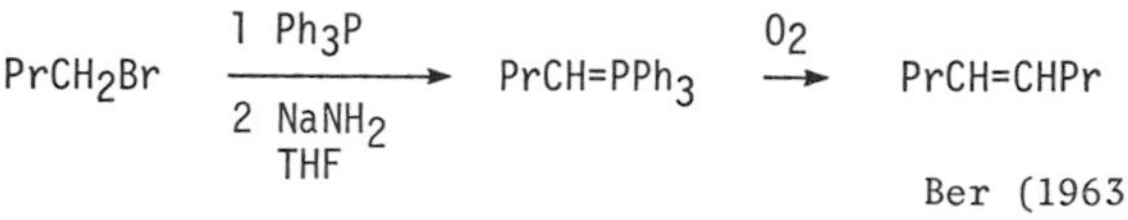

< 66%

Ber (1963) <u>96</u> 1899
(1966) <u>99</u> 2848
(1961) <u>94</u> 1987
Annalen (1969) <u>721</u> 34

$$Ph(CH_2)_2CH_2Br \xrightarrow[\text{2 NaIO}_4]{\text{1 Ph}_3\text{P}} Ph(CH_2)_2CH_2\overset{+}{P}Ph_3 \; \overset{-}{IO_4} \xrightarrow[\text{NH}_3]{\text{NaNH}_2} Ph(CH_2)_2CH=CH(CH_2)_2Ph$$

$$<33\%$$

Ber (1969) <u>102</u> 2259

Org React (1965) <u>14</u> 270

$$CH_2=CHCH_2Br \xrightarrow[\text{C}_6\text{H}_6]{\text{Ph}_3\text{P}} CH_2=CHCH_2\overset{+}{P}Ph_3 \; \overset{-}{Br} \xrightarrow[\text{2 PhCHO}]{\text{1 PhLi   Et}_2\text{O}} CH_2=CHCH=CHPh \qquad 53\%$$

Ber (1954) <u>87</u> 1318
Annalen (1965) <u>682</u> 58

Further examples of the Wittig reaction are included in section 199
(Olefins from Aldehydes) and section 207 (Olefins from Ketones)

36%

JACS (1946) <u>68</u> 1109

$$PhBr \xrightarrow[\text{2 CH}_3\text{COOEt}]{\text{1 Mg   Et}_2\text{O}} \overset{\overset{\displaystyle OH}{|}}{Ph_2CCH_3} \xrightarrow{\Delta} Ph_2C=CH_2$$

Org Synth (1932) Coll Vol 1 226

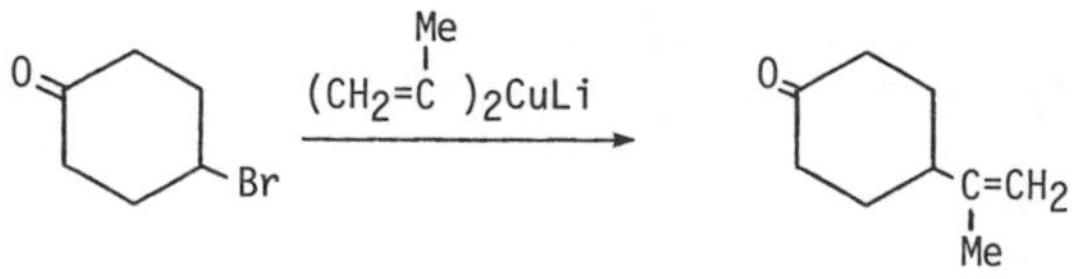

J Indian Chem Soc (1968) 45 1026

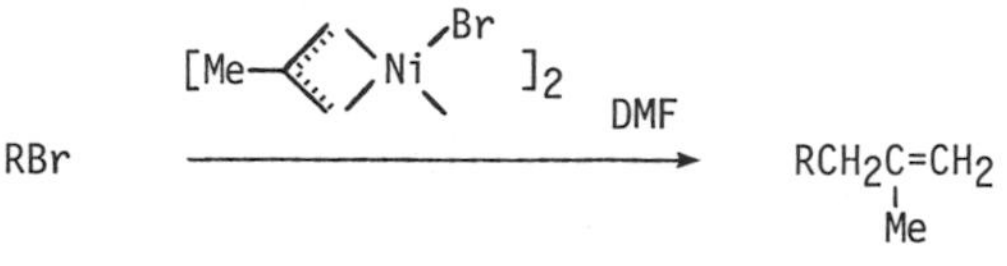

RBr

90% (R= Me)
70% (R= CH_2=CH)
92% (R= $PhCH_2CH_2CH_2$)

JACS (1967) 89 2755

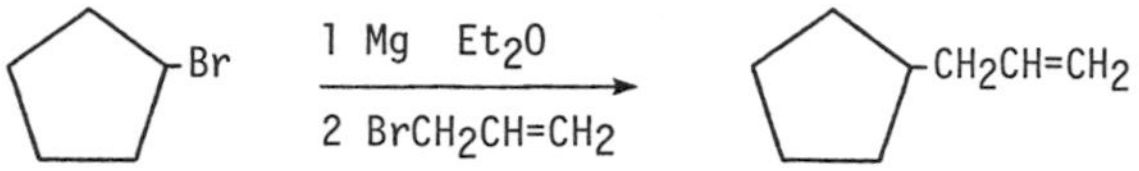

70%

JACS (1946) 68 1101
JOC (1970) 35 22

Further examples of the coupling of halides are included in section 70 (Alkyls and Aryls from Halides and Sulfonates)

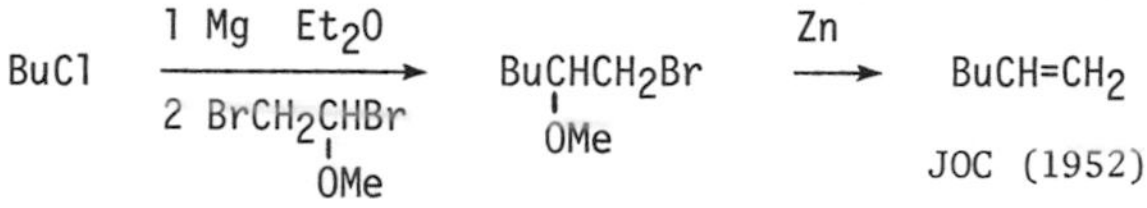

BuCl

JOC (1952) 17 807

BuBr

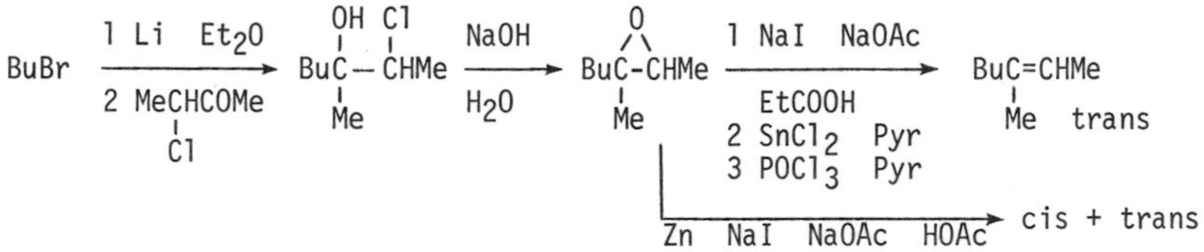

JCS (1959) 112

$$PrCH_2Br \quad \dashrightarrow \quad PrCH_2MgBr \xrightarrow{\quad CF_2Br_2 \quad Et_2O \quad} PrCH=CHCH_2Pr \qquad 72\%$$

Ber (1962) 95 1958

60%

Compt Rend (1965) 260 4535

$$C_5H_{11}CH_2Br \quad \dashrightarrow \quad C_5H_{11}CH_2Li \xrightarrow{\quad Et_2O \quad} C_5H_{11}CH=CHCH_2C_5H_{11} \qquad 33\%$$

trans

Tetr Lett (1964) 2547

$$C_{15}H_{31}CH_2X \xrightarrow{\quad NaH-CH_3SOMe \quad} C_{15}H_{31}CH_2CH_2SOMe \xrightarrow[\quad reflux \quad]{\quad Me_2SO \quad} C_{15}H_{31}CH=CH_2 \qquad <85\%$$

X=Br or OTs

Chem Comm (1965) 29

$$C_5H_{11}CH_2Br \quad \dashrightarrow \quad C_5H_{11}CH_2SH \xrightarrow[\substack{2 \ m\text{-Chloroper-}\\ \text{benzoic acid}}]{\quad 1 \ HCHO \quad HCl \quad} C_5H_{11}CH_2SO_2CH_2Cl$$

$$\downarrow \substack{NaOH \\ H_2O}$$

$$C_5H_{11}CH=CH_2 \qquad <53\%$$

JACS (1964) 86 4383

JOC (1961) 26 2883

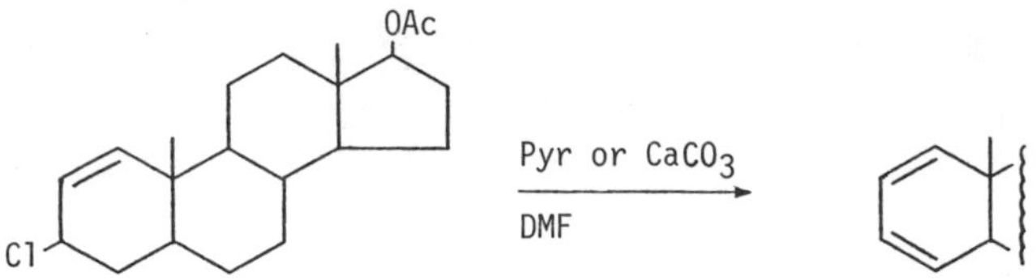

Pyr or CaCO$_3$
———————————→
DMF

80%

JOC (1963) 28 1976

$BuCH_2CH_2Br$ $\xrightarrow{\text{HMPA} \quad 180\text{-}210°}$ $BuCH=CH_2$ 61%

Chem Comm (1971) 113

$\underset{\overset{|}{Br}}{C_3H_7CHCH_2Et}$ $\xrightarrow[\text{Me}_2\text{SO}]{\text{1,5-Diazabicyclo[5.4.0]undec-5-ene}}$ $C_3H_7CH=CHEt$ 91%

Angew (1967) 79 53
(Internat Ed 6 76)
Ber (1966) 99 2012

$C_5H_{11}CH_2CH_2Br$ $\xrightarrow{\text{LiF} \quad \text{Li}_2\text{CO}_3 \quad \text{HMPA}}$ $C_5H_{11}CH=CH_2$ ~51%

Bull Soc Chim Fr (1967) 2455
(1968) 283
JOC (1970) 35 76 1023

$Ph_3CCH_2CH_2Cl$ $\xrightarrow{\text{EtONa} \quad \text{Me}_2\text{SO}}$ $Ph_3CCH=CH_2$ 90%

JCS C (1967) 1115

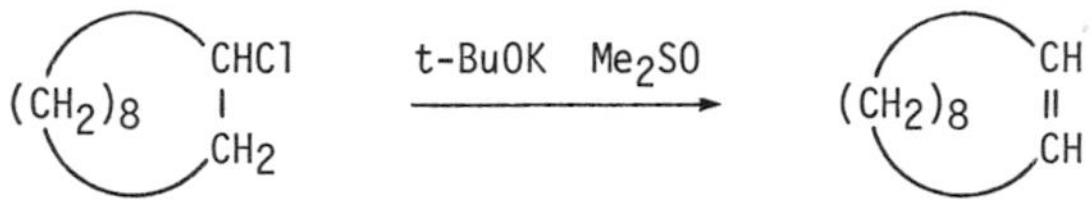

t-BuOK Me$_2$SO
———————————→

JOC (1967) 32 510
(1965) 30 2054
Tetr Lett (1967) 2273

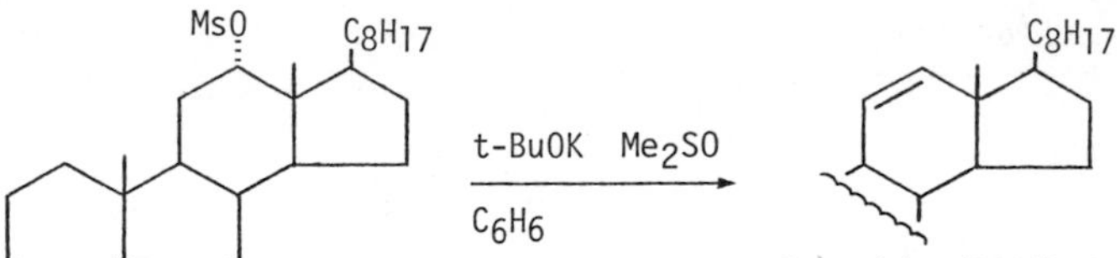

t-BuOK   Me$_2$SO
―――――――――→
C$_6$H$_6$

Steroids (1964) 4 55
JOC (1964) 29 742

65%

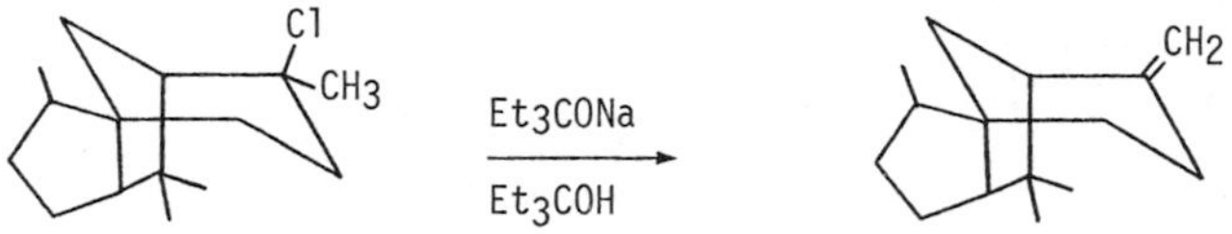

Et$_3$CONa
―――――――→
Et$_3$COH

JOC (1970) 35 196
Chem Comm (1968) 305

52%

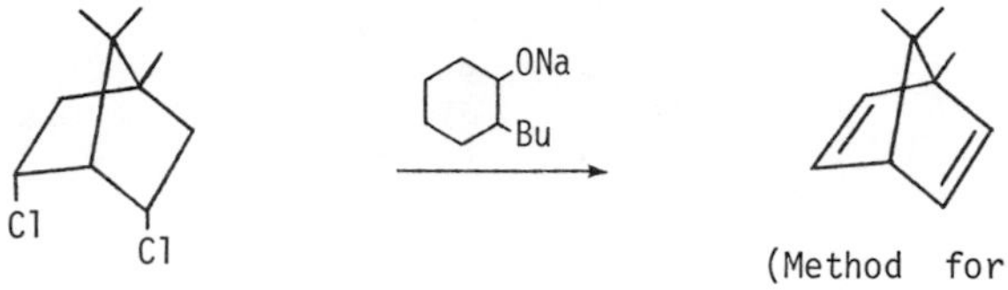

ONa
―――――――→
Bu

(Method   for unreactive halides)

Annalen (1962) 652 96

94%

The following bases/solvents may also be used for the elimination of HX
from halides and sulfonates:

Lithium dicyclohexylamide JOC (1967) 32 510

NaNH$_2$ IIMPA   Bull Soc Chim Fr (1966) 1293

Potassium t-amylate   Helv (1967) 50 2111

Me$_2$SO   JACS (1959) 81 5428
                           Tetr Lett (1968) 4191

(MeO)$_3$P-xylene   Helv (1958) 41 70

C$_{17}$H$_{35}$COOAg-C$_6$H$_6$ Ber (1942) 75B 660

Ag$_2$O-C$_6$H$_6$ Helv (1951) 34 1176

$$AcO(CH_2)_3\underset{Me}{\overset{Cl}{\underset{|}{\overset{|}{C}}}}CH_2Pr \xrightarrow{190\text{-}200°} AcO(CH_2)_3\underset{Me}{\overset{|}{C}}=CHPr \qquad\qquad 80\%$$

JOC (1948) <u>13</u> 239

$$\xrightarrow[\text{(vapor phase)}]{\sim 400°}$$

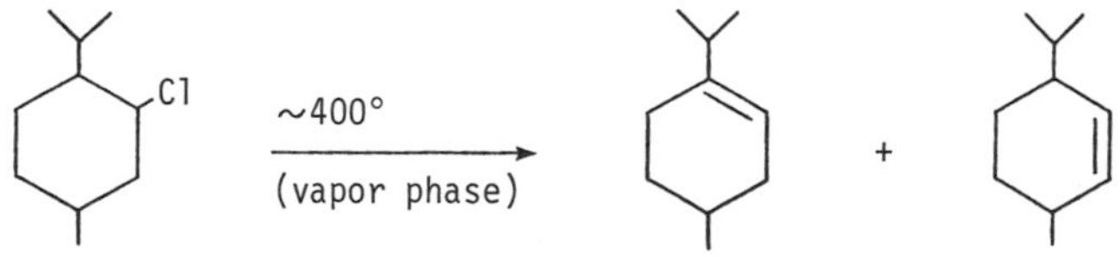

JCS (1952) 453
Chem Rev (1960) <u>60</u> 431

$$Ph_2CX_2 \xrightarrow[\text{or NaI} \\ \text{or Fe(CO)}_5]{Cu} Ph_2C=CPh_2$$

X=Br or Cl

Org Synth (1963) Coll Vol 4 914
JCS (1959) 678
JACS (1961) <u>83</u> 1623

$$\underset{Cl\ \ Cl}{EtCH\text{-}CHEt} \xrightarrow[PrOH]{KOH} \underset{Cl}{EtCH=CEt} \xrightarrow[NH_3]{Na} EtCH=CHEt$$

JACS (1951) <u>73</u> 3329

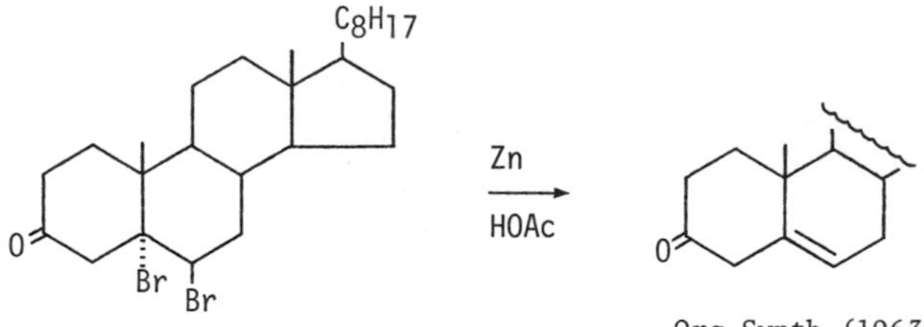

$$\xrightarrow[\text{HOAc}]{Zn}$$

Org Synth (1963) Coll Vol 4 195
JOC (1970) <u>35</u> 1733
JCS (1961) <u>4547</u>

The following reagents may also be used for the conversion of 1,2-dihalides to olefins:

NaI, KI	JOC (1965) <u>30</u> 1658
	(1959) <u>24</u> 143
	Annalen (1967) <u>705</u> 76
CrX_2 (X=Cl, OAc etc.)	JACS (1968) <u>90</u> 1582
	(1967) <u>89</u> 6547
	(1964) <u>86</u> 4603
	JOC (1959) <u>24</u> 1621 1629
	Tetrahedron (1968) <u>24</u> 3503
Bu_3SnH	JOC (1963) <u>28</u> 2165
	Acc Chem Res (1968) <u>1</u> 299
$Na-NH_3$	JACS (1952) <u>74</u> 4590
Sodium dihydronaphthylide	Chem Comm (1969) 78
Phenanthrene disodium	Angew (1964) <u>76</u> 432
PhSNa, EtSNa	Rev Chim (Bucharest) (1962) <u>7</u> 1379
	(Chem Abs <u>61</u> 4208)
	JOC (1959) <u>24</u> 143
Na_2Se	JOC (1966) <u>31</u> 4292
$(NH_2)_2CS$	Chem Ind (1966) 1418
KSCN	Chem Ind (1966) 1418
Potassium cyclohexylphosphide	Ber (1961) <u>94</u> 2664
$(EtO)_3P$	JOC (1970) <u>35</u> 3181
$NaH-Me_2SO$	Chem Ind (1965) 345
$LiAlH_4$	Can J Chem (1964) <u>42</u> 1294
PhLi	Annalen (1967) <u>705</u> 76
PrMgBr	Bull Soc Chim Fr (1946) 604

73%

JCS (1950) 598
JACS (1952) <u>74</u> 4894

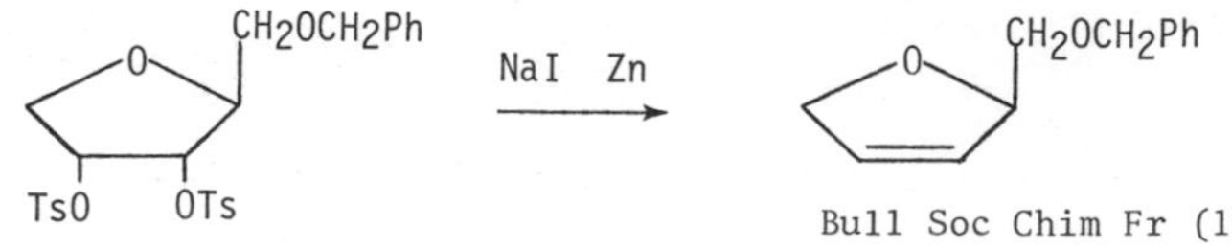

82%

Bull Soc Chim Fr (1967) 4111

KSeCN   DMF

45%

Carbohydrate Res (1967) 5 282

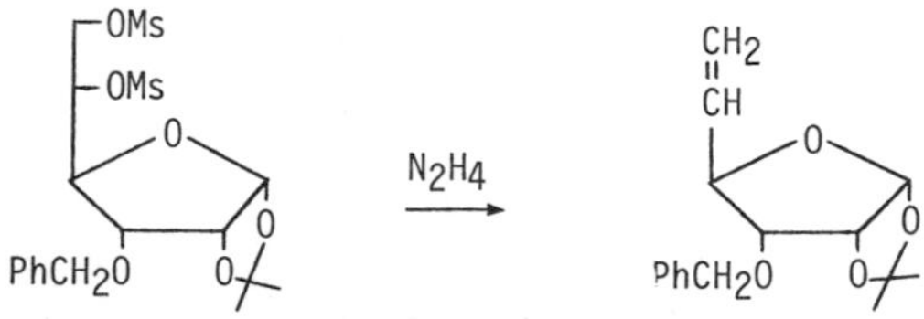

N₂H₄

17%

Ber (1969) 102 820

Section 206   Olefins from Hydrides (RH)

This section lists examples of the replacement of hydrogen by olefinic
groups. For the dehydrogenation of alkyl groups, e.g. $RCH_2CH_2R \longrightarrow RCH=CHR$,
see section 200 (Olefins from Alkyls, Methylenes and Aryls)

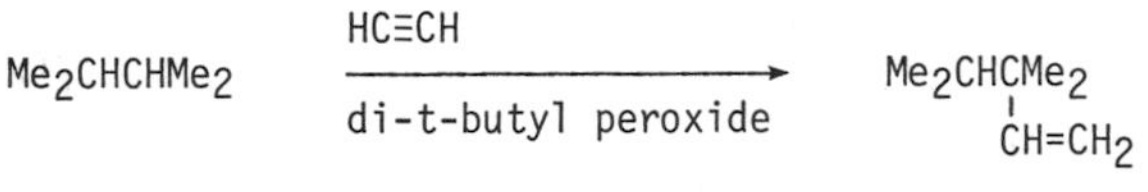

$Me_2CHCHMe_2$
$\xrightarrow[\text{di-t-butyl peroxide}]{HC≡CH}$
Me_2CHCMe_2
$\overset{|}{CH=CH_2}$

JOC (1965) 30 3814

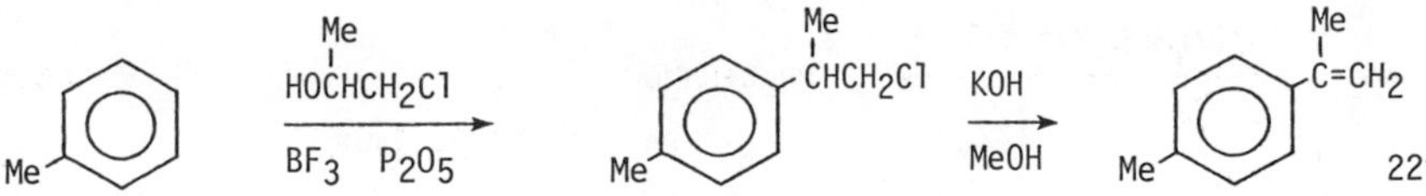

JACS (1948) $\underline{70}$ 1772

Section 207   Olefins from Ketones

Reviews: The Wittig Reaction   Org React (1965) $\underline{14}$ 270
                                        Quart Rev (1963) $\underline{17}$ 406

       New Reactions of Alkylidenephosphoranes and their Preparative Uses

                              Angew (1965) $\underline{77}$ 850
                              (Internat Ed $\underline{4}$ 830)

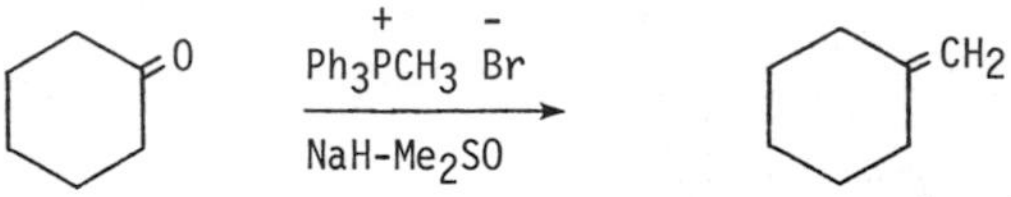

86%

Org React (1965) $\underline{14}$ 270
Ber (1954) $\underline{87}$ 1318

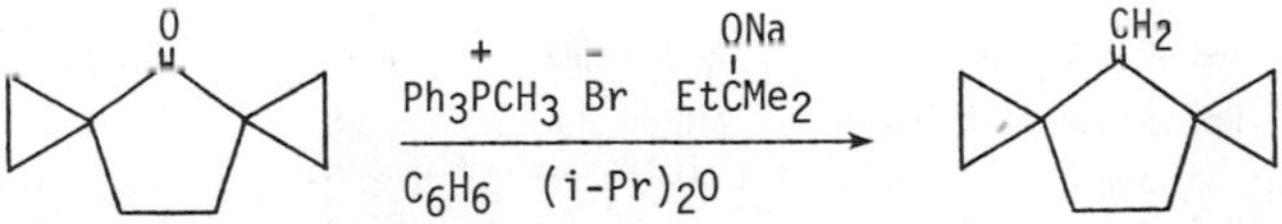

90%

Bull Soc Chim Fr (1967) 1936

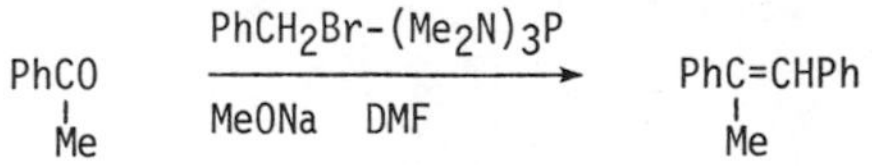

PhCO   PhCH$_2$Br-(Me$_2$N)$_3$P   PhC=CHPh   35%
| ───────────────────────── |
Me   MeONa   DMF   Me

                    Annalen (1965) $\underline{682}$ 58

Ph_2CO $\xrightarrow[\text{toluene}]{Ph_2P(O)CH_2Me \quad t\text{-BuOK}}$ $Ph_2C=CHMe$ 51%

Ber (1959) **92** 2499

Ph_2CO $\xrightarrow[\text{2 Silica gel } C_6H_6 \text{ reflux}]{\overset{\overset{\text{Me}}{|}}{1 \text{ Li}CHPO(NMe_2)_2} \quad THF}$ $Ph_2C=CHMe$ 87%

JACS (1966) **88** 5652 5653
JOC (1969) **34** 3053

$\xrightarrow[\text{2 } C_6H_6 \text{ THF reflux}]{1 \text{ Li}CH_2SO\overset{\overset{\text{Li}}{|}}{N}-\!\!\!\bigcirc\!\!\!-Me \quad THF}$ ~90%

JACS (1966) **88** 5656
(1968) **90** 5548

Ph_2CO $\xrightarrow{\overset{\overset{\text{Me}}{|}}{\text{Li}CHPS(OMe)_2} \quad THF}$ $Ph_2C=CHMe$ 93%

JACS (1966) **88** 5654

The following reagents/solvents may also be used for the generation of Wittig reagents:

$NaH\text{-}Me_2SO$, $NaNH_2\text{-}NH_3$, PhLi, BuLi,

EtOLi-EtOH, EtONa-EtOH	Org React (1965) **14** 270
Na	Chem Abs (1940) **34** 392
t-BuOK-Me$_2$SO	Ber (1965) **98** 604
PhLi-t-BuOK-t-BuOH	Angew (1964) **76** 683 (Internat Ed **3** 636)
Ethylene oxide	Angew (1968) **80** 535 (Internat Ed **7** 536)
Diazabicyclo[3.4.0]non-5-ene	Ber (1966) **99** 2012
Electrolysis (non basic)	JACS (1968) **90** 2728 Tetr Lett (1969) 3523

Further examples of the Wittig reaction are included in section 199 (Olefins from Aldehydes)

> 50%

Tetr Lett (1970) 1137
JOC (1968) 33 780

$(C_5H_{11})_2CO$ $\xrightarrow{C_5H_{11}CHLi_2 \quad THF}$ $(C_5H_{11})_2C=CHC_5H_{11}$　　　　30-35%

Tetr Lett (1966) 4315

$(C_5H_{11})_2CO$ $\xrightarrow{CH_2I_2 \quad Mg \quad Et_2O}$ $(C_5H_{11})_2C=CH_2$　　　　30%

Tetrahedron (1970) 26 1281

Ph_2CO $\xrightarrow[\text{}]{\overset{\displaystyle Li}{\underset{\displaystyle PhCHNC \quad THF}{|}}}$ $Ph_2C=CHPh$　　　　74%

Angew (1968) 80 842
(Internat Ed 7 805)

$Cl(CH_2)_2\underset{\underset{\displaystyle Me}{|}}{C}HCO$ $\xrightarrow[\text{2 Ac}_2O]{1 \ C_5H_{11}MgBr \quad Et_2O}$ $Cl(CH_2)_2\underset{\underset{\displaystyle Me}{|}}{C}H_2\overset{\overset{\displaystyle OAc}{|}}{C}C_5H_{11}$ $\xrightarrow[C_6H_6]{TsOH}$ $Cl(CH_2)_2\underset{\underset{\displaystyle Me}{|}}{C}H=CC_5H_{11}$

JOC (1948) 13 239　　　　　　　　　75%
JACS (1949) 71 819

1 NaC≡CH
2 Acetylation
3 H$_2$ Pd

$\xrightarrow{LiBu_2Cu}$ Et_2O

Chem Comm (1969) 43
JACS (1970) 92 735

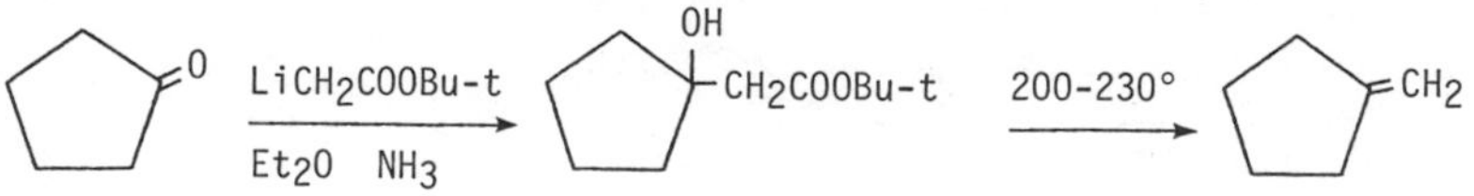

Bull Soc Chim Fr (1960) 1196

Ph_2CO $\xrightarrow{(i\text{-}PrO)_3P}$ $Ph_2C=CPh_2$ low yield

JOC (1964) <u>29</u> 2567

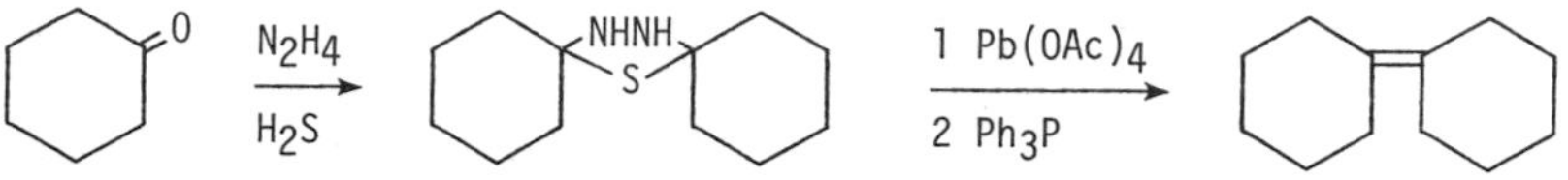

Chem Comm (1970) 1226 1225   73%

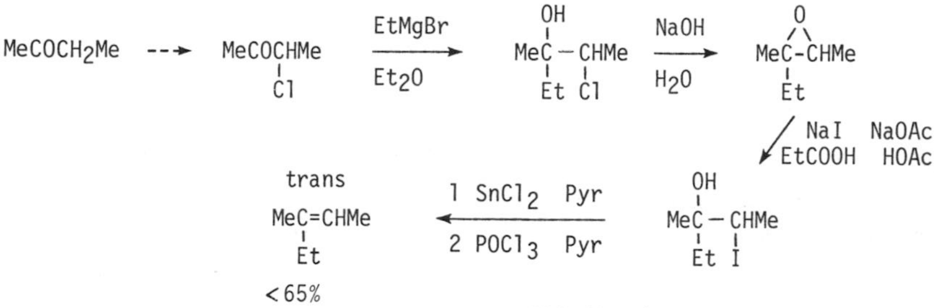

JCS (1959) 112

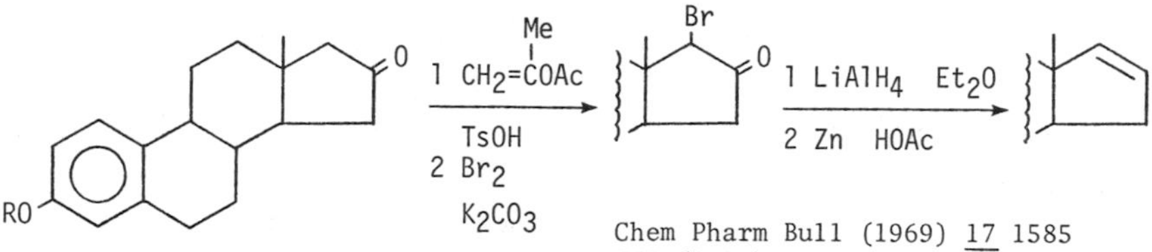

Chem Pharm Bull (1969) <u>17</u> 1585
JCS (1955) 1370

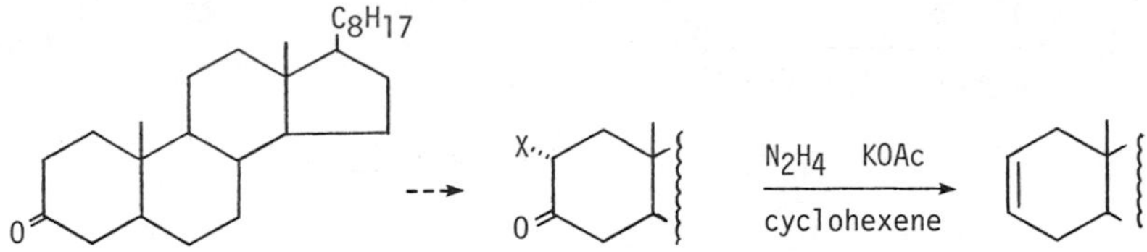
C8H17
X
O
O
N2H4 KOAc
cyclohexene

X=F, Cl, Br or I

54-71%

JOC (1964) 29 958

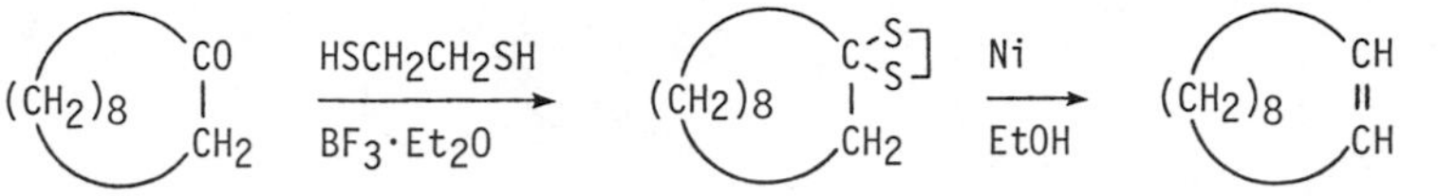
(CH2)8
CO
CH2
HSCH2CH2SH
BF3·Et2O
C
S
S
CH2
Ni
EtOH
CH
CH

55%

Tetrahedron (1969) 25 2823
JOC (1963) 28 1443

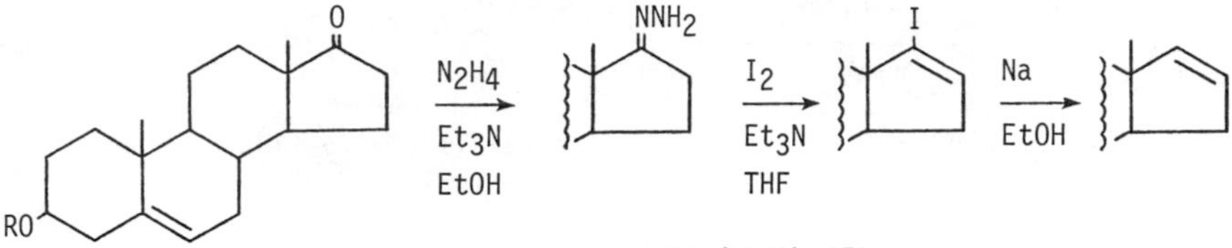
O
RO
N2H4
Et3N
EtOH
NNH2
I2
Et3N
THF
I
Na
EtOH

JCS (1962) 470

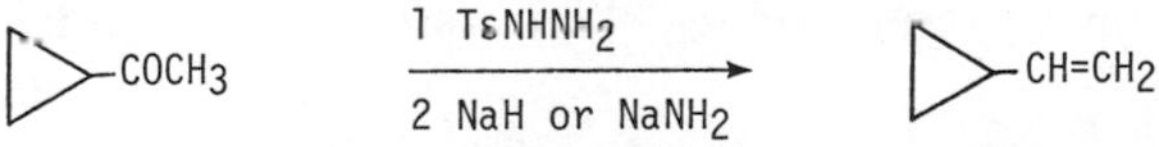
COCH3
1 TsNHNH2
2 NaH or NaNH2
CH=CH2

~100%

Annalen (1966) 691 41

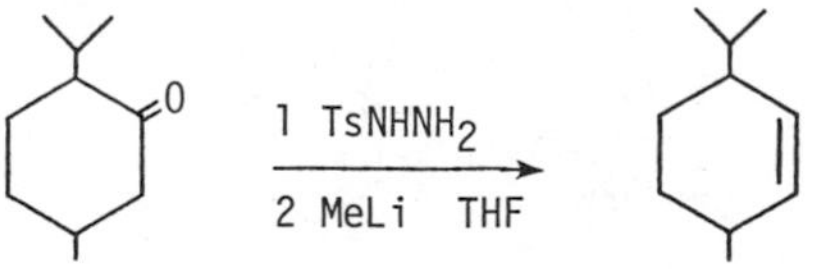
O
1 TsNHNH2
2 MeLi THF

~79%

Synthesis (1970) 595
JACS (1967) 89 5734

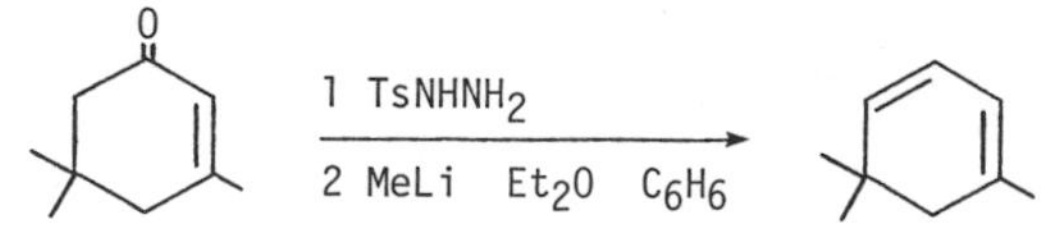

~100%

JACS (1968) 90 4762

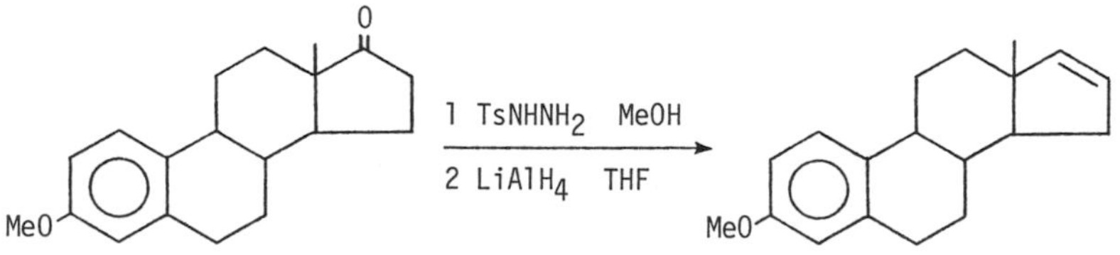

70%

Tetrahedron (1963) 19 1127

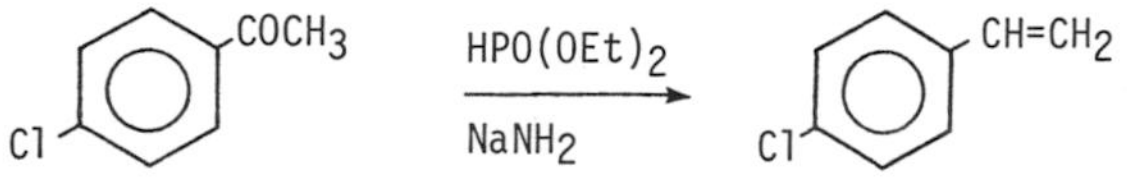

60%

Angew (1963) 75 138
(Internat Ed 2 98)

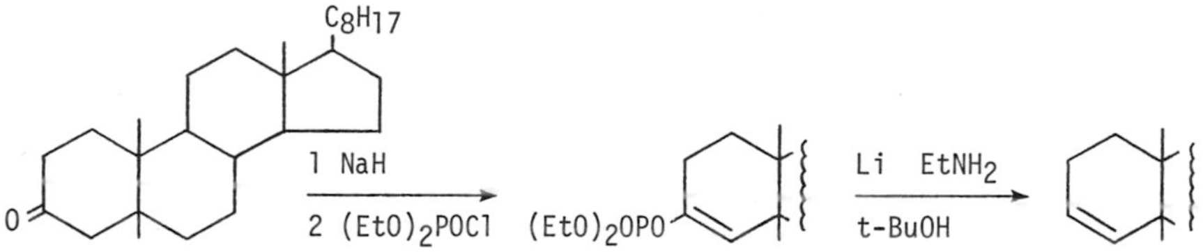

Tetr Lett (1969) 2145
Chem Comm (1969) 112

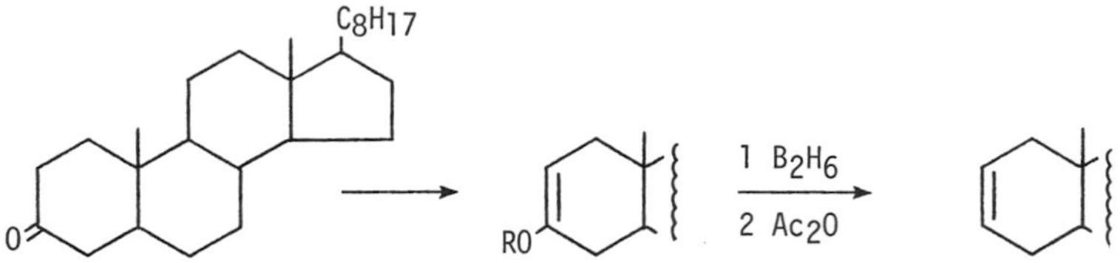

81%

R=Ac or Me

Gazz (1962) 92 309
(Chem Abs 57 12572)

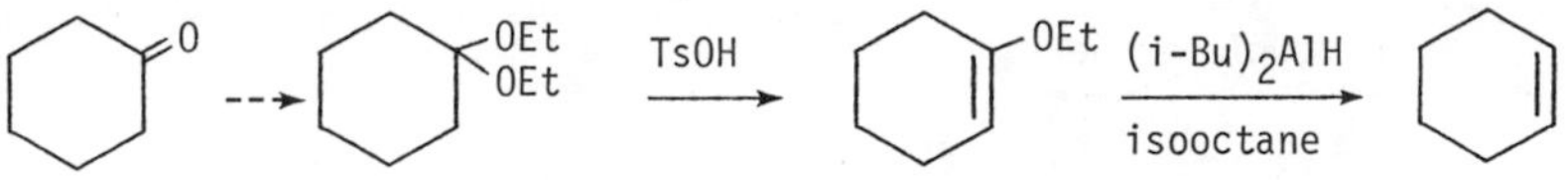
O
OEt
OEt
TsOH
OEt
(i-Bu)2AlH
isooctane

<32%

JOC (1966) 31 329

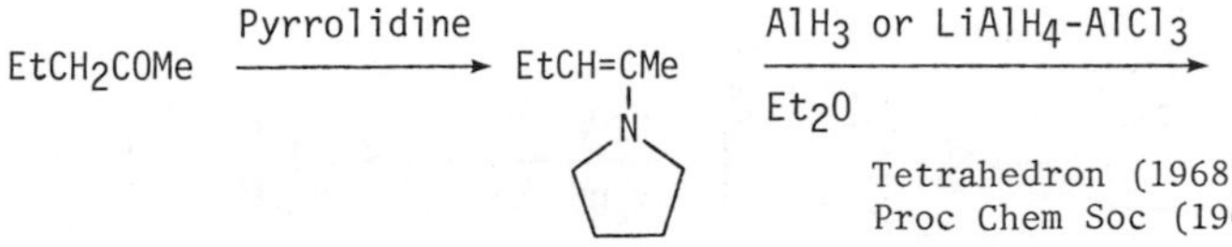
EtCH2COMe
Pyrrolidine
EtCH=CMe
N
AlH3 or LiAlH4-AlCl3
Et2O
EtCH=CHMe

Tetrahedron (1968) 24 4489
Proc Chem Soc (1963) 19

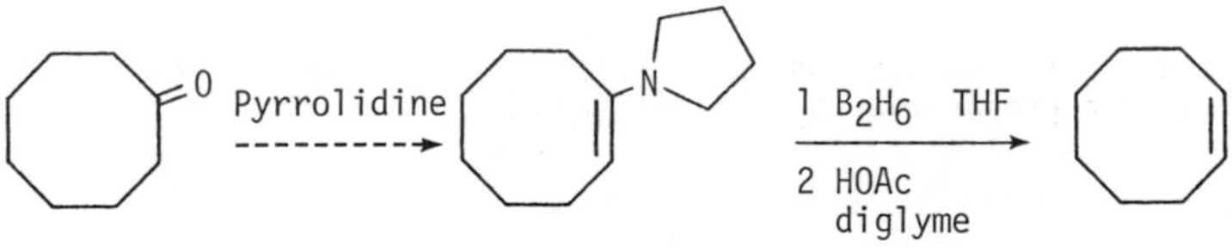
O
Pyrrolidine
N
1 B2H6 THF
2 HOAc
diglyme

< 98%

Tetr Lett (1964) 2039

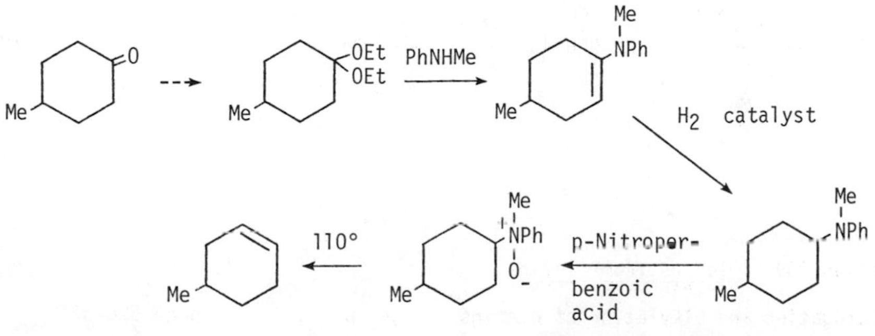
O
Me
OEt
OEt
PhNHMe
Me
NPh
H2 catalyst
p-Nitroper-
benzoic
acid
O-
110°

Tetr Lett (1964) 3853
Compt Rend (1967) C 264 710

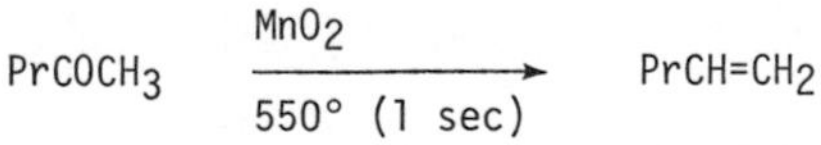
PrCOCH3
MnO2
550° (1 sec)
PrCH=CH2

62%

Chem Comm (1969) 461

JOC (1966) <u>31</u> 1393

Chem Comm (1968) 558

<65%

Some of the methods listed in section 199 (Olefins from Aldehydes) may also be applied to the preparation of olefins from ketones

Section 208   Olefins from Nitriles

$Ph_2CH\underset{CN}{|}$ $\xrightarrow[\underset{Me}{\text{2 PhCHCl}}]{\text{1 KNH}_2\text{ NH}_3}$ $Ph_2C\overset{Me}{\underset{CN}{|}}CHPh$ $\xrightarrow{\text{KNH}_2\text{ NH}_3}$ $Ph_2C\overset{Me}{=}CPh$ 71%

JACS (1956) <u>78</u> 82
       (1960) <u>82</u> 1786

Section 209   Olefins from Olefins

Homologation and alkylation of olefins . . . . page 520-521
Olefin metathesis 521
Migration of double bonds 521-524
Cis-trans interconversion and equilibration 524-526

$BuCH=CH_2$ $\xrightarrow[\text{2 PhHgCCl}_2\text{Br}]{\text{1 B}_2\text{H}_6}$ $BuCH_2CH_2CH=CHCH_2Bu$

JACS (1966) <u>88</u> 1834

$\left(\begin{array}{c}CH=CH \\ (CH_2)_6\end{array}\right)$ $\xrightarrow[\text{t-BuOH}]{\text{CHBr}_3 \quad \text{t-BuOK}}$ $\left(\begin{array}{c}\overset{\text{CBr}_2}{\overset{\wedge}{CHCH}} \\ (CH_2)_6\end{array}\right)$ $\xrightarrow{\text{Mg} \quad \text{Et}_2\text{O}}$ $\left(\begin{array}{c}CH=C=CH \\ (CH_2)_6\end{array}\right)$

$\downarrow \text{H}_2 \quad \text{Pd-C}$

15% $\left(\begin{array}{c}CH_2CH=CH \\ (CH_2)_6\end{array}\right)$

JOC (1961) <u>26</u> 3518

$PhCH_2CH=CH_2$ $\xrightarrow{\text{NaH-Me}_2\text{SO}}$ $PhCH_2\underset{\text{Me}}{CH}CH_2SOMe$ $\xrightarrow{\Delta}$ $PhCH_2\underset{\text{Me}}{C}=CH_2$

JOC (1964) <u>29</u> 2699

$MeCH=CHEt$ $\underset{\xleftarrow{\hspace{2cm}}}{\overset{\text{WCl}_6 \quad \text{EtAlCl}_2 \quad \text{EtOH}}{\xrightarrow{\hspace{2cm}}}}$ $MeCH=CHMe + EtCH=CHEt$

Tetr Lett (1967) 3327
JACS (1968) <u>90</u> 4133

$C_6H_{13}CH=CH_2$ $\xrightarrow[\text{Me}_3\text{Al}_2\text{Cl}_3 \quad \text{PhCl}]{(Ph_3P)_2Cl_2(NO)_2Mo}$ $C_6H_{13}CH=CHC_6H_{13}$ 37%

JACS (1970) <u>92</u> 528

$\xrightarrow[\substack{\text{diglyme} \\ 2 \ C_8H_{17}CH=CH_2}]{1 \ \text{NaBH}_4 \quad \text{BF}_3}$ 62%

JACS (1967) <u>89</u> 567 561
(1960) <u>82</u> 2074

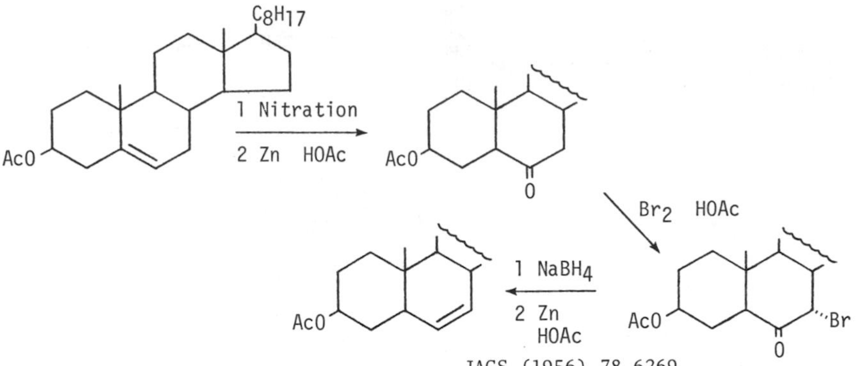

JACS (1956) 78 6269
JOC (1948) 13 424

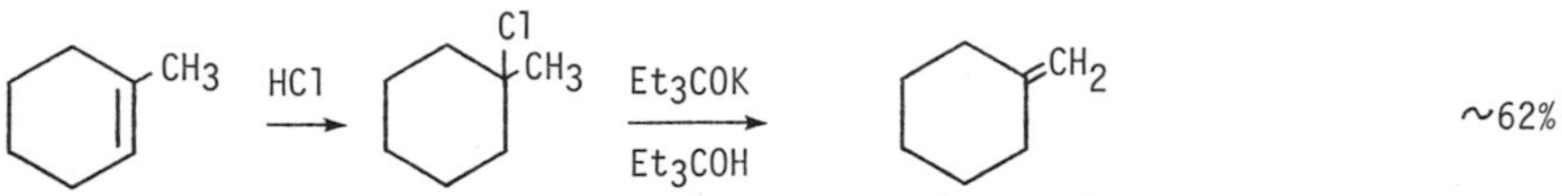

~62%

Chem Comm (1968) 305
Helv (1967) 50 2111

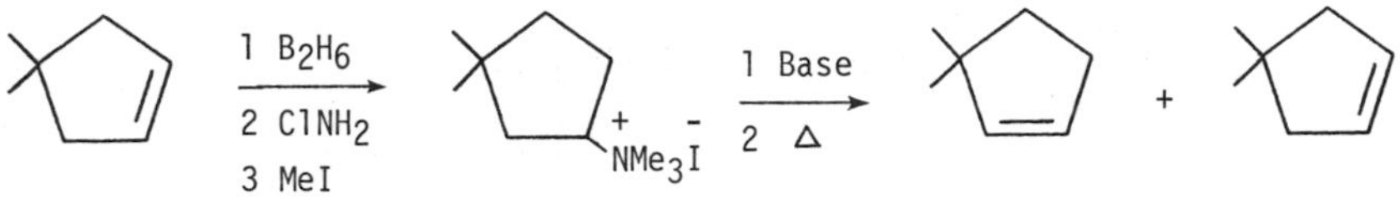

Tetr Lett (1968) 2253

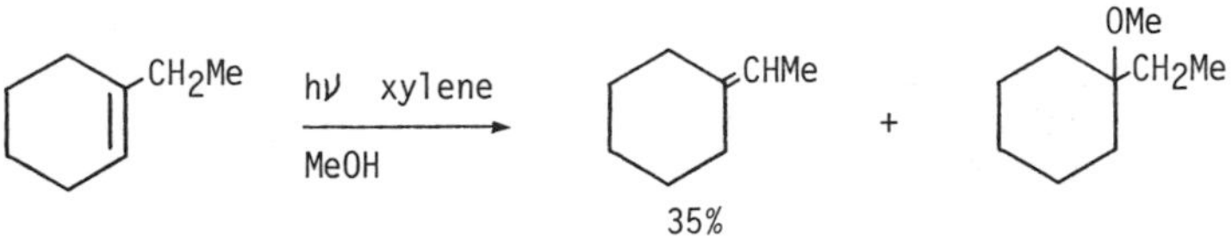

JACS (1967) 89 5199

$$\text{(aryl)}-CH_2CH=CH_2 \xrightarrow[\text{C}_6\text{H}_6]{\text{RhCl(PPh}_3)_3} \text{(aryl)}-CH=CHMe \qquad 60\%$$

Tetr Lett (1968) 3797

$$\text{(ring)}-OMe \xrightarrow{\text{RhCl(PPh}_3)_3 \quad CHCl_3} \text{(ring)}-OMe \qquad 80\%$$

Tetr Lett (1968) 3797

$$EtCH_2CH_2CH=CH_2 \underset{\longleftarrow}{\xrightarrow{RhCl_3 \cdot 3H_2O \quad EtOH}} EtCH_2CH=CHCH_3 \; + \; EtCH=CHCH_2CH_3$$

JACS (1964) <u>86</u> 1776
Helv (1967) <u>50</u> 2445

$$\xrightarrow[\text{2 KCN}]{\text{1 RhCl}_3 \cdot 3H_2O \quad EtOH}$$

JACS (1964) <u>86</u> 2516

$$PrCH_2CH_2CH_2CH=CH_2 \xrightarrow[\text{pet ether}]{\text{Fe}_3(CO)_{12} \quad h\nu} PrCH_2CH_2CH=CHCH_3 \; + \; PrCH_2CH=CHCH_2CH_3$$
$$+ \; PrCH=CHCH_2CH_2CH_3$$

Proc Chem Soc (1964) 408

Review: The Isomerization of Olefins.
         Part I. Base Catalyzed Isomerization of Olefins

Synthesis (1969) 97

COOH
|
(CH₂)₇CH₂CH=CH₂ $\xrightarrow{\text{LiNHCH}_2\text{CH}_2\text{NH}_2}$ COOH
|
(CH₂)₇CH=CHMe

Tetrahedron (1964) <u>20</u> 2911
JOC (1958) <u>23</u> 1136

The following bases may also be used for the migration of double bonds:

LiNHEt	JACS (1964) <u>86</u> 5281
t-BuOK-Me₂SO	Ber (1966) <u>99</u> 1737
	JOC (1968) <u>33</u> 221
	JACS (1961) <u>83</u> 3731
Na-alumina	JACS (1965) <u>87</u> 4107
	JCS <u>C</u> (1967) 2149
	(1966) 260
Na-PhCH₂Na	Tetr Lett (1964) 467

(CH₂)₇   CH
‖
CH
trans $\xrightarrow{\text{2-Naphthalenesulfonic acid}}$ (CH₂)₇   CH
‖
CH
cis

JACS (1952) <u>74</u> 3643

Me   Me
| |
MeCH=C   CH=CCHMe
      CH=CH |
                   OH
cis $\xrightarrow[\text{EtOH}]{\text{I}_2 \ \ h\nu}$ Me
|
MeCH=C   Me
      CH=CH |
            CH=CCHMe   trans
                   |
                   OH

JACS (1954) <u>76</u> 5719

—CH=CH—
(CH₂)₅   (CH₂)₅
—CH=CH—
cis $\xrightarrow{\text{I}_2 \ \ \text{CCl}_4}$ —CH=CH—
(CH₂)₅   (CH₂)₅
CH=CH—
trans

JCS <u>C</u> (1966) 260

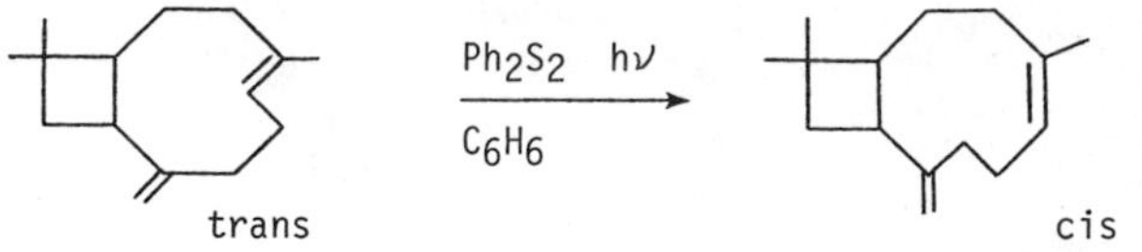

trans

$\xrightarrow[C_6H_6]{Ph_2S_2 \quad h\nu}$

cis

Helv (1968) 51 548
JACS (1967) 89 2758
JCS C (1966) 260

Me   Me
\
   CH=CH

cis

$\underset{\longleftarrow}{\xrightarrow{PhCOMe \quad h\nu \quad C_6H_6}}$

Me
\
   CH=CH

trans   Me

JCS C (1966) 260

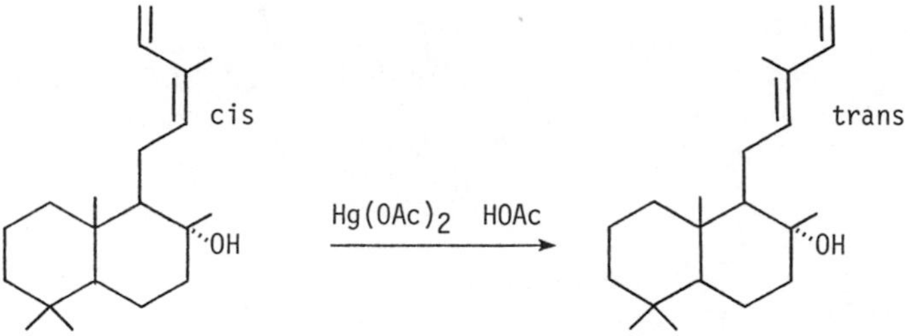

cis

$\xrightarrow{Hg(OAc)_2 \quad HOAc}$

trans

JCS C (1967) 2514

$CH_2=CH$ Me
\   /
  CH=CH

cis

$\underset{\longleftarrow}{\xrightarrow{Ferrocene \quad h\nu}}$

$CH_2=CH$
\
   CH=CH

trans   Me

JACS (1965) 87 1626

Et
\
  CH=CH
trans   Et

$\xrightarrow[CHCl_3]{Cl_2 \quad SbCl_5}$

EtCH-CHEt
| |
Cl   Cl

$\xrightarrow[PrOH]{KOH}$

EtCH=CEt
    |
    Cl

$\xrightarrow[NH_3]{Na}$

Et   Et
\   /
  CH=CH
cis

JACS (1951) 73 3329

C_5H_{11} $\diagdown$ Me
$\diagup CH=CH \diagdown$ 1 INCO   2 MeOH   → $C_5H_{11}CHCHMe$ $\xrightarrow{BuONO}$ $C_5H_{11}\diagdown$
cis   3 180°   4 Base $\underset{NH}{\diagdown/}$ $\diagup CH=CH \diagdown$ 41%
                   5 H_2SO_4 6 Base   trans   Me

Tetr Lett (1969) 4001

$(CH_2)_6 \diagup^{CH}_{\diagdown CH} \overset{\|}{}$ 1 $(SCN)_2$ → $(CH_2)_6 \diagup^{CHS}_{\diagdown CHS} CS$ $\xrightarrow[135°]{(C_8H_{17}O)_3P}$ $(CH_2)_6 \diagup^{CH}_{\diagdown CH} \overset{\|}{}$
cis   2 HBr H_2O trans
                   3 H_2S

JACS (1965) <u>87</u> 934

$\underset{\diagdown (CH_2)_7COOR}{\overset{\diagup C_6H_{13}}{\underset{CH}{\overset{CH}{\|}}}}$ $\xrightarrow[HCOOH]{H_2O_2}$ $\underset{(CH_2)_7COOR}{\overset{C_6H_{13}}{\underset{CHOH}{\overset{CHOH}{|}}}}$ $\xrightarrow[HOAc]{HBr \ H_2SO_4}$ $\underset{(CH_2)_7COOR}{\overset{C_6H_{13}}{\underset{CHBr}{\overset{CHBr}{|}}}}$ $\xrightarrow[EtOH]{Zn}$ $\underset{\diagup (CH_2)_7COOR}{\overset{\diagup C_6H_{13}}{\underset{CH}{\overset{CH}{\|}}}}$
cis   trans

JCS (1954) 4219
    (1953) 2393
    (1951) 1079

Further examples of the conversion of diols and dihalides into olefins are included in section 198 (Olefins from Alcohols) and section 205 (Olefins from Halides and Sulfonates)

Section 210   <u>Olefins from Miscellaneous Compounds</u>
ooooooooooooooooooooooooooooooooooooo

The reduction of vinyl halides, $R_2C=CR \longrightarrow R_2C=CHR$ is included in section
                                $\underset{Cl}{|}$
160 (Hydrides from Halides and Sulfonates)

The hydrogenolysis of unsaturated ketones, e.g. $R_2C=C-CO \longrightarrow R_2CH-C=CH$
                                                  $\underset{R}{|} \underset{R}{|}$ $\underset{R}{|} \underset{R}{|}$
is included in section 72 (Alkyls, Methylenes and Aryls from Ketones)

$C_{15}H_{31}CH_2CH_2SOMe$ $\xrightarrow[\text{reflux}]{Me_2SO}$ $C_{15}H_{31}CH=CH_2$ 85%

Chem Comm (1965) 29
JOC (1964) <u>29</u> 2699

$(Me_2CH)_2SO_2$ $\xrightarrow{t\text{-BuOK} \quad Me_2SO}$ $MeCH=CH_2$ 100%

Chem Ind (1963) 1243
JOC (1967) <u>32</u> 102

PhCS
| $\xrightarrow{Ni \quad xylene}$ $PhC=CPh$ 18%
Me | |
                               Me Me

JACS (1944) <u>66</u> 1136

$\xrightarrow{Ph_3P \quad Et_2O}$ 95%

JOC (1958) <u>23</u> 1767
     (1961) <u>26</u> 3467
Chem Pharm Bull (1960) <u>8</u> 621

$PhCH_2CH_2SO_2Cl$ $\xrightarrow[\text{Et}_2O]{Et_3N \quad CH_2N_2}$ $PhCH_2CH-CH_2$ $\xrightarrow{80°}$ $PhCH_2CH=CH_2$ 96%
                                            \   /
                                            SO_2

Angew (1965) <u>77</u> 41
(Internat Ed <u>4</u> 70)

$BuCHCH_2Br$ $\xrightarrow{Zn \quad PrOH}$ $BuCH=CH_2$ 78%
|
OMe
                              JACS (1932) <u>54</u> 751

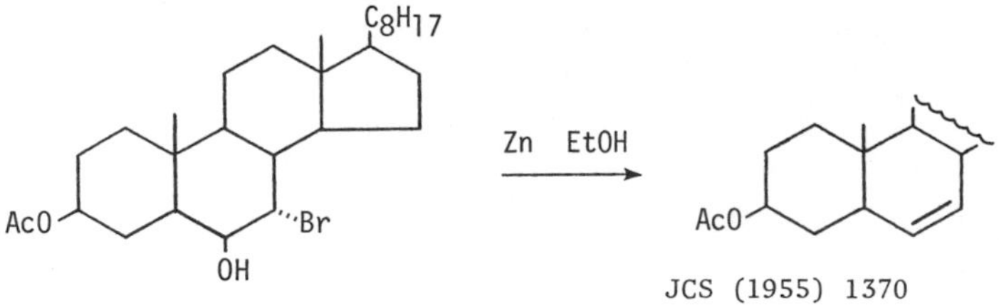

C_8H_{17}

Zn EtOH

AcO

''Br

OH

AcO

JCS (1955) 1370

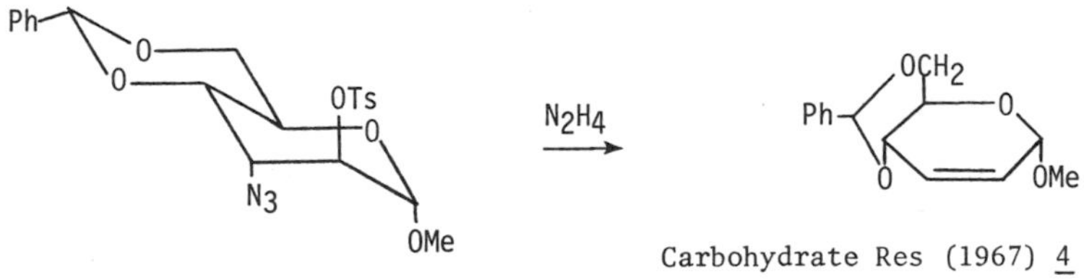

Ph

O

O

OTs

O

N_3

OMe

N_2H_4

OCH_2

Ph

O

O

OMe

72%

Carbohydrate Res (1967) $\underline{4}$ 465

OH

CH_2COOH

Cu quinoline

CH_2

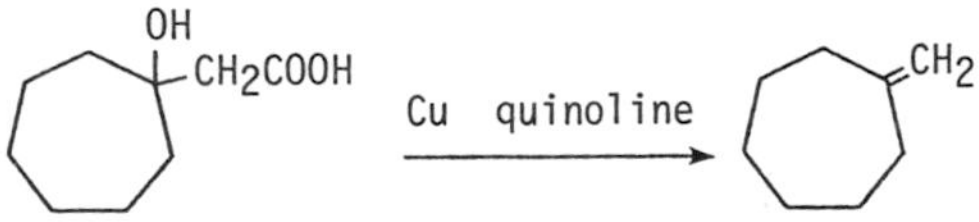

60%

Bull Soc Chim Fr (1960) 1196

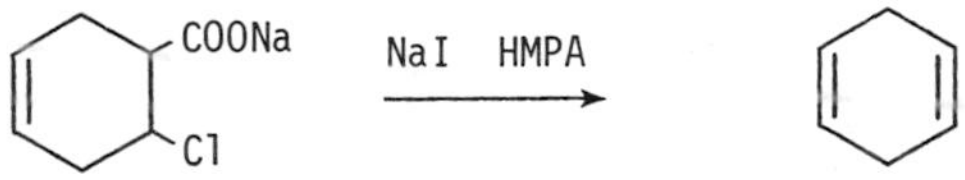

COONa

Cl

NaI HMPA

70%

JOC (1968) $\underline{33}$ 4540

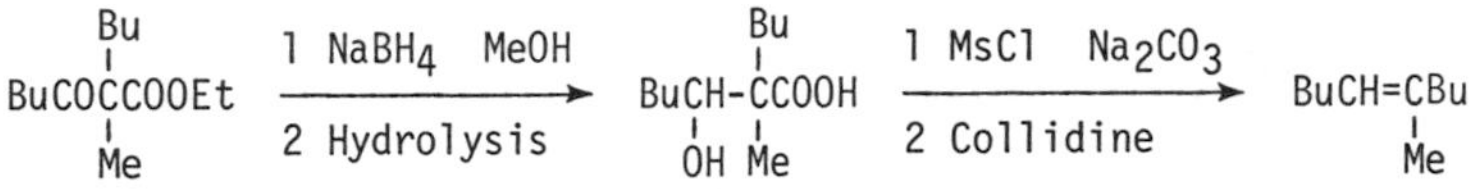

Bu

BuCOCCOOEt

Me

1 NaBH$_4$ MeOH

2 Hydrolysis

Bu

BuCH-CCOOH

OH Me

1 MsCl Na$_2$CO$_3$

2 Collidine

BuCH=CBu

Me

diastereomer A ⟶ trans

diastereomer B ⟶ cis

Tetr Lett (1968) 4569

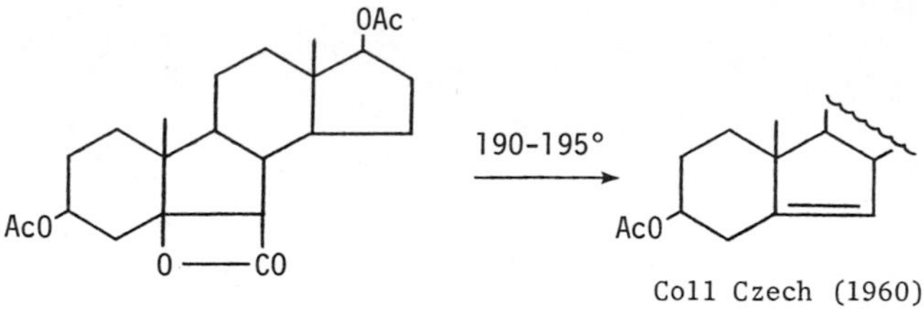

91%

Coll Czech (1960) <u>25</u> 2341

$C_5H_{11}CH_2CH_2CH_2COCOOH$ $\xrightarrow{h\nu \quad C_6H_6}$ $C_5H_{11}CH=CH_2$ 88%

JACS (1968) <u>90</u> 1840

$PhCH=CHCH_2\overset{\underset{OH}{|}}{C}Et_2$ $\xrightarrow{\Delta}$ $PhCH_2CH=CH_2$

JCS (1965) 7242

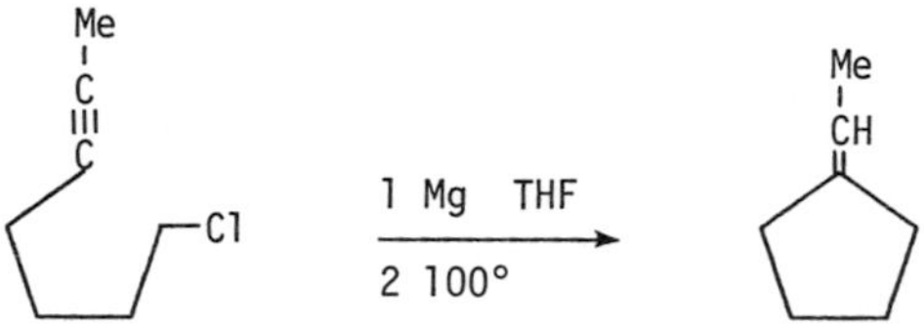

$\sim$90%

Tetr Lett (1968) 1457